CMOS BIOMICROSYSTEMS

CMOS BIOMICROSYSTEMS

Where Electronics Meet Biology

EDITED BY

Krzysztof Iniewski

WILEY

A JOHN WILEY & SONS, INC., PUBLICATION

For general information on our other products and services or for technical support, please contact our Customer Care Department within the United States at (800) 762-2974, outside the United States at (317) 572-3993 or fax (317) 572-4002.

Wiley also publishes its books in a variety of electronic formats. Some content that appears in print may not be available in electronic formats. For more information about Wiley products, visit our web site at www.wiley.com.

Library of Congress Cataloging-in-Publication Data:

Iniewski, Krzysztof, 1960- , author.
CMOS Biomicrosystems : Where Electronics Meet Biology / Krzysztof Iniewski.
p. cm
ISBN 978-0-470-64190-3 (hardback)
1. Medical electronics. 2. Bioelectronics. 3. Metal oxide semiconductors, Complementary.
I. Title.
R856.I385 2011
610.28–dc22

2010042345

Printed in Singapore

oBook ISBN: 978-1-118-01649-7
ePDF ISBN: 978-1-118-01647-3
ePub ISBN: 978-1-118-01648-0

10 9 8 7 6 5 4 3 2 1

CONTENTS

PREFACE

The emerging generation of health care that offers dramatic improvements in disease detection will likely be enabled by bioelectronics, a frontier discipline at the interfaces of electronics, biology, physics, chemistry, and materials science. By integrating these diverse scientific fields, bioelectronics will revolutionize how we interact with, measure, and understand biological systems, enabling emerging technologies from DNA injection to implantable sensors in the human body. This paradigm shift will have enormous impact on improving the quality and hopefully reducing the cost of health care.

CMOS Biomicrosystems provides contemporary coverage of major advances as the well-established CMOS microelectronics technologies are employed to provide innovative solutions in the broad areas of biomedical applications. The book is an overview of the numerous new advancements in this exciting field of microelectronics that is "meeting" biology. It contains many applications and examples of CMOS systems already realized or being developed for providing new tools to interface to biology.

The topic of biomicrosystems is a very active research area worldwide, as the various areas in this field are enjoying considerable popularity. The book contains a broad overview of many different applications of CMOS technology and fabrications, ranging from electrocardiograph and electroencephalogram signals acquisition to molecular and cell detection to *in vivo* imaging systems. The book is more in the style of a reprint book, highlighting individual, self-contained chapters. In this context, the information likely will be of greatest value to those working in the field. However, the chapters are appropriately written to introduce the newcomer to the chapter topic before delving into the detailed technical topics, a benefit for the reader who is from outside the bioelectronics field.

The book is divided into three parts: Human Body Monitoring, Biosensors and Circuits, and Emerging Technologies. The first part on human body monitoring starts with introducing fundamental concepts and performance key metrics, a chapter written by researchers from Imperial College London. This chapter is followed by chapters on neural signal recording written by authors from Politecnico di Milano and National Chung-Cheng University. Researchers from Samsung describe the use of RF technology for health care applications. Finally, a team from Tsinghua University covers design considerations for implantable systems.

The second part on biosensors and biocircuits starts with the fundamentals of biosensors, discussing stochastic modeling and figures of merit, a chapter written by researchers from the University of Texas at Austin. This chapter is followed by a description of CMOS and MEMS biochip technologies written by Dr. Courtois from Circuits Multi-Projets (CMP). The following chapters by authors from Polytechnic Montreal and the University of California, Los Angeles, deal with biointerfaces for lab-on-chip applications and lensfree on-chip imaging, a new tool for telemedicine. Finally, researchers from the University of California, San Diego, and Imperial College London describe monitoring systems for biomicrofluidics and stem cell culture processing.

The third part on emerging technologies starts with an introduction to futuristic biology interfacing written by researchers from Berkeley. They show three examples of the types of interfaces that link CMOS paradigms with biological systems: remote flight control of insects through implantable microsystems, pixelated interfaces to developing cells, and CMOS-compatible very high-density (VLSI) microfluidics. This chapter is followed by a description of technologies for arrayed single-cell biology authored by researchers from the University of Washington. The following chapters by authors from the University of Arkansas and Arizona State University cover an intriguing field of nanoscale engineering systems; both bacterial flagellar motors and gene injection are presented. Finally, an innovative approach to early disease detection based on RF circuits is proposed by researchers from Harvard.

I sincerely hope that you, as reader, will enjoy the book, and I am sure you will learn something new in this exciting field of bioelectronics. If you have any comments or suggestions about the material presented, please contact me at kris.iniewski@gmail.com. I would also love to hear suggestions from you about future books on bioelectronics.

Books like this one would not possible without many creative individuals meeting together in one place to exchange thoughts and ideas in a relaxed atmosphere. I would like to invite you to attend CMOS Emerging Technologies events that are held annually in beautiful British Columbia, Canada, where many topics covered in this book are discussed. See http://www.cmoset.com for presentation slides from the previous meeting and announcements about future ones.

Let electronics meet biology and benefit each other!

KRIS INIEWSKI

CONTRIBUTORS

Editor

Krzysztof Iniewski, CMOS Emerging Technologies, Coqutilam, British Columbia, Canada

Authors

Andrea Bonfanti, Department of Robotic, Brain, and Cognitive Sciences, Italian Institute of Technology, Genoa, Italy

Tommaso Borghi, Department of Electronics and Information, Politecnico di Milano, Milan, Italy

Maide Bucolo, Department of Electrical, Electronics, and Informatics Engineering, University of Catania, Catania, Italy

Alexander J. Casson, Department of Electrical and Electronic Engineering, Imperial College London, London, United Kingdom

Xinkai Chen, Institute of Microelectronics, Tsinghua University, Beijing, China

Chih-Jen Cheng, Department of Electrical Engineering, National Chung-Cheng University, Ming-Hsiung, Chia-Yi, Taiwan

Jung Han Choi, Samsung Advanced Institute of Technology, Yongin, Korea

Daniel Cohen, Department of Bioengineering, University of California, Berkeley, California, USA

Bernard Courtois, CMP, Grenoble, France

Shreepriya Das, Department of Electrical and Computer Engineering, University of Texas, Austin, Texas, USA

Emmanuel M. Drakakis, Department of Bioengineering, Imperial College London, London, United Kingdom

Anthony Erlinger, Electrical Engineering Department, University of California, Los Angeles, California, USA

James R. Etzkorn, Department of Electrical Engineering, University of Washington, Seattle, Washington, USA

Qiang Fang, School of Electrical and Computer Engineering, RMIT University, Melbourne, Victoria, Australia

Ebrahim Ghafar-Zadeh, Department of Bioengineering, University of California, Berkeley, California, USA

Donhee Ham, School of Engineering and Applied Sciences, Harvard University, Cambridge, Massachusetts, USA

Arjang Hassibi, Department of Electrical and Computer Engineering, University of Texas, Austin, Texas, USA

Marcos Intaglietta, Department of Bioengineering, University of California, La Jolla, California, USA

Dong Kyun Kim, Samsung Advanced Institute of Technology, Yongin, Korea

Jin-Woo Kim, Department of Biological and Agricultural Engineering, University of Arkansas, Fayetteville, Arkansas, USA

Andrea L. Lacaita, Department of Electronics and Information, Politecnico di Milano, Milan, Italy

Shuenn-Yuh Lee, Department of Electrical Engineering, National Chung-Cheng University, Ming-Hsiung, Chia-Yi, Taiwan

Shyh-Chyang Lee, Department of Electrical Engineering, National Chung-Cheng University, Ming-Hsiung, Chia-Yi, Taiwan

Xiaowen Li, Institute of Microelectronics, Tsinghua University, Beijing, China

Yong Liu, IBM T. J. Watson Research Center, Yorktown Heights, New York, USA

Michel M. Maharbiz, Department of Electrical Engineering and Computer Science, University of California, Berkeley, California, USA

Sarah C. McQuaide, Department of Electrical Engineering, University of Washington, Seattle, Washington, USA

Jit Muthuswamy, Bioengineering, School of Biological and Health Systems Engineering, Arizona State University, Tempe, Arizona, USA

Aydogan Ozcan, Electrical Engineering Department, University of California, Los Angeles, California, USA

Babak A. Parviz, Department of Electrical Engineering, University of Washington, Seattle, Washington, USA

Ryan Pooran, Microelectronics–Photonics Program, University of Arkansas, Fayetteville, Arkansas, USA

Esther Rodriguez-Villegas, Department of Electrical and Electronic Engineering, Imperial College London, London, United Kingdom

Francesca Sapuppo, Department of Electrical, Electronics and Informatics Engineering, Catania University, Catania, Italy

Hirotaka Sato, Department of Electrical Engineering and Computer Science, University of California, Berkeley, California, USA

Sungkyu Seo, Department of Electronics and Information Engineering, Korea University, Jochiwon, Chungnam, Korea

Arati Sridharan, Bioengineering, School of Biological and Health Systems Engineering, Arizona State University, Tempe, Arizona, USA

Ting-Wei Su, Electrical Engineering Department, University of California, Los Angeles, California, USA

Nan Sun, Department of Electrical and Computer Engineering, University of Texas, Austin, Texas, USA

Steve Tung, Department of Mechanical Engineering, University of Arkansas, Fayetteville, Arkansas, USA

Haris Vikalo, Department of Electrical and Computer Engineering, University of Texas at Austin, Austin, Texas, USA

Zhihua Wang, Institute of Microelectronics, Tsinghua University, Beijing, China

Xiang Xie, Institute of Microelectronics, Tsinghua University, Beijing, China

Xicai Yue, Department of Bioengineering, Imperial College London, London, United Kingdom

Guido Zambra, Department of Electronics and Information, Politecnico di Milano, Milan, Italy

PART I

HUMAN BODY MONITORING

1

INTERFACING BIOLOGY AND CIRCUITS: QUANTIFICATION AND PERFORMANCE METRICS

Alexander J. Casson and Esther Rodriguez-Villegas

1.1 INTRODUCTION

A key aim of bioelectronics is to provide an interface between the biological world (blood pressure, electrocardiogram [ECG], and the like) and the electronics world (analog and digital hardware, software, and the like). This interface allows the characterization and quantification of the biological world, which can be used to gain further understanding of the fundamental biological processes being monitored. Alternatively, long-term monitoring of physiological parameters can lead to new and more effective diagnostic and treatment methods for particular medical conditions.

A typical interface between the biological and electronic worlds is shown in Figure 1.1. Here a suitable sensor or electrode is used to detect a biological parameter, and the resulting signal is then amplified and converted into the digital domain. Once this has been done, the signal can be transferred to a computer for long-term storage and processing. Depending on the application requirements, the data may be transfered over cables or via a wireless link.

The biological world interface system thus includes everything from the sensor to the wired or wireless link. In many applications this system must be as physically small as possible, and capable of operating autonomously over long periods of time. This may be because data is being collected from a lab animal that is physically small, or because a human is being monitored, and they are

CMOS Biomicrosystems: Where Electronics Meet Biology, First Edition. Edited by Krzysztof Iniewski.

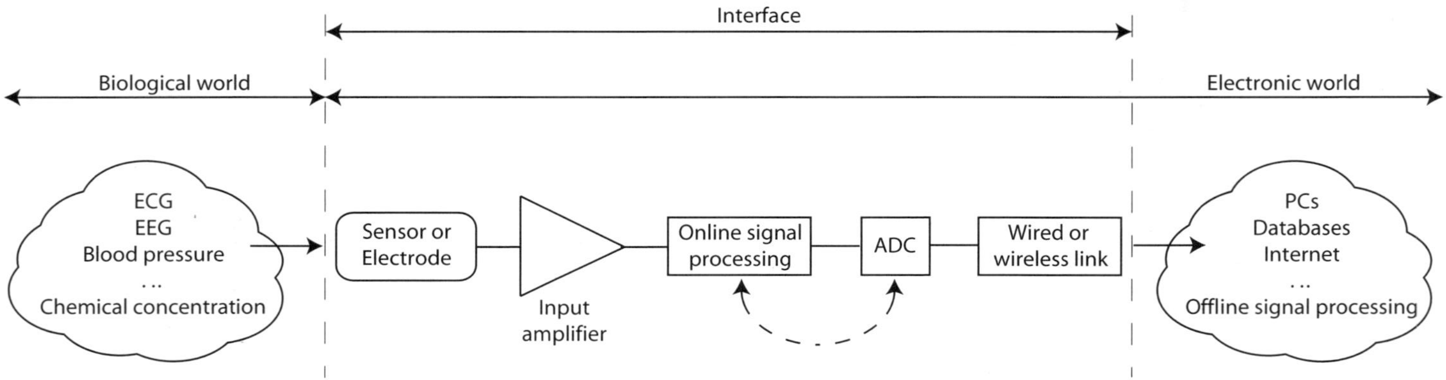

Figure 1.1. A typical interface system between the biological world and the electronic world. Physiological parameters are sensed, amplified, and converted to digital signals, which can be stored and processed on a computer. Online signal processing, implemented in either the analog or digital domain, can be of significant use in enabling device miniaturization.

expected to be going about their normal daily life. For this to be possible, the interface device must be unobtrusive, comfortable, socially acceptable, and long lasting.

Miniaturized, unobtrusive devices imply that only physically small batteries, which have limited energy storage and current sourcing capabilities, are available for use. Simultaneously, long-term monitoring implies that the limited energy capacity of the batteries has to exploited to the maximum by using very low-power electronics. Key for realizing these miniaturized interface systems is thus the optimization of the electronic design. For example, a system using only a 12-bit analog-to-digital converter (ADC) produces much less data to transmit and leads to much lower overall power consumptions than systems using a 24-bit ADC. However, performing such optimizations inevitably requires detailed knowledge of the biological requirements for the given application, and obtaining these requirements is by no means trivial. As an example, consider the electroencephalogram (EEG), which records electrical potentials from the scalp.

Recommendations from the International Federation of Clinical Neurophysiology call for a 12-bit (72 dB) sampling resolution once the direct current (DC) component of the signal has been removed [1]. Most commercial EEG units use 16 or more bits, exceeding this recommendation. Typical analysis of the EEG produced, however, is performed by a human using 16 EEG traces displayed on a screen with 1024 vertical pixels. This gives just 6 bits of resolution [2]. For comparison, traditional paper-based EEG systems had a dynamic range of around 7 bits [2]. Potential room for optimization is thus present, especially if only automated analyses of the EEG are to be performed.

Of course, such uncertainties in the performance requirements are not confined to the biological world alone. For low-power, low-dynamic-range signal processing, analog circuit implementations can potentially significantly outperform their digital counterparts [3]. However, it is then necessary to contend with an amount of *mismatch*: for example, when implemented on a microchip capacitor values may be no more than 20% accurate, and no two transistors will be exactly the same. This leads to a variance in the performance of the analog circuit, and the range of this must be quantified to ensure that such a variance is acceptable. Accurate quantification of both the biological and electronic worlds is thus essential for optimizing the electronic design for device miniaturization.

One further method used to enable device miniaturization is online signal processing, as shown in Figure 1.1. As an example, if the EEG is being monitored, rather than transmitting the entire EEG recording it is possible to detect potential *interesting* sections of data online, and transmit only these sections. This significantly reduces the amount of EEG data to be sent, mitigating the use of high-power transmitters. However, accurate quantification is again necessary. It is essential that the accuracy of this data reduction method is known and acceptable. How many of the interesting sections are missed and how many false detections are made?

Unfortunately, in general bioelectronics applications, the quantification of online signal processing aiming to reduce the system power consumption, and hence the system size, is a problem subject to many constraints that make the algorithm design, implementation, and performance testing far from trivial. On the one hand, the algorithm must achieve acceptable performance accuracy for the given application. On the other hand, this must be done while developing an algorithm that can be implemented in very low-power circuits. There is no benefit in designing an algorithm that, when implemented, requires more power to operate than any potential power savings it provides. In addition to this, the potential for nonidealities in the end implementation, for example, from analog mismatch, must be accounted for.

The procedure for tackling this kind of problem is thus not to optimize any one aspect of it in isolation, but rather to look for a global solution that meets the constraints imposed by both the engineering design (such as the power consumption) and the biological application (such as a clinically acceptable detection accuracy) simultaneously. The overall interface design problem is thus an interdisciplinary one in which the bioelectronics designers must know the aspects of the biology that are going to condition the specifications of the electronic blocks; identify the best metrics to quantify performance for the given application; and devise a rigorous and representative test methodology that characterizes the performance within a certain confidence level.

Accurate characterization of the online signal processing algorithm is an essential part of this. For optimal power performance, these algorithms are best implemented as dedicated circuits, as opposed to in software. The circuit design, however, likely requires man-years of effort. For this not to be wasted on unpromising algorithms, accurate and reliable performance characterization is necessary at the algorithm design stage.

The aim of this chapter is to present the reader with examples of how to design a rigorous test methodology to characterize the performance of online signal processing systems designed to reduce the system-level power consumption. For example, what test factors need to be known and what performance metrics are best used in order to elucidate the most information possible about the performance? What is the impact of using different performance metrics, and how is it possible to ensure that the results are accurate and reliable? To illustrate the results on an actual algorithm, an EEG data reduction algorithm is considered. Nevertheless, although this algorithm is application specific, the methods and characterization routine are similar across many bioelectronics situations.

Section 1.2 thus presents the first part of the problem: the biological application and algorithm aim, in this case data reduction during monitoring of electrical brain activity (EEG). This motivates the need for an online signal processing stage and sets its engineering requirements. Section 1.3 then considers the factors that make representative performance testing nontrivial for this kind of application. Section 1.4 derives different performance metrics that could be used to characterize performance, discussing the advantages and disadvantages of each. Finally, Section 1.5 discusses the statistical testing of the results.

1.2 THE SIGNAL PROCESSING AIM

1.2.1 Introduction

The first step in the design of a rigorous test methodology for an online signal processing algorithm is the precise definition of the algorithm objective. This sets the required specifications for the algorithm, and also sets the objectives of the required test methodology. This Section quantifies the use of online data reduction in bioelectronics interface systems to decrease the system power consumption, and in turn the device size. The analysis allows the required level of data reduction to be found, motivating the algorithm design. The form of the analysis here applies to any general bioelectronics application, although as an illustrative case, the specific numbers here are taken from EEG monitoring.

1.2.2 The Need for Online Data Reduction

EEG recording, where electrodes are placed on the scalp and detect the microvolt-sized signals that result outside the head due to the accumulated neuronal action within the brain [4], is a characteristic bioelectronics problem requiring long-term, miniaturized interface systems connecting the biological and electronic worlds. This is because, although long-term inpatient EEG recordings are ideal for applications such as epilepsy diagnosis [5], they are resource intensive and not universally available [6]. Instead, ambulatory EEG (AEEG) recordings are available, during which the patient has their EEG recorded on a portable unit while undertaking their normal daily life. Such recordings cost approximately 50% of their inpatient counterparts [6] and so are highly desirable, provided the AEEG recording unit is miniaturized.

In general, this miniaturization is limited by the size of the battery required. If the overall device is assumed to have a volume of $1\,cm^3$ (a common aim for long-term ubiquitous recording applications) and half of this space is reserved for a custom-made battery with an energy density of 200 Wh/L, 100 mWh of energy is stored. For operation over 30 days, the average power consumption must be less than $140\,\mu W$ [7].

An input amplifier and ADC system with a measured $25\,\mu W$ power consumption per EEG input channel is presented in Yazicioglu et al. [8], representing the current state-of-the-art performance. Assuming 200 Hz and 12-bit sampling, 300 bytes per second per EEG input channel of data are produced. For a good transmitter that consumes 50 nJ/bit transmitted, including all of the overheads of data buffering, channel selection, and the like, transmitting each channel consumes approximately $120\,\mu W$. With these figures, only EEG systems with one input channel are feasible. To overcome this, online data reduction must be used.

1.2.3 Optimizing the Power–Device Size Trade-Off

The aim of the online signal processing indicated in Figure 1.1 should thus be to reduce the amount of data that is passed through the transmitter stage. From the

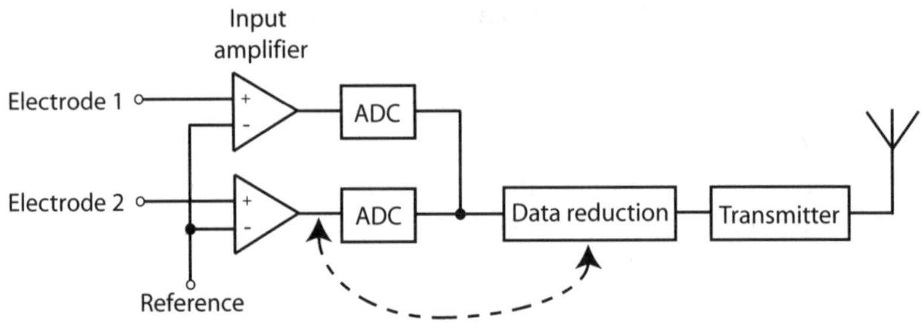

Figure 1.2. A simplified model of a two-input channel wireless interface system based on Yates and Rodriguez-Villegas [9]. The data reduction block can be in either the analog or digital domain.

example above, this transmission consumed 120 μW per channel, compared with only 25 μW for the front-end systems. Transmission thus dominates the system power consumption, and reducing the amount of data to transmit can significantly reduce the overall system power consumption even though the online data reduction will itself require some power. This trade-off is considered here using the framework from Yates and Rodriguez-Villegas [9].

To begin, consider the simplified interface system model in Figure 1.2. This basic architecture contains an input amplifier, an ADC, a data reduction block, and a transmitter. In principle, this architecture could be used to record many different physiological parameters; the form of analysis here is not unique to EEG acquisition.

As a first approximation, the power consumption of the entire system is given by

$$P_{\text{sys}} = \text{NP}_{\text{amp}} + \text{NP}_{\text{ADC}} + P_{\text{c}} + CP_{\text{t}}, \tag{1.1}$$

where n is the number of input channels, C is the percentage of data transmitted giving the ratio of the number of bits that are actually transmitted to the total number of bits if no compression was present, P_{t} is the power consumption of the transmitter, and the other three terms are the power consumptions of the amplifier, ADC and compression, respectively.

If the transmitter has a net power consumption, including overheads, of J joules per bit, P_{t} is given by

$$P_{\text{t}} = Jf_{\text{s}}RN, \tag{1.2}$$

where f_{s} is the sampling frequency and R is the resolution in bits of the ADC.

If the system is operated with no compression stage present, $P_{\text{c}} = 0$ and $C = 1$. Thus, if the inequality

$$P_{\text{c}} < JNf_{\text{s}}R(1-C) \tag{1.3}$$

is satisfied, the online data reduction can be used to decrease the total system power consumption.

Simultaneously, of course, the total system power consumption P_{sys} is governed by the size and capacity of the battery used. For a cell of volume V and energy density D operating over a device lifetime between battery changes T,

$$P_{sys} = \frac{V \times D}{T}. \tag{1.4}$$

For independence from any particular battery technology, the normalized operational lifetime can be defined as

$$T_n = \frac{T}{V \times D}, \tag{1.5}$$

and there is thus a direct three-way trade-off between the amount of data reduction achieved, the power budget available to implement this data reduction, and the normalized operational lifetime that is then possible. This trade-off is illustrated in Figure 1.3 using the EEG system figures from Section 1.2.2 and $n = 2$ for the two-channel system as illustrated in Figure 1.2.

A more comprehensive discussion of this idea can be found in Yates and Rodriguez-Villegas [9], but in brief, Figure 1.3 can be used to quantify the required engineering objectives of the online signal processing algorithm. For example, if from any given algorithm a data reduction of 25% is achieved, using the

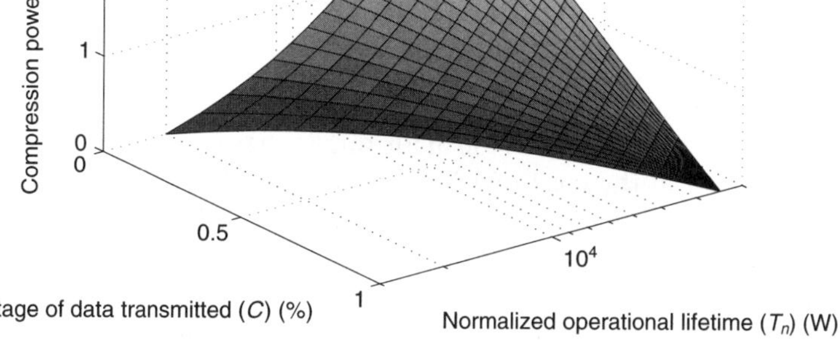

Figure 1.3. The three-way trade-off between the amount of data reduction, the normalized operational lifetime, and the available power budget to implement the compression algorithm.

hypothetical battery from Section 1.2.2, which stores 100 mWh of energy, for operation over 30 days, the online data reduction algorithm must operate using no more than 30 µW of power. Such power consumptions are challenging, but achievable. To avoid wasted effort on the required electronic design, however, the importance of fully verifying that the algorithm operates satisfactorily before attempting the design is clear.

1.3 REPRESENTATIVE TESTING

1.3.1 Introduction

The algorithm aim has now been defined, and it is assumed that a suitable algorithm has been developed. The next step is the design of the test methodology. The performance of any bioelectronics online signal processing algorithm is most suitably assessed prior to hardware implementation by carrying out a range of simulations using a software model of the algorithm. A set of input biological data is passed through the algorithm, and its operation observed. In general, the performance would then be compared with that of a human expert, and performance metrics comparing the two derived. Before considering the necessary performance metrics in detail, it must be ensured, however, that the simulation test methodology is rigorous, representative, repeatable, and accurate. Unfortunately, in real bioelectronics applications, there are a large number of factors that complicate the situation, and need to be identified and controlled if possible. These facts are discussed here for the special case of EEG recording to illustrate the typical factors that must be accounted for.

1.3.2 Data Recording Factors

For the software verification of the algorithm, an amount of biological test data must first be recorded from a subject in order to then be passed through the algorithm. However, one section of data is not necessarily representative of another data section, and is not necessarily representative of the type of data produced in the targeted application population. The data collection recording settings must thus be tightly controlled as they can affect the data traces produced, and hence the algorithm performance. Controlling these factors will also allow the situation under which the algorithm is characterized to be clearly stated.

For example, in the case of EEG collection, different EEG equipment can have different sampling rates, bandwidths, and electrode types, all of which should be specified. Also, the EEG typically records from multiple electrodes placed on the scalp and different *montages* are possible depending on how the channels are interconnected, and these affect the shape of the signals produced. The type of recording must also be controlled. Routine (20–30 minutes), long-term (1–3 days), or AEEG recordings are possible and may use different equipment, settings, channels, and montages. In the case of clinical applications, inpatient and outpatient tests may have very different artifacts present in the EEG recording and such artefacts can be affected by the testing procedure used, for example, whether the eyes are open or closed. In addition, different subject states during testing, such as being awake or drowsy, can affect the EEG traces.

Overall, it is essential that the algorithm is tested using subjects who reflect the anticipated end user population. Examples of this include whether the subject is on medication, and the potential presence of multiple diseases (comorbidity). Factors such as the age of the subject can also be significant in determining the signal conditions, such as the amplitude and the amount of activity. For clinical uses, factors such as the sex, handedness, and many others are also be taken into account and recorded.

1.3.3 The Amount of Data to Test

Once the setup of the data recording has been controlled, a suitable amount of test data must then be collected. For example, to test an EEG spike detection algorithm, Wilson and Emerson [10] recommend that for comprehensive testing 100 subjects, 10,000 spikes and 800 hours of EEG should be used. This seems to be a very high level, especially given the effort required by an expert to mark the events in an EEG trace (see Section 1.3.4). The level, however, is perhaps correct if all of the above recording factors are to be made insignificant purely by the amount of testing done. To our knowledge at this date, only Persyst [11], which uses 18,503 events in 266 hours of data, and Liu et al. [12], who use 145,230 events from 81 patients in 800 hours of data, test anywhere near this amount.

When determining the amount of data to use, note should be made that it is often easy to get good algorithm performance when testing very short (where a human interpreter is essentially perfect) or artifact-free data. Generally, however, neither of these reflect situations where online signal processing algorithms would actually be of use. Long-term data, which is not preselected for the inclusion or exclusion of artifacts, should be used. Of course, to get long-duration recordings, it is often necessary to include multiple recording periods from probably multiple subjects. Due to variations between subjects, testing in multiple different subjects is potentially much more comprehensive than testing the same amount of data in just one subject.

Unfortunately, in most testing situations, it is unlikely that such large data sets are available for use. Even if they are, unless the algorithm being developed operates substantially quicker than real time, it may be impractical to experiment with very many different algorithm setups. To overcome this, it is possible to mathematically gain an insight into how much data should be tested through the idea of confidence intervals. These are considered in Section 1.5.3, once the necessary performance metrics have been defined.

1.3.4 Marker Reliability

Having collected the test data, an expert marker would generally then be used to identify the features of interest. Again, considering the example of an EEG spike detection algorithm, the time locations of the spikes are marked. This then sets the baseline against which the algorithm will be compared. This process can represent a major limitation of the testing procedure as different markers do not always mark the same events, and one marker will sometimes mark differently when reviewing a record for the second time [13]. Marker agreement can be anywhere between 0 and 90%. Any results produced can thus be at most as accurate as this marking procedure. Also, the time markings themselves may only be accurate to the nearest second, minute, or hour, depending on the timescale of the signal being analyzed. This fact should be taken into account when determining how close an expert-marked event and an algorithm-detected event need to be for the detection to be successful.

Ideally, more than one expert marker should be used, and the method for combining the markings of each expert should also be clearly stated. Unfortunately, of course, this significantly increases the amount of time and resources required to prepare the algorithm testing.

1.3.5 Practicability and Ethics

Having collected and marked the biological data accounting for the above issues, suitable testing simulations can now be carried out. In practice, however, it is unlikely that all of the factors identified here will be known, or controllable. This is especially the case when the data is historical, rather than especially collected for the current study, or when the algorithm is developed by engineers, but the data are collected by clinicians, potentially working quite separately. This does not mean that it is impossible to perform good and informative studies without ideal test data. However, the potential limitations of such studies should be appreciated.

It may be ideal if algorithms were tested on a standardized database, and this would allow much more direct comparison between studies. Often, however, ethical approval is not in place to allow individual researchers to share their databases, and this is understandable. Nevertheless, standardized online databases are becoming ever more available.

The very large number of factors, and the potential difficulties in controlling them, can thus be seen. This makes it clear why the generation of accurate performance metrics for a given situation is a nontrivial task and deserves significant attention. Ideally, all of the factors above should be controlled and reported in any algorithm-testing publications. Again, however, there are some ethical considerations that may prevent this from being done. When working with a high-incidence disorder such as epilepsy, reporting that subject 1 has epilepsy and was 29 at the time of the EEG recording (which could have been several years before the first publication based on that test) does not devolve any significant information about subject 1.

It may be, however, that when working with rare diseases, or specific subsets of more prevalent ones, there may only be tens or hundreds of sufferers worldwide. In this case, reporting that subject 2 is male, 29 at the time of test, and is left-handed could provide significant clues to the subject's identity, which could be compromised by a dedicated and resourceful person. Thus, although such information may be relevant, it should not be reported.

1.4 PERFORMANCE METRICS

1.4.1 Introduction

With suitable test data collected, having controlled for factors impacting the test methodology, it is thus now possible to consider which performance metrics should be used, and how these affect the algorithm performance that is reported. Again, a specific EEG data reduction algorithm is considered here as an illustrative case for the similar analysis that should be carried out for each bioelectronics algorithm prior to testing.

1.4.2 Illustration Data Reduction Algorithm

The data reduction strategy investigated here is concisely illustrated in Figure 1.4. Rather than continuously recording the EEG signal, an attempt is made to detect the features of interest, in this case spikes that occur between seizures in epilepsy patients,[1] and to record only a window of data around these automated detections. It can be seen how this can lead to a significant data reduction, even with a number of false detections present [14]. The algorithm aim should be to record all possible events for later interpretation by a human, while still cutting out some background data. Having a relatively high number of false detections does not necessarily compromise this process.

The core performance quantification problem is thus essentially one of signal detection: How many of the spike events are correctly recorded and how many false detections are made? The better the detection performance, the fewer false detections, and therefore more data reduction should be achieved. Reviews of similar EEG spike detection algorithms, although not necessarily for low-power implementation in bioelectronics interface systems, have been given by Frost [15] and Gotman [16] in 1985, and more recently by Wilson and Emerson [10] in 2002. Despite the level of interest illustrated by these, however, a definitive spike detection solution has not been found. It is clear that the task of finding a clinically acceptable trade-off between the number of events correctly detected and the number of false detections is nontrivial, highlighting the importance of accurate performance metrics.

The particular algorithm considered here is a developed version of the one proposed in Casson et al. [17] and is shown at a high level in Figure 1.5. The core of the processing is the extraction of frequency content in two bands by wavelet-based, band-pass filtering (see Casson et al. [18] for details on the filters). The C_5

[1] Here all between-seizure (interictal) events such as spikes, sharp waves, and spike-and-waves are considered under the umbrella term spikes.

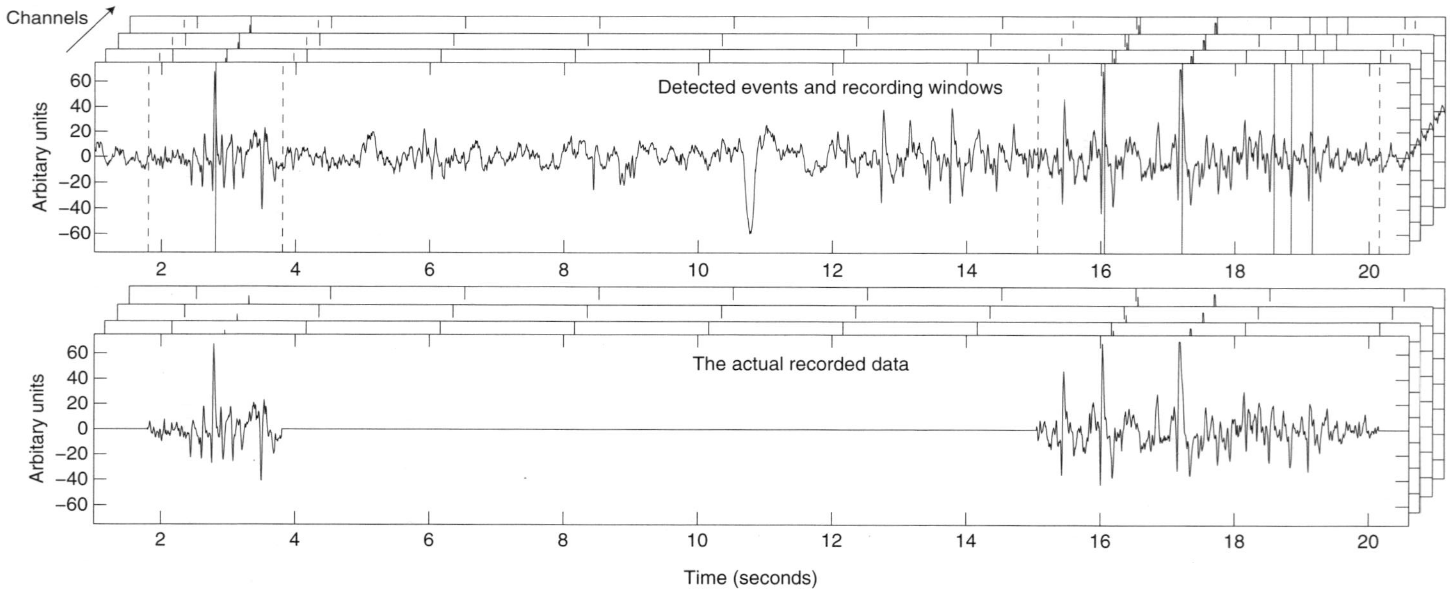

Figure 1.4. A data reduction strategy based on discontinuous recording. Only EEG data for a brief period (dashed vertical lines) on either side of an automated detection of a *candidate spike event* (solid vertical lines) is selected to be recorded. Other data sections are discarded online, significantly reducing the amount of data to be transmitted.

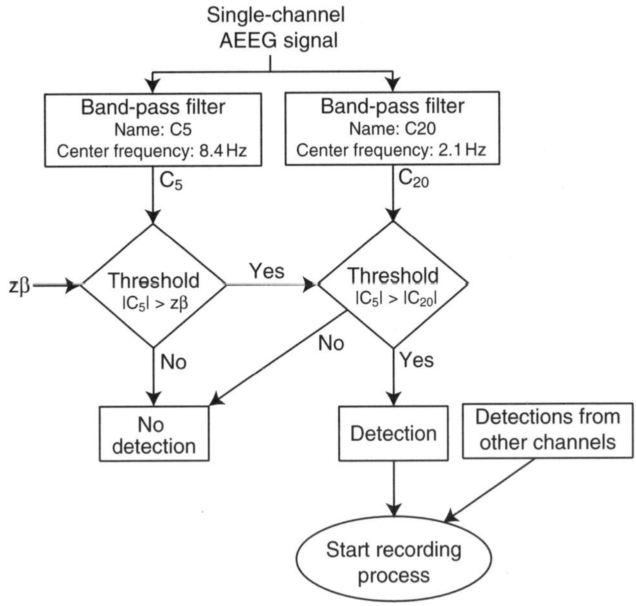

Figure 1.5. A high-level overview of the algorithm to be investigated. Detections in any monitored EEG input channel cause the algorithm to start recording. In the recording process (not shown), a memory buffer must be present to allow sections of EEG data immediately preceding a detection to also be recorded.

information produced is then compared with a threshold value of $z\beta$. z is an automatically generated normalizing parameter to correct for broad level amplitude differences in different EEG traces given approximately by the root mean square (RMS) of the EEG signal. β is a user-set detection threshold. The user is free to sweep β to obtain a range of performances from the algorithm. If all of the comparisons are satisfied, a detection is made, causing a section of EEG data from before (stored in a buffer) and after the detection point to be recorded.

1.4.3 Metrics

In principle, two metrics are required to illustrate the operation of any algorithm. One is the *performance* indicating how well the method operates, and the other is the *cost*, which indicates what undesirable factors are also present. Inevitably, there is some form of trade-off between the two. For event detection algorithms such as the EEG interictal spike case considered here, there are four common measures associated with characterizing the performance compared with that of an expert marker. These all use the following terminology:

- True positives (TP): the number of correct detections of a spike as a spike
- False positives (FP): the number of incorrect detections of a nonspike event as a spike

- True negatives (TN):the number of correct detections of a nonspike as a nonspike
- False negatives (FN): the number of incorrect detections of a spike as a nonspike

and the metrics are given in the following list:

1. *Sensitivity*: the fraction of spikes that are correctly detected:

$$\text{Sensitivity} = \frac{\text{TP}}{\text{TP} + \text{FN}} \times 100\% \tag{1.6}$$

2. *Specificity*: the fraction of nonspikes that are correctly rejected:

$$\text{Specificity} = \frac{\text{TN}}{\text{TN} + \text{FP}} \times 100\% \tag{1.7}$$

3. *Selectivity*: the fraction of correct detections:

$$\text{Selectivity} = \frac{\text{TP}}{\text{TP} + \text{FP}} \times 100\% \tag{1.8}$$

4. *FP rate*: the average number of FP per minute or hour.

Not all papers generate any of these measures explicitly, but it is essentially universal to have some quantification of the *performance* and the *cost*, allowing such measures to be derived if wanted. For example, some articles, such as that by Indiradevi et al. [19], use an *accuracy* metric, although what is meant by this is not always defined, and it is unlikely to be used consistently. Similarly, for the algorithm considered here and, in general, for any algorithm intended to be mapped into hardware, where the aim is data reduction rather than correctly counting the number of spike events, the FP rate is not a suitable metric for use. Instead the percentage of data transmitted (C) introduced in Section 1.2.3, indicating what fraction of the full EEG recording is selected to be transmitted, is a more suitable *cost* metric. However, this is somewhat related to the false detection rate, as the more false detections occur, the larger the amount of data to be transmitted will be.

The sensitivity is the core *performance* metric, and it illustrates how many of the wanted features are correctly found—a high sensitivity is always wanted (although not all papers give a measure of this performance, classically Gotman and Gloor [20]). As noted previously, it is usually calculated by comparing the algorithm detections with those made by an expert marker. When doing this, it is important to specify the detection window, indicating how closely the two must match for a TP to have been found.

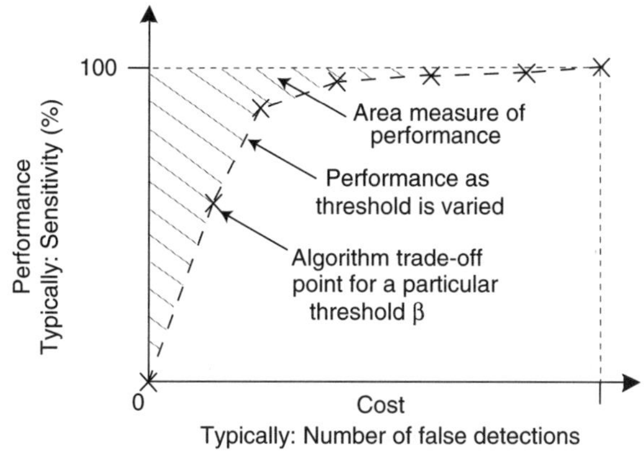

Figure 1.6. A schematic example illustrating how algorithm detection performance can be shown on a trade-off graph.

The specificity, selectivity, and FP rate provide measures of the *cost*. The specificity and selectivity should be high, or the FP rate low, for good results. In the context of spike detection, although the specificity is a well-defined concept, it is not clear what a true negative (TN) EEG event is. It is presumably a measure of how much background activity is present, but it is not easily reduced to an integer as the TP, FP, and FN measures are, and so the specificity is perhaps not an optimal measure to use, although again some articles, such as that by Exarchos et al. [21], attempt it. One method of calculating it is to use a time domain approach. Say for review by a human, 10 seconds is recorded in response to each false detection. A false detection rate of one per hour, which is quite poor if accurate counting of the number of events is wanted, results in a specificity of 99.7%. A specificity value over 99% thus does not necessarily correspond to a particularly good performance, and again other metrics such as the FP rate may be more illustrative.

Given these metrics, it is generally found that there is a trade-off between the sensitivity of an algorithm and the number of false detections that it makes. It is possible to achieve a high sensitivity (correctly detecting lots of events) if lots of false detections are tolerated. Most algorithms take this aim, although some, such as Ramabhadran et al. [22], aim for no false detections at the cost of a reduced sensitivity.

To display the results, as Wilson and Emerson [10] note, if an algorithm detection parameter, called say β, can be easily varied, the most natural way to display this trade-off is on a graph, as illustrated in Figure 1.6. For each value of β, a particular *performance* (sensitivity) is obtained at a certain *cost* (e.g., number of false detections). As β is varied, this pair of points defines a curve, which is essentially a receiver–operator curve (ROC) and the area to the top left of the line can be used as a high-level, dimensionless measure of the algorithm performance.

1.4.4 Illustrating the Average Performance

When using large numbers of subjects or tests to characterize the algorithm performance, it is not feasible to present the above metrics for each individual test, and it is desirable to have an overall headline performance figure. Methods for displaying the average performance of the algorithm over different tests are thus required. Different methods for calculating the average are possible, for example, by weighting the sensitivity values found to correct for nonideal test cases, and these are investigated below.

One example of a nonideal test case is that some data records may contain hundreds of expert-marked events, while others will contain only one. Thus, in different records, the detection or nondetection of one event can have a very different effect on the sensitivity found for the record, and this affects the average sensitivity found. A quantitative treatment showing the effect of the averaging method is very insightful for showing how results can be inadvertently weighted.

Here, four different methods of calculating the average sensitivity between different records are investigated, although other methods are undoubtedly possible, and will likely provide different properties to the methods considered here. The calculation procedures are detailed below using the terminology that there are M records and the ith record has a duration T_i, with N_i marked events and D_i correctly detected events. The sensitivity for any one record is given by $100\% \times D_i/N_i$. Sensitivities for different records can then be combined in the following ways:

1. Arithmetic mean:

$$\text{Sensitivity} = \frac{1}{M}\sum_{i=1}^{M}\frac{D_i}{N_i}\times 100\% \tag{1.9}$$

2. Time-weighted average:

$$\text{Sensitivity} = \frac{1}{\sum_{i=1}^{M} T_i}\sum_{i=1}^{M}\frac{D_i}{N_i}T_i \times 100\% \tag{1.10}$$

3. Total sensitivity average:

$$\text{Sensitivity} = \frac{1}{\sum_{i=1}^{M} N_i}\sum_{i=1}^{M} D_i \times 100\% \tag{1.11}$$

4. Time/event-weighted average:

$$\text{Sensitivity} = \frac{1}{\sum_{i=1}^{M} T_i / N_i}\sum_{i=1}^{M}\frac{D_i}{N_i}\frac{T_i}{N_i}\times 100\% \tag{1.12}$$

TABLE 1.1. A Synthetic Algorithm Results Set with Tests of Variable Lengths and Differing Numbers of Events and Correct Detections

Data Set (i)	Duration (T_i) (minutes)	Events (N_i)	Detections (D_i)	Duration/ Events (T_i/N_i)	Record Sensitivity (D_i/N_i) (%)
1	20	2	1	10	50.0
2	20	400	385	0.05	96.3
3	30	6	6	5	100
4	30	25	15	1.2	60.0
5	60	28	28	2.1	100
6	60	500	463	0.12	92.6
7	60	5	4	12	80.0
8	1440	40	19	36	47.5
9	1440	16	15	90	93.8
10*a*	60	3	1	20	33.3
10*b*	60	3	2	20	66.7

TABLE 1.2. Results of the Different Calculation Methods on Data Sets ***a*** and ***b***

	Sensitivity (%)	
Averaging method	1–9 and 10*a*	1–9 and 10*b*
Arithmetic mean	75.3	78.7
Time weighted	71.3	71.9
Total sensitivity	91.4	91.5
Time/event weighted	74.1	77.9

The arithmetic mean method treats all of the different test cases equally, and so any one record that has, say a low sensitivity, can significantly affect the overall value found. As a result, it is potentially weighted by records that are very short or contain very few events. In contrast, the total sensitivity measure treats all of the records as if they were one long record concatenated together. It is, thus, potentially weighted by records with large numbers of spikes or ones where the detection rate is particularly good. The time-weighted average and time/event-weighted average weight the individual sensitivities to make longer records more significant and long records with few events significant, respectively, in an attempt to overcome the limitations noted above.

The effect of the different averaging methods is illustrated, using purely numerical arguments, in Tables 1.1 and 1.2. Table 1.1 illustrates a typical set of 10 data records available for algorithm testing, with variable test lengths and numbers of events present in each recording. In addition, an assumed example level of algorithm performance is shown, with the number of correct detections made being arbitrarily chosen to illustrate a range of performance cases. For each data set, given the number of detections and the number of events, the sensitivity within each record is calculated and shown.

For record 10, however, two different cases, *a* and *b*, are considered. In case *a*, only one of the three events in the recording is correctly detected, while in case *b*, two events are correctly detected. The effect of this slight change is illustrated in Table 1.2. In this table, the sensitivity is calculated using $M = 10$, the four different sensitivity methods discussed previously, and the figures for N_i and D_i from Table 1.1. In case *a* (the middle column in Table 1.2), data sets 1–9 and 10*a* are analyzed, while in case *b* (the right-hand column), data sets 1–9 and 10*b* are analyzed.

It can be seen how the detection or nondetection of just one event from over 1000 events in 32 hours of data appreciably affects the averages found. Furthermore, the different averages change by noticeably different amounts. For example, the total sensitivity hardly changes (one event in 1033 is a very small percentage), while the arithmetic mean changes noticeably (one event in the three in record 10 is a large percentage). It is also interesting to note the spread of average values that are present in Table 1.2: more than 20% in case *a*. When devising an algorithm testing methodology, it is essential to take these potential weightings into account.

To illustrate how the ROC-like curves proposed in Section 1.4.3 are affected, an example testing situation using the algorithm from Section 1.4.2 is considered. For testing, three different data sets are analyzed: *A*, *B*, and *C*, detailed in Table 1.3. Data set *A* contains nine tests ($M = 9$), each of which is approximately an hour or more in length, with a total duration of 16 hours. There are 120 expert-marked events present. Data set *B* is the same as *A*, but with an extra 37-minute data set, making little difference to the total time analyzed ($M = 10$). This record, however, contains 644 events. Data set *C* is again the same as *A*, but with an extra 5-minute record, which contains seven events ($M = 10$). The algorithm is run on these three data sets, and the results produced using the four different metrics are shown in Figure 1.7. Results for both axes are weighted according to the averaging method used.

For the analysis, 19 values for the threshold β between 1 and 0.32 are used to produce the trade-off curve, and 5 seconds of EEG are recorded in response to each detection. (A typical spike has a duration of 140 msec [23].) For a detection to be considered correct, it must occur no more than 2 seconds away from the expert-marked position. In addition to the calculated result points, the *known* end points that if no data is sent the sensitivity must be 0% and, similarly, that if all of the data is sent the sensitivity must be 100% are also included. Finally, the *chance performance* line is also drawn. If, as a first approximation, spike events are assumed to occur at random times, randomly transmitting 10% of the raw EEG trace should result in a 10% sensitivity. The $y = x$ line thus corresponds to the performance of a chance detection scheme, and any algorithm performance should always be above this.

Considering Figure 1.7, data set *A* illustrates the baseline level of performance. Although the performance of the algorithm will always depend on the data analyzed to some extent, in principle the sensitivity is a property of the algorithm and should not be weighted by less representative test cases. This is

TABLE 1.3. Data Available for Analysis with the Algorithm Detailed in Section 1.4.2

Record (i)	Data Set	Duration (T_i) (HH:MM:SS)	Events (N_i)
1	A, B, and C	00:59:08	4
2	A, B, and C	00:58:56	4
3	A, B, and C	02:00:11	41
4	A, B, and C	02:00:11	7
5	A, B, and C	02:00:11	3
6	A, B and C	02:00:11	21
7	A, B and C	02:00:11	28
8	A, B and C	02:00:11	9
9	A, B and C	02:00:11	3
10	B	00:36:55	644
11	C	00:05:00	7

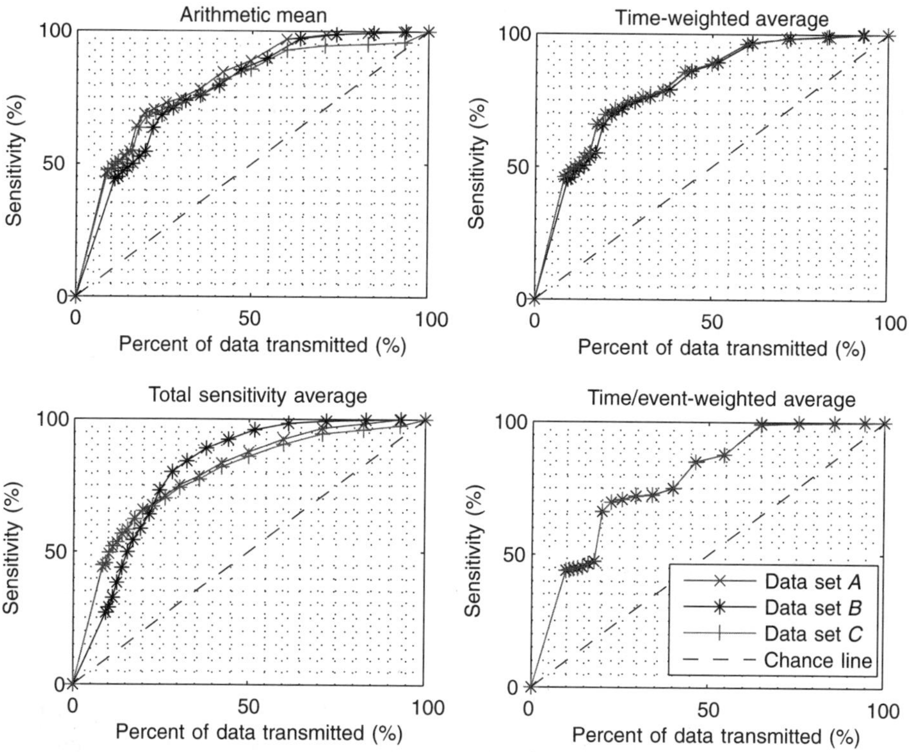

Figure 1.7. Performance results for the data reduction algorithm when different data sets are analyzed using the four different averaging methods.

clearly not the case, however, for the arithmetic mean or total sensitivity averaging methods where the results are visually different depending on which data is analyzed. The time-weighted average and time/event-weighted average have less sensitivity to the data analyzed, although less skew is present in the time/event averaging case, particularly when the high event count record is analyzed (data set *B*).

Given these results, it is thus essential to appreciate how the testing procedure can potentially be weighted. Assuming that the same test setup is used in all cases (see Section 1.3), the two core parameters that can vary between tests from a signal processing point of view are the test duration and the number of events present. The test duration can be controlled to a certain extent, although longer tests require more resources to carry out. The number of events present cannot be controlled. (Although only a set number can be considered if all records contain this minimum number.)

Thus, it does not seem unreasonable to normalize for both these parameters, giving the time/event-weighting method. At the same time, however, weighting the results can distort the sensitivity figure that is seen. For example, for the data set *B* case from Figure 1.7, at one threshold level 611 of the 764 events are correctly detected, giving a total sensitivity of 80%, but the time/event-weighted sensitivity is only 72%. In some cases, the time/event-weighted figure thus does not well represent the number of correct detections made. It is thus likely that reporting both the total sensitivity and time/event-weighted sensitivity is advisable when attempting to draw as much information from the algorithm performance as possible.

Finally, when interpreting the average performance lines, consideration must be given to the fact that although an algorithm may achieve a certain average performance level, there is not necessarily an equal probability of different events being detected. Wilson et al. [13] puts it as

> a particular spike has a probability of detection, which may be high or low depending on its size, morphology, background activity and other attributes.

As a result, the average performance is not necessarily easily related back to the performance of any one test: It can be used only as a general guideline of overall performance.

1.4.5 Illustrating the Variance in Performance

Methods for presenting the algorithm's average performance have been discussed in detail above. It is equally important, however, to illustrate how the performance results obtained vary from test to test, and person to person. Given the differences between people, it is not necessarily unexpected if an algorithm performs very well on one person, but only moderately well on another person. This would indicate the some subject-dependent tuning is required. It may even be found that different recordings in the same person produce noticeably differ-

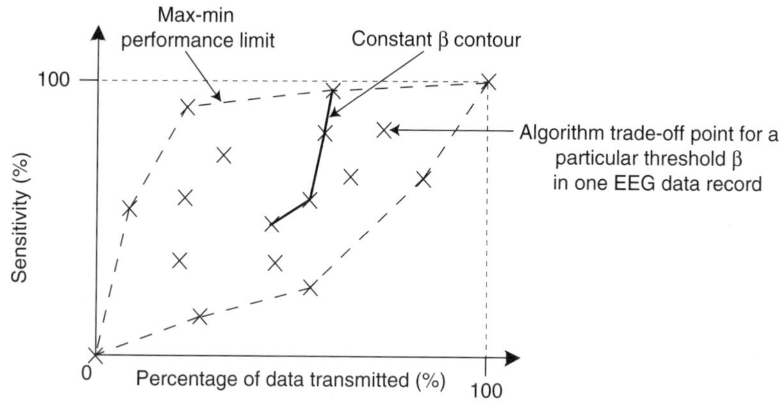

Figure 1.8. An example method for illustrating the performance variance by showing the max-min performance limit and the performance at the same threshold β in each record.

ent performances, for example, due to different equipment setups or awareness states during the data recording. Some variance in the performance is thus entirely reasonable, but for a successful algorithm it must be at an acceptable level, and so needs quantifying.

Of course, for any arbitrary algorithm, it is unlikely that the variance follows a normal distribution about the average. Simply plotting the mean result and the standard deviation is thus not a viable option. Instead, one potential method is illustrated in Figure 1.8 and discussed here, although as with the averaging methods considered above, other methods are undoubtedly possible.

First, on the same fundamental ROC-like curve, the *individual results* are shown. If the data set i is being analyzed and contains one or more marked events, each time the algorithm is run at a particular threshold β, individual values for the sensitivity ($S_i = 100\% \times D_i/N_i$) and percentage of data transmitted (C_i) are found. This pair of values (C_i, S_i) can then be plotted, giving a large number of data points if multiple tests and thresholds are used.

Then, the convex hull joining the outer most individual results can be drawn. This thus illustrates the max-min performance limit of the algorithm. Given the test data used, the algorithm performance can always be expected to lie in this region, and ideally, this region should also lie above the chance performance line. The generation of this line though does assume that only *sensible* values for the threshold β are used. Otherwise, the region can be artificially expanded by using thresholds that would never be considered in practice.

Finally, a constant threshold line is drawn. This is a line connecting all of the individual results that are calculated at the same β value. This reflects the system performance when used in practice where β would be set in advance and the performance will vary along this line. In the graphs here, this is drawn from the highest sensitivity point to the lowest sensitivity point, and so cannot *double back* in this direction.

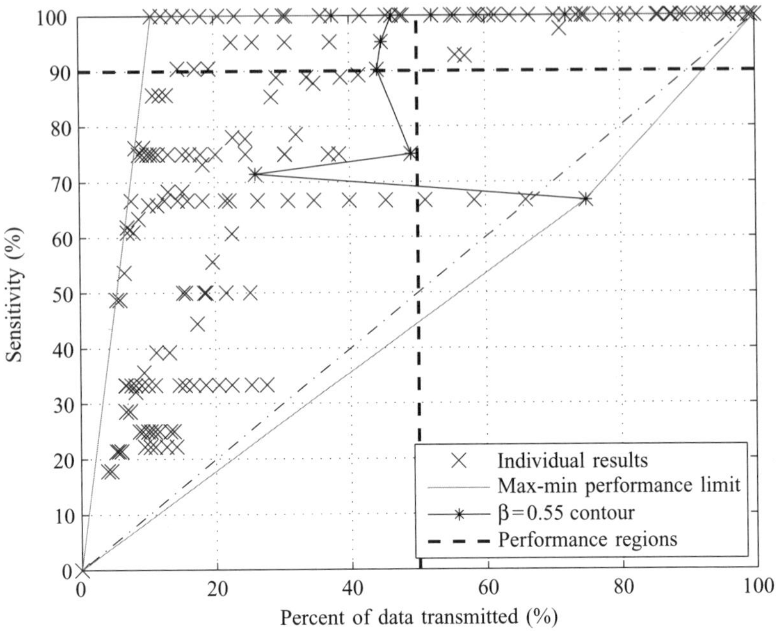

Figure 1.9. Results showing the variance in performance for the algorithm from Section 1.4.2 using data set *A* and a 5-second recording window.

Figure 1.9 thus shows such a variance graph using the data set *A* detailed in Table 1.3. The value of β for the contour line is arbitrarily chosen to show a particular operating point. In addition to the lines identified above, the graph has also been split into four regions. These come from the assumed specifications that 90% sensitivity with more than 50% data reduction lead to acceptable performance. These are arbitrarily selected in this case, and in reality, must be carefully chosen based on the particular application requirements.

It can thus be seen that for four tests (44%), the sensitivity is over 90% and less than 50% of the raw data is transmitted. The algorithm is thus operating well for these cases. For two cases (22%), high sensitivities are achieved, but a lot of data has to be transmitted. This could correspond either to tests that have a very large number of events relative to their length or simply to poor algorithm performance. It is straightforward to investigate how much data would be sent by an ideal algorithm for these tests to differentiate between these two possibilities.

For the other three cases present (33%), the sensitivity is too low and it is likely that further algorithm development is necessary. For the two cases where the amount of data transmitted is below 50%, the performance is not necessarily disastrous. From the system-level point of view (Section 1.2), provided that the algorithm implementation consumes very little power, turning off the high-power transmitter stage can lead to significant overall power reductions. This will increase the operational lifetime and it may be possible to *recover* some of the lost sensitivity by recording potentially more events in this longer lifetime.

The single point in the lower right-hand quadrant, however, clearly represents poor algorithm performance: the performance is below that of the chance line. Events that should be recorded are missed, and still large amounts of data are selected for transmitting. Whether such a situation is tolerable at all, and for what fraction of cases if it is, is an open question.

The presented method thus clearly demonstrates all aspects of the algorithm variance in one graph. It is still up to the user, however, to determine whether the level of variance is acceptable, and whether there is any pattern between the observed results. For example, subject 3 may always obtain a low sensitivity, indicating that some subject-specific modifications may be necessary, or the algorithm may perform much better in young subjects, but not old ones. Overall, it can be seen how the illustration of the algorithm variance is nontrivial, and leads to the generation of very complicated plots.

1.5 STATISTICAL VALIDATION

1.5.1 Introduction

Ideally, the results for the average and variance in the performance of an algorithm need to answer the following question: Is the algorithm performance statistically *good enough*? Unfortunately, this is not easily done. It depends on the overall aim of the algorithm and the performance necessary for acceptable operation. For example, in EEG spike detection, Gotman [16] suggests that it is unlikely that all spikes in the EEG record need to be correctly recorded and identified in order to allow accurate diagnosis by a human. The aim of diagnosis is to pool all of the available information to enable a decision based on the balance of probabilities to be made. The presence or absence of a small number of spikes should not be a critical factor in this decision process. Thus, sensitivities of 90% or 80% may be readily acceptable. The true performance can only be made in terms of the *aid to diagnosis* that miniaturized and long-lasting systems offer. Of course, however, this can only be measured once systems are in place, and it does not help in determining whether a particular algorithm is of use, and suitable for detailed investigation, optimization, and hardware implementation. In reality, such work may take several years, so we do not want to waste effort on unpromising algorithms.

1.5.2 Statistical Significance Testing

Instead, for most algorithms, it is likely that the closest possible test is showing that the average algorithm performance is statistically better than chance performance. This would illustrate that the algorithm does have some skill and acts as a basis point for further improvement. To do this for the ROC-like result curves such as the ones considered here, the Mann–Whitney U-test can be used. This is a nonparametric test and is frequently used for testing the areas under ROC curves [24].

The test works as follows. For each ROC-like average performance curve, y-axis points corresponding to sensitivity values are used to form the set s, and x-axis points corresponding to percentage of data transmitted values are used to form the set c. The following null hypothesis is then considered [25]:

"*There is no tendency for members of set* s *to exceed members of set* c."

That is, if the sensitivity values tend to be the same as the data transmited values, the algorithm performance would be along the chance line ($y = x$). If they are statistically different, and the ROC curve is seen to lie above the chance performance line, the performance must be statistically better than chance. Note that the Mann–Whitney U-test assumes that samples in s and c are independent. For the specific situation illustrated here, this seems reasonable provided that spikes are rare, short events: With any level of transmission, it is possible to get any sensitivity; it is not necessary to send large amounts of data to get a large sensitivity value.

As an example, the statistics for the 12 result lines (four different averages with three data sets) in Figure 1.7 are calculated. To do this, the x and y values from each result curve are extracted to form sets s and c, with the known end points at 0 and 100% excluded. Each set thus has 19 entries ($n_1 = n_2 = 19$) corresponding to the 19 thresholds used in the algorithm. U-values are then calculated using the formula [25]

$$U = \sum_{i=1}^{n_1} r_i - \frac{n_1(n_1+1)}{2}, \tag{1.13}$$

where r_i is the *rank*. The rank is calculated by concatenating sets s and c and then sorting into numerical order. The rank is then the end positions of the members of set s. Note that this formulation assumes that each value in s and c are unique. If repetitions are present, a slightly modified procedure should be used [24, 25].

The U-value is calculated for each result line and is shown in Table 1.4. To perform the statistical test, each value is then compared with the critical U-value, tabulated in Bland [25]. For a $p = 0.05$ two-tailed test, $U_{crit} = 113$. All of the U-values in Table 1.4 are below the critical U-value, and so the null hypothesis is rejected ($p = 0.05$ two-tailed test, $n_1 = n_2 = 19$, $U_{crit} = 113$, $U_{max} = 61$). It is thus concluded that the average performance of the algorithm is statistically superior to that of a chance classifier. Again, this does not provide any insight into whether the algorithm performance is acceptable, but it is undoubtedly a good result.

1.5.3 Confidence Intervals to Determine Test Sample Size

Once the performance metrics of interest for the algorithm have been defined and investigated, it is then possible to go back and use them to help guide the design of the algorithm testing methodology. This can be done, as suggested in

TABLE 1.4. *U*-Values from the Mann–Whitney *U*-Test Applied to the 12 Result Curves from Figure 1.7

Data Set	Arithmetic Mean	Time-Weighted Average	Total Sensitivity	Time/Event-Weighted Average
A	48	48	50	61
B	55	50	59	61
C	49	48	51	61

Section 1.3, by using statistical information to guide the amount of data that should be tested. This procedure is based on statistical confidence intervals. Note that it has to be first assumed that the biological test data is representative, containing the features of interest in all of their likely morphologies from all of the potential test setups and user populations. Confidence intervals only illustrate how much of this representative data should be analyzed. Beyond this, the results below depend only on the performance metrics and amount of data used, not on the particular algorithm being investigated.

The method is based on the idea that at a particular operating point, the algorithm has a certain *true* performance. The algorithm testing attempts to find this by analyzing a set of data and obtaining an estimate, or *reported* performance. The aim, of course, is that the true and reported performances should match. Confidence intervals calculate the range of values that, given the reported result and amount of data tested, the true result could actually lie in and reasonably produce the reported result by chance alone. This is clarified in an illustration here. As in Sackellares et al. [26], confidence intervals for the *performance* (sensitivity) and *cost* (percentage of data transmitted) are calculated separately. In principle, these values could be plotted on the ROC-like result curves to show the uncertainties at each trade-off point, but it is likely that this will lead to overcomplicated graphs. Instead they are plotted in isolation in Figure 1.10.

The starting point for confidence interval generation is the assumption of a suitable probability distribution that describes the underlying performance metric. For the sensitivity metric, a binomial distribution, which quantifies the probability of detecting a certain fraction of events from a total number, is suitable [26]. If there are a total of n events and the reported sensitivity is S, a distribution $B(n, S/100)$ is used. The 95% two-tailed confidence intervals can then be either simply read from tables [27] or generated using the MATLAB binofit function.

The intervals are illustrated in Figure 1.10a for any algorithm with a reported sensitivity of 80% [26]. Given a reported sensitivity S, these intervals show the range in which the true value of S could reasonably occur. Note that the binomial distribution assumes that the detection of any one event is independent from the detection of any other. This seems reasonable when testing a large amount of data from multiple people. In general, without detailed modeling of the probability distribution of the algorithm detections, it is not possible to avoid such assumptions entirely. Instead, however, rough values can be generated, which, while approximate, are very informative.

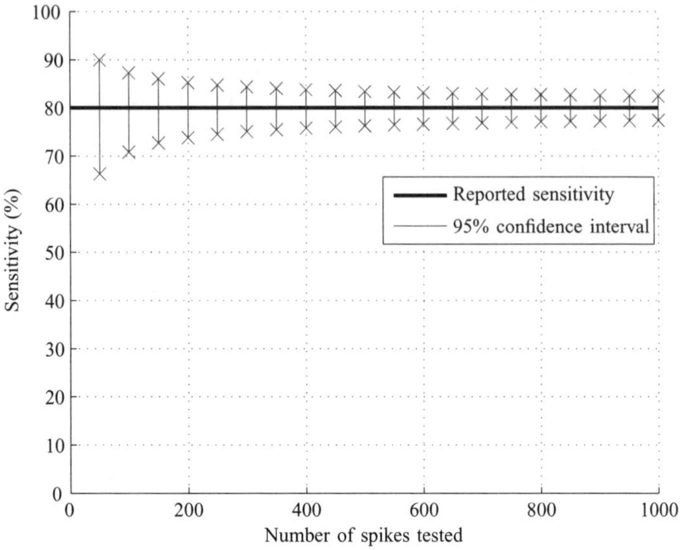

(a) Sensitivity intervals for 80% performance

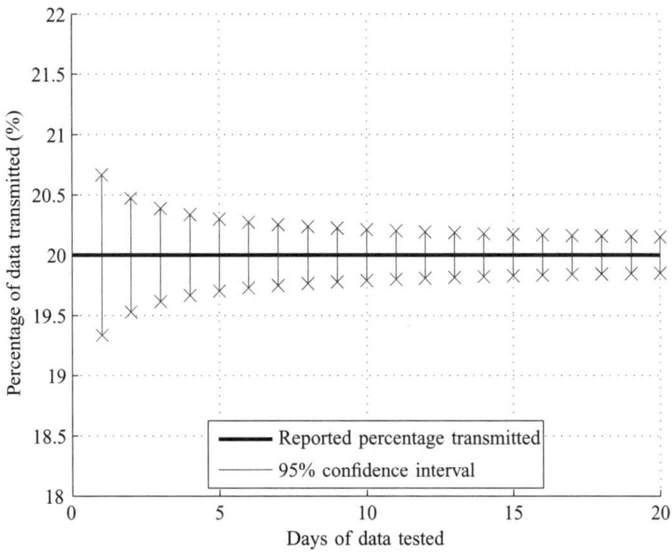

(b) Percentage of data reduction intervals for 20% performance

Figure 1.10. Estimated 95% two-tailed confidence intervals showing the range that, given the observed result, the true value could reasonably lie within. There is a 95% chance that the true sensitivity is within the interval shown. (a) Sensitivity intervals for 80% performance; (b) percentage of data reduction intervals for 20% performance. (a) Based on Reference 26.

The graph from Figure 1.10a can be replicated for different reported sensitivity values if so desired. The largest confidence intervals are found when the reported sensitivity is 50%. When testing 120 events as in data set A, it is found that the reported sensitivity may be overestimating the true sensitivity by up to 9.26%. Insufficient data is being analyzed to produce statistically confident results. For most sensitivity values, however, the confidence intervals will be smaller than this. Also, if 10,000 events, as recommended by Wilson and Emerson [10], are analyzed, the sensitivity should not be overestimated by more than 0.98% in the worst case.

If confidence intervals are wanted for the percentage of data transmitted, they can be calculated similarly. In this case, however, a binomial distribution is less suitable for use as there is no analog of the value n. Instead, in lack of a better probability distribution suitable for use, a procedure based on the false detection rate can be used to approximate the confidence intervals.

False detection rate confidence intervals are found from a Poisson distribution [26]. This distribution arises from a binomial distribution when the number of tests is very large, and the events (false detections) are rare and occur at a fixed average rate. It is thus equivalent to testing at each instantaneous time point to observe whether a false detection has occurred, and 95% two-tailed confidence intervals can thus be generated for an assumed false detection rate based on confidence interval tables [27]. For the EEG data reduction algorithm considered previously, assuming that false detections are rare, when recording the signal around one false detection, it will not *overlap* with the signal recorded from another false detection. There is thus a fixed ratio between the false detection rate and the amount of data transmitted, allowing the wanted confidence intervals to be calculated.

Results for a percentage of data transmitted value of 20% are shown in Figure 1.10b. For the approximately 16 hours of data in data set A, the confidence intervals indicate that the percentage of data transmitted should not be underestimated by more than 1%. For 800 hours, as suggested by Wilson and Emerson [10], the value should not be overestimated by more than 0.12%.

It is thus possible to use confidence interval figures to show that sufficient data is being tested in order to have reasonable confidence that the results produced are representative of the algorithm performance, and are unlikely to have occured by chance. Given the form of the intervals seen in Figure 1.10, large amounts of data must be tested in order to get significant improvements in the confidence intervals, but a guide to a reasonable amount of data to test can be made.

1.6 CONCLUSIONS

In order to optimize the interface between the biological world and the electronic world, it is essential that the performance of the interface system is accurately quantified. This will allow sensible design decisions to be taken so that effort is not wasted on the implementation of algorithmic methods that will not result in satisfactory performance. Methodological and rigorous testing, however, is not a trivial task. It is often difficult to collect sufficient data for comprehensive testing of such algorithms, and when such data is available, it is often not completely controlled and all of the recording parameters known. It is still possible, and often necessary, to perform insightful and useful testing using such data, but the limitations should be taken into account.

Using a data reduction algorithm for EEG monitoring as an example, this chapter has shown how the algorithm results can potentially be weighted to compensate for nonideal test cases. When selecting a method to present the average performance of an algorithm, care must be taken to select the appropriate averaging scheme, given the overall application aims. Otherwise, it is possible to inadvertently weight the results produced, which could lead to incorrect conclusions and design decisions being made. A potential method for quantifying the variance in the performance has also been outlined.

Furthermore, given the limitations of algorithm testing, is it essential that appropriate statistical validation be used where possible. For ROC-like result cases, as illustrated here, the Mann–Whitney U-test can be used to determine whether the algorithm performance is statistically better than that of a chance classifier. Such algorithms can then act as a starting point for algorithm improvements, although they do not help answer the question of whether a particular performance is good enough. Confidence intervals can also be used to help determine how much data should be tested, and the benefit of spending large amounts of time in collecting more data for testing. The performance metrics considered here are inevitably application specific to some extent, but it is likely that similar metrics and weightings are suitable for many different applications.

Finally, for future low-power online signal processing algorithms implemented in the analog domain, the mismatch associated with analog circuits will lead to an amount of uncertainty about the exact performance of any one chip before it is explicitly tested. For statistical algorithm testing, there is also an amount of uncertainty as the performance will never be known in advance, and will depend on how representative the test data is. Thus, it may be possible to *couple* these uncertainties in order to get improved performance. For example, the mismatch of a filter may lead to slight differences in the performance of an algorithm. If this has a large effect on the performance, mismatch robust filter topologies must be used, and this may come at the cost of other factors, such as power consumption, linearity, and noise performance. Alternatively, if the filter mismatch makes no statistically significant change to the algorithm performance, alternative topologies can be sought, potentially easing the design constraints elsewhere. The investigation of such trade-offs is essential for future interface system research.

REFERENCES

[1] M. R. Nuwer, G. Comi, R. Emerson, A. Fuglsang-Frederiksen, J.-M. Guerit, H. Hinrichs, A. Ikeda, F. J. C. Luccas, and P. Rappelsberger, "IFCN standards for digital recording

of clinical EEG," in *Recommendations for the Practice of Clinical Neurophysiology: Guidelines of the International Federation of Clinical Physiology (Electroencephalography and Clinical Neurophysiology Supplement 52)*, 2nd ed., G. Deuschl, and A. Eisen, Eds. Amsterdam: Elsevier, 1999, pp. 11–14.

[2] G. L. Krauss and R. S. Fisher, *The Johns Hopkins Atlas of Digital EEG: An Interactive Training Guide*. Baltimore: Johns Hopkins University Press, 2006.

[3] E. A. Vittoz, "Future of analog in the VLSI environment," *IEEE ISCAS*, New Orleans, May 1990.

[4] C. D. Binnie, A. J. Rowan, and T. Gutter, *A Manual of Electroencephalographic Technology*. Cambridge, U.K.: Cambridge University Press, 1982.

[5] P. E. M. Smith and S. J. Wallace, *Clinicians' Guide to Epilepsy*. London: Arnold, 2001.

[6] E. Waterhouse, "New horizons in ambulatory electroencephalography," *IEEE Eng. Med. Biol. Mag.*, 22(3), pp. 74–80, 2003.

[7] B. Gyselinckx, C. Van Hoof, J. Ryckaert, R. Yazicioglu, P. Fiorini, and V. Leonov, "Human++: Autonomous wireless sensors for body area networks," *IEEE CICC*, San Jose, September 2005.

[8] R. F. Yazicioglu, P. Merken, R. Puers, and C. Van Hoof, "A 200 μW eight-channel EEG acquisition ASIC for ambulatory EEG systems," *IEEE J. Solid-State Circuits*, 43(12), pp. 3025–3038, 2008.

[9] D. C. Yates and E. Rodriguez-Villegas, "A key power trade-off in wireless EEG headset design," *IEEE EMBS NER*, Hawaii, May 2007.

[10] S. B. Wilson and R. Emerson, "Spike detection: A review and comparison of algorithms," *Clin. Neurophysiol.*, 113(12), pp. 1873–1881, 2002.

[11] Persyst Development Corporation, "Persyst application notes: Optimizing spike and seizure detection for long-term monitoring and event notifications II," Insights newsletter, Spring 2004.

[12] H. S. Liu, T. Zhang, and F. S. Yang, "A multistage, multimethod approach for automatic detection and classification of epileptiform EEG," *IEEE Trans. Biomed. Eng.*, 49(12), pp. 1557–1566, 2002.

[13] S. B. Wilson, R. N. Harner, F. H. Duffy, B. R. Tharp, M. R. Nuwer, and M. R. Sperling, "Spike detection. I. Correlation and reliability of human experts," *Electroencephalogr. Clin. Neurophysiol.*, 98(3), pp. 186–198, 1996.

[14] A. J. Casson and E. Rodriguez-Villegas, "Data reduction techniques to facilitate wireless and long term AEEG epilepsy monitoring," *IEEE EMBS NER*, Hawaii, May 2007.

[15] J. D. Frost, "Automatic recognition and characterization of epileptiform discharges in the human EEG," *J. Clin. Neurophysiol.*, 2(3), pp. 231–249, 1985.

[16] J. Gotman, "Automatic recognition of interictal spikes," in *Long-Term Monitoring in Epilepsy*, J. Gotman, J. R. Ives, and P. Gloor, Eds. Amsterdam: Elsevier, 1985, pp. 93–114.

[17] A. J. Casson, D. C. Yates, S. Patel, and E. Rodriguez-Villegas, "Algorithm for AEEG data selection leading to wireless and long term epilepsy monitoring," *IEEE EMBC*, Lyon, August 2007.

[18] A. J. Casson, D. C. Yates, S. Patel, and E. Rodriguez-Villegas, "An analogue bandpass filter realisation of the continuous wavelet transform," *IEEE EMBC*, Lyon, August 2007.

[19] K. P. Indiradevi, E. Elias, P. S. Sathidevi, S. Dinesh Nayak, and K. Radhakrishnan, "A multi-level wavelet approach for automatic detection of epileptic spikes in the electroencephalogram," *Com. Biol. Med.*, 38(7), pp. 805–816, 2008.

[20] J. Gotman and P. Gloor, "Automatic recognition and quantification of interictal epileptic activity in the human scalp EEG," *Electroencephalogr. Clin. Neurophysiol.*, 41(5), pp. 513–529, 1976.

[21] T. P. Exarchos, A. Tzallas, D. I. Fotiadis, S. Konitsiotis, and S. Giannopoulos, "EEG transient event detection and classification using association rules," *IEEE Trans. Inform. Technol. Biomed.*, 10(3), pp. 451–457, 2006.

[22] B. Ramabhadran, J. D. Frost, J. R. Glover, and P. Y. Ktonas, "An automated system for epileptogenic focus localization in the electroencephalogram," *J. Clin. Neurophysiol.*, 16(1), pp. 59–68, 1999.

[23] G. E. Chatrian, L. Bergamini, M. Dondey, D. W. Klass, M. Lennox-Buchthal, and I. Petersén, "A glossary of terms most commonly used by clinical electroencephalographers," *Electroencephalogr. Clin. Neurophysiol.*, 37(5), pp. 538–548, 1974.

[24] S. J. Mason and N. E. Graham, "Areas beneath the relative operating characteristics (ROC) and relative operating levels (ROL) curves: Statistical significance and interpretation," *Q. J. R. Meteorol. Soc.*, 128, pp. 2145–2166, 2002.

[25] M. Bland, *An Introduction to Medical Statistics*, 3rd ed. Oxford: Oxford University Press, 2000.

[26] J. C. Sackellares, D.-S. Shiau, K. M. Kelly, and S. P. Nair, "Testing a prediction algorithm: Assessment of performance," in *Seizure Prediction in Epilepsy*, B. Schelter, J. Timmer, and A. Schulze-Bonhage, Eds. Weinheim: Wiley, 2008, pp. 249–259.

[27] R. A. Fisher and F. Yates, *Statistical Tables for Biological, Agricultural and Medical Research*, 6th ed. Edinburgh: Oliver and Boyd, 1963.

2

FULLY INTEGRATED SYSTEMS FOR NEURAL SIGNAL RECORDING: TECHNOLOGY PERSPECTIVE AND LOW-NOISE FRONT-END DESIGN

Andrea Bonfanti, Tommaso Borghi, Guido Zambra, and Andrea L. Lacaita

2.1 INTRODUCTION

Neurological disorders like epilepsy, migraines, multiple sclerosis, and Parkinson's disease represent 35% of all diseases in Europe [1], accounting for about 46 million cases over 466 million EU residents, for an aggregate cost of more than 80 billion euros. On a more global scale, 450 million people are affected by neural disorders and 6.8 million people are estimated to die every year as a result of these pathologies worldwide [2]. Moreover, this toll is expected to quickly rise since disorders such as Parkinson's disease and Alzheimer's are forecasted to massively spread in our aging society.

In this frame, electronics can play an important role contributing to developing new monitoring and diagnostic devices and new rehabilitation and therapeutic tools to restore lost skills or to deliver effective treatment of neural diseases. For example, in the last decade, advances in electroencephalogram (EEG) acquisition and processing made it possible to introduce neural activity monitors in

CMOS Biomicrosystems: Where Electronics Meet Biology, First Edition. Edited by Krzysztof Iniewski.

the clinical practice [3], and a number of therapies and prosthetic systems gradually entered hospitals and clinics to help patients live a better life: deep brain stimulation for Parkinson's disease [4] and cochlear implants [5] are just some examples of neurotechnologies today available in health-care centers worldwide. Despite their widespread adoption, these tools are still in an early stage of development and significant room for improvement is left [6]. This chapter offers an introductory overview of microelectronic systems for neural probes, pointing out the major issues met in deciding their architectures and function partitioning. Then the discussion will focus on the integration of low-power low-noise multichannel amplifiers for extracellular action potential (AP).

2.2 NEURAL SIGNALS AND FRONT-END REQUIREMENTS

Interfacing brain and electronic circuits implies the effective transduction of ionic-driven potential spikes into electronic signals. Requirements on the electronic side are basically set by the anatomy and physiology of the brain cells. A comprehensive description of the neuronal cells, of their morphology, of their working principles, and of their interaction dynamics would certainly be an intriguing journey through a wide spectrum of different topics and disciplines, but it clearly falls outside of the scope of the chapter. Therefore, in following, we will just recall some few key facts and figures, ultimately setting the specs of an implantable neural probe and the requirements for any system for neural monitoring and therapeutics.

Neurons comprise three different parts: a cell body called soma, a number of input branches (or processes) called dendrites that collect information coming from other neurons, and an output process called axon that connects each neuron to other nerve cells or muscles. The soma contains everything, is usually inside a cell body (nucleus, Golgi apparatus, endoplasmic reticulum, mitochondria, liposome, and ribosome), and has a characteristic diameter of 20 µm. Dendrites and axons can reach lengths up to hundreds of centimeters and their diameter defines the speed of information propagation along the nerves. As a matter of fact, neurons can communicate with each other by means of electrical pulses called APs or spikes. As all other cells, they contain high concentrations of K^+, while keeping high concentrations of Na^+ and Ca^{2+} outside the soma. The unbalance of charge across the cell membrane generates the so-called resting potential of –60 mV.

Such a distribution of positive and negative ions is at the basis of cell-to-cell communication: The reason behind this arrangement dates back to the origin of life when the first small single-cell organisms arose in the sea, a Na^+-rich environment. In such a hostile habitat, the development of a Na^+-proof membrane was the only choice nature had to defend the living cells from the environment. On the other hand, Na^+ abundance suggested using this ion charge for signaling. As a matter of fact, the generation of each spike starts with a flow of Na^+ into the cell. The flow is controlled by ion channels, proteins that fold in small pore structures and regulate the flow of different ions across the cellular membrane.

The dynamics of spike generation can be divided into three steps: the initial inflow of Na^+ that depolarizes the neuron (i.e., the inner potential of the cell becomes positive by a few millivolts); an outflow of K^+ ions that hyperpolarizes (i.e., makes the inside potential negative again); and finally, a refractory period, preventing the neuron to fire a new spike for a few milliseconds [7]. Na^+ ions have mobility six orders of magnitude lower than electrons in Si. Their intrinsic low speed as charge carriers has impact on both the treelike morphology of the neurons and the way they connect to each other to give rise to intelligent behavior. Nature chose to link these slow devices in complex redundant networks. If the details of the brain processing algorithm are still to be fully understood, its architecture and morphology were clear since the early experiments at the beginning of the last century: The brain is a highly parallel processor with a high density of slow devices. It follows that a deep investigation of neural mechanisms requires simultaneous sensing of many neurons from a variety of different brain regions.

To really appreciate the dynamics of a network and the role of each neuron, investigators need different tools to zoom in and out at different temporal and spatial resolutions. The overall activity of the neuron population can be readily detected using a few-centimeter-wide electrode on the surface of the skull. As a matter of fact, this was the first kind of information ever recorded from the brain [8]. Unfortunately, these signals (called EEG) are loosely representative of the activity of each individual cell since they are a superposition of thousands or millions of signals from firing neurons filtered and attenuated by the five cortical layers, the skull, the skin, and finally, the sensor, or electrode. Even if EEG signals have been successfully used to study important cerebral processes (such as epileptic seizure recognition [9], anesthesia monitoring, and sleep disorders analysis [3]), they cannot provide key information on smaller-scale phenomena. The highest temporal and spatial resolutions can be achieved with an invasive recording technique referred to as single-unit extracellular AP recording. This method was first pioneered by David Hubel in the late 1950s and has led to the discovery of a lot important properties [10]. In these measurements, a needlelike electrode with a small conductive tip is inserted into the cortex. The tip diameter is about 20 μm close to the size of a single cell. The electrode is usually fixed to a microdrive that precisely controls its position: The closer the tip to a cell, the larger the signal collected as the neuron fires.

2.2.1 Signals and Noise

Extracellular APs are biphasic pulses with a typical duration ranging from 750 and 1000 μsec and a power spectrum residing in a frequency band from a few hertz to 5–10 kHz (see Fig. 2.1). The shape and amplitude of each pulse depend on the relative tip-to-cell position. Biophysical arguments and experimental recordings demonstrated that spikes collected near the soma show a negative trough followed by a positive peak, while extracellular APs collected in proximity of dendrites have peaks followed by a trough [7]. The peak amplitude of extracellular APs ranges from several tens to few hundreds of microvolts, the specific

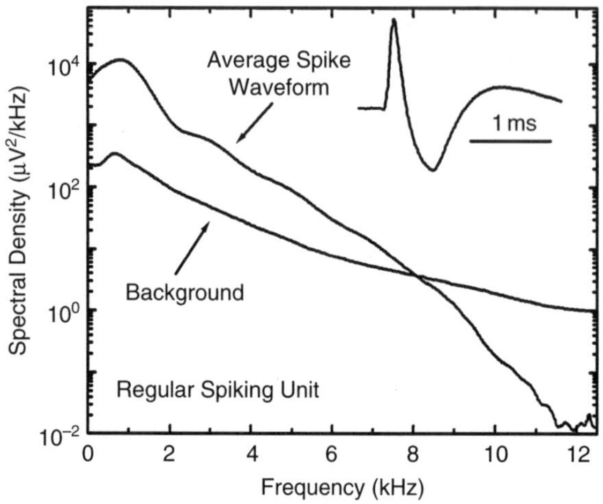

Figure 2.1. Spectral components of the average action potential waveform and power spectrum of background signals and noise collected by neural electrodes. *Inset*: Schematic waveform of action potential pulses.

shape being a complicated function of the electrode materials and geometry as well as the tip-to-cell distance.

This variable is of extreme importance since the spike amplitude rapidly decays as a function of the distance from the cell. In Henze et al. [11] and Pettersen and Einevoll [12], a $1/r^2$ decay was estimated. It follows that a single electrode can collect signals from a small volume around the tip. An experiment performed by Henze et al. [11], using simultaneous recording of intracellular and extracellular signals, demonstrated that the largest spikes are fired at distances <50 µm from the electrode, while at distances >140 µm no distinguishable waveforms are collected.

Given these numbers, each single electrode probing a thin cortical layer (60-µm thick) should receive signals from all the neurons within a 50-µm radius. Taking into account the typical neuron density, the theoretical total number of cells sensed by each electrode should be more than 140. This number is much higher than what is sensed in practice. As a matter of fact, a single electrode usually receives distinctive signals from just three to four neurons. The remaining hundreds or even thousands in its proximity contribute with their tiny APs to the background noise that is the most important source of noise in extracellular recordings [13].

2.2.2 Electrode, Its Noise Source, and General Specs for Front-End Design

In contrast to the broad information content of EEG signals, extracellular recordings offer a window on single cell behavior. The two approaches lay at the opposite, looking either at too big or too small cell populations. A possible solution to this mismatch was proposed in the mid-1960s by Prof. Moll at Stanford who envisioned the use of highly scalable manufacturing techniques such as lithography to produce array of extracellular electrodes [14]. Such an invention offered not only high resolution but also the valuable possibility to simultaneously record from a number of different cells. Since then, silicon-based microelectrode array evolved in two different architectures, the so-called Michigan probe and the Utah array.

The Michigan probe is the natural evolution of the original idea incubated at Stanford. Deep boron diffusion defines the electrode shank, and recording spots are exposed along the shank, thus sensing at different depths in the tissue. Each electrode can be coated with different materials (iridium, gold, and platinum), and up to 64 channels can be packed in a single device. The Utah array is instead a three-dimensional (3D) array made of a matrix of needlelike 1.5-mm-long electrodes [15]. In contrast to the Michigan probe, the recording tips of this array can be either placed all on the same plane or in a 3D arrangement depending on the size of each needle. The structure is obtained by dicing and chemically etching an n-type silicon wafer. The conductive tips are coated with Pt and then insulated with polyamide. Whatever the electrode structure is, the ultimate goal in the design of neural recording systems is the integration of these sensors with receiving and processing circuits on the same substrate in order to reduce the lengths of interconnects that can collect additional noise and interferences. A reasonable volume for such a system can be $4 \times 4 \times 1.5\,\text{mm}^3$. The contact surface with the surrounding tissues is an important parameter. Many studies [16, 17] showed that the power dissipation density of an implantable neural system should not exceed $800\,\mu\text{W/mm}^2$ to prevent tissue overheating and necrosis. For a 16-mm^2 surface, this corresponds to about a 12.5-mW power budget for a complete system.

Electrodes are the physical interface between ionic currents and integrated electronics, and thus their specific characteristics influence the front-end design in terms of noise performance, input impedance, and filter sizing. Only a small number of materials are available for neural electrodes applications: They have to be neurocompatible and conductive and just a handful of materials can match these requirements—stainless steel, platinum and iridium are the most commonly used. Often the exposed tip is processed with other materials (black Pt, iridium oxide, polymers) to increase the surface and thus reducing the overall impedance.

The impedance of the electrodes is not directly defined by the electrical properties of the materials. It rather depends on the particular chemical reactions taking place at the interface between the tip and the electrolyte filling the extracellular matrix, the so-called electrode–electrolyte interface. When a metal is immersed in a conductive solution, it triggers the formation of the Helmholtz layer: a double layer of charged molecules that can be approximately modeled as a resistance with a capacitor placed in parallel [18] (see Fig. 2.2).

The water molecules are attracted to the metal surface (usually negatively charged) forming the Helmholtz inner layer, while another layer of hydrated ions

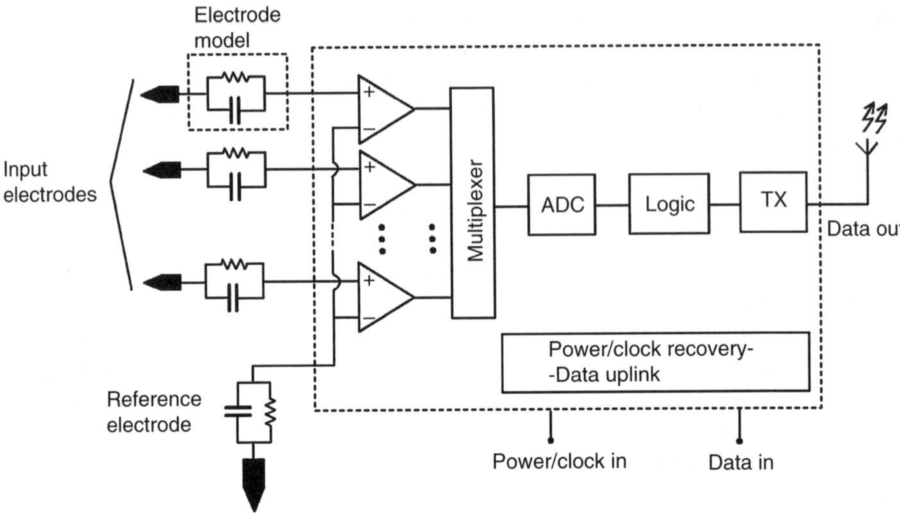

Figure 2.2. Reference architecture of a fully integrated neural recording system. The low-noise analog front end, interfaced with the electrode array, is followed by analog to digital conversion and high-throughput uplink data transmission steps. The power source is typically provided by induction link electromagnetic coupling, which also supports a low bit rate downlink for control purposes.

forms the outer Helmholtz layer. The two opposed charged layers cause a capacitive response, while the small leakage currents flowing through them are accounted by a resistive impedance, $\mathrm{Re}(Z_e(\omega))$, which is dependent on frequency. In the frequency band of interest, the electrode impedance model can be mostly approximated by a single capacitor. Commercially available electrodes for single-unit extracellular recordings feature impedance at 1 kHz ranging from several hundreds of kiloohms up to more than 1 MΩ. These values correspond to a capacitance in the 150–350-pF range. The front-end amplifier should be properly designed to interface such large capacitive impedance without signal losses. On the other hand, the real part of the electrode impedance contributes to thermal noise like an equivalent resistance [19]. It turns out that, for a typical electrode, both the thermal noise due to the electrode impedance and the background fluctuations generate 5–10 μV_{rms} over a 10-kHz bandwidth. These values suggest setting a conservative 5 μV_{rms} as the upper limit for the input-referred rms noise of the front-end amplifier within the signal band.

In addition to noise, the physiological fluctuations of charged carriers in the extracellular matrix as well as additional reactions (binding and unbinding of extracellular proteins on the tip surface) generate a DC bias on the tip that can be as high as hundreds of millivolts. Since neural extracellular signals are 1000 times smaller, a DC removal technique will be required to be able to record the APs. In summary, this introductory discussion defines four key requirements for a neural probe that have impact on the design of the complete system:

1. The parallel strategy of cortical information processing makes multichannel recording mandatory for both monitoring tools and neuroprosthetic devices. An array of electrodes is needed.
2. The neural signal spectrum and the characteristics of the background signals collected by the electrodes immersed in the extracellular matrix require band-pass filtering to remove both DC offset and high-frequency noise.
3. The noise of the electrode and the background noise due to distant firing neurons set the minimum noise requirements of the front-end electronics.
4. The thermal conduction properties of neural tissues determine the total power budget for a complete neural recording system (amplifiers, analog to digital converter [ADC], processing unit, receiver [Rx] and transmitter [Tx]).

These specs, together with additional limitations set by the available technologies, guide system partitioning and force the careful design of all the functional blocks.

2.3 SYSTEM ARCHITECTURE AND POWER BUDGET PARTITIONING

In principle, a neural recording system is a conventional signal acquisition chain comprising an array of sensors and amplifiers, analog-to-digital conversion, a processing unit, and a wireless link (Fig. 2.2). Moreover, a power management block is required to provide stable power supply to the whole system. Most of the design challenges come out from the objective to bring the system to a single chip solution, meeting tight noise specs and power budget limitations. As seen in the previous section, the noise floor of 5 μV is basically set by disturbs picked up from the environment and by the electrode impedance. The tight power budget of 10–15 mW is set by the limited system size and by brain physiology. This Section is now devoted to describe the system architecture, the partitioning of the power budget, and the target performance achievable by the different building blocks.

2.3.1 System Architecture

The small amplitude of the cellular neural signals and the peculiar capacitive impedance of the probe suggest selecting and amplifying the signals immediately, before performing sampling and analog to digital (A/D) conversion. The large potential variations due to background local field potentials (LFPs) and disturbs force the use of differential stages. A reference electrode immersed into the same extracellular environment provides the reference potential. After amplification and filtering, the signals are sampled, digitalized, and transmitted to a backbone receiver. In addition, a power source is required to provide stable power supply to the whole system. This stage may also provide a low-rate downlink for setting and control signals. Most of the design challenges derive from the goal of squeezing a complex system on a single die with close volume constraints, tight noise specs, and power budget limitations. Increasing pressure on performance is also provided by the expectation to handle a large number of array electrodes. The number of analog channels increases following the array size, thus requiring more power budget. A similar trend is also shared by sampling, A/D conversion, and transmission stages. As the number of input channels increases the overall information throughput, the generated bit rate, will rise, thus demanding more power. It is therefore obvious that full integration of neural recording systems interfacing those large electrode arrays required for high-density neuron-by-neuron interfacing needs to minimize down to the ultimate limits the energy required by analog and digital processing and transmission. Tough challenges are also met in providing energy to the system, which can be easily characterized by current values of several milliamperes, needed under mostly continuous operation. Since use of implanted batteries has to be avoided due to their size, potential toxic composition, and finite lifetime, power transfer by inductive link is seen as the only viable solution to power the implanted circuit [20–24]. This choice makes efficient energy coupling and management an additional requirement for these systems. In summary, matching very low-power consumption while meeting low-noise performance of each analog front-end and parallel operation of a large number of channels in an implantable system is a highly challenging design task. Before going into the details of the analog front-end design, the following paragraphs offer an overview of the state of the art of the different system blocks, pointing out the most recent performance, the contribution to energy savings provided by the long-lasting downsizing of the silicon technologies, and the key points to be addressed by future advances.

2.3.2 Power and Noise Constraints in Amplifier Design

Power consumption of the analog front end is essentially set by the noise requirements and by the g_m/I transistor efficiency. A noise efficiency factor (NEF) was proposed in Steyaert et al. [25] to compare the noise performance of different designs. The starting point is the expression of the total equivalent input noise of a bipolar differential amplifier. Taking into account only thermal noise of the input transistors and neglecting base resistance contribution, the input-referred rms noise, $V_{\text{in,rms}}$, is given by

$$V_{\text{in,rms}} = \sqrt{\text{BW} \cdot \frac{\pi}{2} \cdot \frac{4kT}{g_m}}, \tag{2.1}$$

where BW is the frequency bandwidth. The NEF of the system can be therefore defined as

$$\text{NEF} = V_{\text{in,rms}} \sqrt{\frac{2I_{\text{tot}}}{\pi U_T 4kT\text{BW}}}, \tag{2.2}$$

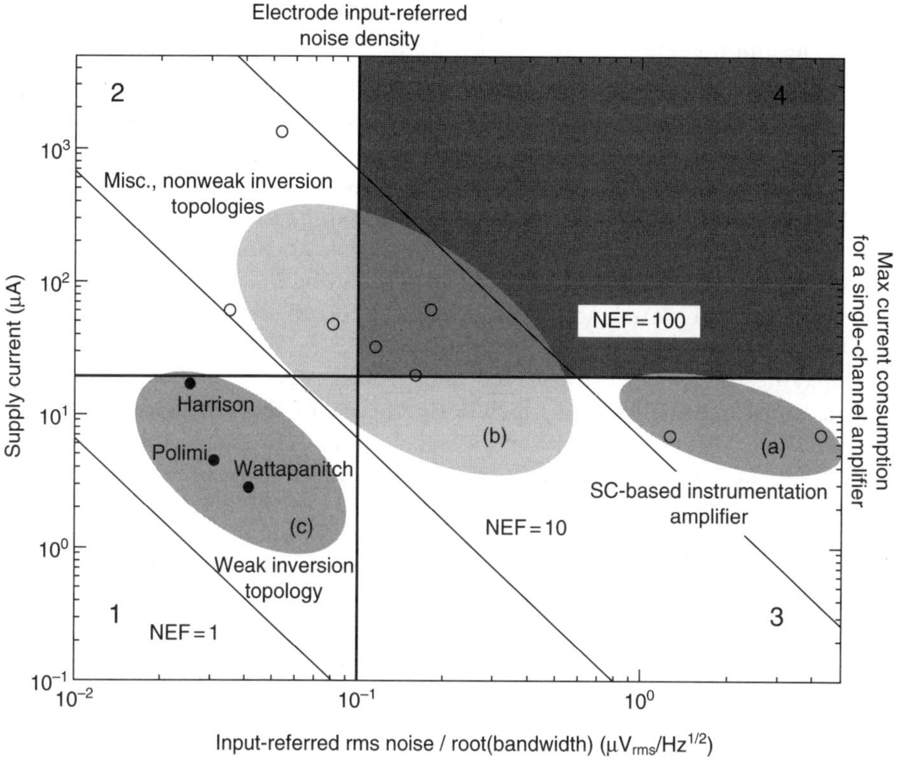

Figure 2.3. Noise performance and current consumption of different OTA topologies. Circuits within Region 1 comply with both the noise and the low-power requirements of neural recording front ends. Constant NEF lines are also drawn to highlight the power–noise trade-off. Circuits exploiting subthreshold operation of CMOS transistors guarantee the best in class performance.

where I_{tot} is the total supply current, $U_t = kT/q$ is the thermal voltage, and BW is the signal bandwidth of the amplifier. Equation 2.2 suggests a theoretical limit of $\sqrt{2}$ for a bipolar differential pair. Since CMOS transistors have a g_m/I ratio lower than the bipolars, the NEF values for CMOS amplifiers, like those considered in the following, is expected to be higher, Figure 2.3 shows the performance of state-of-the-art designs.

The horizontal axis quotes the input noise spectral density, while the current consumption of the amplifying stages runs on the vertical axis. The vertical line drawn at $0.1\,\mu V_{rms}/\sqrt{Hz}$ roughly corresponds to the input noise from the neural probes. The horizontal line at $20\,\mu A$ defines an upper limit of the current budget allowed to these stages. Constant NEF lines highlight instead the noise/consumption trade-off reached by the different solutions. The design plane results clearly divided into four different regions:

- Region 1 includes designs with both noise performance and power consumption matching the application requirements

- Region 2 includes design with good noise performance but high power consumption
- Region 3 comprises amplifiers with very low power consumption but noise floor exceeding the noise of typical neural electrodes
- Region 4 includes designs not matching neither noise nor power specs

The results taken from the literature fall into three different clusters: cluster (a) mainly includes switched capacitor-based instrumentation amplifiers; they operate at reasonable low power but their noise is dominated by the white noise sources of the differential inputs and $1/f$ noise is added when sampling frequency is lowered close to the signal bandwidth (about 8 kHz). For their specific features, they may be used in LFP/EEG recordings where electrode noise poses more relaxed constraints. Cluster (b) includes a variety of low-power design: JFET input buffers, current-feedback single-stage amplifiers, or simple folded cascode topologies where little attention was posed to minimize the white noise contribution of the input stages. Proper transistor sizing and operation in the subthreshold regime, where the g_m/I efficiency of MOS transistors peaks, are instead the key point addressed by designs in cluster (c), which are the best in class results so far reported. It is worth noting that the ultimate NEF value for a differential CMOS analog front end is 2.02. This value is already close to the published results. It follows that even if technology downsizing will proceed further, the current consumption of the analog front end is expected to remain in the order of 5 μA per channel, while power dissipation may marginally benefit from further scaling of the voltage supply.

2.3.3 Analog-to-Digital Conversion

Analog-to-digital conversion follows low-noise amplification. As already pointed out in the previous section, the maximum amplitude for an extracellular AP is about 1 mV, while the minimum signal is approximately 20 μV. Therefore, the ratio between the full-scale range (FSR) and the ADC least significant bit should be better than 1 mV/20 μV ≅ 50, leading to a minimum number of 6 bits. In addition, the ADC resolution should be large enough to prevent quantization noise to impair sensitivity. In other terms, the quantization noise level should be kept well below the electrode noise. This requirement translates into

$$\frac{\text{FSR}}{2^n G\sqrt{12}} << 5\mu\text{V}_{\text{rms}}, \tag{2.3}$$

where G is the amplifier gain and n is the bit number. The obvious result is that the higher the gain, the lower the noise added by the ADC quantization. On the other hand, denoting as $A_{\max}$ the maximum amplitude of the extracellular AP, the maximum gain is given by

$$G_{max} = \frac{\text{FSR}}{A_{max}}, \tag{2.4}$$

leading Equation 2.3 to be written as

$$\frac{A_{max}}{2^n\sqrt{12}} = \frac{1\text{mV}}{2^n\sqrt{12}} << 5\mu\text{V}_{\text{rms}} \tag{2.5}$$

$$n >> \log_2 \frac{A_{max}}{\sqrt{300}\mu\text{V}_{\text{rms}}} \tag{2.6}$$

that gives again a minimum number of 6 bits. In practice, taking into account that the effective number of bit, ENOB, is always lower than the nominal figure, the adoption of an 8-bit ADC will provide a safe margin to the design.

Let us now consider the power consumption (P), which is definitely affected by the requested throughput. The ADC sampling rate is set by the channel parallelism and the signal bandwidth; serving 64 electrodes sampled at 30 kHz requires a sampling rate, F_S, of about 2 Msps. In this frame, an 8-bit resolution translates into a total throughput of about 20 Mbps. Both these values for resolution and sampling frequency are quite moderate and suggest for minimum power, to rely on the effective charge redistribution process of a capacitive successive approximation register (SAR) architecture.

In more general terms, the ability of an ADC to fulfill throughput requirements with minimum power consumption is summarized by a Figure of merit (FOM) defined as

$$\text{FOM} = \frac{P}{2^{\text{ENOB}} \cdot F_S}. \tag{2.7}$$

Such an FOM has empirical ground and takes into account that the ADC power consumption scales with the dynamic range and throughput requirements. It basically gives the energy dissipated per quantization level. Figure 2.4 shows its historical trend for Nyquist rate and oversampled topologies [26], the dashed line corresponding to a twofold efficiency improvement every 18 months. It is clear that technology downsizing is making possible steady improvements of ADC efficiency, with the Nyquist rate architectures benefiting the most [27]. This result may appear somehow surprising since it is well known that, contrary to digital circuits, analog circuits may require larger power dissipation as the voltage supply scales with the technology. In fact, to keep constant the signal-to-noise ratio, the noise level should be improved at the same pace of the voltage headroom reduction. Since noise trades with power dissipation, larger power consumption should be expected. The reason for the opposite trend is essentially that performance of moderate resolution ADCs are not limited by thermal noise but by process constraints (i.e., minimum capacitance that can be integrated) and component

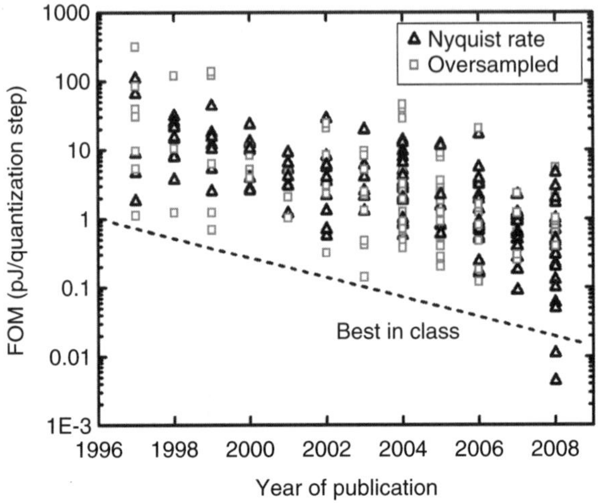

Figure 2.4. A/D Figure of merit: historical trend of Nyquist rate and oversampled A/D. The dashed line corresponds to a twofold efficiency improvement every 18 months.

mismatches [27]. Both these limitations improve with technology scaling, thus making it possible to attain better efficiency. In addition, as the energy cost for digital operations improves with scaling, a new analog design style is emerging. Power-hungry, high-performance amplifiers tend to be replaced by simplified analog stages and more power-efficient digital circuits are introduced to compensate for nonidealities and mismatches. The relentless technology downsizing, the emerging digitally assisted analog design style, the accrual of design experience, and the judicial use of all process options are pushing up ADC efficiency. In this frame, two recent results have clearly step out of the trend line in Figure 2.4 as examples of highly efficient design.

The work of Craninckx and Van der Plas [28] presents a SAR architecture in a 90-nm technology with 65 fJ/quantization step FOM, while the best result in the literature has been achieved using a 65-nm technology, demonstrating a 10-bit, 10-Msps SAR ADC with 4.4 fJ/quantization step [29]. On the other hand, the adoption of less scaled technologies is the reason for the less aggressive results so far reported in neural recording. The systems in Wise et al. [14], Harrison et al. [30], and Gosselin et al. [31] show much higher FOM values, ranging from 2 to 10 pJ/quantization step, the best performance being 2.72 pJ/quantization in a 0.18-µm technology. However, a lot of room is left for energy savings. Even taking a conversion energy of 0.1 pJ/quantization step, well above the latest results in Figure 2.4, the complete digitization of a channel requiring 8-bit resolution and 30-ksps sampling rate translates into a power budget less than 1 µW per channel, in agreement with the target proposed by Zumsteng et al. [32].

It should be also noted that throughput requirements could be relaxed using spike detection and sorting compression algorithms. Just a simple threshold detector can be indeed used to identify neural spikes [30], lowering to a mere 1–2 kbit/s the data rate needed for each electrode (i.e., less than 150 kbit/s for 64 channels). The drawback of this simple scheme is that no information is given on which neuron is firing, which is a serious limit in neural prosthesis systems where the activity of each neuron has to be isolated to better predict the intended movement [33]. In this perspective, researchers in Olsson and Wise [34] demonstrated a digital circuit that is able to catch the spike features needed for spike sorting; namely, the amplitudes of the peak and the trough as well as the width of the spike [35]. A similar approach, but implemented with very low-power analog circuits, is reported in Bonfanti et al. [36], demonstrating a throughput reduction to 5 kbit/channel and a capability to isolate single neuron activity. However, although featuring interesting performance, spike-sorting approaches have never been used so far in real neurophysiology experiments where careful spike clustering is still mandatory. Spike sorting therefore remains an area where major breakthroughs will be essential to improve performance and reliability of fully implantable neural systems.

2.3.4 Data Link

Data transfer from or to implantable devices is one of the most crucial functional blocks of a brain–machine interface (BMI). The inductive link adopted to supply power may be also used to download signals to control the implantable device. In some cases, this channel has been used for the uplink, even if at a low data rate [37]. Referring to the reference system in Figure 2.2, the most natural choice is therefore to use a high-throughput, low-power TX to transfer the large amount of data generated by the array electrodes, while the downlink, with much lower capacity, is provided through the inductive link. This section is devoted to give an overview of the viable solutions for the high-capacity TX. The next section will instead shortly deal with the inductive links.

The design of high-capacity, low-power data links is a very challenging and hot design issue. For short-range applications, two options are being considered: wireless ultrawide band (UWB) and infrared (IR) transmission. Due to the tight power budget, the key figure is the energy spent per transmitted bit. Figure 2.5 shows the performance of different wireless and IR TX systems taken from the literature and makes possible to appreciate the state of the art in this rapidly evolving field. Note that transmitters for WLAN (IEEE 802.11a/b) and short-range communication protocols (Zigbee and Bluetooth) are very power hungry and do not match the high-throughput requirements. As an example, Bluetooth transmitters need about 100 nJ/bit, most of the energy being spent in baseband processing and in the coherent radio frequency (RF) circuitry. Picoradio solutions are therefore implemented reducing to the essentials the transmitter structure, avoiding coherent transmissions, and using simple communication protocols. This is the case of UWB TXs that works by transmitting nanosecond pulses with frequency components in the 3.1–10.6-GHz band. The transmitter structure has digital pulse generators, while no phase-locked loop (PLL) or tuned filters are

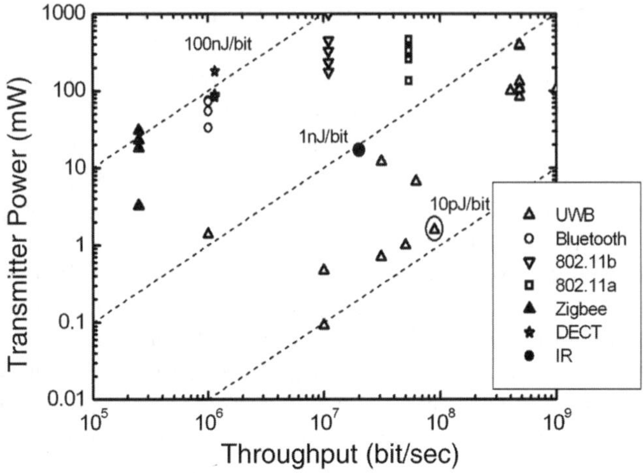

Figure 2.5. Power consumption versus throughput dependence of wireless and IR fully integrated transmitters. The dashed lines highlight the energy needed per bit transmission. The circled triangle refers to the only integrated transmitter for neural recording applications [40].

involved. The power amplifier (PA) has an intermittent operation, thus reducing the power dissipation. Throughput up to 10 Mbit/s has been demonstrated for an RF identification (RFID) tag with less than 10 pJ/bit energy consumption and a communication range of 13.9 m [38].

Note that taking as a reference the 240 kbps generated by each neural recording channel (30-ksps sampling rate, 8-bit resolution), a 10 pJ/bit transmission energy corresponds to an impressive target of just 2.5 μW per channel. It is true that possible shortfalls of UWB transmissions may result from the tissue absorption at such high frequencies, but preliminary simulation studies suggest that such an issue could be fixed in the future [39]. Moreover, encouraging results have been already demonstrated. The UWB transmitter in Chae et al. [40], designed for neural recording systems in a 0.35-μm technology, reaches a 90 Mbit/s throughput with a power consumption of only 1.6 mW (i.e., 17 pJ/bit) even if using an off-chip antenna, which may pose issues in implanted systems. Unfortunately, no details have been given on the antenna size and on the communication range to elaborate further.

In addition to wireless approaches, IR signals are being considered as an alternative. In this solution, the implanted chip is interfaced with a laser positioned under the head skin to exploit the low skin absorption in the IR light spectrum. Figure 2.5 shows the energy efficiency achieved with the IR systems reported in the literature [41]. A power consumption of 20 mW is needed to reach a throughput of 20 Mbps, corresponding to 1 nJ/bit, two orders of magnitude higher than wireless TXs. Moreover, IR transmission requires precise alignment between the laser source and the receiver, and the range is limited to less than 1 cm. At the present stage, we may therefore conclude that, even if further developments may improve this performance, wireless solutions seem to be more promising.

2.3.5 Power Source

Intensive research is also ongoing on inductive coupling systems for powering implanted circuits. These systems are basically composed of an external transmitter (typically a class E amplifier) driving a coil and an on-chip receiver that picks up the signal reaching a secondary magnetically coupled coil. This coil is tuned to the transmitter frequency with a proper shunt capacitor. The transmitter generates an RF magnetic field in the megahertz range, thus inducing a harmonic voltage in the receiver tank. An RF limiter is adopted to clamp the amplitude of the induced voltage and to protect the implanted circuit from overvoltages. A wideband rectifier converts the AC signal into an unregulated DC supply voltage, and finally, a voltage regulator provides a stable and regulated power supply to the implanted device.

One of the most critical parameters of those systems is the carrier frequency. High frequencies have the advantage of transferring more energy per unit of time and to allow a smaller tank inductor and capacitor. Unfortunately, increasing the carrier frequency determines more current consumption in the receiver and more power absorption in the biological tissue, degrading the efficiency. Typically, the frequency is chosen in the 1–10-MHz range, considering also the need of a particular clock frequency on the chip (e.g., A/D converter clock).

The inductive link is also used for back telemetry, that is, to transmit data to the implantable system. These data are exploited to set and adjust some parameters of the recording system: for example, to select a bank of amplifiers, to control the gain of the amplification chain, or to set the A/D converter resolution. A low-rate control channel is enough for these purposes and can be easily obtained by modulating either the amplitude [20] or the frequency [23] of the power carrier. A transfer rate of 16 kbit/s has been demonstrated in Harrison et al. [24] using an amplitude shift keying modulation of a 2.56-MHz carrier. It is interesting to note that the overall power management system, in a 0.6-μm technology, needed 1 mW to deliver 8 mW of power, while an additional 0.5 mW was required by clock and data recovery circuits. These numbers highlight that the impact of supply coupling and regulation on the overall power budget is very significant. High-efficiency, low-power voltage converters without off-chip components are essential enabling blocks for future implanted biomedical devices. Research breakthroughs in this field are still needed.

2.3.6 System Perspectives

In summary, technologies and circuit designs are opening the way to high-density neural recording systems capable of handling high-throughput transmission of sampled signals derived from hundreds of simultaneously recording sites. The technology trends discussed above show that the 240-kbps throughput generated by each channel of a neural recording system (30-ksps sampling rate, 8-bit resolution) may require just 5 μW for low-noise amplification and filtering, 1 μW for digitalization, and about 2.5 μW for wireless transmission, leading to an estimate of overall power allocation of about 10 μW per channel. Spike detection and sorting may add further room for system flexibility. In the future, technology scaling is expected to further improve the energy performance of A/D converters and RF transmitters. Advances are certainly needed to improve the overall efficiency of the induction link power source, which otherwise may need up to 10% of the overall power budget. Packaging technologies may be asked to allow the heterointegration of some off-chip components (inductors, antenna, and filters).

Neural recording systems share with other systems on chip most of all these blocks. Therefore, it can be expected that design advances made in the frame of other applications will be also beneficial to these systems. On the other hand, the analog front end is quite peculiar and needs a custom approach. This is due to the large capacitive input impedance of the electrode, the tight input noise floor, and the need to decouple the large DC input offset. The next sections will be therefore devoted to cover the most relevant issues met in the design of these stages.

2.4 PREAMPLIFIER AND FILTER

2.4.1 Preamplifier and Filter Requirements

The small amplitude of extracellular neural signals, the high impedance of the electrode–tissue interface, and the need to reduce noise and disturbs caused by clock artifacts and aliasing impose the performance of amplification and filtering before sampling and converting the signals into the digital domain. In these systems, an integrated preamplifier for neural signals must

1. Have much higher input impedance than the electrode–tissue interface (i.e., much larger than 1 MΩ at 1 kHz of frequency) and negligible DC input current
2. Block the DC offset generated at the electrode–tissue interface (up to 300 mV) to prevent the saturation of the amplifier
3. Select the band where most of the signal power lies, that is 300 Hz–10 kHz [34]; such a band-pass filtering removes both low-frequency signals such as LFP that lays in the 1–100-Hz band and can be as large as several millivolts, and high-frequency noise that can be aliased back into the signal band due to sampling performed by analog multiplexing
4. Have sufficient dynamic range to convey spikes having a peak-to-peak amplitude of up to 500 μV
5. Have a high common-mode rejection to minimize interference from 50–60-Hz power line noise and other common-mode disturbs such as human breath, brain pulsing, heartbeat, and movement artifacts, and a high-power

supply rejection if the power supply noise is significant, for example, for an AC inductive power link
6. Feature an input-referred noise lower than the electrode/background noise that can be as high as $5\,\mu V_{rms}$
7. Have minimum power consumption so as not to impact the overall power budget

This Section is devoted to discussing the design options and the most relevant choices made to comply with these requirements. The result is the best in class low-noise amplifier for neural recording so far reported, which is an improved version of the neural amplifier described in Borghi et al. [42]. Numerical values are given referring to the actual design that was implemented in a 0.35-μm AMS CMOS process with a 3-V power supply.

In order to comply with the common-mode rejection requirements, there is no other choice than using a differential input stage. Its gain has to be a significant fraction of the overall amplification required from the electrode to the A/D converter. Considering a power supply and an FSR for the A/D converter of 3 V, the whole FSR is fully exploited, adopting a total gain of 5000–6000, the best partition being to assign a 1000–2000 gain to the preamplifier and leaving the remaining to the variable-gain amplifier (VGA) placed before the ADC. A higher VGA gain has to be avoided for two reasons: (1) the large bandwidth, in the megahertz range, of this amplifier, which processes signals coming from the different multiplexed channels; (2) so as not to amplify too much clock signal spurs and artifacts as well as power supply noise.

2.4.2 Preamplifier and Filter Architecture

In principle, three possible strategies can be followed in designing the continuous-time high-gain architecture of the preamplifier/filter stage. The stage has to provide enough gain (60–70 dB), and select the signal band removing the DC electrode offset (Fig. 2.6).

The solution depicted in Figure 2.6a shows a high-pass filter, with a cut-off frequency f_{HP}, set to the lower side of the signal band (i.e. $f_{HP} \approx 300\,Hz$), followed by a high-gain preamplifier that fixes the upper side of the amplifier band, cutting off noise at frequencies higher than f_{LP}. Figure 2.6b shows an alternative approach where a first amplifier provides moderate gain to prevent saturation of the overall chain. Offset and low-frequency signals at frequencies below f_{HP} are removed at the amplifier output. A second stage is used to boost the gain to the desired level. The third solution (see Fig. 2.6c) features a first amplifier with AC-coupled inputs that implements a high-pass filter with a cutoff frequency f_1 well below f_{HP}, a second high-pass filter that correctly sets the low-frequency cutoff band at f_{HP}, followed by a second gain stage, which also acts as low-pass filter. The architecture in Figure 2.6a is interesting since the limited number of stages may save power. Unfortunately, as it will be explained in Section 2.4.3.3, the noise introduced by

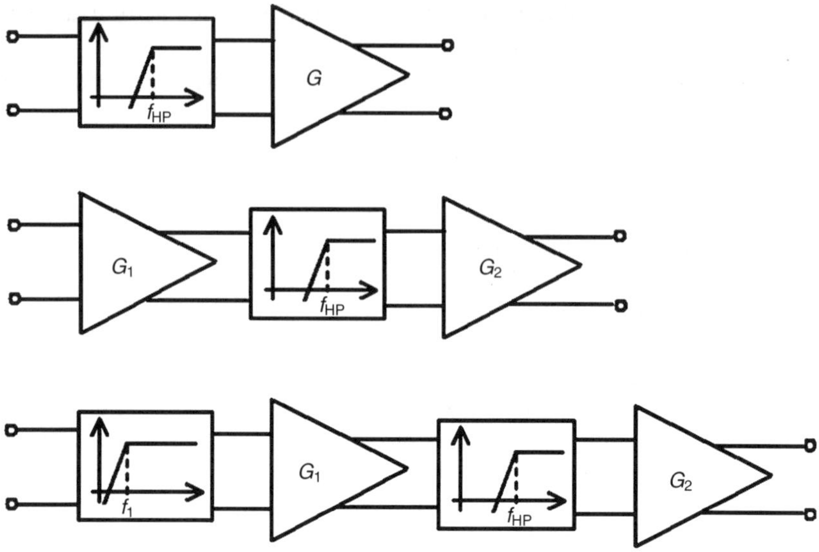

Figure 2.6. Different architectures of the low-noise preamplifier performing signal amplification, offset removal, and cutoff of the high-frequency components exceeding the band of interest.

the high-pass filter impairs its performance. Also, the solution in Figure 2.6b is critical since the possible level of electrode offset limits the first-stage gain to values as low as 2–3, thus preventing the suppression of the noise contributions arising from the following stages. The best solution for the input stage of a neural recording system is therefore the architecture in Figure 2.6c. The very low-frequency high-pass filter does not significantly contribute to the equivalent input noise that is mainly set by the first amplifier. Figure 2.7 shows the schematic of the selected three-stage circuit.

The first stage is an AC-coupled high-pass filter, using two MOS-bipolar pseudo-resistors as feedback elements [43]. This solution enables the synthesis of high-value resistance without using large area components. Figure 2.8 shows the experimental resistance of the pseudobipolar elements used in the first stage.

A peak value close to 0.5 TΩ is reached. It is true that the resistance dependence on the voltage swing may causes distortion; however, in the following, some solutions will be described to circumvent this potential issue. The midband amplifier gain is given by $G_1 \cong -C_1/C_2$. Its value is set to about –67 by taking $C_1 = 10\,\text{pF}$ and $C_2 = 150\,\text{fF}$. The high-pass pole frequency is instead placed below 10 Hz just to reject the offset and the slow voltage drift of the electrode. A $G_{M\text{-}C}$ high-pass filter with a cutoff frequency of about 300 Hz is introduced after the first stage [30]: it properly sets the low-frequency cutoff of the signal band. In addition to the beneficial impact on noise, which will be discussed in the following section, the stage filters out the signals from the background LFPs that can prevent a correct detection of the potential spikes or even saturate the amplifier. A G_M cell

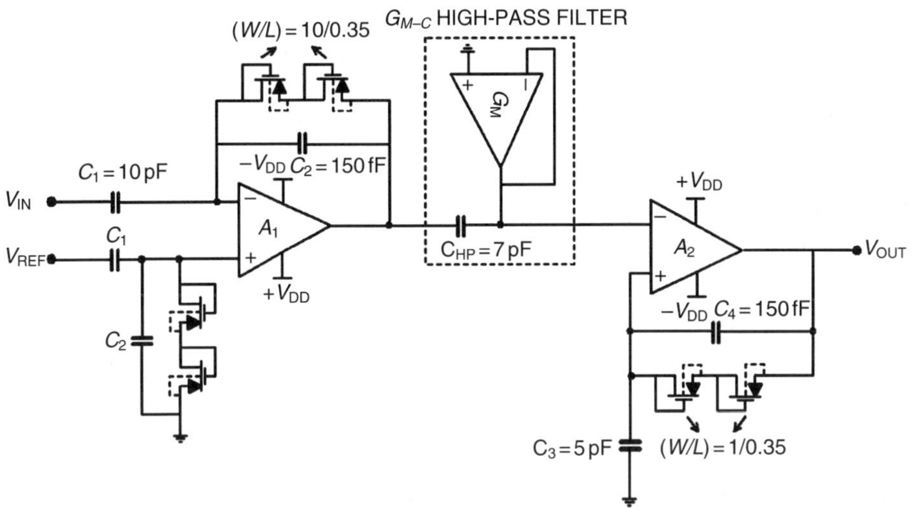

Figure 2.7. Schematic structure of the low-noise neural amplifier, implementing the architecture in Figure 2.6c.

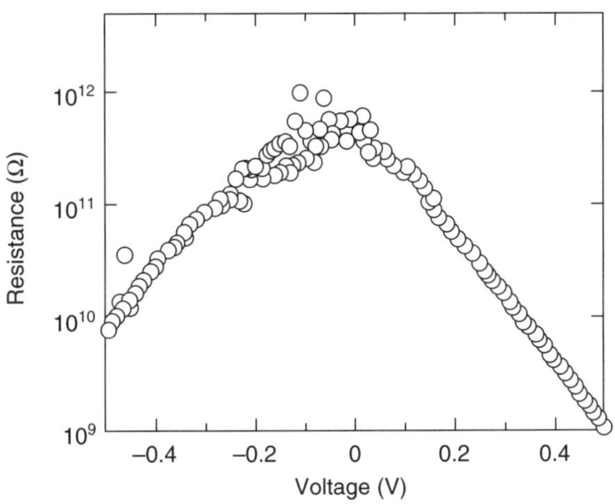

Figure 2.8. Experimental incremental resistance of the pseudoresistor elements implemented in the first stage of the preamplifier in Figure 2.7. Reprinted with permission from Borghi et al. [44]. (Copyright © 2007 IEEE)

was preferred to a tunable subthreshold MOS resistor as in Wattanapanitch et al. [45] because these latter components feature a large spread of the resistance value, even up to one decade over the same die [46]. After the selective high-pass filter, a second noninverting gain stage is added to provide further signal amplification and to define the high-frequency cutoff, f_{LP}. The stage is a single-ended capacitive-coupled voltage amplifier with a gain G_2 of about 33, achieved by using

$C_3 = 5\,\text{pF}$ and $C_4 = 150\,\text{fF}$. A capacitive-coupled structure was preferred to a purely resistive feedback amplifier to minimize the current drawn by the operational transconductance amplifier (OTA) output stage. The DC voltage at the amplifier input is set by the pseudoresistor element in the feedback path. The low-pass cutoff frequency is determined by the gain-bandwidth product of the second operational amplifier (GBWP_2) and is set to about $\text{GBWP}_2/G_2 = 15\,\text{kHz}$. Note that at very low frequencies, this noninverting stage has a unity gain in order not to amplify the offset of the operational amplifier.

2.4.3 Low-Power Low-Noise Amplifier Design

The noise performance of the whole front-end amplifier depends on the design of the first-stage OTA, its input noise power density being mainly determined by a peculiar compromise between the thermal and the flicker noises of the first stage. To explain this point, let us denote as C_p the input parasitic capacitance to ground of each input terminals of the first-stage OTA. (C_p is not drawn in Fig. 2.7. It can be easily verified that these strays do not affect, in principle, the differential gain of the first stage. Only their mismatch causes a pole-zero doublet, which, however, has no practical impact since it lays at very low frequency, beyond the band of interest.) To simplify the noise analysis, let us now neglect the noise contribution due to the two pseudoresistors. This term will be addressed in a following step. Under this assumption, the input-referred noise of the overall amplifier can be written as

$$\overline{E^2_{n_{\text{eq}}}} = \left(\frac{C_p + C_1 + C_2}{C_1}\right)^2 \cdot \overline{E^2_{n_{\text{eq}}\,\text{OTA}}} = \left(1 + \frac{1}{|G_1|} + \frac{C_p}{C_1}\right)^2 \cdot \overline{E^2_{n_{\text{eq}}\,\text{OTA}}}, \tag{2.8}$$

where $\overline{E^2_{n_{\text{eq}}\,\text{OTA}}}$ is the input-referred voltage noise of the first-stage OTA.

Equation 2.8 highlights that the input parasitic C_p, which is mainly determined by the input transistor dimensions, cannot be increased too much without impairing the overall input-referred noise. On the other hand, it is well known that large area input transistors are required to minimize the OTA flicker noise. Therefore, a judicious trade-off has to be reached to get the best overall performance. Furthermore, Equation 2.8 suggests increasing the gain of the first stage, G_1, maximizing C_1. The limit to its value is set by the source impedance that is in the 150–350-pF range. A C_1 value of 10 pF therefore guarantees high-enough input impedance. To get the target gain of $G_1 = -C_1/C_2 \approx -67$, C_2 is taken equal to 150 fF. For the parasitic C_p, an upper limit of 0.5 pF guarantees enough margins to keep under control the $1/f$ noise at a cost of a marginal 10% increase of the overall amplifier with respect to the OTA input-referred noise (see Eq. 2.8). Due to these design choices $\overline{E^2_{n_{\text{eq}}}}$ and $\overline{E^2_{n_{\text{eq}}\,\text{OTA}}}$ become almost the same. From here, both terms will be therefore denoted with the same symbol, $\overline{E^2_{n_{\text{eq}}}}$.

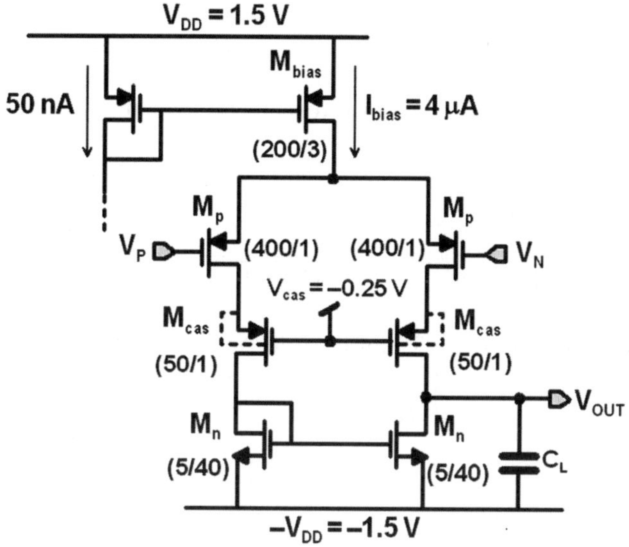

Figure 2.9. Schematic of the telescopic cascode op-amp used in the first stage of the presented amplifier. Reprinted with permission from Borghi et al. [42]. (Copyright © 2008 IEEE)

2.4.3.1 ***OTA Noise Analysis and Optimization.*** In order to better exploit the noise–power trade-off, the first OTA stage has been synthesized following a telescopic cascode approach (Fig. 2.9).

This well-known configuration guarantees excellent noise performance, thanks to its very few transistors, while cascoding and proper transistor sizing ensure enough gain.

The input-referred noise power spectral density of the circuit in Figure 2.9 is given by

$$\overline{E_{n_{eq}}^2} \cong \frac{8kT\gamma}{g_{mp}} + \frac{8kT\gamma}{g_{mp}}\left(\frac{g_{mn}}{g_{mp}}\right) + \frac{2K_p^{(1/f)}}{C'_{ox}W_pL_p}\frac{1}{f} + \frac{2K_n^{(1/f)}}{C'_{ox}W_nL_n}\left(\frac{g_{mn}}{g_{mp}}\right)^2\frac{1}{f}. \tag{2.9}$$

The result has been derived taking into account that M_{bias} and M_{cas} transistors do not significantly contribute to the overall noise. To minimize the noise, the condition $g_{mp} >> g_{mn}$ must be fulfilled, that is, $(W/L)_p >> (W/L)_n$. Under this assumption, the total noise power density reads

$$\overline{E_{n_{eq}}^2} \cong \frac{8kT\gamma}{g_{mp}} + \frac{2K_p^{(1/f)}}{C'_{ox}W_pL_p}\frac{1}{f}, \tag{2.10}$$

γ being equal to 2/3 for transistors working in strong inversion or $1/(2\kappa)$ in weak inversion ($\kappa \approx 0.7$ [47]). In order to minimize the thermal noise without increasing the current consumption, the M_p transistors have to work in weak inversion, where the transconductance reaches the maximum value. We can estimate the transconductance using the EKV model [48], valid in all regions of inversion, as

$$g_m \cong \frac{\kappa I_D}{U_T} \cdot \frac{2}{1+\sqrt{1+4\times \mathrm{IC}}}, \tag{2.11}$$

where IC is the inversion coefficient [47], defined as

$$\mathrm{IC} = \frac{\kappa I_D}{2\mu C'_{ox}\left(\frac{W}{L}\right)U_T^2}. \tag{2.12}$$

In the subthreshold or weak inversion region, this coefficient is much less than 1. Using Equation 2.11, the overall-amplifier input-referred thermal noise power spectral density results in

$$\overline{E_{n\text{th}}^2} = \frac{8kTU_T}{\kappa^2 I_{\text{bias}}}, \tag{2.13}$$

where I_{bias} is the bias current of the telescopic cascode amplifier. Assuming a first-order roll-off of the frequency response, the noise bandwidth is $(\pi/2)\cdot 15\,\text{kHz} \approx 23.56\,\text{kHz}$. Setting an upper limit of $3\,\mu V_{\text{rms}}$ for the thermal noise contribution in this noise band, a minimal I_{bias} value of about $4\,\mu A$ is required. In order to assure that input pair transistors work in the weak inversion region ($\mathrm{IC} < 0.1$), their form factor has to be less than about 300; we set $(W_p/L_p) = 400$ to be conservative.

The flicker noise contribution can be reduced by increasing the PMOS transistor area ($W_p \cdot L_p$). Setting a noise corner frequency lower than 100 Hz and considering that for the adopted technology $K_p^{(1/f)} \approx 2\times 10^{-26}\ \text{V}^2\text{F}$ and $C'_{ox} \approx 5\ \text{fF}/\mu\text{m}^2$, the input PMOS transistors were sized with $W_p \cdot L_p = 400\,\mu\text{m}^2$ (thus $W_p = 400\,\mu\text{m}$ and $L_p = 1\,\mu\text{m}$). This sizing leads to a stray OTA input capacitance C_p of approximately 500 fF that does not excessively impair the equivalent input noise.

Moreover, in order to fulfill the requirement $(W/L)_p >> (W/L)_n$, it was set that $W_n = 5\,\mu\text{m}$ and $L_n = 40\,\mu\text{m}$, thus forcing M_n transistors to work in strong inversion (see Table 2.1).

This choice makes it possible to avoid the adoption of a cascoded mirror not to degrade the amplifier gain, while the introduction of M_{cas} guarantees an output resistance given $R_{\text{out}} = r_{0n} \| g_{m\text{cas}} r_{0\text{cas}} r_{0p} \approx r_{0n} \approx 108\,\text{M}\Omega$ and an amplifier gain $g_{mp} r_{0n} \approx 75\,\text{dB}$. In summary, the noise within the amplifier band may be reduced by careful transistor sizing to the sole thermal noise. In this limit, the minimum

TABLE 2.1. Transistor Operating Points

Transistor	$V_{GS} - V_T$ (mV)	IC	g_m (A/V)	r_0 (MΩ)
M_p	110	0.075	52.3	5.1
M_n	490	51	7.76	108
M_{cas}	24	0.36	43.7	8.2

NEF achievable with a single differential stage can be estimated from Equations 2.2 and 2.13 leading to a value of $\sqrt{2}/\kappa = 2.02$. Actually, the contribution to the thermal noise of the current mirror transistors cannot be completely neglected. In fact, even if M_p and M_n transistors work in weak and strong inversion regions, respectively, the transconductance ratio is only about 7 (see Table 2.1) in the present design. Considering the general expression of the transconductance given by Equation 2.11 and the total input referred thermal noise given by the first two terms of Equation 2.9, the theoretical NEF for this preamplifier becomes

$$\mathrm{NEF_{th}} = \frac{\sqrt{2}}{\kappa} \cdot \sqrt{1 + \frac{2\gamma\kappa}{\sqrt{\mathrm{IC}_n}}} \cong 2.15, \tag{2.14}$$

being $\gamma \approx 2/3$, $\kappa \approx 0.7$, and IC_n the inversion coefficient of current mirror transistors. As a reference, consider that the folded cascode amplifier described in Wattanapanitch et al. [45] targets a minimum theoretical NEF = 2.47, while a limit of 2.9 was found in Harrison and Charles [43] for a mirrored cascode op-amp. Note also that in Holleman and Otis [49], the authors propose an amplifier with an NEF = 1.8, but the topology is not differential and thus it has an intrinsic very low power supply rejection ratio (PSRR).

However, the achievable NEF of the proposed amplifier is slightly larger compared with the previously valuated theoretical limit. Three factors combine to raise the *NEF*:

1. The transconductance of the input transistors is slightly lower than $\kappa I_D/U_T$ since their inversion coefficient is larger than zero. For the present design, the input transistor IC is 0.075 (see Table 2.1), that means an increase of the rms input noise, and also of the NEF, of about 1.034. Note that the technology scaling is beneficial from this point of view since, for a given bias current, the inversion coefficient tends to be lower due to a larger specific oxide capacitance (see Eq. 2.12).
2. The input referred noise of the overall amplifier is a factor $(1 + 1/|G_1| + C_p/C_1)^2$ larger than the input-referred noise of the first operational amplifier, as stated by Equation 2.8. This factor determines an increase of the NEF of about 1.065, considering a parasitic input capacitance of about 0.5 pF, mainly due to the OTA input transistors. A small contribution derives from the strays associated to the input capacitor plates. This contribution can be drastically reduced connecting the capacitor bottom

plate (which has the largest parasitism) to the amplifier input and the top plate to the OTA terminal. In this way, a parasitic capacitance larger than 1.5 pF is avoided.

3. The current drawn in the second amplifying stage contributes to the total current in Equation 2.2 but does not contribute to lowering the input-referred thermal noise. NEF increases by a factor of $\sqrt{1+(I_2/I_1)}$, I_1 and I_2 being the current drawn by the first and the second operational amplifier, respectively. This corresponds to an increase of about 1.037 of the NEF. Considering these three factors, the estimated NEF of the overall amplifier is about 2.45. The adoption of a telescopic cascode stage has to be reconsidered in the frame of very low-voltage technologies. In these cases, a folded cascode topology is more appropriate even if with some penalty on power dissipation and NEF value. Moreover, telescopic cascode is known to have a small output swing. However, most of the time, the values are large enough to accommodate the dynamics of the first amplifying stage. In the 0.35-μm design in Figure 2.9, the positive voltage swing is ≈450 mV ($|V_{T,p}| + V_{cas}$) while the negative swing is about 1 V (1.5 V – $V_{ov,n}$), posing no issue to the amplifier operation.

2.4.3.2 Second Amplifying Stage. A second gain stage is needed to increase the dynamic range of the input signal before multiplexing. Since this stage requires a relative large output swing, two-stage OTA (not shown in the figures) can be used. Provided that the first stage has enough gain, the impact of this second amplifier on the input-referred noise is negligible; therefore, its power dissipation can be reduced without affecting the noise performance. In fact, the noise of the second OTA is dominated by input transistors, designed to operate in the weak inversion region. Imposing a contribution to the equivalent input noise added by the second-stage OTA less than 0.3 μV_{rms} (i.e., 1/10 of the dominant contribution), a minimum current of 100 nA is needed in the input differential pair of the second op-amp. To be conservative, we set a bias current of 200 nA in the first stage, while a current of 100 nA is drawn by the second stage.

Finally, nonlinear distortion may instead be of some concern. Distortion arises from the nonlinear high-resistance pseudoresistor placed in the feedback path, which is driven by a large output voltage swing. For this reason, the subthreshold MOS transistors have been sized to have resistance values an order of magnitude higher than those in the first stage. This choice makes the signal current flowing through them always orders of magnitudes lower than the signal through the capacitors. This way, even if the resistance value is modulated under a 1.5-V peak-to-peak output swing, the total harmonic distortion (THD) generated driving the stage with a 1-kHz sinusoidal input was verified to remain less than 5%. Moreover, the application to neural recording is intrinsically more robust to minimal distortion of the impulse response.

2.4.3.3 High-Pass Filter Design and Optimization. Let us now consider the noise due to the MOS-bipolar pseudoresistors that was neglected in Equation

2.9. Denoting as R the small signal resistance of the two PMOS subthreshold transistors, it was experimentally verified that their current noise spectral density complies with the usual equation

$$\overline{S_R^2} = \frac{4kT}{R}, \tag{2.15}$$

taking for R the peak incremental resistance (Fig. 2.8). Their contribution to the OTA input-referred noise power spectral density follows a $1/f^2$ law and is given by

$$\overline{E_{nR}^2} = 2\frac{4kT}{R}\frac{1}{(2\pi C_1 f)^2} = \frac{8kTf_1}{2\pi G_1^2 C_2 f^2}, \tag{2.16}$$

f_1 being the cutoff frequency of the high-pass filter set by the first amplifier stage. The factor 2 in Equation 2.16 accounts for the two pseudoresistors, one in the feedback path and one connected to the positive input terminal of the operational amplifier. For the sake of simplicity, let us now neglect the presence of the selective high-pass filter following the first stage. By integrating Equation 2.16 from f_1 to the low-pass cutoff frequency of the overall amplifier, f_{LP}, we get

$$\int_{f_1}^{f_{LP}} \overline{E_{nR}^2}\, df \cong \frac{8kT}{2\pi G_1^2 C_2}, \tag{2.17}$$

which is independent of the cutoff frequency f_1. From Equation 2.17, the contribution of the pseudoresistors to the OTA input-referred noise results about $2.75\,\mu V_{rms}$. A similar result can be obtained considering the total noise due to the two pseudoresistors at the output of the first stage, $2kT/C_2$, and dividing it by the squared midband gain, $(G_1)^2$. This means an input-referred noise of $3.44\,\mu V_{rms}$. This value is not negligible and explains why it is not convenient to implement the selective high-pass filter (i.e., the 300-Hz cutoff frequency high-pass filter) in the first stage. This solution would add the noise of the pseudoresistors to the other contributions, thus degrading the noise performance. The high-pass filter in the topology of Figure 2.7 is instead able to cut off most of the noise from the pseudoresistors of the first stage. However, a careful choice of the filter capacitor is mandatory to reduce the noise contribution of the $G_{M\text{-}C}$ filter. The point will be discussed in the following. Let us now suppose that $f_1 << f_{HP}$, f_{HP} being the cutoff frequency of the $G_{M\text{-}C}$ high-pass filter. The pseudoresistors of the first stage and the $G_{M\text{-}C}$ filter determine the $1/f^2$ component of the input-referred noise, which can be denoted as $\overline{E_{eq}^2}^{(1/f^2)}$. For frequency larger than f_1, it results

$$\overline{E_{eq}^2}^{(1/f^2)} = \frac{8kT}{R}\frac{1}{(2\pi C_1 f)^2} + \overline{S_{nI}^2}\frac{1}{(2\pi G_1 C_{HP} f)^2}, \tag{2.18}$$

$\overline{S_{nI}^2}$ being the output current noise of the high-pass filter. Let us now assume, for the sake of simplicity, that the current noise generated by the G_M cell ($G_M = 1/R_{HP}$) can be written as $4kT/R_{HP}$. Equation 2.18 becomes

$$\overline{E_{eq}^2}^{(1/f^2)} = \frac{8kTf_1}{2\pi G_1^2 C_2 f^2} + \frac{4kTf_{HP}}{2\pi G_1^2 C_{HP} f^2}. \tag{2.19}$$

Integrating Equation 2.19 in the overall amplifier band, that is from f_{HP} to f_{LP}, it turns out:

$$\int_{f_{HP}}^{f_{LP}} \overline{E_{eq}^2}^{(1/f^2)} df \cong \frac{8kT_1}{2\pi G_1^2 C_2} \frac{f_1}{f_{HP}} + \frac{4kT}{2\pi G_1^2 C_{HP}}. \tag{2.20}$$

The result suggests reducing the first term by minimizing the f_1/f_{HP} ratio. A large capacitor value C_{HP} is also needed to reduce the second term. In practice, the G_M cell is a simple differential stage in which the bias current can be externally tuned, as depicted in Figure 2.10.

Due to the small bias current of 1 nA needed to synthesize a high-value resistance, all the transistors work in the weak inversion region and their transconductance is $g_m = G_M \approx \kappa I/U_T$. The current noise of this configuration is therefore

$$\overline{S_{nI}^2} = 4\frac{2kT}{\kappa} g_m = \frac{8kT}{\kappa R_{HP}}. \tag{2.21}$$

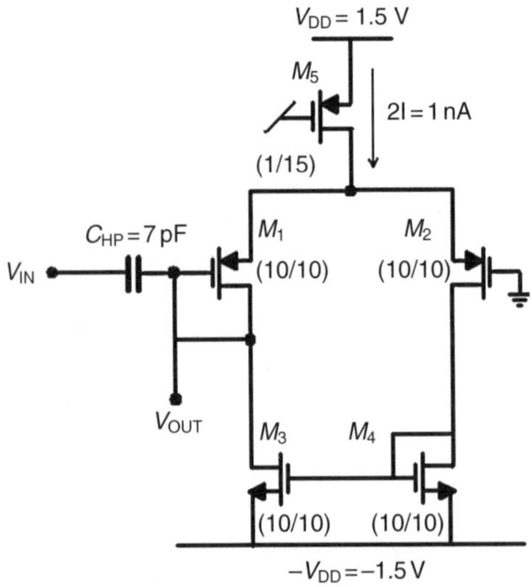

Figure 2.10. Schematic structure of the $G_{M\text{-}C}$ high-pass filter in Figure 2.7. Note the very low bias current needed to synthesize a high-value equivalent resistance.

The current noise of the G_M cell is therefore a factor of $2/\kappa$ larger than that of a resistor R_{HP}, and this factor has to be added in the second term of Equation 2.20. In summary, in order to keep the input-referred noise due to the first-stage pseudoresistors and the G_M cell in the amplifier band below $1\,\mu V_{rms}$, f_1 has to be at least a factor of 10 smaller than f_{HP} and, at the same time, C_{HP} must be greater than 2 pF. The two pseudoresistors in the circuit of Figure 2.7 were designed to have $f_1 < 10\,Hz$ (with $f_{HP} = 300\,Hz$) and $C_{HP} = 7\,pF$.

2.5 SYSTEM INTEGRATION AND RESULTS

The front end was integrated in a wireless neural recording system. Figure 2.11 shows the block diagram of the system, implemented in a 0.35-μm AMS (austriamicrosystems) CMOS process with a 3-V power supply delivered by two coin batteries.

The system was designed to sense the signals provided by an array of 16 neural probes. After amplification, sampling, and conversion, the data are transmitted using an IR laser transmitter. The adoption of an IR transmitter was decided as first step (1) to avoid the broadband external antenna needed in UWB systems and (2) so as not to face skull and skin absorption at high frequencies [39]. The circuit front end features an array of 16 preamplifiers, connected to an analog multiplexer operating at a 40-kHz sampling rate. Multiplexing is performed connecting the output of each preamplifier to a pass-gate switch. A shift register, controlled by a 320-kHz clock (ck_{mux} in Fig. 2.11), sequentially enables each amplifier to access the common data lead; each pass gate is switched on both the rising and the falling edge of multiplexer clock, avoiding clock transitions in the middle of the sampling window. In this topology, each amplifier has to drive a long routing line connecting all the amplifier outputs to the multiplexer. The resulting load capacitance is large (about 1 pF) and has to be driven in a 1.56-μsec time frame. An AB-class buffer is therefore needed to boost each line. Its design was based on a differential flipped voltage follower [50]. To save power, each buffer is switched off when the corresponding amplifier is not selected and turned on one clock edge before, that is, when the previous amplifier is selected. This way, at any time, only two of the 16 buffers are on, draining about $10\,\mu A$ each from the power supply. Further amplification is provided before sampling by a VGA added to the amplification chain at the multiplexer output. This stage is a noninverting amplifier whose feedback resistor is digitally controlled by 3 bits in order to change the VGA gain from 1 to 8. The amplified signal is then digitized by an 8-bit SAR ADC operating at a clock frequency (ck_{adc} in Fig. 2.11) of 10.24 MHz (640 kS/s) for a 5.12 Mbit/s output stream. The conversion cycle is composed by a sample phase, lasting seven clock cycles (683.6 nsec), followed by an evaluation phase, eight-clock-cycles long. A single bit is evaluated within each clock cycle starting from the most significant bit (MSB). The 16th clock period is devoted to the end of conversion operations and to reset the ADC. The parallel outputs of the converter are then processed by a logic block before to be sent to the laser driver.

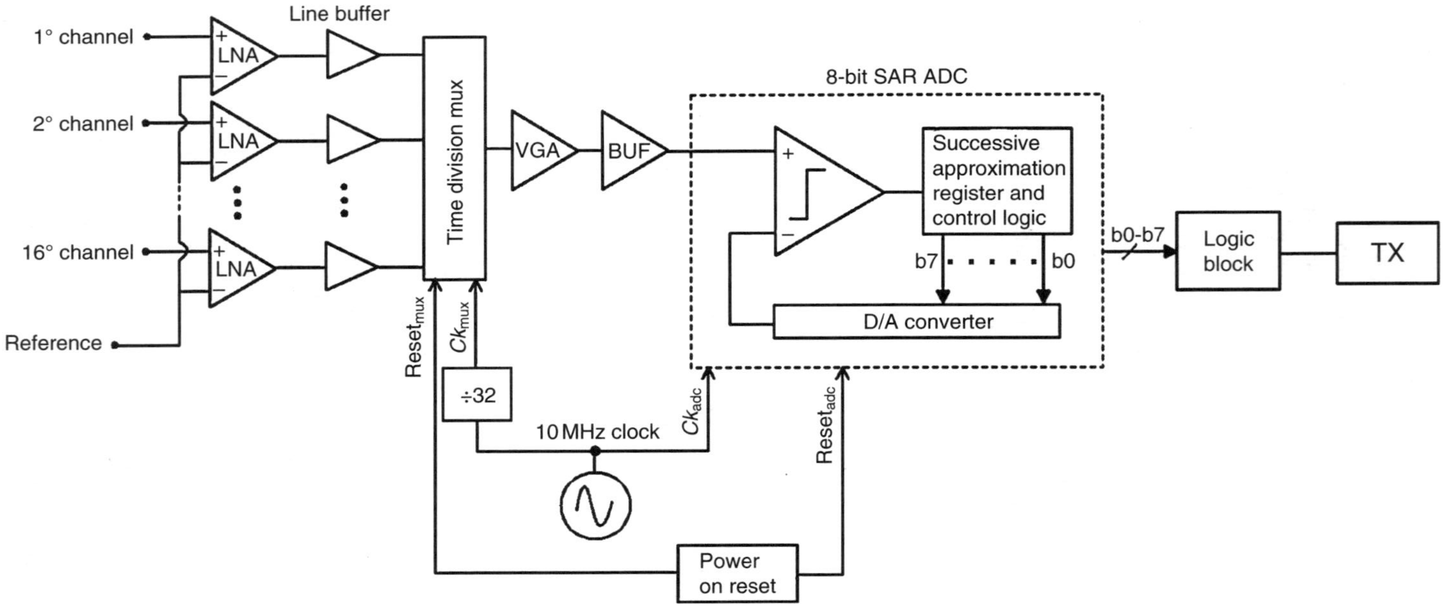

Figure 2.11. Block diagram of the implemented system.

This digital stage provides the serialization of the binary data and the inclusion of service bits (e.g., to indicate the start and stop of each digital word and to recover the information of which channel a byte belongs to). A low-power on-chip relaxation oscillator generates the ck_{adc} signal. After a frequency divider by 32 it also provides the multiplexer clock, ck_{mux}. At the power-up, the negative pulse from a power on reset (POR) stage resets the multiplexer serial register and starts the digital conversion. This reset signal is mandatory to assure proper synchronization between the multiplexer and the sampling phase of the A/D converter since each channel has to be sampled at the end of the corresponding multiplexing period. The system was successfully tested for data transmission at 5.12 Mbit/s through pigskin, about 4-mm thick. The power consumption was 7.7 mW, dominated by the laser driver. A 4-mA current was needed to drive an off-chip 850-nm vertical cavity surface emitting laser (VCSEL) packaged in a TO-18. Transmission was performed using an on-off keying (OOK) modulation. The 6-mW power needed by the transmitter, leading about 1.2 nJ/bit, is in line with the performance of similar transmitters (Fig. 2.5). Before discussing in more detail the performance of the low-noise front end, it is worth noting that the ADC required 570 µW for a sampling rate of 640 kS/s, a differential non-linearity (DNL) < 0.4 least significant bit (LSB), an integral non-linearity (INL) < 0.8 LSB, and an ENOB of 7.3. These values correspond to a power budget of about 35 µW per channel and to an FOM of 5.86 pJ/quantization step. It favorably compares with the best value of 2.72 pJ/quantization step, reported so far in neural recording systems [51].

2.5.1 Amplifier and Filter Characterization

Figure 2.12 shows a die microphotograph of the implemented system with the 16-channel amplifier array, the A/D converter, the logic block, and the laser driver. Each amplifier occupies an area of 0.066 mm^2, line buffer included. Figure 2.13 shows the transfer function of a typical amplifier. The midband gain is 67.1 dB, while the signal band extends from 290 Hz to 15 kHz, close to the expected values. The gain deviation measured on different amplifiers of different dies is less than 0.2 dB, while the high-pass and low-pass frequencies have a standard deviation less than 9 and 4%, respectively. Figure 2.14 shows the measured input-referred noise power spectrum compared with the result obtained simulating the extracted view. It was obtained by dividing the output noise power spectral density by the square of the overall transfer function.

The input-referred thermal noise is about 25 nV/√Hz close to the value expected from the bias current value. Following Wattanapanitch et al. [45], the equivalent input noise is evaluated by dividing the total output noise for the midband gain. Using this procedure for the implemented amplifier and considering the total output noise in the 10 Hz–50 kHz band, the input-referred noise results 3.53 µV, while the measured current consumption is 4 µA for the first-stage OTA and 0.3 µA for the operational amplifier of the second gain stage (the consumption of the $G_{M\text{-}C}$ high-pass filter is negligible). The NEF of the entire amplifier is about 2.41, which is the best result reported to date (see Table 2.2) and close to the value estimated in the previous section for the present design.

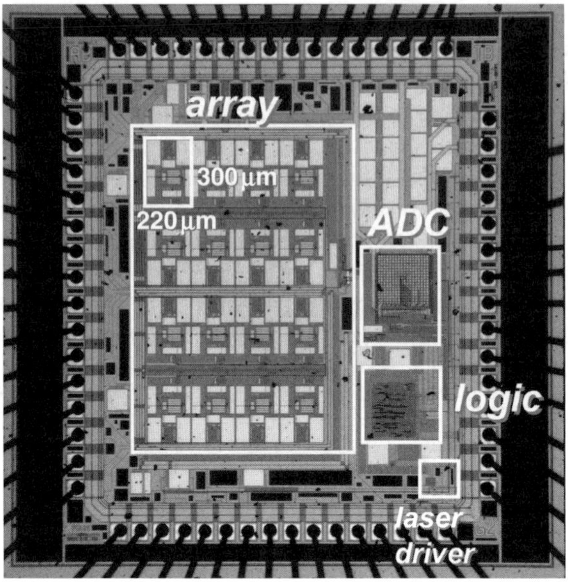

Figure 2.12. Die photo of the integrated system with 16-channel amplifier array, A/D converter, logic block, and laser driver.

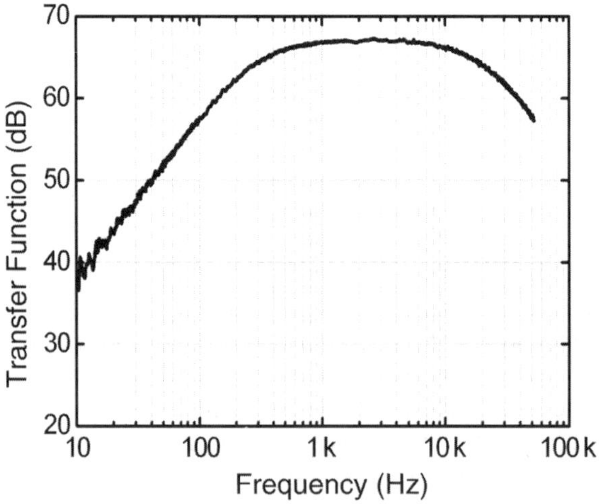

Figure 2.13. Transfer function of the implemented low-noise amplifier. The band ranges from approximately 290 Hz to 15 kHz.

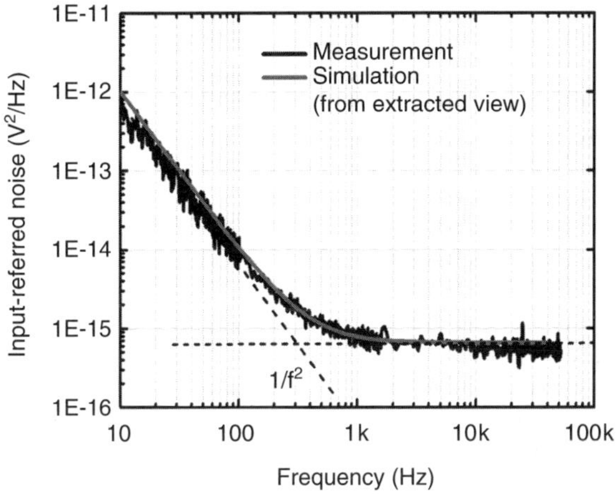

Figure 2.14. Measured and simulated input-referred noise spectra of the low-noise amplifier. The spectrum is obtained dividing the amplifier output noise by the overall transfer function.

TABLE 2.2. Comparison of Low-Power Low-Noise Neural Amplifier Characteristics

Amplifier	Technology (μm)	Bandwidth	Input Noise (μV_{rms})	Supply Current (μA)	Gain (dB)	NEF
Reference [45]	0.5	45 Hz–5.32 kHz	3.06	2.7	40.85	2.67
Reference [43]	1.5	0.025–7.2 kHz	2.2	16	39.5	4.0
Reference [31]	0.18	102–9.1 kHz	5.6	4.67	49.52	4.9
Reference [52]	0.5	94–8.2 kHz	1.94	8	39.6	2.9
Reference [53]	1.5	0.015–4 kHz	3.6	8	39.3	4.9
This work	0.35	290–15 kHz	3.53	4.3	67.1	2.41

To validate the operation of the amplifier under real conditions, the PSRR (defined as the ratio of the differential-mode gain to the gain from the power supply to the output) was measured to be larger than 50 dB in the amplifier band, sufficient to enable proper working even in the most challenging conditions. Finally, the common-mode rejection ratio (CMRR, defined as the ratio of the differential-mode gain to the common-mode gain) was measured to be greater than 65 dB in the signal band (see Table 2.3 for the amplifier performance summary). The amplifier was also tested in *in vivo* experiments on an anesthetized rat. Recordings were performed using an array of electrodes (Tucker Davis) implanted in the primary motor cortex (M1) and with an impedance of about 1 MΩ at 1 kHz. Figure 2.15 shows a trace recorded with the amplifier, while in Figure 2.16 a collection of APs from the same electrode is presented.

TABLE 2.3. Performance Summary of the Low-Power Low-Noise Amplifier

Technology	0.35-μm AMS
Midband gain	67.1 dB
High-pass cutoff frequency	290 Hz
Bandwidth	15 kHz
Input-referred noise	$3.53\,\mu V_{rms}$
Noise efficiency factor	2.41
THD	5%
CMRR (10 Hz–15 kHz)	65 dB
$PSRR^+$ (30 Hz–15 kHz)	55 dB
$PSRR^-$ (30 Hz–15 kHz)	50 dB
Supply voltage	3 V
Supply current	4.3 μA
Area	$0.066\,mm^2$

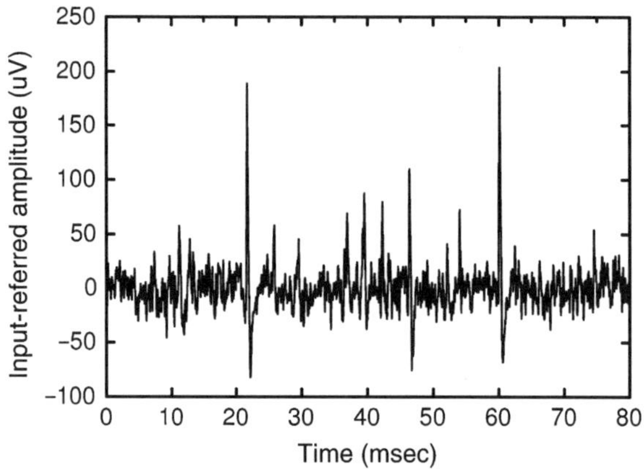

Figure 2.15. Extracellular recording from the primary motor cortex of a rat obtained with the low-noise amplifier.

2.6 REFINEMENTS OF AMPLIFIER DESIGN

The low-noise front end presented here (Fig. 2.7) and those in Harrison and Charles [43] and Wattanapanitch et al. [45] follow the same three-stage scheme of Figure 2.6c: the main difference being the topology of the first OTA stage. All these amplifiers can be classified as passive low-frequency suppression schemes. They use a capacitive feedback network with an AC-coupled input and high-impedance pseudoresistors. For this reason, they need very large integrated capacitors to achieve satisfactory gains. Moreover, nonlinearities arising from

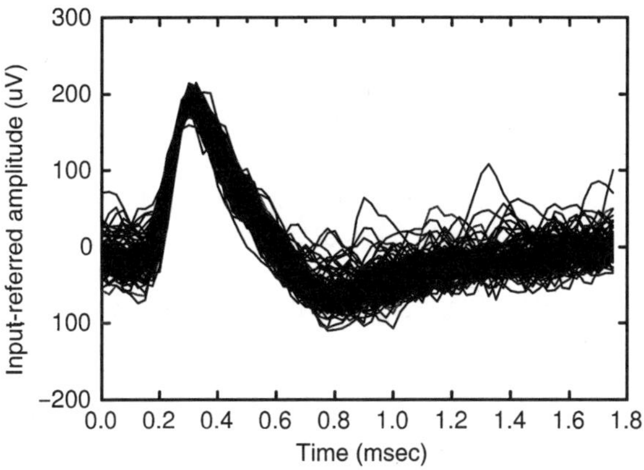

Figure 2.16. Action potential pulses from the trace in Figure 2.15 plotted starting form the same time origin.

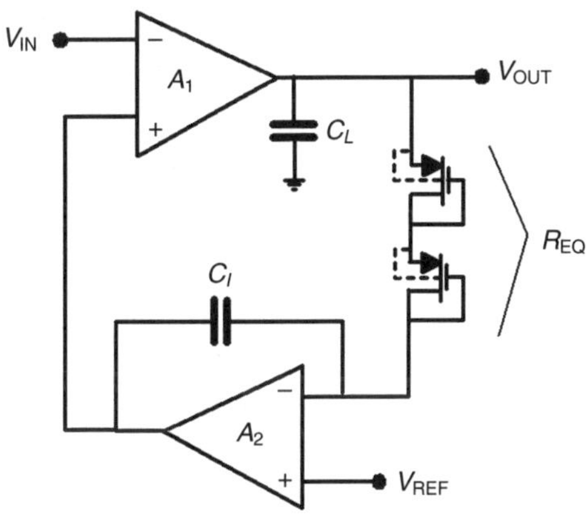

Figure 2.17. Schematic structure of an amplifier performing active low-frequency suppression for minimum area consumption [31].

pseudoresistors must be properly minimized by sizing or additional circuit solutions. This Section is therefore devoted to cover alternative solutions proposed in the literature to reduce the area occupation and to manage pseudoresistor nonlinearities.

2.6.1 Active Low-Frequency Suppression Amplifier

Reduced area occupations can be achieved by adopting the active scheme based on closed-loop control of the DC level proposed by Gosselin et al. [31] and depicted in Figure 2.17.

This system consists of a low-noise OTA (A_1) featuring an inverting Miller integrator within its feedback path. The active integrator is made by a second OTA (A_2), a capacitor (C_I), and high-value resistor (R_{EQ}) implemented with two subthreshold PMOS transistors. Its time constant (τ) sets the –3-dB high-pass cutoff frequency (f_{HP}) of the amplifier, being $\tau = R_{EQ}C_I$. The midband gain of the amplifier equals the low-frequency open-loop gain of the first OTA, A_1, while the –3-dB low-pass cutoff frequency f_{LP} is set by the A_1 dominant pole. The Miller integrator integrates the amplifier output voltage and applies it to the positive input terminal of the first OTA. As a result, this node tracks the DC level at the V_{IN} node, which effectively cancels out any DC offset, drift, and other low-frequency input voltages. The active integrator yields a loop gain greater than unity from the output node to the positive input terminal of the first OTA at low frequency while it provides a strong attenuation at higher frequency, that is, in the neural signal band. This means that in the frequency band of interest the gain of the amplifier is determined by the first OTA that sets the in-band gain and the low-pass cutoff frequency. Considering A_2 as an ideal OTA, the transfer function of the overall amplifier can be written as

$$T(s) = \frac{-s\tau A_{01}}{s\tau + A_{01}} \cdot \frac{1}{1+s\tau_L}, \tag{2.22}$$

A_{01} being the gain of the first OTA and $\tau_L = 1/(2\pi f_{LP})$ the time constant determined by its output resistance and capacitor C_L. This transfer function yield a –3-dB high-pass cutoff frequency that is set by the gain A_{01} and τ:

$$f_{HP} = \frac{1}{2\pi}\frac{A_{01}}{R_{EQ}C_I}. \tag{2.23}$$

Clearly, the DC and the low-frequency suppression performance of the amplifier are determined by the second OTA gain, A_{02}, which was chosen by the authors to be 20 kV/V. Other parameters of their implementation are: $A_{01} = 400$, $C_L = 400$ fF, $C_I = 1$ pF, $f_{HP} = 100$ Hz, and $f_{LP} = 9$ kHz. The main advantages of this amplifier topology are the high input impedance that is not determined by a large capacitor as in traditional AC-coupled amplifier and the small area (0.050 mm^2, while the amplifier described in the previous Section is 0.066 mm^2); only two relatively small capacitors are required just to set the high-pass and the low-pass poles of the band-pass filter. The amplifier gain is determined by the open-loop gain of an OTA, which is the major drawback of this topology since it cannot be well controlled. Gosselin et al. [31] measured a gain of 50.1 dB with a standard deviation of 0.6 dB in a same die and about 2 dB considering different batches. Another drawback of this topology is related to the noise-power efficiency of the overall amplifier. In fact, both operational amplifiers determine the equivalent input noise of the scheme in Figure 2.17. Since in the neural signal band the second OTA can be considered closed in a buffer configuration, the input-referred noise power spectral density can be approximated as

$$\overline{E^2_{n_{eq}}} = \overline{E^2_{nA1}} + \overline{E^2_{nA2}}. \tag{2.24}$$

Considering γ as the ratio between the A_1 and A_2 input stage bias current, the achievable NEF with this active low-frequency suppression scheme is a factor of $(1+\gamma)/\sqrt{\gamma}$ larger than the one evaluated in Section 2.4.3. For $\gamma = 1$, that is, with OTAs biased with the same current, the minimum NEF is about 4.04, while Gosselin et al. [31] get an NEF of 4.6 biasing the second OTA at a current two times larger than the first OTA for loop stability reasons. In conclusion, an active suppression of the low-frequency components makes it possible to reduce the silicon real estate, avoiding the adoption of large capacitors, but at the cost of doubling the target NEF.

2.6.2 Low Signal Distortion Amplifier

The amplifier presented in Section 2.4.2 uses a MOS-bipolar pseudoresistor in the feedback path to provide a very low cutoff frequency. However, as seen in Figure 2.8, the pseudo-MOS resistance is highly dependent on the output signal level, which may result in signal distortion and high-pass filter cutoff frequency variation. The same problems arise in the solution adopted by Olsson et al. [54] where an NMOS pseudoresistor with controlled gate voltage is used to achieve a tunable high-pass cutoff frequency, or in Aziz et al. [55] where a PMOS is used instead of an NMOS transistor. A solution for synthesizing a high-value resistance independent of the signal level is presented in Yin and Ghovanloo [53] and described in Figure 2.18.

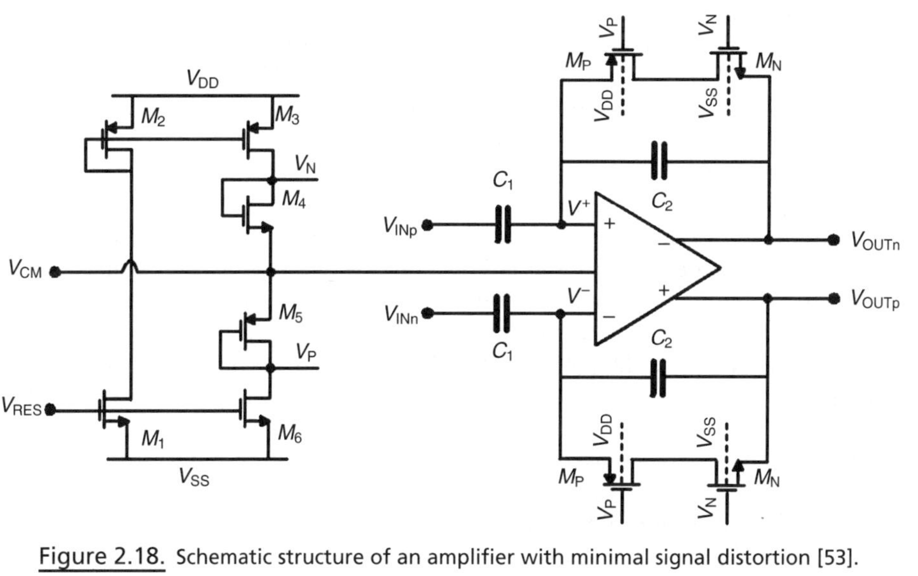

Figure 2.18. Schematic structure of an amplifier with minimal signal distortion [53].

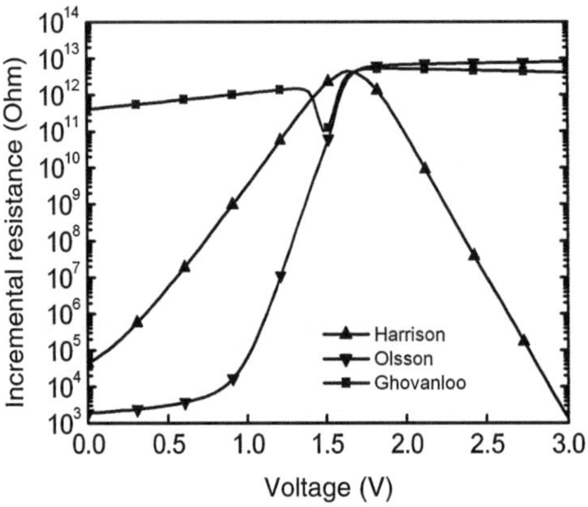

Figure 2.19. Comparison between the incremental resistance of different pseudoresistor elements. Harrison refers to the pseudoresistor used in the presented design and in Harrison and Charles [43], Olsson to a gate-controlled NMOS [54], and Ghovanloo to the scheme presented in Yin and Ghovanloo [53].

Yin and Ghovanloo [53] used a voltage-controlled PMOS and NMOS pseudoresistor to create a stable low cutoff frequency in a fully differential amplifier. In this configuration, an external control voltage, V_{RES}, biases M_1 in the deep subthreshold region; the same current passes through M_4 and M_5, which are diode-connected and sized with large W/L. Noting that the input terminal of the amplifier are kept around V_{CM} by the common feedback network, it can be shown that M_4 and M_5 keep the current passing through the PMOS-NMOS pseudoresistors (M_P and M_N) around k times $I_{D4} = I_{D5}$, k being the ratio $(W/L)_N/(W/L)_4 = (W/L)_P/(W/L)_5$, regardless of the output voltage variations. In fact, when the voltage at the positive output node, V_{OUTp}, is lower than the voltage at the negative input terminal of the OTA, V_n, a current kI_{D5} passes from V^- to V_{OUTp} because M_P and M_5 have the same gate to source voltage; when $V_{OUTp} > V^-$, the current passing from V_{OUTp} to V^- keeps the voltage between M_P and M_N, V_X, slightly higher than V^-, and so $V_{GSn} = V_N - V_X \approx V_N - V^- = V_{GS4}$. Therefore, the current passing through the pseudoresistor $M_4 - M_5$ is k times the current in M_4 and M_5. This explains how the resistance of the NMOS-PMOS pseudoresistor is almost constant even for the large voltage signal applied to M_P and M_N transistors. Figure 2.19 shows a comparison between simulated incremental resistances of different MOS pseudoresistors used in Yin and Ghovanloo [53], Olsson et al. [54], and Harrison et al. [30], the latter being the same pseudoresistor adopted in the first stage of the amplifier described in Section 2.4.2. The comparison is made setting transistor sizes and gate voltages in order to have the same resistance at small voltage. The NMOS-PMOS pseudoresistor in Yin and Ghovanloo [53] shows the minimum resistance deviation for the whole range of applied signals, while the other topologies present a variation of different decades. This solution can be therefore used to lower signal distortion in high-gain amplifiers with capacitive-coupled topologies such as the one in Figure 2.7, instead of using a second stage with pure linear resistors.

2.7 CONCLUSIONS

Since the dawn of microelectronic industry, integrated technologies have been fueling tremendous advances in science, engineering, and applications, leading to an increasing inclusion of intelligence in infrastructures, equipment, and products. This trend, leveraging on silicon device miniaturization, is still ongoing and is having a profound impact in all fields, medical science and therapeutics included. In the forthcoming years, the availability of decananometer silicon technologies, and the advances in micromechanical and packaging manufacturing, energy-conversion techniques, and material engineering are expected to provide the solutions needed to develop fully miniaturized, low-power, energy-autonomous smart systems. These systems will promote a more intimate smart link between humans, from a high-level interaction down to the cellular level, "things," and environment. Implantable recording systems are a challenging test field for deeply scaled technologies since demanding performance is required for its application and by the tight constraints imposed by the surrounding environment, that is, the body. But big challenges translate in big opportunities: The potentials of this trend are already clearly visible, neurotechnology being one of the leading examples. Technological advances are enabling innovative interfaces between neurons and electronics, opening the way to new therapeutic devices for neurological diseases as well as to detailed investigation tools of the cognitive processes. The chapter reviewed the performance requirements and the perspectives of fully integrated neural recording systems, pointing out the issues faced in the definition of optimal architectures and function partitioning. In this frame, energy efficiency and low-noise design are key ingredients. The fundamental metrics to quantitatively judge the trade-off between noise, power consumption, and processing speed have been introduced and adopted to compare the most recent system implementations. It has been shown that a 10-μW power budget target per sensing channel is attainable by using cutting-edge technologies and careful design. Finally, we focused on the particular issue of neural amplifier design: Leveraging on a detailed breakdown of the noise sources and by means of an insightful design strategy, we addressed the problem of noise–power trade-off and we presented a neural amplifier that achieved the best performance so far reported.

REFERENCES

[1] P. Andlin-Sobocki, B. Jonsson, H. Wittchen, and J. Olesen, "Costs of disorders of the brain in Europe," *Eur. J. Neurol.*, 12, pp. 1–27, 2005.

[2] WHO, *Neurological Disorders: Public Health Challenges*. Geneva: WHO Press, 2007.

[3] S. Tong and N. Thakor, *Quantitative EEG Analysis Methods and Clinical Applications*. Norwood, MA: Artech House, 2009.

[4] S. Breit, J. Schulz, and A. Benabid, "Deep brain stimulation," *Cell Tissue Res.*, 318, pp. 275–288, 2004.

[5] R. Briggs, H. Eder, P. Seligman, R. Cowan, K. Plant, J. Dalton, D. Money, and J. Patrick, "Deep brain stimulation," *Otol. Neurotol.*, 29(2), pp. 114–119, 2008.

[6] D. Di Lorenzo and J. Bronzino, *Neuroengeneering*. Boca Raton, FL: CRC, 2007.[56]

[7] E. Kandel, J. Schwartz, and T. Jessell, *Principles of Neuroscience*. 4th edition, chapter 9. New York: McGraw-Hill, 2000.

[8] G. Buzsaki, *Rhythms of the Brain*. New York: Oxford University Press, 2006.

[9] A. Geva and D. Kerem, "Forecasting generalized epileptic seizures from the EEG signal by wavelet analysis and dynamic unsupervisioned fuzzy clustering," *IEEE Trans. Biomed. Eng.*, 45(10), pp. 1205–1216, 1998.

[10] G. Buzsaki, "Large-scale recording of neuronal ensembles," *Nat. Neurosci.*, 7, pp. 446–451, 2004.

[11] D. Henze, Z. Borhegyi, J. Csicsvari, A. Mamiya, K. Harris, and G. Buzsaki, "Intracellular features predicted by extracellular recordings in the hippocampus *in vivo*," *J. Neurophysiol.*, 84, pp. 390–400, 2000.

[12] K. Pettersen and G. Einevoll, "Amplitude variability and extracellular low-pass filtering of neuronal spikes," *Biophys. J.*, 94, pp. 784–802, 2008.

[13] K. Harris, D. Henze, J. Csicsvari, H. Hirase, and G. Buzsaki, "Accuracy of tetrode spike separation as determined by simultaneous intracellular and extracellular measurements," *J. Neurophysiol.*, 84, pp. 401–414, 2000.

[14] K. Wise, D. Anderson, J. Hetke, D. Kipke, and K. Najafi, "Wireless implantable microsystem: High density electronic interfaces to the nervous system," *Proceedings of the IEEE*, pp. 76–97, 2004.

[15] P. Campbell, K. Jones, R. Hubert, K. Horch, and R. Normann, "A silicon-based, three dimensional neural interface: Manufacturing processes for an intracortical electrode array," *IEEE Trans. Biomed. Eng.*, 38(12), pp. 758–768, 1991.

[16] T. Seese, H. Harasaki, G. Saidel, and C. Davies, "Characterization of tissue morphology, angiogenesis, and temperature in the adaptive response of muscle tissue to chronic heating," *Lab. Invest.*, 78, pp. 1553–1562, 1998.

[17] K. Silay, C. Dehollain, and M. Declercq, "Numerical analysis of temperature elevation in the head due to power dissipation in a cortical implant," *Proceedings of the 2008 International Conference of the IEEE Engineering in Medicine and Biology Society*, pp. 951–956, 2008.

[18] T. Jochum, T. Denison, and P. Wolf, "Integrated circuit amplifiers for multi-electrode intracortical recording," *J. Neural Eng.*, 6(1), p. 012001 (26pp), 2009.

[19] R. Gesteland, B. Howlandt, J. Y. Lettvin, and W. H. Pittsk, "Comments on microelectrodes," *Proceedings of the IRE*, pp. 1856–1862, 1959.

[20] N. Neihart and R. Harrison, "Micropower circuits for bidirectional wireless telemetry in neural recording applications," *IEEE Trans. Biomed. Eng.*, 52(11), pp. 1950–1959, 2005.

[21] M. Ghovanloo and S. Atluri, "A wide-band power-efficient inductive wireless link for implantable microelectronic devices using multiple carriers," *IEEE Trans. Circuits Syst.*, 10, pp. 2211–2221, 2007.

[22] A. Sodagar, K. Najafi, K. Wise, and M. Ghvanloo, "Fully-integrated CMOS power regulator for telemetry-powered implantable biomedical microsystems," *Proceedings of the CICC*, 2006, pp. 659–662.

[23] M. Ghovanloo and K. Najafi, "A wideband frequency-shift keying wireless link for inductively powered biomedical implants," *IEEE Trans. Circuits Syst.*, 51(12), pp. 2374–2383, 2004.

[24] R. Harrison, R. Kier, B. Greger, F. Solzbacher, C. Chestek, V. Gilja, P. Nuyujukian, S. Ryu, and K. V. Shenoy, "Wireless neural signal acquisition with single low-power integrated circuit," *Proceedings of the ISCAS*, pp. 1748–1751, 2008.

[25] J. Steyaert and W. Sansen, "A micropower low-noise monolithic instrumentation amplifier for medical purposes," *IEEE J. Solid-State Circuits*, 22, pp. 1163–1168, 1987.

[26] B. Murmann, *ADC performance survey 1997–2008*." Available: http://www.stanford.edu/~murmann/adcsurvey.html.

[27] B. Murmann, "A/D converter trends: Power dissipation, scaling and digitally assisted architectures," *Proceedings of the IEEE Custom Integrated Circuits Conference (CICC)*, pp.105–112, 2008.

[28] J. Craninckx and G. Van der Plas, "A 65fJ/conversion-step 0-to-50 Ms/s 0-to-0.7 mW 9b charge-sharing SAR ADC in 90 nm digital CMOS," *Proceedings of the IEEE International Solid-State Circuits Conference*, pp. 246–247, 2007.

[29] M. Elzakker, E. Tuijl, P. Geraedts, D. Schinkel, E. Klumperink, and B. Nauta, "A 1.9 nW 4.4fJ/conversion-step 10b 1 Ms/s charge-redistribution ADC," *Proceedings of the IEEE International Solid-State Circuits Conference*, pp. 244–245, 2008.

[30] R. R. Harrison, P. T. Watkins, R. Kier, R. Lovejoy, D. J. Black, B. Greger, and F. Solzbacher, "A low-power integrated circuit for a wireless 100-electrode neural recording system," *IEEE J. Solid-State Circuits*, 42(1), pp. 123–133, 2007.

[31] B. Gosselin, M. Sawan, and C. A. Chapman, "A low-power integrated bioamplifier with active low-frequency suppression," *IEEE Trans. Biomed. Circuits Syst.*, 1(3), pp. 184–192, 2007.

[32] Z. Zumsteg, C. Kemere, S. O'Driscoll, G. Santhanam, R. Ahmed, K. Shenoy, and T. Meng, "Power feasibility of implantable digital spike sorting circuits for neural prosthetic systems," *IEEE Trans. Neural Syst. Rehabil. Eng.*, 13, pp. 272–279, 2005.

[33] L. Hochberg, M. Serruya, G. Friehs, J. Mukand, M. Saleh, A. Caplan, A. Branner, D. Chen, R. Penn, and J. Donoghue, "Neuronal ensemble control of prosthetic devices by a human with tetraplegia," *Nature*, 442, pp. 164–171, 2006.

[34] R. Olsson III and K. Wise, "A three-dimensional neural recording microsystem with implantable data compression circuitry," *IEEE J. Solid-State Circuits*, 40, pp. 2796–2804, 2005.

[35] J. Vibert and J. Costa, "Spike separation in multiunit records: A multivariate analysis of spike descriptive parameters," *Electroencephalogr. Clin. Neurophysiol.*, 47, pp. 172–182, 1979.

[36] A. Bonfanti, T. Borghi, R. Gusmeroli, G. Zambra, A. Oliyink, L. Fadiga, A. S. Spinelli, and G. Baranauskas, "A low-power integrated circuit for analog spike detection and sorting in neural prosthesis systems," *Proceedings of the BioCAS*, pp. 257–260, 2008.

[37] Z. Tang, B. Smith, J. Schild, and P. Peckham, "Data transmission from an implantable biotelemeter by load-shift keying using circuit configuration modulator," *IEEE Trans. Biomed. Eng.*, 42, pp. 524–528, 1995.

[38] M. Baghaei-Nejad, D. Mendoza, Z. Zou, S. Radiom, G. Gielen, L.-R. Zheng, and H. Tenhunen, "A remote-powered RFID tag with 10 Mb/s UWB uplink and –18.5 dBm

sensitivity UHF downlink in 0.18 μm," *IEEE Int. Solid-State Circ. Conf.*, pp. 198–199, 2009.

[39] M. Klemm and G. Troester, "EM energy absorption in the human body tissues due to UWB antennas," *Prog. Electromagnetics Res.*, 62, pp. 261–280, 2006.

[40] M. Chae, W. Liu, Z. Yang, T. Chen, J. Kim, M. Sivaprakasam, and M. Yuce, "A 128-channel 6 mW wireless neural recording IC with on-the-fly spike sorting and UWB transmitter," *Proceedings of the IEEE International Solid-State Circuits Conference*, pp. 146–148, 2008.

[41] K. Guillory, A. Misener, and A. Pungor, "Hybrid RF/IR transcutaneous telemetry for power and high-bandwidth data," *Proceedings of the IEEE EMBS Conference*, pp. 4338–4340, 2004.

[42] T. Borghi, A. Bonfanti, R. Gusmeroli, G. Zambra, and A. S. Spinelli, "A power-efficient analog integrated circuit for amplification and detection of neural signals," *Proceedings of the EMBC 2008*, pp.4911–4914, Vancouver, 2008.

[43] R. Harrison and C. Charles, "A low-power low-noise CMOS amplifier for neural recording applications," *IEEE J. Solid-State Circuits*, 38(6), pp. 958–965, 2003.

[44] T. Borghi, A. Bonfanti, G. Zambra, R. Gusmeroli, A. S. Spinelli, and G. Baranausks, "An integrated low-noise multichannel system for neural signals amplification," *Proceedings of the IEEE ESSCIRC*, Munich (Germany), pp.456–459, 2007.

[45] W. Wattanapanitch, M. Fee, and R. Sarpeshkar, "An energy-efficient micropower neural recording amplifier," *IEEE Trans. Biomed. Eng.*, 1(2), pp. 136–147, 2007.

[46] Y. Perelman and R. Ginosar, "An integrated system for multichannel neuronal recording with spike/LFP separation, integrated A/D conversion and threshold detection," *IEEE Trans. Biomed. Eng.*, 54(1), pp. 130–137, 2007.

[47] E. Vittoz and J. Fellrath, "CMOS analog integrated circuits based on weak inversion operations," *IEEE J. Solid-State Circuits*, 12(3), pp. 224–231, 1977.

[48] C. Enz, F. Krummenacher and E. Vittoz, "An analytical MOS transistor model valid in all region of operation and dedicated to low-voltage and low-current applications," *Analog Integr. Circuits Process.*, 8, pp. 83–114, 1995.

[49] J. Holleman and B. Otis, "A sub-microwatt low-noise amplifier for neural recording," *Proceedings of the 2007 International Conference of the IEEE Engineering in Medicine and Biology Society*, pp. 3930–3933, 2005.

[50] R. Carvajal, J. Ramirez-Angulo, A. Lopez-Martin, A. Torralba, J. Galan, A. Carlosena, and F. Chavero, "The flipped voltage follower: A useful cell for low-voltage low-power circuit design," *IEEE Trans. Circ. Syst.*, 52(7), pp. 1276–1291, 2005.

[51] B. Gosselin, A. Ayoub, J.-F. Ry, M. Sawan, F. Lepore, A. Chaudhuri, and D. Guitton, "A mixed-signal multichip neural recording interface with bandwidth reduction," *Proceedings of the BioCAS*, pp. 129–141, 2009.

[52] M. Mollazadeh, K. Murari, G. Cauwenberghs, and N. Thakor, "Micropower MOS integrated low-noise amplification, filtering and digitization of multimodal neuropotentials," *IEEE Trans. Biomed. Circuits Syst.*, 3(1), pp. 1–10, 2009.

[53] M. Yin and M. Ghovanloo, "A low-noise preamplifier with adjustable gain and bandwidth for biopotential recording applications," *Proceedings of the ISCAS*, pp. 321–324, 2007.

[54] R. H. Olsson, D. L. Buhl, A. M. Sirota, G. Buzsaki, and K. D. Wise, "Band-tunable and multiplexed integrated circuits for simultaneous recording and stimulation with microelectrode arrays," *IEEE Trans. Biomed. Eng.*, 52, pp. 1303–1311, 2005.

[55] J. N. Y. Aziz, K. Abdelhalim, R. Shulyzki, R. Genov, B. Bardakjian, M. Derchansky, D. Serletis, and P. L. Carlen, "256-channel neural recording and delta compression microsystem with 3D electrodes," *IEEE J. Solid-State Circuits*, 44(3), pp. 995–1005, 2009.

3

VLSI IMPLEMENTATION OF WIRELESS NEURAL RECORDING MICROSYSTEM FOR NEUROMUSCULAR STIMULATION

Shuenn-Yuh Lee, Chih-Jen Cheng, Shyh-Chyang Lee, and Qiang Fang

3.1 REVIEW OF THE RECORDING MICROSYSTEM

With the advancements in the semiconductor industry, the very large-scale integrated (VLSI) technology has made impressive improvements in the past 10 years. Integrating more and more functionalities into a chip has become feasible.

System-on-chip (SoC) is a mainstream in which the whole circuit and all digital and mixed-signal components are integrated into a single substrate of silicon [1]. SoC can be used in many branches of microelectronics, such as biomedical applications (e.g., health monitoring, medical diagnosis, microsurgery, nanochemistry, and environmental monitoring), which have been a prominent topic. Due to the highly integrated and single-chip characteristics of SoC, biomedical devices have benefited the most from SoC, especially when they are intended to be implantable. Hence, an increase in the number biomicrosystem-related research is expected in the future.

CMOS Biomicrosystems: Where Electronics Meet Biology, First Edition. Edited by Krzysztof Iniewski.

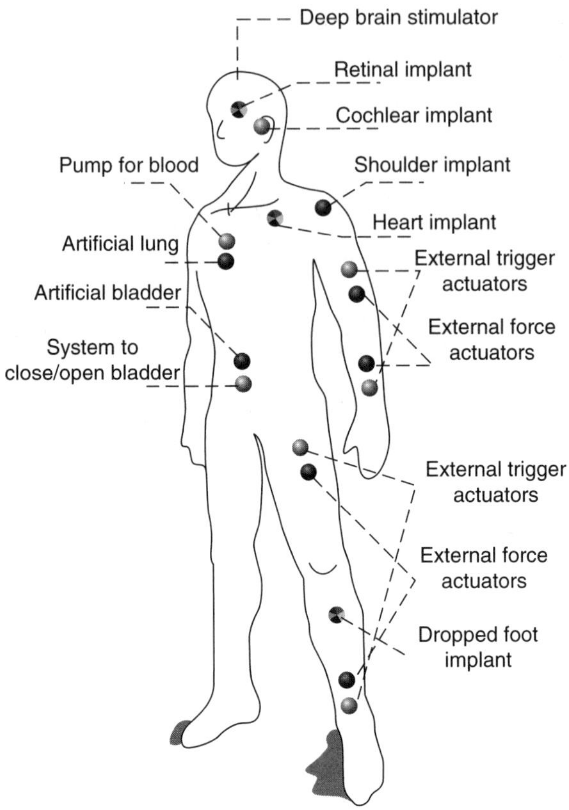

Figure 3.1. The different parts of the human body to be monitored or regulated.

Following the mainstream, many researchers have devoted their time to the development of biomedical devices. As Figure 3.1 shows, almost every aspect of human health can be monitored or regulated by an implanted device. Hence, studies on biomedical devices and systems being implanted into the human body have increased rapidly. A number of published research have stressed that it is efficient to generate neural action potential electrically and to control dysfunctional organs further [2]. Therefore, various implantable microstimulators have been designed for various clinical applications, such as cardiac pacemakers, cochlear implants, retinal prostheses, functional neuromuscular stimulation (FNS) systems, and deep brain stimulators [3–8]. In the following subsections, an overview of the different microstimulators will be presented.

3.1.1 Cardiac Pacemaker

Figure 3.2 shows a cardiac pacemaker system that consists of three components: battery, pulse generator, and pacing leads. The battery, which is used to supply

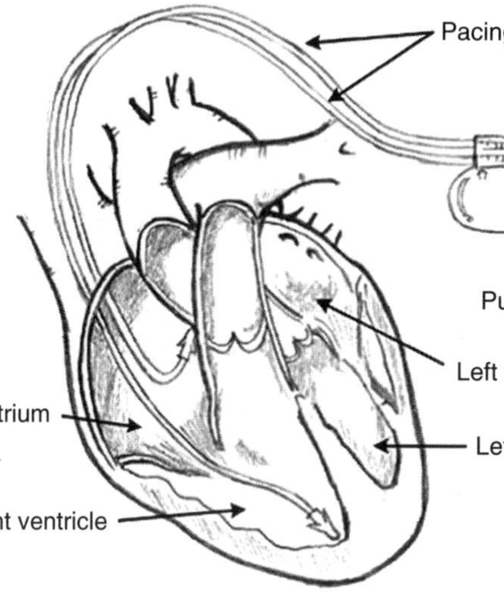

Figure 3.2. The cardiac pacemaker system.

power to the pacemaker, is strategically designed to be small and flat to fit into the pacemaker case. The pulse generator is the brain of the cardiac pacemaker system and is placed under the skin below the collarbone. It checks the heart rate and produces tiny electrical pulses that keep the heart beating at a normal pace. Usually, a normal heart rate is restored in this unit so that the heart can pump properly. The insulated pacing leads connect the pulse generator and the chambers of the heart. The leads can carry impulses from the pulse generator to the electrode, which is fixed against the heart, and stimulate the heart to beat. At the same time, the leads can also carry information from the heart back to the pulse generator, which enables the physician to monitor the stimulation.

3.1.2 Cochlear Implant

Figure 3.3 shows a cochlear implant system consisting of two main parts: an external part, which is like a hearing aid; and an internal part to be implanted by a surgeon. The external part (headpiece) includes a microphone, a transmitter, and a speech processor. The internal part (implant) includes a receiver, a stimulator, and an electrode array. A cochlear implant functions as follows. The microphone picks up the sound from the environment and the processor codes the sound input into electrical signals that are sent back to the transmitter. The receiver receives a digital code from the transmitter, and the stimulator outputs the stimulus pattern to the auditory nerve via the electrode array. Finally, the auditory nerve picks up this stimulus signal and transmits it to the brain (auditory cortex), where it is perceived as sound.

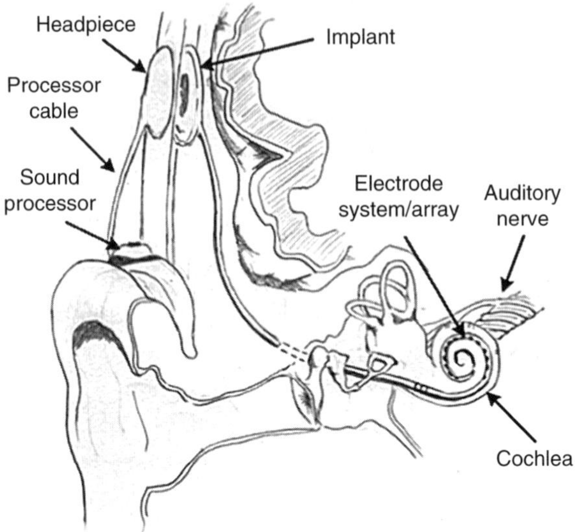

Figure 3.3. A cochlear implant system.

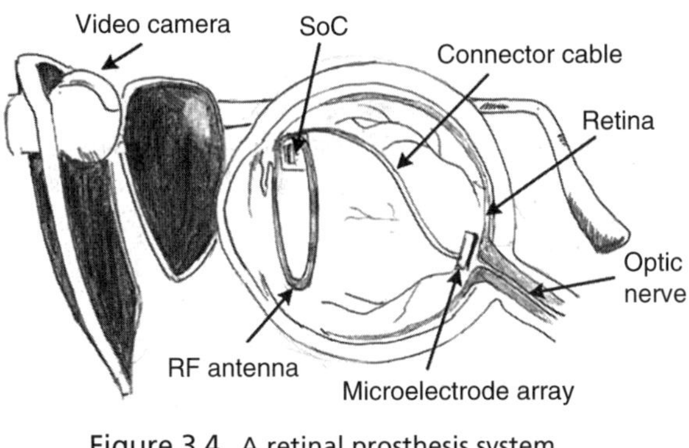

Figure 3.4. A retinal prosthesis system.

3.1.3 Retinal Prosthesis

A retinal prosthesis system is shown in Figure 3.4. The system consists of an external unit, a glasses-like device with a tiny camera embedded in the lens; and an implanted unit, which includes a receiver, a stimulator, and an electrode array. The external camera captures an image that is converted into digital data by an image processor. Through a wireless link, the data is coupled to the implanted unit. The implanted unit receives the signal, recovers the power and data, and controls the output potential of the stimulator. Finally, the stimulus pattern can be applied to the retinal nerve via the electrode array. As there are over 100 million photoreceptors in the retina, the channel number of the electrode array should be enough to store the image in the visual scene.

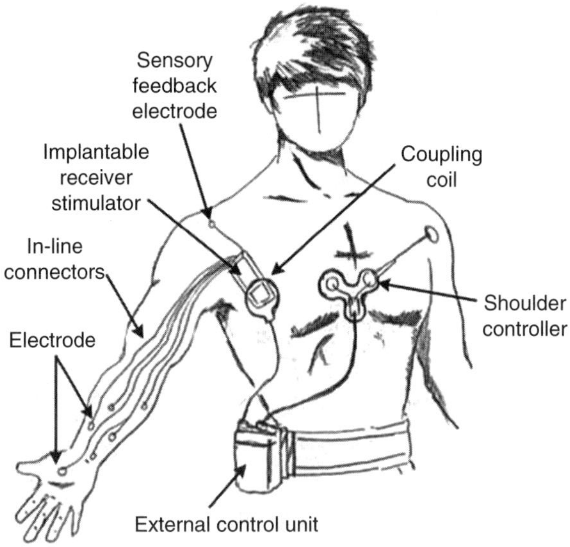

Figure 3.5. The functional neuromuscular stimulation system.

3.1.4 Functional Neuromuscular Stimulation

Figure 3.5 shows a common FNS system. It is divided into the implanted and the external part. The external components include a PC-based controller that regulates the stimulus. The implanted components include the electrodes, the leads, and a stimulator to generate the electrical pulse. All the implanted components are powered by the wireless coupling link. By triggering and modulating the intensity of the electrical stimulus, functional movements, such as reaching, grasping, releasing, standing, and walking, can be achieved.

3.1.5 Deep Brain Stimulator

Figure 3.6 shows the deep brain stimulation (DBS) system, which implants electrodes deep into the thalamus, a part of the brain that handles movement-related communication. It is composed of three components: lead (electrode), extension, and stimulator. The electrodes deliver mild electrical pulses to the thalamus, blocking the brain signals that cause muscle tremor. The extension is an insulated wire passed under the skin of the head, neck, and shoulder, connecting the lead to the stimulator. The stimulator is implanted under the skin near the collarbone, and it generates electrical signals that are delivered to the targeted structures in the brain via the extension.

DBS does not cure Parkinson's disease (PD), but it can effectively treat many of its symptoms and help improve motor functions even though it tends to increase the risk of infection.

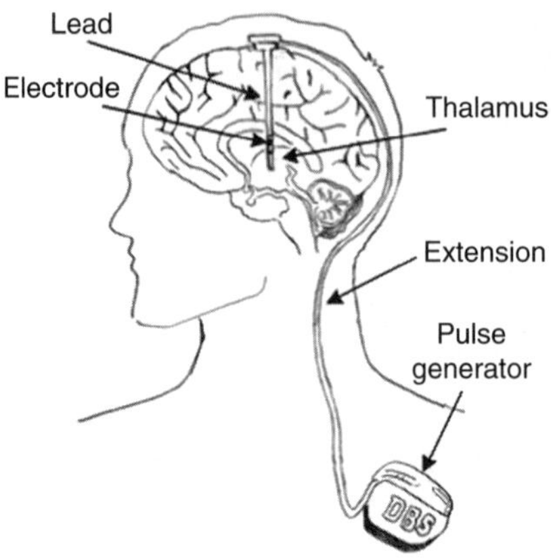

Figure 3.6. The deep brain stimulation system.

3.2 WIRELESS POWER AND DATA TRANSMISSION MICROSTIMULATOR SYSTEM

Figure 3.7 shows the overall system block diagram including the wireless power and data transmission microstimulator system and recording microsystem. The external unit includes a control module, a transmitter coil, a high-efficiency power amplifier, and a receiver. In general, a class E power amplifier is employed in an external transmitter for high-efficiency transmission [9]. The digital data and commands are encoded and streamed in a series of pulses, which are sent to the high-efficiency radio frequency (RF) transmitter and are inductively coupled to the implantable device. The internal unit is a fully integrated SOC, except for the RF coupling link. The wireless power and data transmission microstimulator system includes a receiver coil, RF powering circuits, a control circuit, and a microstimulator. Through the receiver coil and RF powering circuits, power and data from the external transmitter are extracted for the whole-system operations. The decoded data are then employed to control the potential of the microstimulator. All system blocks will be introduced completely in the following subsections.

3.2.1 Radio Frequency Powering Circuits

The integrated full-wave rectifiers are composed of diode-connected devices commonly used in radio frequency identification (RFID) and implantable systems [10, 11]. However, the threshold voltage associated with the p-type metal-oxide-silicon (pMOS) transistor of the diode-connected pMOS drastically decreases the available rectified output current with small $V_{ace}(=V_{rf+} - V_{rf-})$, posing a headroom limitation in low-voltage design. As a result, a self-driven synchronous rectifier can be adopted to avoid the additional voltage drop, as shown in Figure 3.8a, wherein devices M_{p1-2} and M_{n3-4} make up the conventional four-transistor

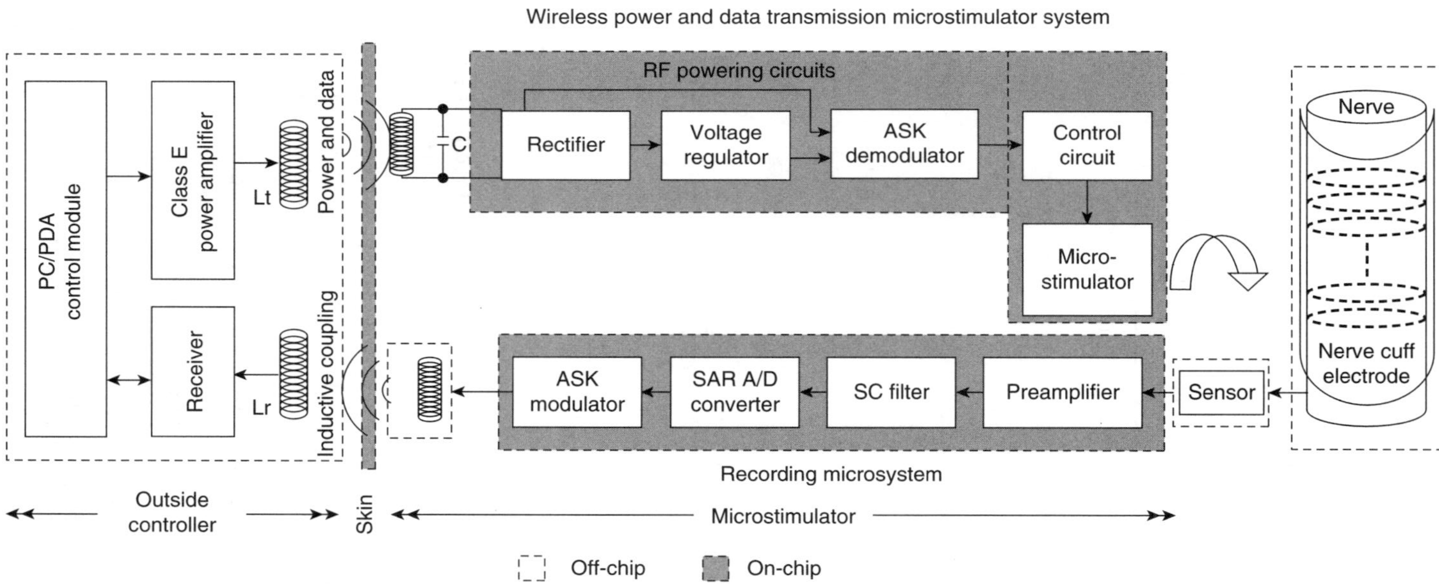

Figure 3.7. Block diagram of the bidirectional microstimulator system.

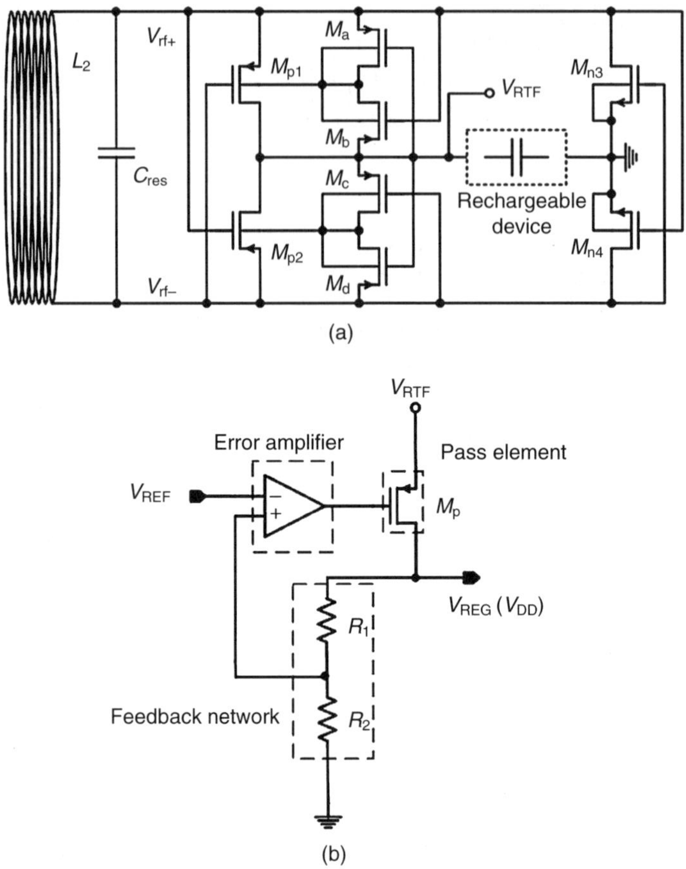

Figure 3.8. (a) Self-driven bridge rectifier; (b) LDO regulator.

cell. The voltage regulator receives the output voltage of the rectifier (V_{rtf}) and regulates the voltage to a stable and precise direct current (DC) voltage source (V_{REG}) for the implant stimulator. The precise output voltage and the low quiescent current are important issues in designing voltage regulators. Figure 3.8b shows a typical low-dropout (LDO) regulator, which consists of a metal-oxide-silicon (MOS) pass element, an error amplifier, and one set of feedback network [12, 13].

Amplitude shift keying (ASK) is a form of modulation representing digital data as variations in the amplitude of a carrier wave. It is commonly used in many microsystems that are powered by an inductive link [14, 15]. The major limitation is the high-bandwidth data transmission. Figure 3.9 shows the ASK demodulator including a pair of unity-gain amplifiers, an OR logic gate, a low-pass filter, and a comparator. The unity-gain amplifiers are used to make the RF coupling signal different. Through the OR logic gate, the voltages, which are larger than the threshold voltage, will be pulled to logic high (V_{DD}, 3V). Hence, the voltage dis-

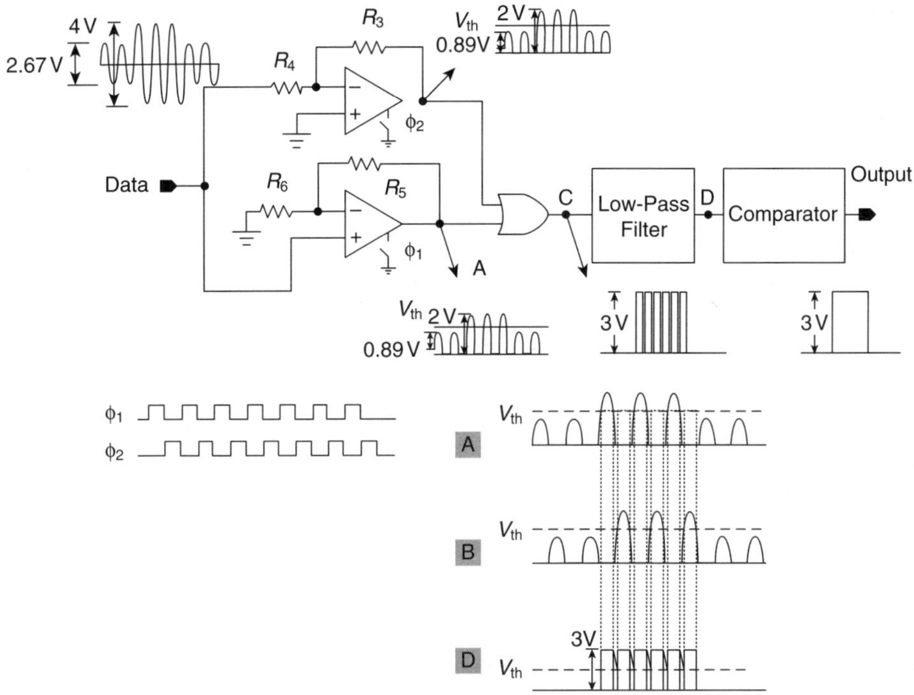

Figure 3.9. The ASK demodulator block and waveform in each of the output stages.

charge time between the adjacent peaks will be considerably reduced. Moreover, the low-pass filter is employed to regulate the noise and to extract the envelope from the carrier. The comparator, with a threshold voltage of $V_{DD}/2$, is then used to detect the digital data precisely. The detailed waveforms for each of the stages are also shown in Figure 3.9.

3.2.2 Control Circuit

Figure 3.10 shows the functional blocks [14, 15] of the control circuit, including a Manchester decoder, a serial-to-parallel converter, a finite state machine, and the relative protocol. The main functions of the Manchester decoder are to decode the Manchester code from the ASK demodulator and to restore its data and clock. The different time bases between the external transmitter and the internal receiver can cause error due to the sampling of the data at a wrong time. Hence, the Manchester code is often used to resolve the synchronization between the two blocks because of its strong timing information. The data extracted by the Manchester decoder are in serial sequence, whereas the parallel-type data are more convenient for later logical operations. Therefore, a serial-to-parallel converter is necessary. The output data of the serial-to-parallel converter are used

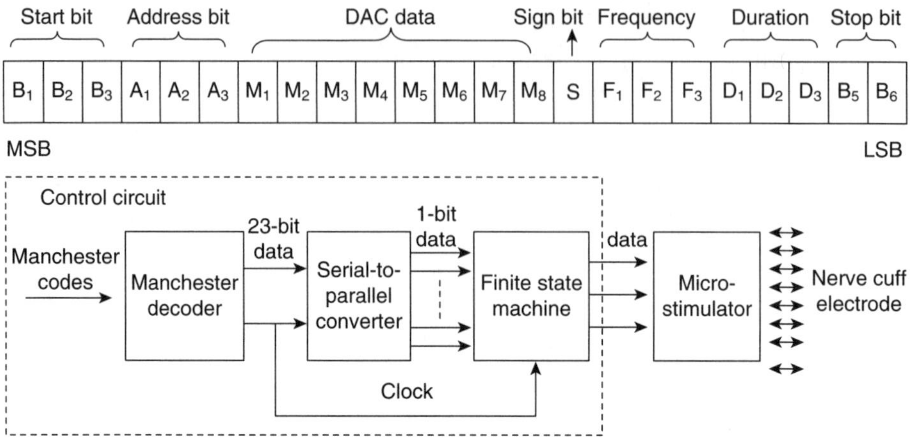

Figure 3.10. Format of the 23-bit command frame and functional blocks of the control circuit.

to control the microstimulator through the control center, a finite state machine. The function of the finite state machine is to determine the proper stimulation channel, the scale of the stimulation magnitude, and the frequency, all of which are controlled by the previously decoded data.

3.2.3 Microstimulator

The stimulator is used to pass the current through the tissue and to generate useful action potentials. According to the output type, the stimulator is categorized into three types: voltage mode, current mode, and charge mode. In the voltage-mode stimulator, the output voltages can be generated precisely. However, the current passing through the tissue is dependent on the impedance of the tissue, which is highly variable. Hence, the passing current cannot be properly controlled even though its output voltages are controlled efficiently. The charge-mode stimulator has similar capacity with the voltage-mode stimulator. Therefore, the current-mode stimulator is preferred more than the voltage-mode and the charge-mode stimulators in the microstimulator system. Figure 3.11 shows the architecture of the current-mode microstimulator [15]. It is composed of three components, namely, the 8-bit segment current-mode digital-to-analog converter (DAC), the current-to-voltage (I–V) converter, and the control logic. Maintaining the stability and linearity of the output current is the main issue of the microstimulator design [16]. Hence, the thermometer code scheme is employed in the DAC design because of its small glitch error. As using the biphasic pulse could avoid ion charge accumulation in tissues, two pairs of switches, S_1 and S_2, which are controlled by the different clock phases, are employed to produce biphasic electrical stimulation pulses.

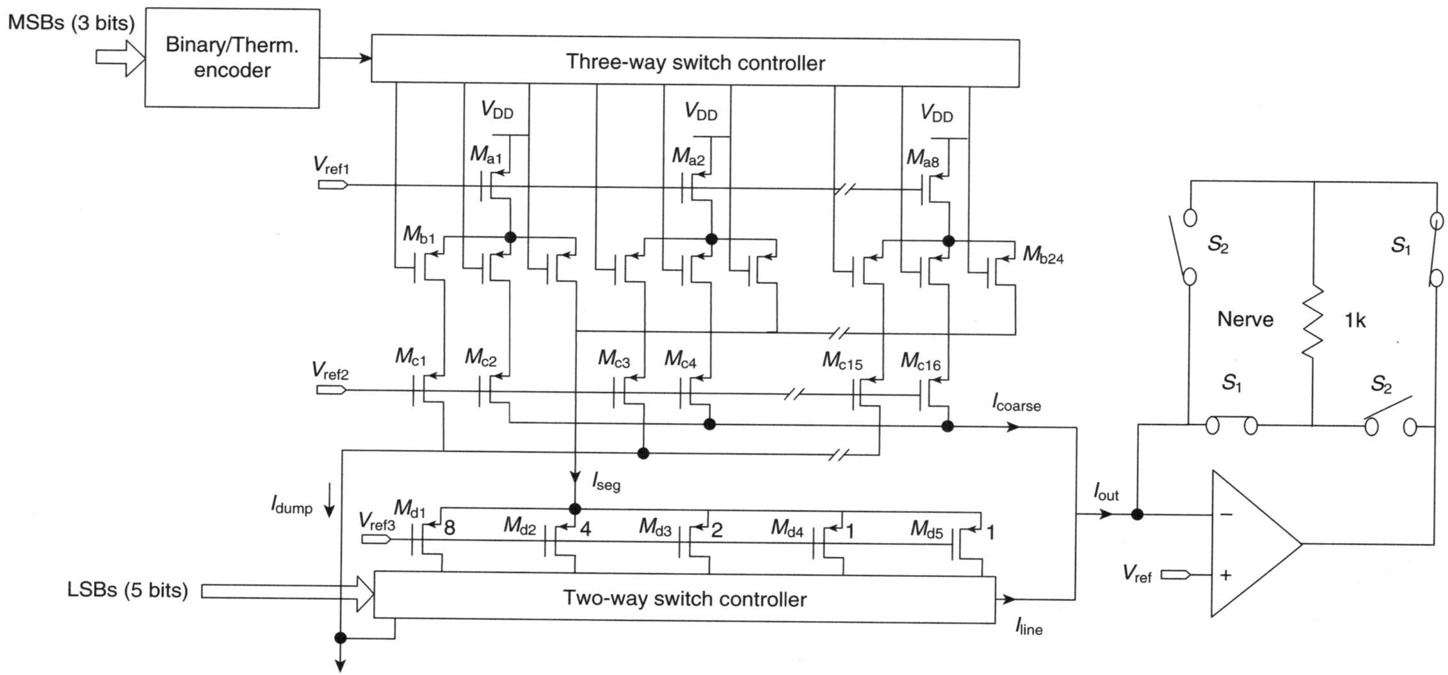

Figure 3.11. The 8-bit segment current-mode microstimulator.

3.3 VERY LARGE-SCALE INTEGRATED CIRCUITS IN THE RECORDING MICROSYSTEM

Bidirectional communication is an important issue in all kinds of implants due to the backward information of said implants, which have the following purposes: making adjustments in accordance with the data sent out from the implants, and monitoring or recording biosignals at the external devices. As shown in Figure 3.7, the recording microsystem can monitor the biosignals from the neural cell. It consists of a programmable gain preamplifier, an operational transconductance amplifier-capacitor (OTA-C) filter, a successive approximation (SA) analog-to-digital converter (ADC), and an ASK modulator, which will be described in the following subsections.

3.3.1 Programmable Gain Preamplifier

The schematic of the closed-loop preamplifier is shown in Figure 3.12, wherein the DC feedback by utilizing two MOS-bipolar pseudoresistors, M_{r1} and M_{r2}, is used to diminish the loading effect on the output stage [17]. To perform the variable gain, the structure of the differential difference amplifier (DDA) is adopted in this continuous circuit, as the second input stage associated with $V_{21(22)}$ has the ability to set the gain without deteriorating the signal when using switches in the signal path.

3.3.2 Antialiasing OTA-C Filter

Switched capacitor (SC) is popularly used in integrating this long-term biosignal monitor system [18]. With the switch-based circuits, the sampling frequency of up to kilohertz must be adopted for the filter to avoid leakage problems, whereas the switching behavior consumes additional dynamic power. Thus, it needs to use the continuous-time OTA-based filters in which the devices operate in the subthreshold region to save more power that would result to a very low transconductance (G_m, typically of the order of a few nanoamperes per volt) [19]. In OTA-based circuits, the OTA will dominate the performance of the filter circuit,

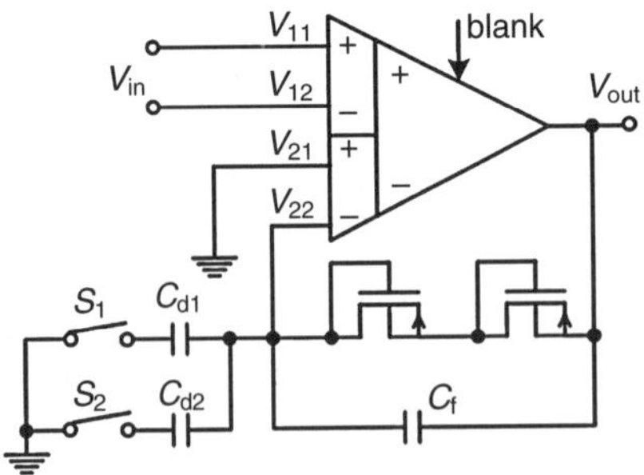

Figure 3.12. Diagram of the DDA preamplifier.

and the ratio of the capacitor to the small transconductance determines the time constant of OTA-C integrators.

A design example of the 5th-order ladder-type Butterworth filter with a maximum flat response and a cutoff frequency of 250 Hz has been selected for cardiac signal processing [20] (Fig. 3.13). The ladder solution has the advantage of having transfer function robustness and is inherently insensitive to component variations. Furthermore, the signal flow graph (SFG) mapping method can be utilized to verify that this filter possesses a nearly constant group delay below 150 Hz where the most cardiac signals are located [21]. Referring to Figure 3.13, the 5th-order OTA-C filter with common-mode feedback circuits is realized. The overall circuit consists of two grounded resistors, G_{m0} and G_{m6}; five capacitors, C_1–C_5; and two gyrators, A and B, which are used to implement the equivalent inductors, L_2 and L_4, respectively.

3.3.3 Successive Approximation Analog-to-Digital Converter

The architecture of the SA ADC, along with the sample-and-hold (S/H) circuit, is shown in Figure 3.14a. The SA ADC operation is based on the binary search algorithm. This algorithm searches the closest digital word to match the input signal from most significant bit (MSB) to least significant bit (LSB). Hence, the latency is n for an n-bit conversion. It consists of an S/H, a voltage comparator, a successive approximation register (SAR)-based controller, and an 8-bit DAC. A dummy MOS capacitance is placed adjacent to the sampling switch to reduce the charge injection and clock feedthrough. To save power, the charge redistribution structure is applied to the DAC whose unit capacitor is 25 fF to suppress the inband kT/C noise. The sampling voltage and the output of DAC are then fed into the regenerative comparator [22] to execute the comparison operation and produce a resulting signal for the SAR controller. Unlike the conventional designs, the nonredundant structure as shown in Figure 3.14b [23] can be used to reduce the usage of the registers [24]. At the beginning of the conversion, the MSB is set to 1, and the other bits are set to 0. If the comparator output is low, the MSB will be 0 and saved at the output of the SAR. The residue bits perform the same operations until the LSB is determined.

3.3.4 Amplitude Shift Keying Modulator

As the ASK modulation is characterized by simplicity and low power consumption, the digital output of the SA ADC will be modulated by the ASK scheme and transmitted to the external receiver for data analysis and recording. Figure 3.15 shows the architecture of the ASK modulator [25]. A carrier frequency is generated from the ring oscillator and the divider, which is used to lighten the stage requirement of the ring oscillator. Through the AND logic gate, digital data from the SA ADC are modulated. Through an active transmitter, a common-source class A power amplifier, and an inductor-capacitor (LC) tank, the modulated ASK signal can be amplified and transmitted.

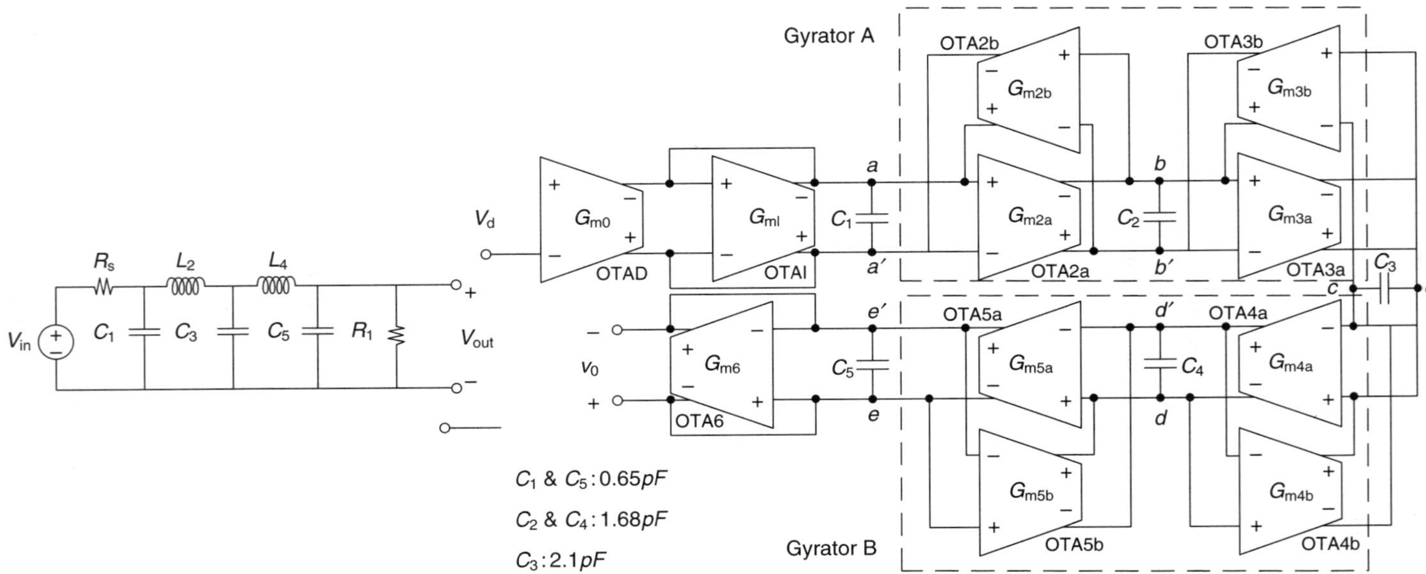

Figure 3.13. Passive and active circuit realization of the 5th-order Butterworth filter.

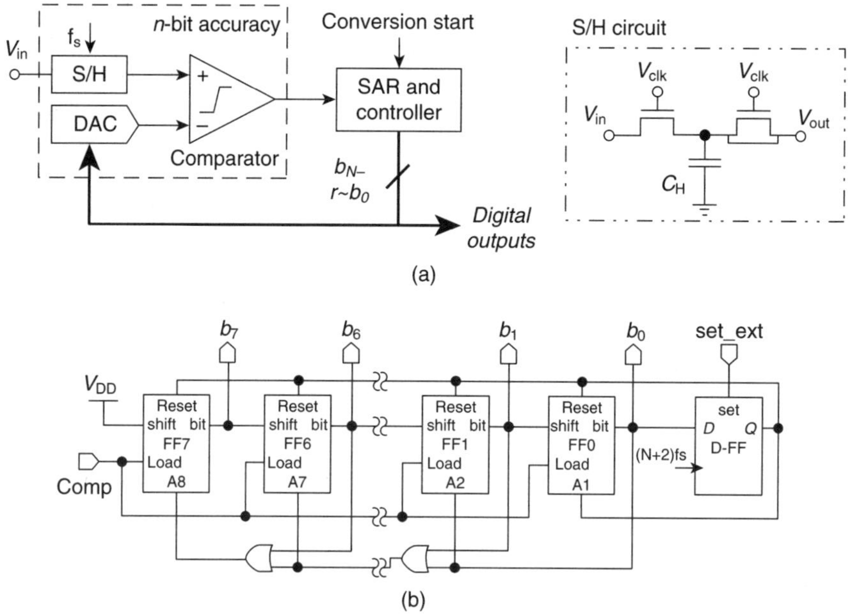

Figure 3.14. (a) Block diagram of the SA ADC with S/H circuit; (b) *n*-bit SAR controller based on the nonredundant structure.

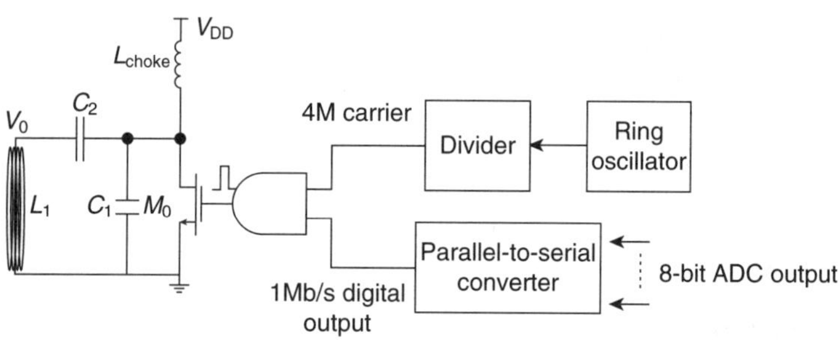

Figure 3.15. Architecture of the ASK modulator and power amplifier.

3.4 CONCLUSION

The concept of the SoC has been applied to the microstimulator system. The different applications of the microstimulator are introduced in Section 3.1. The reviews indicate that the microstimulators are indispensable for patients who suffer neural disorders. With regard to the requirements, a wireless telemetry using the near-field coupling technique for the implanted devices is presented. The presented system possesses an external powering amplifier and an internal bidirectional microstimulator. All circuitries associated with the implantable stimulator are operated normally by the coupling power interface that includes an efficient rectifier and a fully integrated regulator. A miniature digital processor, along with the ASK demodulator and a system controller with a Manchester decoder, is in charge of decoding the transmission data and handling the operation of the whole system, respectively. To acquire the biosignal, an analog front-end circuit, together with a preamplifier, an OTA-C filter, an SA ADC, and an ASK modulator, performs monitoring, data conversion, and data transmission out of the body for external recording and analysis. All functional blocks for the bidirectional implantable microstimulator with the aid of the VLSI implementation of the implantable devices have been introduced in this chapter.

REFERENCES

[1] R. Saleh, S. Wilton, S. Mirabbasi, A. Hu, M. Greenstreet, G. Lemieus, P. P. Pande, C. Grecu, and A. Ivanov, "System-on-chip: Reuse and integration," *Proc. IEEE*, 94(6), pp. 1050–1069, 2006.

[2] F. T. Hambrecht and J. B. Reswick, Eds., *Functional Electrical Stimulation, Application in Neural Prosthesis*. New York: Marcel Dekker, 1977.

[3] T. W. Dawson, M. A. Stuchly, K. Caputa, A. Sastre, R. B. Shepard, and R. Kavet, "Pacemaker interference and low-frequency electric induction in humans by external fields and electrodes," *IEEE Trans. Biomed. Eng.*, 47, pp. 1211–1218, 2000.

[4] P. C. Loizou, "Introduction to cochlear implants," *IEEE Eng. Med. Biol. Mag.*, 18, pp. 32–42, 1999.

[5] W. Liu, K. Vichienchom, M. Clements, S. C. DeMarco, C. Hughes, E. McGucken, M. S. Humayun, E. De Juan, J. D. Weiland, and R. Greenberg, "A neuro-stimulus chip with telemetry unit for retinal prosthetic device," *IEEE J. Solid-State Circuits*, 35, pp. 1487–1497, 2000.

[6] K. Arabi and M. A. Sawan, "Electronic design of a multi-channel programmable implant for neuromuscular electrical stimulation," *IEEE Trans. Rehabil. Eng.*, 7, pp. 204–214, 1999.

[7] S. Boyer, M. Sawan, M. Abdel-Gawad, S. Robin, and M. M. Elhilali, "Implantable selective stimulator to improve bladder voiding: Design and chronic experiments in dogs," *IEEE Trans. Rehabil. Eng.*, 8(4), pp. 464–470, 2000.

[8] H. A. Shill and A. G. Shetter, "Reliability in deep brain stimulation," *IEEE Trans. Device Mater. Reliab.*, 5, pp. 445–448, 2005.

[9] J. C. Mandojana, K. J. Herman, and R. E. Zulinski, "A Discrete/continuous time-domain analysis of a generalized class E amplifier," *IEEE Trans. Circuits Syst.*, 36, pp. 1057–1160, 1990.

[10] L. A. Glasser, A. C. Malamy, and C. W. Selvidge, "A magnetic power and communication interface for a CMOS integrated circuit," *IEEE J. Solid-State Circuits*, 24, pp. 1146–1149, 1989.

[11] M. Ghovanloo and K. Najafi, "Fully integrated wideband high-current rectifiers for inductively powered devices," *IEEE J. Solid-State Circuits*, 39, pp. 1976–1984, 2004.

[12] G. A. Rincon-Mora and P. E. Allen, "A low-voltage, low quiescent current, low drop-out regulator," *IEEE J. Solid-State Circuits*, 33, pp. 36–44, 1998.

[13] C. J. Cheng, C. J. Wu, and S. Y. Lee, "Programmable pacing channel with fully on-chip LDO regulator for cardiac pacemaker," *IEEE Asia Solid State Circuit Conference*, pp. 285–288, 2008.

[14] S. Y. Lee, S. C. Lee, and J. J. J. Chen, "VLSI implementation of implantable wireless power and data transmission micro-stimulator for neuromuscular stimulation," *IEICE Trans. Electron.*, E87-C(6), pp. 1062–1068, 2004.

[15] S. Y. Lee and S. C. Lee, "An implantable wireless bidirectional communication microstimulator for neuromuscular stimulation," *IEEE Trans. Circuits Syst. I*, 52(12), pp. 2526–2538, 2005.

[16] S. Bourret, M. Sawan, and R. Plamondon, "Programmable high-amplitude balanced stimulus current-source for implantable microstimulators," *Proceedings of the 19th IEEE/EMBS Conference*, pp. 1938–1941, October 1997.

[17] R. R. Harrison and C. Charles, "A low-power low-noise CMOS amplifier for neural recording applications," *IEEE J. Solid-State Circuits*, 38, pp. 958–965, 2003.

[18] K. Lasanen and J. Kostamovaara, "A 1-V analog CMOS front-end for detecting QRS complexes in a cardiac signal," *IEEE Trans. Circuits Syst. I*, 52, pp. 2584–2594, 2005.

[19] C. D. Salthouse and R. Sarpeshkar, "A practical micropower programmable bandpass filter for use in bionic ears," *IEEE J. Solid-State Circuits*, 38(1), pp. 63–70, 2003.

[20] S. Y. Lee and C. J. Cheng, "Systematic design and modeling of a *OTA-C* filter for portable ECG detection," *IEEE Trans. Biomed. Circuits Syst.*, 3(1), pp. 53–64, 2009.

[21] J. D. Bronzino, *The Biomedical Engineering Handbook*, 2nd ed. Boca Raton, FL: CRC, 2000.

[22] S. Y. Lee, C. J. Cheng, C. P. Wang, and S. C. Lee, "A 1-V 8-bit 0.95 uW successive approximation ADC for biosignal acquisition systems," *IEEE International Symposium on Circuits and Systems*, pp. 649–652, May 2009.

[23] C. J. Cheng, S. Y. Lee, C. P. Wang, and W. C. Kao, "Low-power signal processing devices for portable ECG detection," *30th Annual International Conference of the IEEE Engineering in Medicine and Biology Society (IEEE EMBS 2008)*, pp. 1683–1686, 2008.

[24] A. Rossi and G. Fucili, "Nonredundant successive approximation register for A/D converters," *Electron. Lett.*, 32(12), pp. 1055–1056, 1996.

[25] G. Wang, W. Liu, M. Sivaprakasam, and G. A. Kendir, "Design and analysis of an adaptive transcutaneous power telemetry for biomedical implants," *IEEE Trans. Circuits Syst. I*, 52(10), pp. 2109–2117, 2005.

4

HEALTH-CARE DEVICES USING RADIO FREQUENCY TECHNOLOGY

Jung Han Choi and Dong Kyun Kim

4.1 INTRODUCTION

In recent times, the health-care market has been growing rapidly in advanced countries. In the United States, the total health-care expenditure was estimated at $2.72 trillion in 2010. The compound annual growth rate (CAGR) of the health-care industry is about 7% [1]. The health-care market includes various sorts of services, appliances, and products. Considerable effort has been made to develop the products associated with the health-care industry. Mobile technology can be incorporated into health-care services and products to innovate conventional medical services and expand opportunities in the health-care market.

Radio frequency (RF) technology is developing in a challenging era during which RF systems will be applied to health care. RF technologies have been used in mobile phones for customer mobile services and in military radar systems. With wireless technology penetrating daily life (e.g., mobile phones and wireless Internet), mobile health-care sensors can be thought of as an application of mobile technology. In this environment, the use of health-care sensors for the remote monitoring of human heartbeat and respiration rates is gaining attention [2–6]. Most commercial health-care sensors have to be directly attached to the human body. It is definitely inconvenient to carry the device daily. Remote monitoring of human cardiac and respiratory activities through a mobile device is therefore desirable.

CMOS Biomicrosystems: Where Electronics Meet Biology, First Edition. Edited by Krzysztof Iniewski.

In this chapter, the use of health-care sensors for the monitoring of human cardiac and respiratory activities is addressed in detail. We show the feasibility of using the sensors for mobile health-care products.

4.2 REMOTE DETECTION OF HUMAN VITAL SIGNS

4.2.1 Operating Principle

The operating principle of a health-care device using RF technology is fundamentally the same as that used in radar systems. Most radar systems use a linearly polarized electromagnetic (EM) wave for transmission (TX) and reception (RX). When the EM wave is linearly polarized, the sensors are unfavorably vulnerable to interference if they cross each other. The linear polarization of the EM wave degrades the sensitivity of the receiver since the received power is a function of $\cos^2 \theta$, where θ is the rotation angle of the reflected signal from the target [7]. To overcome these problems and to design a robust health-care sensor, a circularly polarized EM wave can be utilized. However, although the circular polarization technique provides invulnerability to interference, it requires two separate antennas with different polarization. Therefore, the size of the sensor is inevitably very large. A sensor using circular polarization and a single antenna, which would be ideal for compact design, still remains to be developed.

Figure 4.1a shows the principle of the remote detection of human vital signs with an RF radar sensor. If the RF transmitter of the sensor continuously emits a sinusoidal RF signal with a single frequency (f_o), then the reflected signal will have the same frequency as that of the TX signal with a phase delay. The TX and RX signals can be expressed, respectively, as follows:

$$\text{TX signal: } V_{\text{TX}}(t) = A \cdot \Re\left\{e^{j2\pi f_o t}\right\} \tag{4.1}$$

$$\text{RX signal: } V_{\text{RX}}(t) = A' \cdot \Re\left\{e^{j2\pi f_o t} \cdot e^{j\Phi(t)}\right\}, \tag{4.2}$$

where A and A' are the amplitude of the TX and RX signals, respectively. $\Re(\cdot)$ denotes the real part of the complex value and $\Phi(t)$ represents the phase delay. The periodic displacement by the regular movement of the heart causes the phase delay in the received signal. If a right-hand circularly polarized (RHCP) EM wave is transmitted, the reflected signal from the target is a left-hand circularly polarized (LHCP) wave due to the boundary condition of the EM field [7] (Fig. 4.1b). The received signal has a time-varying phase, $\Phi(t)$. In other words, the reflected signal is modulated by the periodic displacement by the heart and respiration. The time-varying phase can be calculated using

$$\Phi(t) = 2\pi \cdot \frac{2d(t)}{\lambda_0}, \tag{4.3}$$

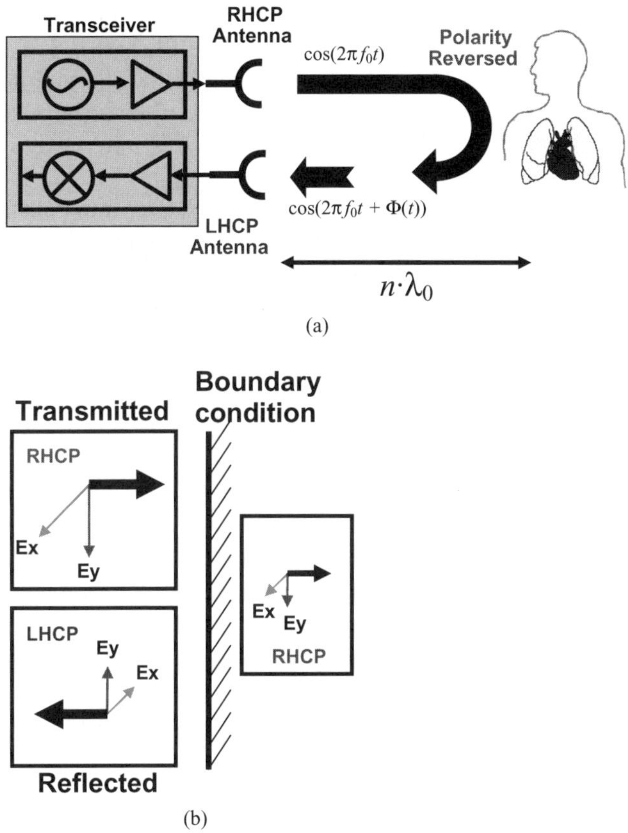

Figure 4.1. (a) The operation principle of the mobile health-care device. (b) The polarization of the EM wave at the interface.

where $d(t)$ is the distance from the antenna to the human body as given below:

$$d(t) = n \cdot \lambda_0 + x(t). \tag{4.4}$$

In Equation 4.4, n is an integer and $x(t)$ is the periodic displacement of the heart. Since the change in the displacement is small compared with the wavelength, 12.5 mm at 24 GHz, the demodulated signal is proportional to the periodic displacement of the object given by

$$V_{OUT}(t) = \cos\left(\frac{4\pi x(t)}{\lambda_0}\right) \approx \frac{4\pi x(t)}{\lambda_0}. \tag{4.5}$$

4.2.2 Radio Frequency Architecture

There have been several reports about using RF technology for *ex vivo* detection of vital signs. In previous reports, 2.4, 5, and 10 GHz have been used for detection

[8–11]. The 5 GHz RF architectures in Xiao et al. [9] and Li et al. [10] have utilized the double-sideband TX and separate antennas for TX and RX. The advantage of the double-sideband TX is discussed in Li et al. [12]. The architectures have used two separate free-running oscillators and utilized linearly polarized waves. The use of two separate antennas operating at 5 GHz inherently increases the size of the radar system.

The design of a mobile health-care sensor should take into consideration size (including the digital signal part); power; maintaining sensitivity; invulnerability to interference signals; and reliable operation under conditions of change in the outer environment. First, in order to incorporate a compact-sized antenna, the operating RF frequency has to be increased because antenna size is inversely proportional to operating frequency. The frequency band of 24 GHz, which is reserved internationally for industrial, scientific, and medical (ISM) purposes, other than for communications [13], can be considered,. It is appropriate to use such a high-frequency band for the development of a compact sensor system. Also, in order to reduce power consumption, a single voltage-controlled oscillator (VCO) signal is generated and used for two purposes. One is to transmit the RF signal without using a power amplifier (PA), and the other is to drive the mixer circuit. Apart from the VCO, the active circuits used in RF front-end circuits are a low-noise amplifier and a mixer. The other circuits are passive and do not consume power. The block diagram of the developed RF front-end architecture is shown in Figure 4.2.

Maintainance of device sensitivity and invulnerability to interfering signals can be achieved by using the circularly polarized EM wave. The antenna structure in Figure 4.2 is designed for this purpose. The antenna is connected to both the TX and the RX. As shown in Figure 4.2, the LHCP antenna is formed at the transmitter, whereas the RHCP antenna is formed at the receiver. Current feeding ports with a 90° phase difference to the antenna excite two orthogonal modes [14]. When the transmitted signal is polarized left-handed in the antenna, the same antenna only receives the right-handed polarized EM wave. This isolates the RX signals from the TX signals, and renders the device invulnerable to interference. A 2 × 1 patch antenna is connected in parallel in order to increase sensitivity. The received power increases since the antenna array shows higher gain than the single patch antenna.

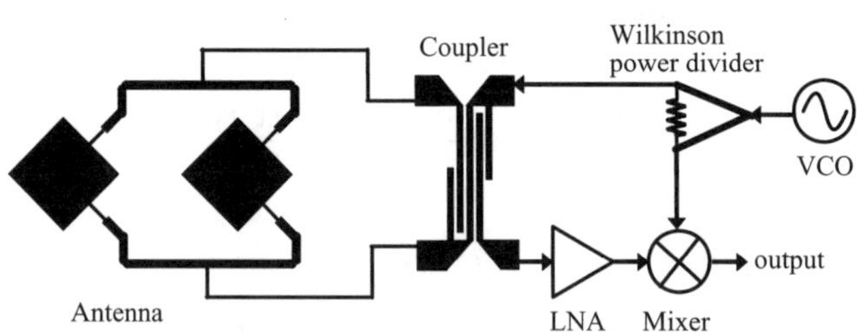

Figure 4.2. The RF front-end architecture. (Copyright © 2009 IEEE)

4.3 HEALTH-CARE SENSOR USING RADIO FREQUENCY TECHNOLOGY

4.3.1 Circularly Polarized Electromagnetic Wave

In general, in a linearly polarized radar sensor, there are two antennas for the TX and the RX. In order to reduce size, a single antenna should be used. A Lange coupler or directional coupler can be considered for the separation of the TX and the RX signals [7]. Unfortunately, power losses are unavoidable when a coupler is used. As shown in Figure 4.3, an overall power loss of 6dB is expected when transmitting and receiving signals.

To overcome the loss introduced by a coupler and to achieve a compact size, a circularly polarized RF sensor composed of circularly polarized antennas, a Lange coupler, a Wilkinson power divider, a VCO, an LNA, and a mixer is considered here. The Lange coupler simultaneously operates both as a polarizer and a duplexer [14]. The transmitter block is implemented at one input port of the Lange coupler, and the receiver is constructed at the other one. The two feeding ports in the antenna are isolated from each other by the Lange coupler. As a result, two differently polarized antennas can be established with the Lange coupler. Because the sensor system uses circular polarization, it is not subject to the interference problem encountered in the linearly polarized radar system. Moreover, even though it uses a Lange coupler to separate the TX and the RX paths, there is no power loss of 6dB. The input signal is radiated without any power loss, and also the received signal is combined at the receiver port through the Lange coupler.

A free-running oscillator provides less complex RF architecture and consumes less power than the phase-locked loop (PLL)-based synthesizer. It produces a highly stable oscillating signal, thus providing highly accurate transmission of heartbeat and respiration signals.

4.3.2 Radio Frequency Circuit Design

The RF front-end integrated circuit (IC) of the sensor was fabricated using the 6-in. InGaP/GaAs heterojunction bipolar transistor (HBT) technology. The large signal model of the transistor was obtained using the vertical bipolar

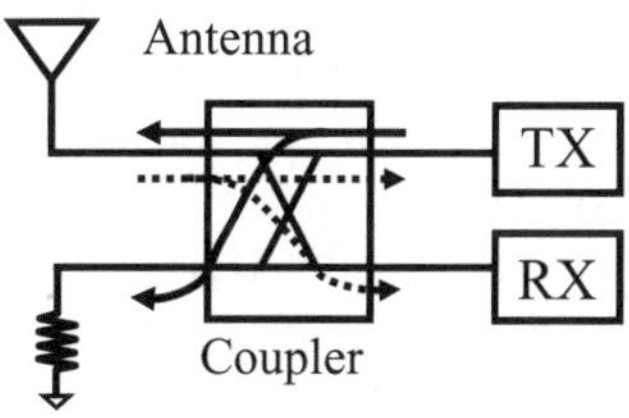

Figure 4.3. A single antenna transceiver employing a coupler. (Copyright © IEEE 2009)

intercompany (VBIC) model [15]. The HBT with one finger ($1 \times 10\,\mu m^2$) shows a cutoff frequency (f_T) of 55 GHz and a maximum oscillation frequency (f_{MAX}) of 100 GHz. The turn-on voltage between emitter and base in the HBT is 1.21 V. This technology provides a SiNx metal-insulator-metal (MIM) capacitor of 600 pF/mm^2, a NiCr resistor of 50Ω/□, and two metal layers with thicknesses of 1.3 and 4 μm. All circuits are passivated with polyimide material. The wafer with backside via holes is thinned to 95 μm.

4.3.2.1 Voltage-Controlled Oscillator. The VCO is designed utilizing negative resistance with capacitive feedback. For the tuning of the oscillator frequency, the base-collector junction of the HBT is used. The designed schematic of the VCO circuit is shown in Figure 4.4a. The spectrum of the single-ended VCO was obtained using an HP8564E spectrum analyzer (Agilent) and a phase noise measurement kit. Figure 4.4b,c show the output spectrum of the VCO and the measured performance of the phase noise, respectively. A free-running oscillation frequency of 27.345 GHz was achieved. It provides an output power of −9.67 dBm. A phase noise of −91.17 dBc/MHz at 1 MHz offset frequency was measured. The residual phase noise after down-conversion should be smaller than Equation 4.5, in which the phase noise of the VCO is assumed to be negligible. The tuning range of the oscillation frequency was from 25.5 to 27.81 GHz.

4.3.2.2 Low-Noise Amplifier. The LNA was implemented using the three stages of gain block shown in Figure 4.5. At the first stage, the transistor is degenerated with an inductive microstrip line in order to match the gain and the optimum noise point at the same time. The second and third stages do not include degeneration so that the overall gain will increase. The designed LNA satisfies the need for unconditional stability. The frequency range of the operation was measured from 19.0 to 24.5 GHz. The small signal gain of the LNA was over 17.0 dB from 20 to 25 GHz. The return losses at the input and output ports were below −10 dB at 24 GHz. The 1-dB compression point of the output power (P_{1dB}) was 10.3 dBm. The power-added efficiency (PAE) was 8.4%, and the power consumption of the LNA was 126 mW.

4.3.2.3 Mixer. The received signal is amplified in the LNA and down-converted to the intermediate frequency (IF), 100 KHz, in the mixer circuit. The mixer was designed with a single-ended input and balanced outputs. The schematic of the mixer circuit is shown in Figure 4.6a. The differential outputs of the VCO are used to drive the mixer circuit. In order to suppress high-frequency components such as local oscillator (LO) and RF leakage signals at the output port, a capacitor is connected in parallel with the load resistor. An emitter follower is used as a buffer amplifier to reduce the loading effect. The performance of the single-balanced mixer is characterized by two signal sources (Anritsu MG369XA and Agilent 83650A) and a spectrum analyzer (HP8564E). Figure 4.6b shows the measured conversion gain and the IF output power as the RF input power is swept from −28 to 2 dBm. A maximum conversion gain of 7 dB

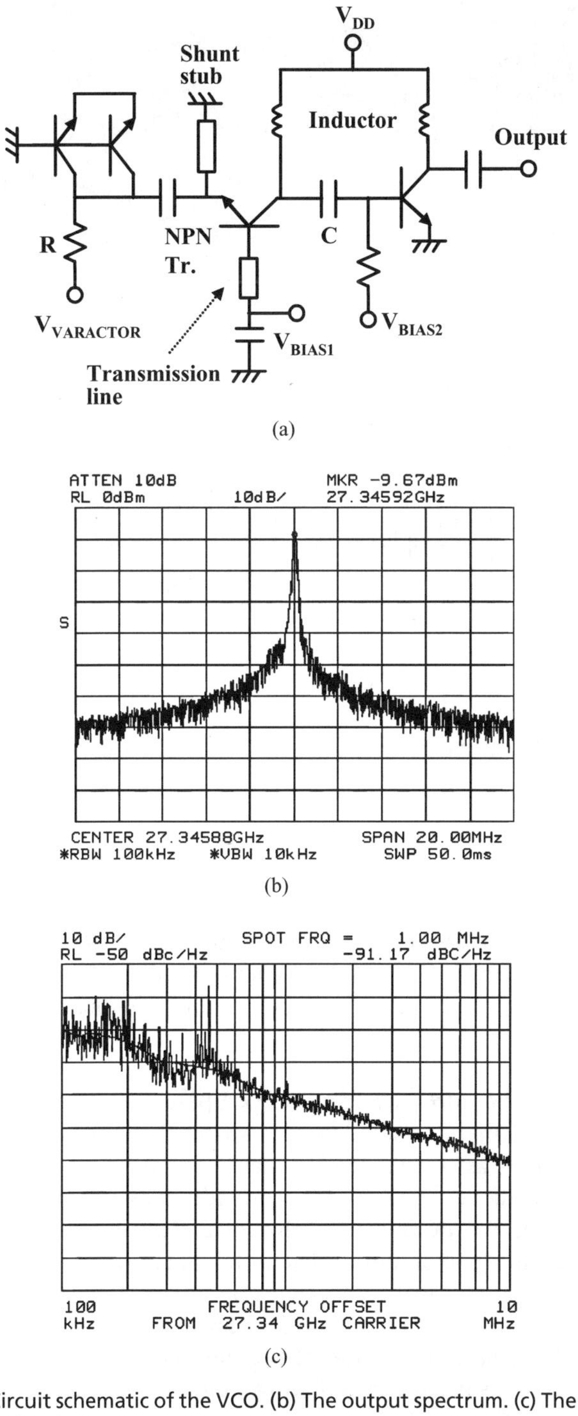

Figure 4.4. (a) Circuit schematic of the VCO. (b) The output spectrum. (c) The measured phase noise. (Copyright © 2009 IEEE)

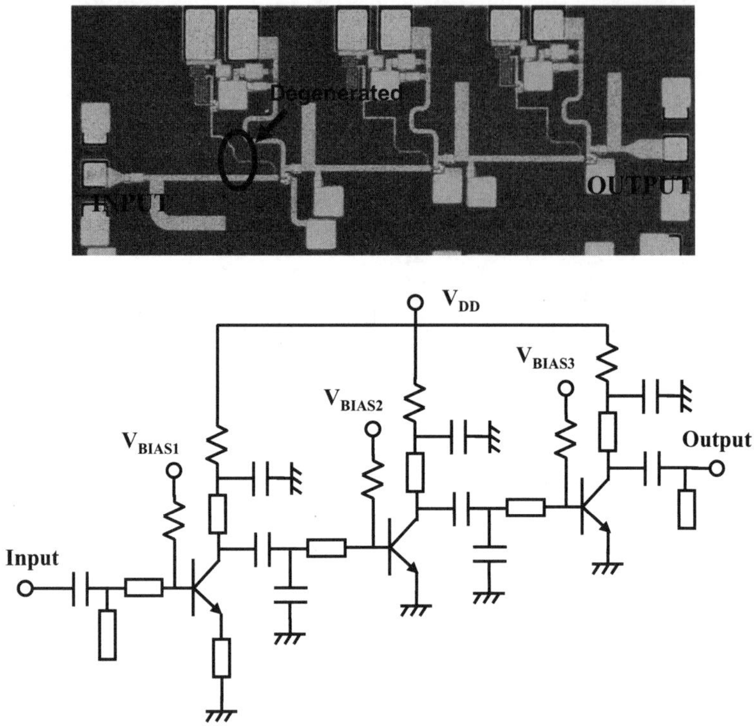

Figure 4.5. Circuit schematic of LNA and the fabricated photo.

was obtained. The isolation performances of LO to IF and RF to IF were 17.3 and 17.4 dB, respectively, and the DC power consumption was about 140 mW.

4.3.2.4 Lange Coupler. The Lange coupler was designed using the EM wave simulator, momentum software (Agilent). Figure 4.7 shows the result measurements. The measured coupling and the isolation coefficients were 4 and 19 dB at 24 GHz, respectively.

4.3.2.5 Antenna and Package. A 2×1 patch antenna array was designed using the 2.5 dimensional EM simulator by Ansoft Designer. A multilayer printed circuit board (PCB) was considered for the design. A simulated antenna gain of about 11 dBi was achieved. The 10-dB bandwidth was 1.2 GHz (23.4–24.6 GHz). For the fabrication, two PCBs with different dielectric constants were bonded together using an adhesive material. The patch antenna array was fabricated on the front side of the multilayer PCB. The substrate material of the front side has a low dielectric constant of 2.17 and a thickness of 787 μm (see Fig. 4.8a). The RF IC was attached on the bottom side of the multilayer PCB and interconnected with the electrical lines using wire bonding technology (see Fig. 4.8b). The material of the bottom side was RO3003, which has a high dielectric constant of 3 and

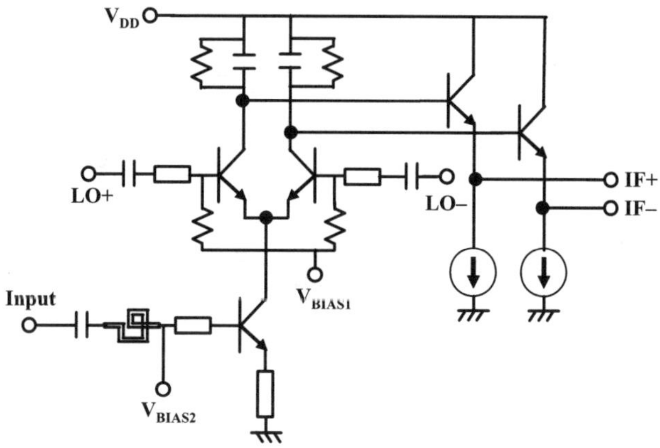

(a)

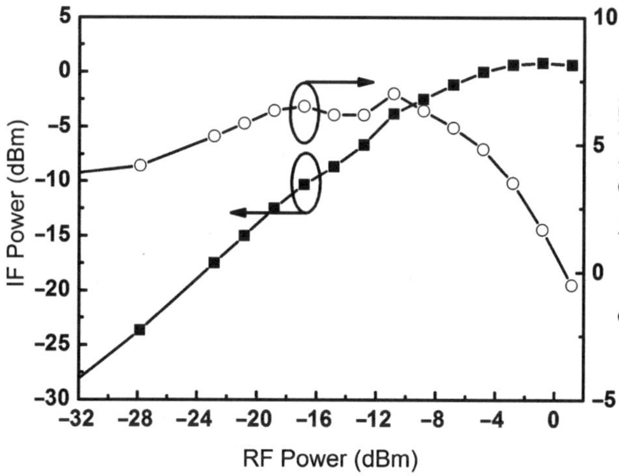

(b)

Figure 4.6. (a) The circuit schematic of the single-balanced mixer. (b) The IF power and the conversion gain versus RF power. (Copyright © 2009 IEEE)

a thickness of 254 μm. The patch antenna array and the RF IC are electrically connected using via holes. In Figure 4.8, the developed circularly polarized radar sensor is presented.

4.3.3 Signal Conditioning and Data Acquisition

The electrical signal output from the mixer circuit is processed by the signal processing block. The signal processing block is composed of a signal conditioning block (SCB), a data acquisition (DAQ) unit, and a digital signal processing

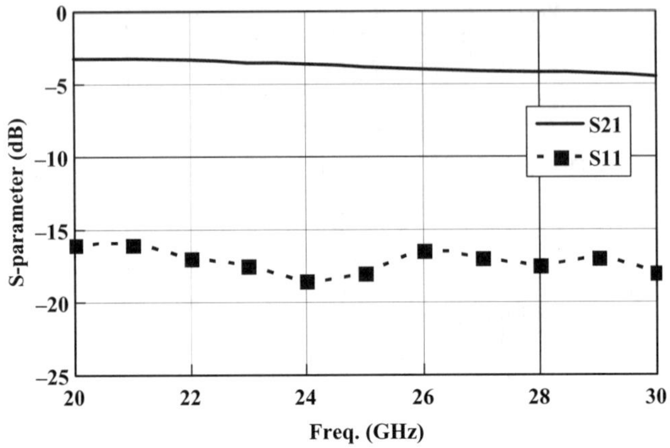

Figure 4.7. Measured S parameter of the Lange coupler. (Copyright © 2009 IEEE).

(DSP) program. Figure 4.9 shows the block diagram of the SCB. It consists of a high-pass filter (HPF), a low-pass filter (LPF), and a baseband amplifier.

The HPF was employed to remove the DC component of the IF signal and block the DC leakage from the mixer. The HPF has a cutoff frequency of 0.02 Hz. The LPF has to follow after the HPF due to the aliasing effect. The antialiasing LPF has a cutoff frequency of 50 Hz. The HPF and the LPF in the SCB were designed using 4th-order Sallen–Key filter topology [16] and are shown in Figure 4.10. To achieve the proper voltage range for the DAQ card (6024E by National Instrument), a postamplifier had to be used. The characteristics of the SCB were simulated and are shown in Figure 4.11. The overall gain can be adjusted to 46 dB, and the bandwidth of the transfer function is from 0.02 to 50 Hz. The output signal of the SCB is then converted to digital data with the DAQ card. The DAQ card has an input resolution of 12 bits and a sampling rate of 200 kS/s. The sampling rate used for measurements is 100 Hz. The collected digital data on the DAQ card is then processed by the DSP program.

4.3.4 Digital Signal Processing

4.3.4.1 Signal Processing Algorithms. Basically, the detection of heartbeat and the respiration rates from the radar signal is based on the principle of the Doppler shift. The precise detection of the wanted signal is subject to the analysis of the wanted signal and noise components. The distinctive nature of the heartbeat and the respiratory signals is represented by difference in the occupied frequency spectrum. The frequency of the heartbeat of human beings ranges from 0.9 to 3 Hz (54–180 beats/min), whereas that of respiration ranges from 0 to 0.5 Hz (0–30 beats/min). Using these characteristic frequencies, two signals can be discriminated using a band-pass filter.

Most transmitted signals are reflected back from the human skin. Furthermore, the strength of the heartbeat signal is about 10 times smaller than that of the

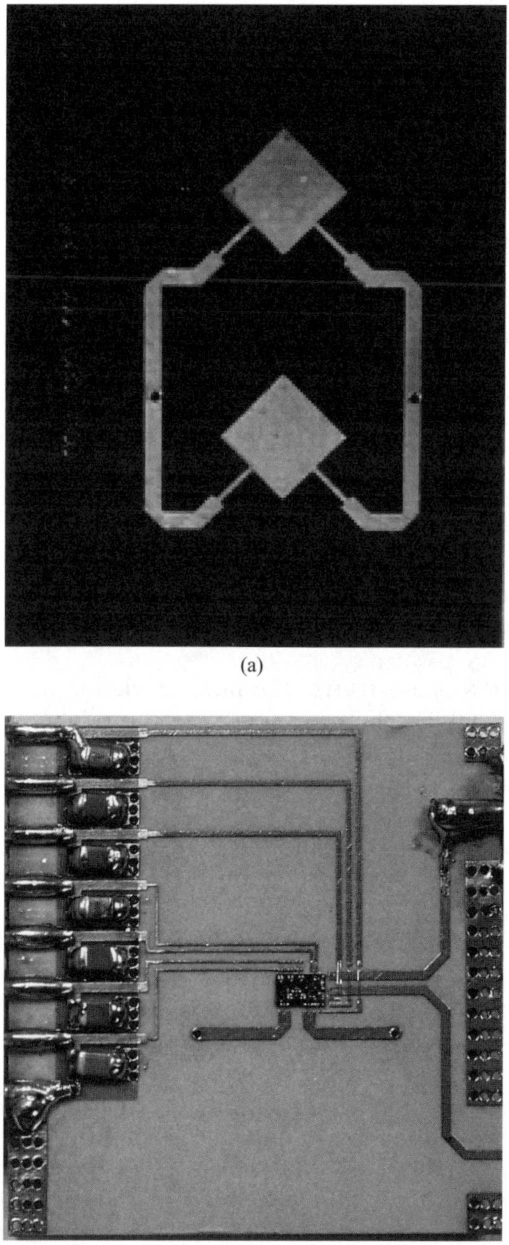

(a)

(b)

Figure 4.8. (a) A 2 × 1 patch array antenna was patterned (front side). (b) The developed RF chip is attached to the substrate and electrically connected using the wire bonding technology (bottom side). (Copyright © 2009 IEEE)

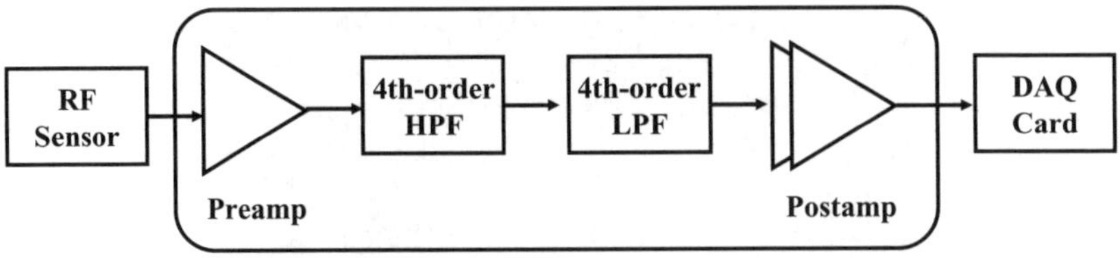

Figure 4.9. The block diagram of the SCB.

respiration signal. This is because the heart is located inside the human body and the movement of the heart relative to the movement of the lungs is quite small. Figure 4.12 depicts a received radar signal from a human subject in motion. As shown in the figure, one of the principle sources of noise is spontaneous movement of the body. Figure 4.13 shows the heartbeat and respiratory signals as recovered by the band-pass filter

The relative signal strength of the heartbeat is about 0.1–0.2 and the respiration signal is around 2; the noise signals in Figure 4.13 are 10 times larger than the wanted signal, that is, the heartbeat. Therefore, the primary purpose of the signal processing algorithm is to remove the noise signal from the wanted signal, resulting in precise detection of, the heartbeat and the respiration rates.

AUTOCORRELATION ALGORITHMS. The autocorrelation method estimates the periodicity of the signal. The flowchart for the autocorrelation algorithm is shown in Figure 4.14. First of all, the detected signal is band-passed in order to discriminate the heartbeat and the respiratory signal. As explained before, two signals occupy different frequency bands. For example, for the extraction of the respiration signal, a 4th-order Butterworth LPF is used, which has a cutoff frequency of 0.7 Hz; for the heartbeat, a 4th-order Butterworth band-pass filter is used, which has cutoff frequencies of 1 and 3 Hz.

In the next step, the band-passed signals go through a sliding window. This window represents the fixed time interval in which the numerical data is processed. In the algorithm, the time duration for the window is 10 seconds. Using the window, sufficient data can be provided for signal processing, and abrupt fluctuations of the wanted signal can be traced enough to result in precise extraction. After numerical computation in the window, it moves to the following time and captures data again for 10 seconds of the timing window. After the window sliding, the *Hanning* window is introduced in order to prevent spectral loss during the sliding window. This technique improves signal quality; however, it leads to a distortion of the autocorrelation function that multiplies the inverse function of the *Hanning* window.

After the preprocessing of raw signals, autocorrelation is carried out. One of the eminent characteristics of the autocorrelation function is to reveal periodicity (frequency) once the input signal shows periodic behavior. The autocorrelation function obtains the maximum value in every periodic interval. The maximum values are found by searching for the maximum value of the second derivative

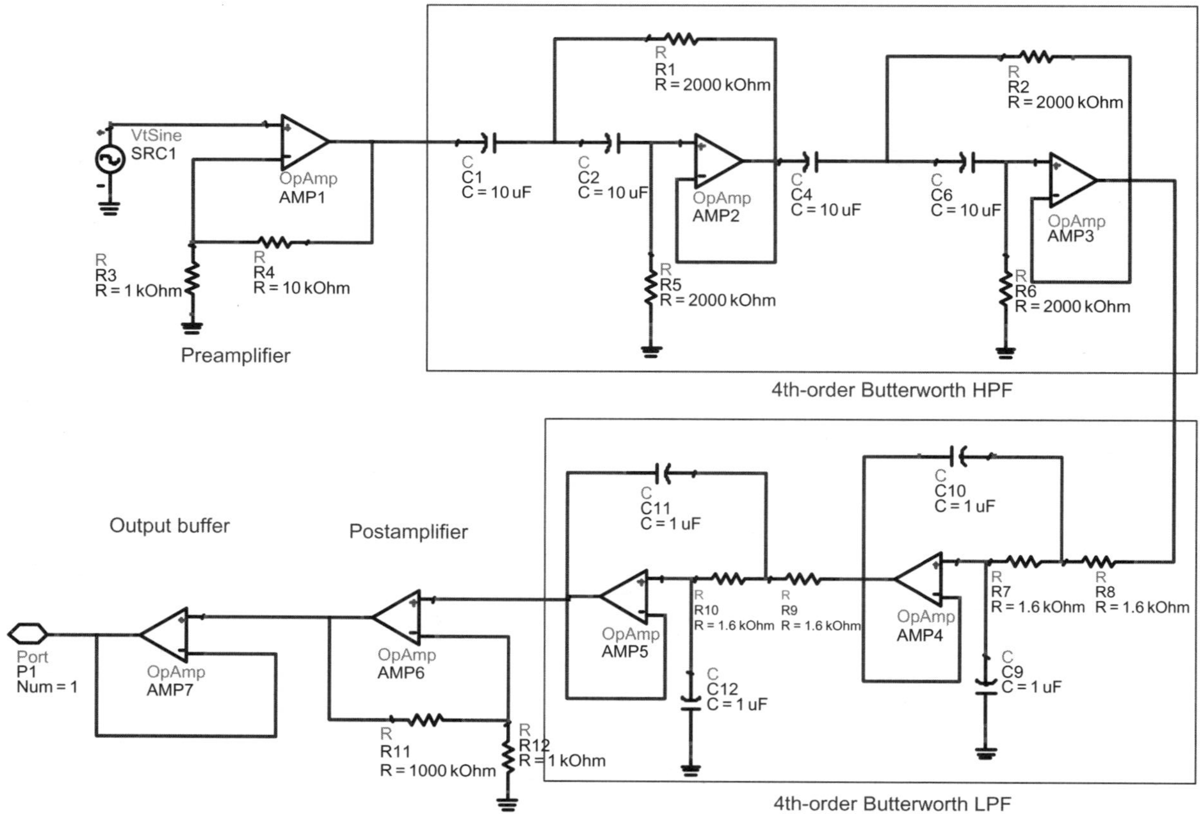

Figure 4.10. Circuit schematic of the signal conditioning block.

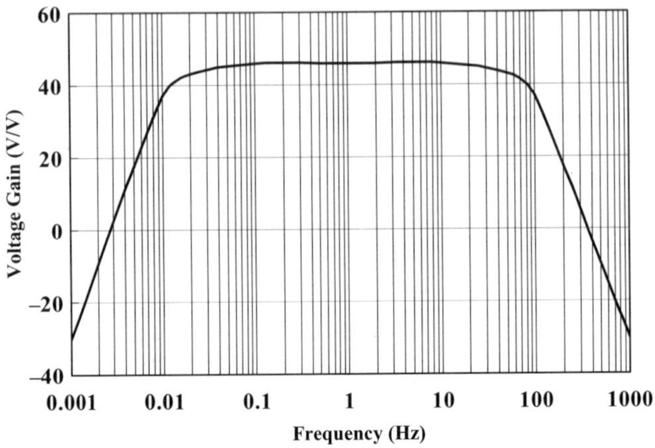

Figure 4.11. Frequency response of the SCB. (Copyright © 2009 IEEE)

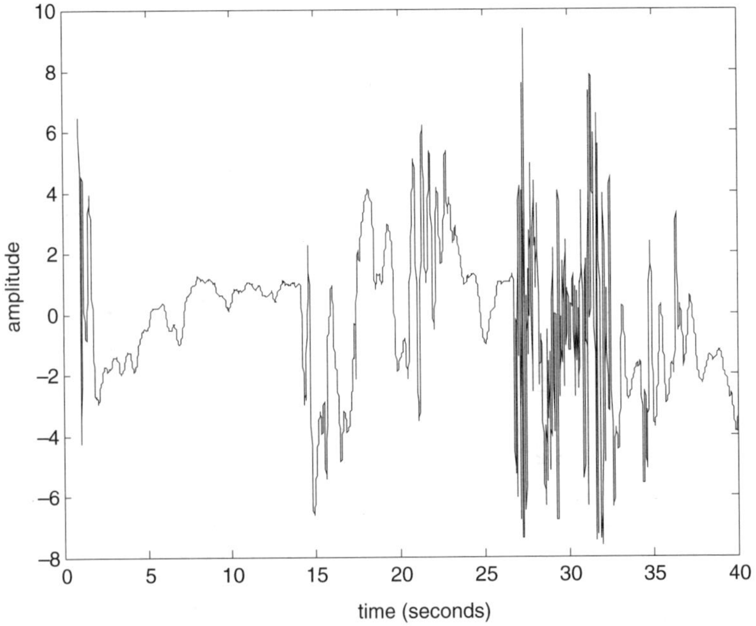

Figure 4.12. Received radar signal in the time domain.

of three consecutive samples. The time interval between maximum values is then calculated, resulting in the periodicity. In order to improve the result, the center clipping technique is employed before the calculation of the period (see Fig. 4.14). This is often used to remove the unwanted maximum values of the signals. The center clipper is defined as

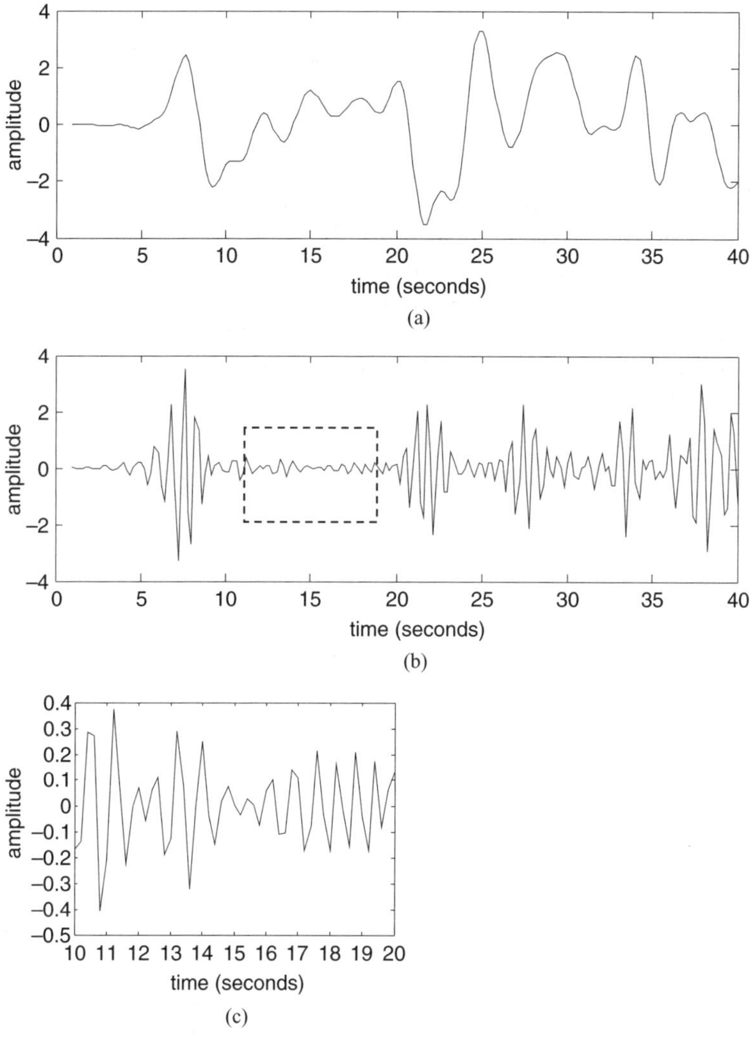

Figure 4.13. Band-passed waveform of the received radar signal. (a) Respiration signal. (b) Band-passed raw signal of heartbeat including noise. (c) Heartbeat signal (enlarged view of the dotted box in panel b).

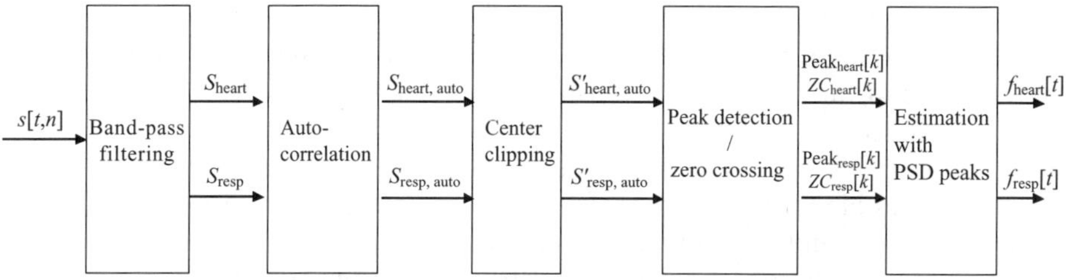

Figure 4.14. Flowchart for the autocorrelation algorithm.

$$c(n) = \begin{cases} 0 & \text{if} \quad |s(n)| \le k \cdot a_{\max} \\ s(n) & \text{if} \quad |s(n)| > k \cdot a_{\max} \end{cases}, \tag{4.6}$$

where $c(n)$ is the output signal, $s(n)$ is the source signal, and $a_{\max}$ is the maximum value within the window. k is determined to define the boundary of the window.

In the following step, the periodicity of the signal is determined. Two methods are simultaneously considered: peak detection and zero crossing algorithms. The zero-crossing rate (ZCR) is determined from the frequency of the crossings of 0. Each peak has two zero crossings, so only one crossing is accounted for when a signal changes from minus to positive value. The time interval between peaks (zero crossing points) can be calculated using the following equations:

$$\begin{aligned} \Delta T_{\text{heart}} &= \{\text{Peak}_{\text{heart}}(k + D/2) - \text{Peak}_{\text{heart}}(k - D/2)\} / D \\ \Delta T_{\text{resp}} &= \{\text{Peak}_{\text{resp}}(k + D/2) - \text{Peak}_{\text{resp}}(k - D/2)\} / D \end{aligned} \tag{4.7}$$

$$\begin{aligned} \Delta ZC_{\text{heart}} &= \{ZC_{\text{heart}}(k + D/2) - ZC_{\text{heart}}(k - D/2)\} / D \\ \Delta ZC_{\text{resp}} &= \{ZC_{\text{resp}}(k + D/2) - ZC_{\text{resp}}(k - D/2)\} / D \end{aligned}, \tag{4.8}$$

where D is the number of peak/zero crossings. Using the above equations, we can determine the heartbeat and the respiration rates:

$$\begin{aligned} f_{\text{heart,peak}} &= \frac{1}{\Delta T_{\text{heart}}}, \quad f_{\text{resp,peak}} = \frac{1}{\Delta T_{\text{resp}}} \\ f_{\text{heart},ZC} &= \frac{1}{\Delta ZC_{\text{heart}}}, \quad f_{\text{resp},ZC} = \frac{1}{\Delta ZC_{\text{resp}}} \end{aligned}. \tag{4.9}$$

Power Spectrum Density Algorithms. Since the heartbeat and respiratory signals from the human body have a periodic quality, the algorithm of the DSP program is mainly dedicated to detect the periodic signal component from the conditioned signal. To fulfill this objective, the power spectral density (PSD) of the signal is also examined. The procedural block diagram of the PSD is illustrated in Figure 4.15.

First, the output signal from the SCB at time t, $s[t,n]$, is convolved with the band-pass filters of the heartbeat rate ($h_{\text{heart}}[n]$) and the respiration rate ($h_{\text{resp}}[n]$) in the time domain, respectively. Here, $s[t,n]$ means the signal of the t-th block with the size of the fast Fourier transformation (FFT), N ($n = 0, 1, 2, \ldots, N-1$). Since the normal respiration rate of humans is within less than 0.9 Hz (54 beats/min) and the heartbeat rate is between 0.9 (54 beats/min) and 3 Hz (180 beats/min), it is possible to discriminate the heartbeat and the respiration signals with the appropriate band-pass filtering, as discussed in Section 4.3.4.1. Then, the FFT and the PSD calculations are carried out for each output signal:

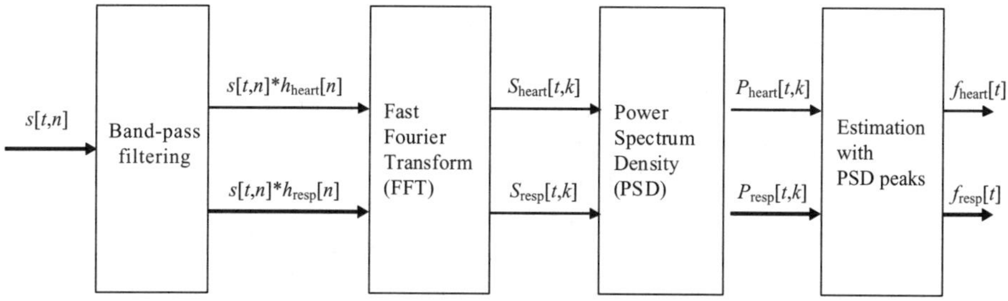

Figure 4.15. Flowchart for the power spectrum density algorithm.

$$P_{\text{heart}}[t,k] = \text{FFT}\left\{\left(s[t,n] * h_{\text{heart}}[n]\right)^2\right\} \tag{4.10}$$

$$P_{\text{resp}}[t,k] = \text{FFT}\left\{\left(s[t,n] * h_{\text{resp}}[n]\right)^2\right\}, \tag{4.11}$$

where $P_{\text{heart}}[t,k]$ and $P_{\text{resp}}[t,k]$ stand for the PSD at time t with the frequency index k of the band-passed heartbeat $(s[t,n] * h_{\text{heart}}[n])$ and the respiratory signals $(s[t,n] * h_{\text{resp}}[n])$, respectively. The index k is a positive integer and smaller than N. We note that "*" is a convolution given by

$$x[n] * y[n] = \sum_{k=-\infty}^{\infty} x[k]y[n-k]. \tag{4.12}$$

After calculating the PSDs for the heartbeat and respiration rates, respectively, the frequency at which the power of the PSD is highest for each is chosen as the heartbeat or the respiration rate.

4.3.4.2 Proposed Signal Processing Algorithm and Comparisons.

Power Spectral Density with Tracking Algorithm. Basically, the proposed algorithm improves the PSD algorithm. In order to enhance the quality of the detected signal even when a person is in motion, a tracking algorithm can be applied. In the tracking algorithm, first the PSD results of the heartbeat and the respiration rates are averaged over the data size, L. $\overline{P_{\text{heart}}[t,k]}$ and $\overline{P_{\text{resp}}[t,k]}$ represent the averaged PSDs of the heartbeat and respiration signals, respectively. When the algorithm is introduced for the first time, the initial tracking values are required. These values are obtained by averaging the PSD over the time interval $[t_1, t_2]$. Within the predefined tracking range of M, the algorithm is repeated unless the heartbeat and the respiration rates are out of the range. The tracking of the peak value of the averaged PSD results in the heartbeat rate (f_{heart}) and the respiration rate (f_{resp}) as below:

$$f_{\text{heart}}[t] = \arg\max_k \overline{P_{\text{heart}}[t,k]} \quad f_{\text{heart}}[t-1] - M/2 \le k \le f_{\text{heart}}[t-1] + M/2 \tag{4.13}$$

$$f_{\text{resp}}[t] = \arg\max_{k} \overline{P_{\text{resp}}[t,k]} \quad f_{\text{resp}}[t-1] - M/2 \le k \le f_{\text{resp}}[t-1] + M/2. \tag{4.14}$$

The heartbeat and the respiration rates correspond to the frequency indexes of k, at which the averaged PSDs of the heartbeat and the respiration rates are highest. If the tracking has failed, the heartbeat algorithm in Figure 4.16 is initialized only if the present PSD result is smaller than 0.8 times the peak value among the past PSD values. Since the heartbeat does not change abruptly in a short time, it is assumed that the present PSD result is larger than 80% of the maximum of the past PSD results. However, this assumption cannot be applied to the respiration rate. Since humans can halt their breathing willingly, the respiration rate can be zero. So, if the output power of the PSD for respiration is below the criteria value, the respiration rate is considered to be zero. By repeating the above processes, the respiration rate and the heartbeat rate can be calculated. In Figure 4.16, the procedural block diagram is depicted.

Comparisons of the Signal Algorithms. In order to extract human vital signs, various algorithms have been developed [17–22] and typical examples of algorithms have been discussed. In the case of human heartbeat and respiration rates, it is assumed that the heartbeat does not change quickly. In fact, both rates show periodicity. Even when heartbeat and respiration signals are corrupted by noise, the periodic spectral components can be accurately extracted. Numerical simulations were carried out for the peak detection of the power spectral density (PDP) with the tracking algorithm, the PDP alone, the ZCR, and the autocorrelation method. Comparison of the simulation results shows that using the PDP with the tracking algorithm is the superior method for capturing heartbeat and respiration rates.

Because of the periodic characteristic of the wanted signals, the DSP program is primarily focused to detect the periodic spectral component in the frequency domain. For the simulation, the additive white Gaussian noise (AWGN) environment is assumed. With the autocorrelation method, the calculation result can be different from the result without the noise. In particular, when the signal-to-noise ratio (SNR) becomes low, the autocorrelation method obviously fails to detect vital signs. In Figure 4.17, the autocorrelation, the ZCR, and the PDP methods are compared varying the SNR of the signal. As the SNR decreases, the PDP method shows superior performance to the autocorrelation and the ZCR methods. Over the whole range of the SNR, the minimum mean square error (MMSE) of the PDP with the tracking algorithm is the smallest among the three and is almost constant. On the basis of this comparison, it can be said that the PDP with the tracking algorithm is optimally tolerant to the noise environment.

It should be noted that even when a person is in motion around the steady position, the DSP algorithm should be able to render the heartbeat and the respiration rates without significant errors. The motions of the person are considered the origin of the noise. When the heart beats periodically, the relative displacement of the heart is only a few millimeters. Even relatively small movements of the body can deteriorate the frequency spectrum of the heartbeat signal.

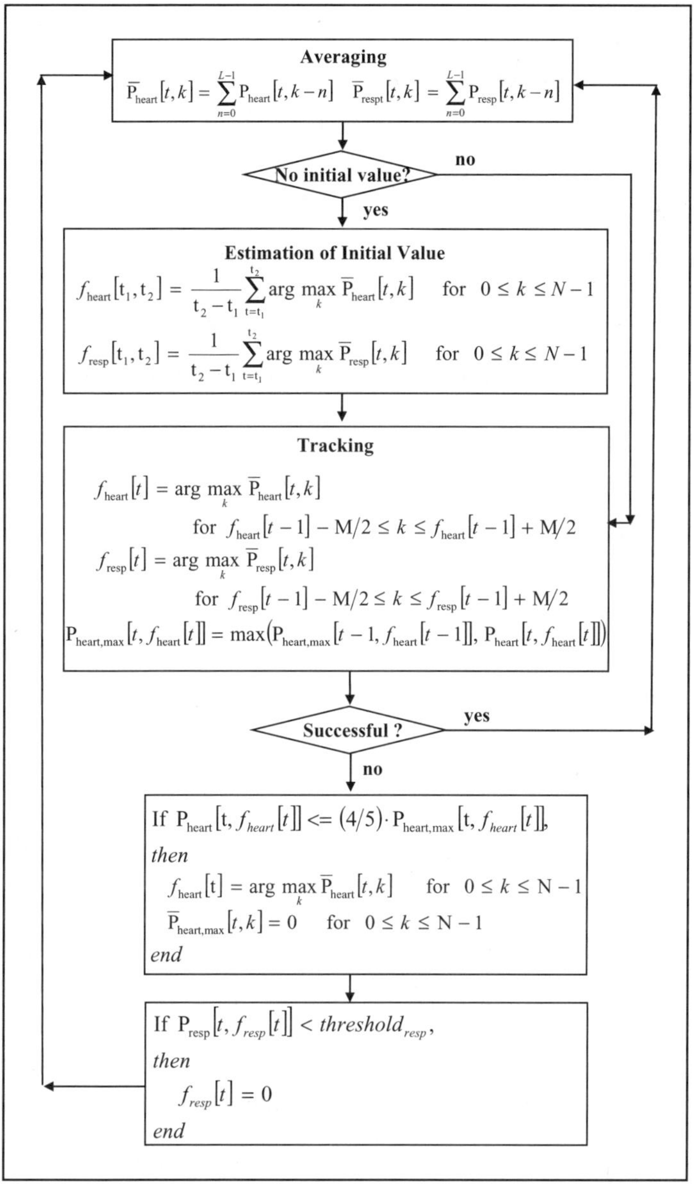

Figure 4.16. The algorithm of the peak detection of the power spectral density with the tracking method. (Copyright © 2009 IEEE)

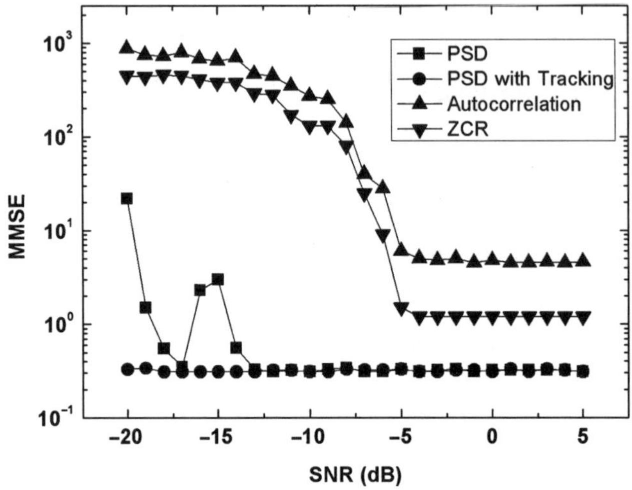

Figure 4.17. Comparison of the DSP algorithms. (Copyright © 2009 IEEE)

Nevertheless, the sensor should have the capability to predict human vital signs accurately with the DSP algorithm.

4.4 MEASUREMENTS

In Figure 4.18, the photo of the developed sensor module is shown. The antenna of the RF front-end is located at the top layer, and the SCB circuit is at the second layer. The DC-DC converter circuit is also integrated to provide stable DC biases to the module. The size of the module is 2.5 cm × 2.5 cm, with a height of 2.3 cm.

The sensor is located at a distance of 0.5 m from the human chest for the experiment. The transmitted power is about 60 µW. Using the LabView program (National Instrument), the DSP algorithm is coded and executed. The signals of the heartbeat and the respiration rates are displayed on a computer screen in the time domain and the frequency domain, respectively, as shown in Figure 4.19. In the first row of each figure, the raw data without any signal processing is depicted. It contains the information of the respiration rate and the heartbeat rate, including noise; the corresponding FFT results are depicted in the second column. The band-passed signals, that is, the respiration rate and the heartbeat rate, are shown in the 2nd and 3rd rows, respectively.

Whether a person is breathing or holding his breath, clear vital signals can be obtained. When the subject for this experiment was breathing, the heartbeat and the respiration rates were 69 and 21 beats/min, respectively, as illustrated in

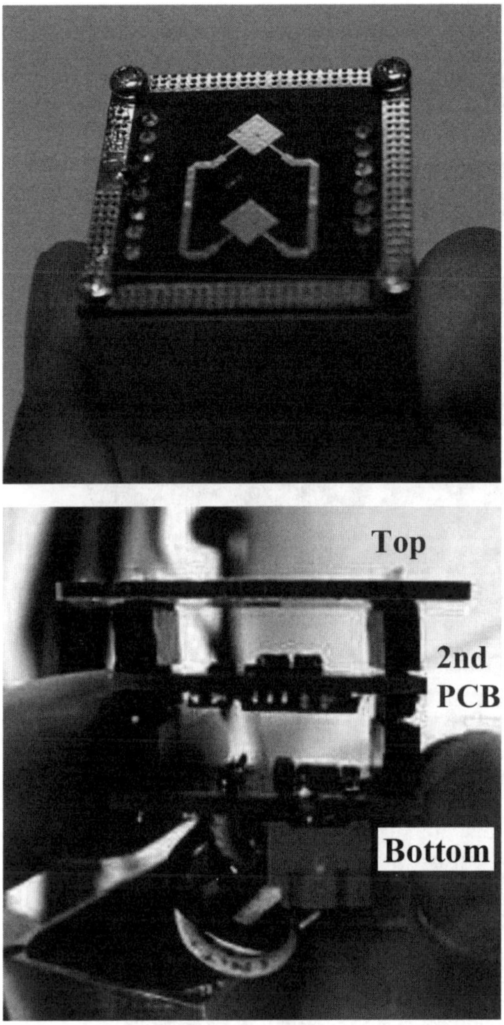

Figure 4.18. The patch antenna of the sensor. The RF front-end (top), the SCB (2nd layer), and the DC-DC converter circuit (bottom) are integrated.

Figure 4.19a. However, when the breath was held, the respiration rate abruptly dropped to 6 beats/min, with the heartbeat rate at 63 beats/min (Fig. 4.19b). Although the subject did move a little, the heartbeat rate did not change much, as would have been expected even had the subject held his breath. The developed sensor was compared with the commercial fingertip sensor, which only measures heartbeat. The measured heartbeat detected by the two sensors was the same, 84 beats/min. This result of this comparison validates the performance of the developed sensor system [23] as shown in Figure 4.20. The developed sensor system shows an accuracy of 100%.

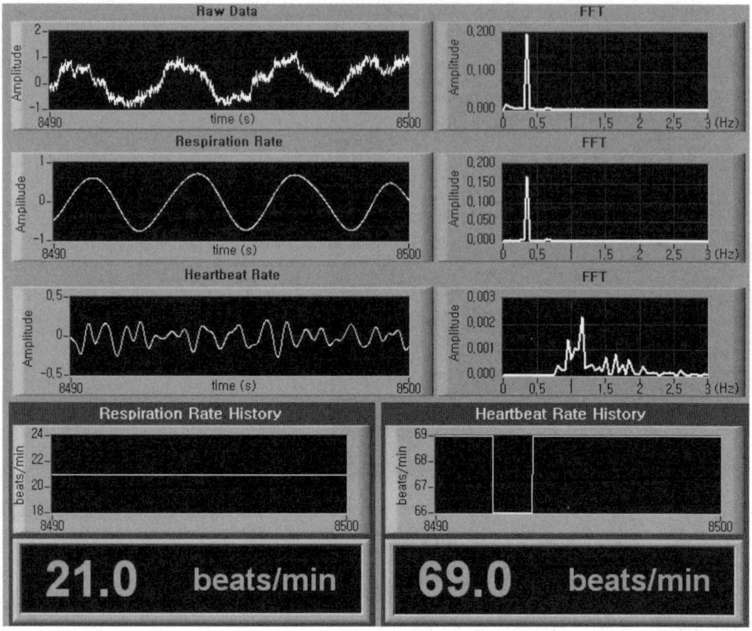

(a)

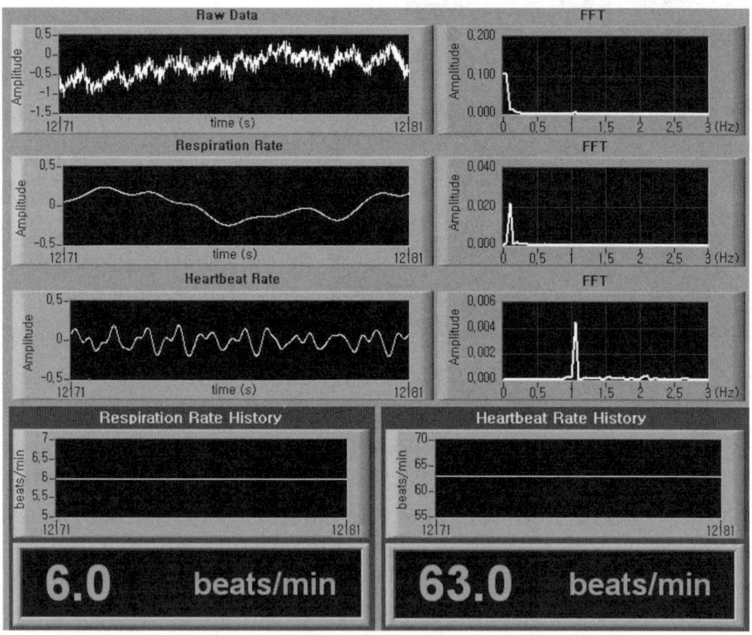

(b)

Figure 4.19. The measurement results of the respiration rate (*r*) and the heartbeat rate (*h*): (a) when breathing ($r = 21.0$; $h = 69.0$ beats/min); and (b) when holding the breath ($r = 6.0$; $h = 63.0$ beats/min) at the distance of 0.5 m. (Copyright © 2009 IEEE)

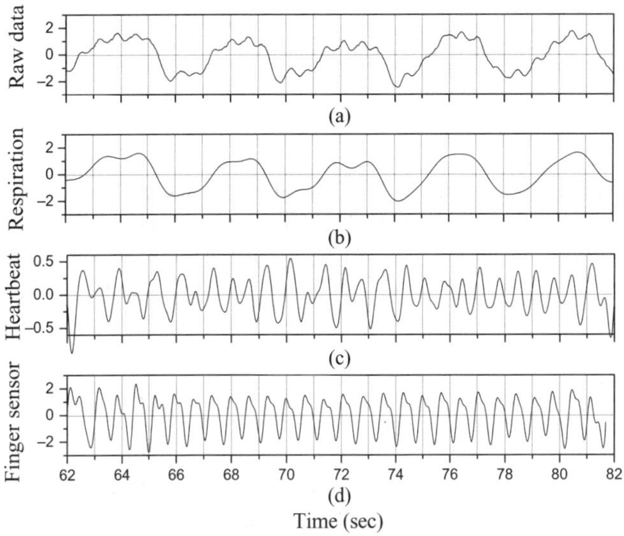

Figure 4.20. The comparison of the developed sensor with the commercial fingertip sensor. (a) Measured raw data using the developed sensor; (b) extracted respiration rate signal from the raw data; (c) extracted heartbeat signal; and (d) measured signal of the fingertip sensor. (Copyright © 2009 IEEE)

4.5 CONCLUSION

A mobile health-care device using RF technology for the health-care industry has been discussed. A sensor for the detection of human vital signs, such as the heartbeat and the respiration rates, has been developed employing the circularly polarized EM wave. The guidelines for sensor development have been given: size, power, sensitivity, invulnerability to interference, and reliable operation under conditions of change in the outer environment. A single antenna has been developed to keep the size of the device compact. Utilizing RF architecture, which uses a single free-running VCO, power consumption can be reduced considerably. The circular polarization results in invulnerability to interference, as well as reliable operation.

For the extraction of human vital signs, a baseband signal processing unit, which is composed of an SCB, a DAQ unit, and a DSP program, has been developed. In the SCB, the baseband signal is preprocessed using the HPF, the LPF, and the postamplifier. In order to extract the heartbeat and the respiration rates, a digital signal algorithm has been developed. The algorithm used in the sensor depends on the signal characteristic we want to obtain. To sense heartbeat and respiration rates, we have utilized the the periodicity of the signals. The PSD with tracking algorithm method has been developed. Even when the human body moves irregularly, the algorithm can extract the wanted signal accurately. The experimental result of the sensor was compared with that of a commercial sensor.

The comparison shows an excellent agreement of results, which validates the performance of the developed sensor system.

ACKNOWLEDGMENTS

The authors would like to thank Prof. Songcheol Hong and Prof. Hyung-Myung Kim at the Korea Advanced Institute of Technology (KAIST) for their fruitful support for this work. They are also grateful to Dr. Jeong-geun Kim at the University of California at San Diego for assistance with measurement and circuit design.

REFERENCES

[1] "Introduction to the health care industry." Available: http://www.plunkettresearch.com/Industries/HealthCare/HalthCareTrends/tabid/294/Default.aspx.

[2] M. Steffen, A. Aleksandrowicz, and S. Leonhardt, "Mobile noncontact monitoring of heart and lung activity," *IEEE Biomed. Circuits Syst.*, 1(4), pp. 250–257, 2007.

[3] Y. Lin, I. Jan, K. P. C.-I. Ko, Y. Chen, J. Wong, and G. Jan, "A wireless PDA-based physiological monitoring system for patient transport," *IEEE Trans. Inf. Technol. Biomed.*, 8(4), pp. 439–447, 2004.

[4] H.-C. Yu and S.-M. Tseng, "A wireless based sensor for patient monitoring system with remote diagnostic," *Proceedings of the International Conference on Networking and Services 2007*, p. 92, Athens, June 2007.

[5] V. Lubecke, O. Boric-Lubecke, and E. Beck, "A compact low-cost add-on module for Doppler radar sensing of vital signs using a wireless communications terminal," pp. 1767–1770, IEEE MTT-S Int. Microwave Symp., Dig., Seattle, WA, June 2002.

[6] J. H. Oum, S. E. Lee, D.-W. Kim, and S. Hong, "Non-contact heartbeat and respiration detector using capacitive sensor with Colpitts oscillator," *Electron. Lett.*, 44(2), pp. 87–88, 2008.

[7] J.-G. Kim, S.-H. Sim, S. Cheon, and S. Hong, "24 GHz circularly polarized Doppler radar with a single antenna," *European Microwave Conference 2005*, pp. 4–7, Paris, France, October 2005.

[8] Q. Zhou, J. Liu, A. Høst-Madsen, O. Boric-Lubecke, and V. Lubecke, "Detection of multiple heartbeats using Doppler radar," *IEEE ICASSP, '06*, pp. 1160–1163, Toulouse, France, May 2006.

[9] Y. Xiao, C. Li, and J. Lin, "A portable noncontact heartbeat and respiration monitoring system using 5-GHz radar," *IEEE Sens. J.*, 7(7), pp. 1042–1043, 2007.

[10] C. Li, Y. Xiao, and J. Lin, "A 5 GHz double-sideband radar Sensor chip in 0.18 μm CMOS for non-contact vital sign detection," *IEEE Microw. Wireless Compon. Lett.*, 18(7), pp. 494–496, 2008.

[11] K. M. Chen, D. Mirsa, H. Wang, H. R. Chuang, and E. Postow, "An X-band microwave life detection system," *IEEE Trans. Biomed. Eng.*, 33, pp. 697–700, 1986.

[12] C. Li, Y. Xiao, and J. Lin, "Experiment and spectral analysis of a low-power Ka-band heartbeat detector measuring from four sides of a human body," *IEEE Trans. Microwave Theory Tech.*, 54(12), pp. 4464–4471, 2006.

[13] "ISM band." Available: http://en.wikipedia.org/wiki/ISM_band.

[14] C. A. Balanis, *Antenna Theory: Analysis and Design*, 2nd ed. New York: John Wiley & Sons, 1996.

[15] C. McAndrew, J. Seitchik, D. Bowers, M. Dunn, M. Foisy, I. Getreu, M. McSwain, S. Moinian, J. Parker, D. Roulston, M. Schroter, P. van Wijnen, and L. Wagner, "VBIC95: The vertical bipolar intercompany model," *IEEE J. Solid-State Circuits*, 31, pp. 1476–1483, 1996.

[16] W.-K. Chen, *The Circuits and Filters Handbook*, 1st ed. Boca Raton, FL: CRC Press, 1995.

[17] B. Widrow, J. R. Glover, J. M. McCool, J. Kaunitz, C. S. Williams, R. H. Hearn, J. R. Zeidler, E. Dong, and R. C. Goodlin, "Adaptive noise cancelling: Principles and applications," *IEEE Proc.*, 63(12), pp. 1692–1716, 1975.

[18] N. V. Thakor and Y. Zhu, "Applications of adaptive filtering to ECG analysis: Noise cancellation and arrhythmia detection," *IEEE Trans. Biomed. Eng.*, 38(8), pp. 785–794, 1991.

[19] M. Richter, T. Schreiber, and D. T. Kaplan, "Fetal ECG extraction with nonlinear state-space projections," *IEEE Trans. Biomed. Eng.*, 45, pp. 133–137, 1998.

[20] L. De Lathauwer, B. De Moor, and J. Vandewalle, "Fetal electrocardiogram extraction by blind source subspace separation," *IEEE Trans. Biomed. Eng.*, 47, pp. 567–572, 2000.

[21] P. P. Kanjilal, S. Palit, and G. Saha, "Fetal ECG extraction from single-channel maternal ECG using singular value decomposition," *IEEE Trans. Biomed. Eng.*, 44, pp. 51–59, 1997.

[22] J. S. Paul, M. R. Reddy, and V. J. Kumar, "A transform domain SVD filter for suppression of muscle noise artifacts in exercise ECGs," *IEEE Trans. Biomed. Eng.*, 47, pp. 645–663, 2000.

[23] J. H. Choi and D. K. Kim, "A remote compact sensor for the real-time monitoring of human heartbeat and respiration rate," *IEEE Trans. Biomed. Circuits Syst.*, 3(3), pp. 181–188, 2009.

5

DESIGN CONSIDERATIONS OF LOW-POWER DIGITAL INTEGRATED SYSTEMS FOR IMPLANTABLE MEDICAL APPLICATIONS

Zhihua Wang, Xiang Xie, Xinkai Chen, and Xiaowen Li

5.1 INTRODUCTION

Implantable medical electronic devices (IMEDs) are implanted into human beings for the purposes of long-term measurement and surveillance of physiological and biochemical parameters within the human body, and are also used for therapy and replacement of disabled organs. Their unique interface with the human body makes it possible to learn more about the natural behavior or function of certain body parts and organs, and to design more effective prosthetic devices. Compared with other medical devices, IMEDs have the following distinct advantages: more accurate and intact measurement [1, 2], long-term monitoring [3–5], direct control of living organs or tissues [6–8], distinct therapeutic capability [9–11], and replacement of lost biological functionality [12–16]. The demand for implantable medical devices is growing every year because of these unique merits. For example, in the United States, the demand will rise 7.7% to $40.2 billion in 2011 [17]. IMEDs are playing an increasingly important role in medical applications. The field of implantable medical electronics has become one of the most important research fields in biomedical engineering.

CMOS Biomicrosystems: Where Electronics Meet Biology, First Edition. Edited by Krzysztof Iniewski.

The unique environment within the human body leads to many challenges and constraints in designing IMEDs, which mainly include limited size and low power. Reducing power consumption of IMEDs has two beneficial effects: prolonging battery life, thus reducing the number of needed recharges in an external battery-operated system, and reducing heat dissipation in the tissue around the implant. Reducing the size can make IMEDs easier to implant, reduce side effects to living tissue, and enable more accurate control and measurement. These strict requirements applying to IMEDs are met only by the use of very large-scale integration (VLSI) design. In VLSI design, a digital integrated circuit (IC) design plays a key role because of its low power and small size. Considering that the complementary metal-oxide-semiconductor (CMOS) process technology is ideally suited for ultralow-power and system on chip (SOC) designs in implant applications, low-power, digital CMOS IC design for IMEDs will be discussed in detail in this chapter. The chapter is organized as follows. The general model of IMEDs is analyzed in Section 5.2. Design considerations for implantable medical applications using low-power digital IC are discussed in Section 5.3. A design case, a low-power digital IC design for wireless capsule endoscopy, is given in Section 5.4. The chapter closes in Section 5.5 with some conclusions.

5.2 GENERAL MODEL OF IMPLANTABLE MEDICAL ELECTRONIC DEVICES

In the modern implantable medical electronics field, typical implantable medical applications can be summarized as the following: (1) heart pacemakers [15, 18, 19]; (2) implantable defibrillators [20–22]; (3) cochlear prostheses [23, 24]; (4) visual prostheses [25, 26]; (5) implantable pain controllers [27, 28]; (6) urinary incontinence prostheses [29, 30]; (7) neural recording microsystems [31–33]; (8) implantable measurement microsystems of physiological parameters [3, 4, 34], such as pressure, temperature, and bioimpedance; (9) implantable measurement microsystems of biochemical parameters [1, 35, 36], such as pH, glucose, and manner of ion concentrations; (10) drug delivery microsystems [11, 37]; and (11) wireless capsule endoscopy [2]. From the elaborate analyses of these typical implantable applications, a general model of IMEDs can be extracted, as illustrated in Figure 5.1.

The implantable medical electronic system in this model is partitioned into two parts: the implant unit (the IMED) and the external unit. The implant unit is the key part of the whole system [38].

The external unit mainly includes an antenna, a power transmitter, a wireless transceiver, and a digital signal processor (DSP) and controller. The basic functions of the external unit are to transmit the necessary data to the implant unit, to receive the bioinformation captured by the implant unit via the outside wireless transceiver and antenna, and to refine and upgrade the DSP algorithm, functions that cannot be integrated into the implant unit owing to stringent power and heat dissipation limitations. If batteries are used in the implant unit, because of its limited size and lifespan, a surgical operation into a living body is required

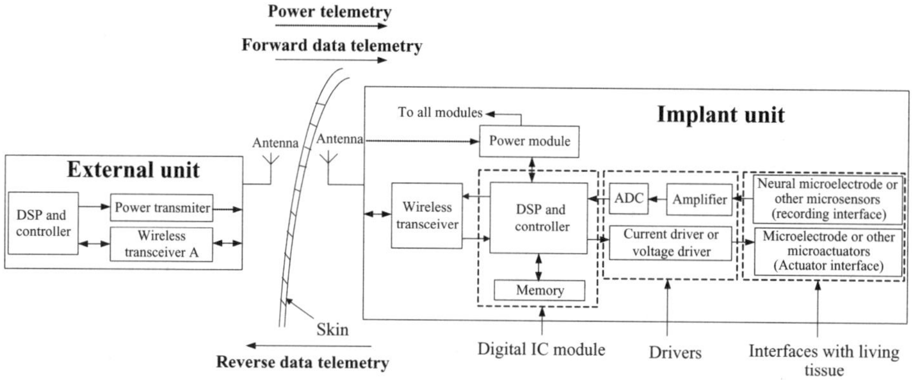

Figure 5.1. The model of implantable medical electronic system.

to exchange the batteries in the implant unit. To avoid these invasive operations, a power transmitter is included in this model for wireless energy transmission. Currently, two major wireless power transmission methods have been developed for implant units. One makes use of a near infrared laser [39] to transmit power, and the other method utilizes electromagnetic induction between two inductors [40, 41]. The benefits of external units include a decreased risk of adverse reactions to implanted materials, a reduction in internal heat dissipation, the ability to place the power source externally, ease of access for refinement and upgrade of signal processing algorithms, and the functionality of the external components.

The implant unit consists of an antenna, a wireless transceiver, a digital IC module, a power module, and interfaces with living tissue, as well as its drivers. Each module is discussed next.

5.2.1 Interfaces with Living Tissue

The interface with living tissue is the bridge between the machine and the human body. It can be divided into two types. One type is called the recording interface, which acquires the bioinformation directly from the living tissues, such as neural electric signals, pH, ion concentrations, pressure, and temperature. The recording interface of an implant unit includes a neural microelectrode, a physiological microsensor, and a biochemical microsensor. This interface can also convert the nonelectric biosignals into electric signals. The other type is an actuator interface. It acts directly on the living tissue under the control of the digital IC module. The most common actuator interface in a modern implantable device is the microelectrode, which is applied to stimulate nerves for all kinds of functional prostheses, such as cochlear, visual, and urinary incontinence prostheses. Another typical actuator interface is the microactuator, which can convert electrical energy into mechanical energy to operate micromechanical motors, pumps, or valves that meter delivery from a larger-scale reservoir [37]. Two key factors should be considered in designing the interfaces with living tissue. The first is safety and reliability because the interface is in direct and long-term contact with living tissue [42, 43]. The second is low power and small size. Low-power and small-size design of the image sensor is vital in the wireless capsule endoscopy system [44].

5.2.2 Drivers

Drivers have two types. The first consists of an amplifier and an analog-to-digital converter (ADC). The biosignal from the recording interface is amplified and converted into a digital signal by this driver. The amplifier is the key module of this type of driver. Low-power neural amplifiers are extremely important in recording brain–machine interfaces (BMIs) since one such amplifier is needed for each microelectrode [31]. Three challenges exist in amplifier design:

5.2.2.1 Small-Input, Low-Noise, and High-Gain Design. Biosignals are very weak.. For example, neural electric signals range from 1 to 500 μV, and have bandwidths ranging from several millihertz to 10 kilohertz [45, 46]. Hence, the weak biosignal will be submerged in the noise of a conventional CMOS two-stage op-amp topology. By design, an implantable amplifier should overcome 1/f noise first, then amplify the signal with high gain (about 10,000 times [47]) and must then reject the common mode signal.

5.2.2.2 Rejection of Large DC Offsets. Due to electrochemical effects at the electrode–tissue interface, DC offsets of 1–2 V are common across differential recording electrodes [46]. These large offsets should be rejected when the biosignal with bandwidth increases from millihertz to kilohertz.

5.2.2.3 Low-Power Design. An implantable amplifier must dissipate little power so that surrounding tissues are not damaged by the heat. A heat flux of as low as 80 mW/cm can cause necrosis in muscle tissue [48]. The power consumption of bioamplifiers for neural electric signals is about 100 mW [46, 49].

In summary, an amplifier with low noise and high gain and that uses low power and rejects large DC offsets is needed for most IMEDs. In typical implantable medical applications, the sample rate of the ADC is less than 1000 kS/s and its resolution is no more than 10 bits [50–52]. The design of the ADC in this type of driver focuses on ultralow power consumption and small area [50–52].

The second type of driver includes current or voltage drivers to drive the actuator interface. These provide all kinds of current or voltage signals required for the actuator interface under the control of the digital IC module. For example, in implantable pacemakers, there is a voltage driver, called a high-voltage generator, that provides a high-voltage pulse to stimulate the heart muscle (e.g., to initiate a heartbeat) through the pacing electrodes. These high-voltage pulses can be 5 V or more [18]. The current driver applied in cochlear implants consists of a group of current sources. It generates a stimulating current according to the amplitude information from the digital IC module. It usually consists of a digital-to-analog converter (DAC) and current mirrors [23].

5.2.3 Wireless Transceiver and Antenna

A wireless transceiver in an implant unit receives control commands or data from an external unit and transmits collected biosignals to the external unit via a microantenna. It eliminates the problems associated with percutaneous connectors while supporting data requirements of an implanted prosthesis. Percutaneous connectors may require continued medical supervision to guard against infections. The wireless transceiver design solutions are driven by two primary requirements, namely low power and small size. In some implantable medical applications, such as wireless endoscopic capsules, the transmitter of the implant unit dissipates a significant amount of power [53]. This module must be designed to consume very little power so that it can last on battery power for a long time, and it must often achieve this in a small area by using few, external, small-sized components [54]. The wireless communication distance is ultrashort in implantable medical applications, so the simple modulation/demodulation methods, such as amplitude-shift keying (ASK), frequency-shift keying (FSK), or phase-shift keying (PSK), are usually used for low power and small areas in the wireless transceiver [55]. To save transceiver power, channel coding is usually applied in the digital IC module for the specified bit error rate (BER) [56] and half-duplex transmission is used in the wireless transceiver [57, 58]. In most medical applications, the implant unit does not have to communicate continuously with the external unit. Therefore, it makes sense to wake up the implanted transceiver only when there is a need to exchange information. Battery life can be saved by employing an optimal wake-up protocol. Balancing security and privacy with safety and efficacy will become increasingly important as IMED technologies evolve [59]. The encryption of telemetry data in implant units may therefore be considered in the near future.

The implantable antenna plays a major role in implantable systems since it provides communication between the implant and the external unit, and provides power from the external unit. Designing an efficient implanted antenna is a key requirement in the creation of reliable medical implanted communication systems [60]. However, implantable antenna design is quite challenging with regard to antenna size, impedance matching, low-power requirements, and biocompatibility with the body's physiology [61]. The designs have been limited to *in silico* and *in vitro* studies. Traditionally, antennas at frequencies lower than 10 MHz are used for both power and data telemetries. To increase the bandwidth of the data telemetry system to accommodate a large number of data communications, where the data are transmitted using miniature antennas operating at microwave frequencies and the power is transmitted at lower radio frequencies (RFs) [62].

5.2.4 Power Module

A power module provides direct current and voltage energy for all other modules in an implant unit. At present, there are two typical power sources applied in power modules. One is a portable battery, and the other is wireless transmission power from the outside. Implantable devices have traditionally used batteries as the energy source for the implantable system [63]. Despite advances in miniaturization, batteries are still relatively large, occupying a considerable proportion of the volume of the implant, and they pose a risk to the patient if they leak. Wireless power transfer technology offers a route to further miniaturization of the implant unit while increasing the lifespan of the unit and eliminating the health risk of battery leaks. So, wireless transmission of power appears to be more attractive than portable batteries for IMEDs. More and more IMEDs will make use of wireless power as their energy source.

The wireless power harvesting part of a power module consists of a power receiver that recovers the transmitted power and a power regulator ensures stable voltages for the operation of implant electronics. In the design of wireless power transfer systems, power efficiency is one of the most important parameters that can be used to quantitatively evaluate an inductive link [41]. Tissue safety should also be considered in this design because high electromagnetic field intensities may cause adverse effects on the body [64].

5.2.5 Digital Integrated Circuits

The digital IC module plays a key role in an implant unit. It is the control core and manipulates all the other modules in the implant unit. The DSP required for an implant unit is also merged into the digital IC module in all kinds of medical clinic applications. A digital IC module consists of a DSP and controller module and a memory module in this IMED model. The DSP and control module provides baseband signal processing and system control functions, such as communication baseband signal processing, data encoding/decoding, and power management. A memory module is needed here to store some temporal data or program data for a processor in the DSP and control module.

The generic architecture and design considerations of the digital IC in an implant unit are discussed in detail in the next section.

5.3 DESIGN CONSIDERATIONS OF LOW-POWER DIGITAL INTEGRATED CIRCUIT DESIGN INSIDE AN IMPLANT UNIT

According to the above analysis for the general model of IMEDs, the generic architecture of a low-power digital IC in an implant unit can be designed as illustrated in Figure 5.2. The digital module includes a sensor interface, data compression, random-access memory (RAM), read-only memory (ROM), baseband communication processing, power management, clock management, a controller, a low clock controller, and an actuator interface. Low-power design for each part of the digital IC module will be discussed in the following section.

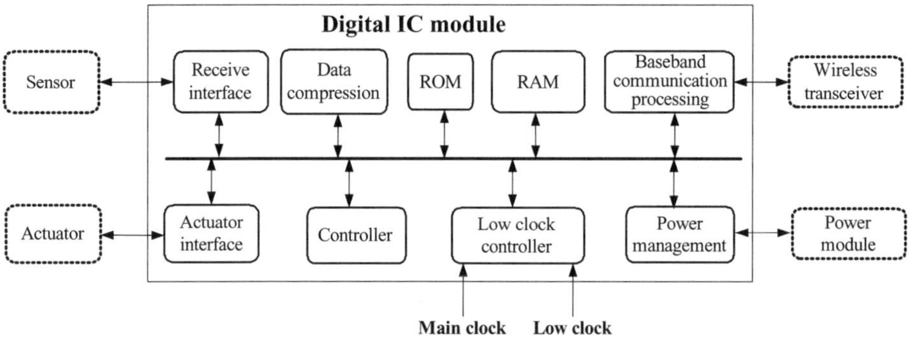

Figure 5.2. A generic architecture of low-power digital IC in the implant unit.

5.3.1 Sensor Interface and Actuator Interface

The sensor interface controls the front-end sensor's work state and acquires the electric biosignals from the sensor, while the actuator interface controls the actuator's work state and sends the operational data to the actuator. In general, there are two data buses. One is between the sensor interface and the sensor, and the other is between the actuator interface and the actuator.

Three strategies should be considered for low-power design in these two interfaces. The first is that when the sensor cannot be integrated into one chip with the digital module, the bus encoder technology [65] should be used to reduce signal activity for low power. In the same way, this technology should also be applied for the data bus between the actuator interface and the actuator. The second strategy is that low clocks should be provided to the two interfaces for configuring the sensor and actuator. The third strategy is to shut off the power or the working clocks of the sensor and actuator to put them into sleep mode when they are idle.

5.3.2 Data Compression

In fact, most IMEDs can be viewed as complex data acquisition, processing, and communication agents that have been designed for many years with portable batteries. Data storage and transmission represents one of the major challenges in designing such devices, requiring dedication of substantial memory and power resources. For example, in BMI applications, a typical recording experiment with a 100-electrode array sampled at 25 kHz per channel with 12-bit precision yields an aggregate data rate of 30 Mbps, which is beyond the reach of state-of-the-art telemetry links for biological applications [66], and consumes beyond the power afforded by the implant unit. To meet this challenge, data within the device must be sufficiently compressed [67]. The compression algorithm suitable for IMEDs must have low computational complexity, high quality, and a high compression ratio, which is more challenging. So, the known highly efficient compression algorithms such as Huffman and arithmetic encoding, generally, cannot be directly used for the implant unit [67]. In typical implantable neuroprosthetics, some new compression algorithms based on discrete wavelet transform (DWT) are proposed to provide efficient data compression for neural records without compromising the information content [66, 67]. For data reduction, some new, noncompression processing techniques are also applied to eliminate nonuseful or redundant data. The compressive-sensing technology [68] has been used for real-time spike train decoding of IMEDs [69]. Adaptive, learnable, multi-input–multi-output neural decoding techniques that project firing-rate neuronal data onto several motor output parameters are used to decode movement intentions from neural signals [70].

Generally, a data compression algorithm and its application-specific integrated circuit (ASIC) design for IMEDs should take into account three main factors, namely, low complexity for low power, high quality for data precision, and a high compression ratio for small memory, suitable wireless bandwidth, and low power of data transmission. Trade-offs among these factors have to be made. The order of importance of these three factors is high quality, low power, and high compression ratio.

5.3.3 Baseband Communication Processing

Baseband communication processing is an interface between the digital module and the wireless transceiver in an implant unit. It includes two communication channels. One is the transmitting channel, and the other is the receiving channel. The transmitting channel mainly performs two functions, namely channel coding, and generating the training codes and frame synchronous codes. The receiving channel realizes clock and data recovery, and channel decoding. In general, the baseband communication processing in an implant unit is an asymmetrical communication interface with a wireless transceiver. In some implantable applications, such as neural recording microsystems and wireless capsule endoscopy, the uplink (from the implant unit to the external unit) data rate is higher than the downlink because capturing the bioinformation is the main work of IMEDs and only some simple commands are required to be transmitted from the outside to the implant unit. In other implantable applications, such as cochlear prostheses, visual prostheses, and implantable pain controllers, the downlink (from the external unit to the implant unit) data rate is much higher than the uplink. For example, a large amount of image data needs to be transmitted from the outside to the implanted visual prosthetic device. So, asymmetrical communication should be considered for the design of the baseband communication processing part. Two work clocks must be used for saving power. A low clock is used for the low data rate communication channel, and a high clock is applied for the high data rate communication channel. Moreover, when one communication channel is idle, it should enter into sleep mode.

5.3.4 Controller

The working states of all modules in an implant are controlled by the controller, which is the control core of the implant unit. Its work clock is the main clock

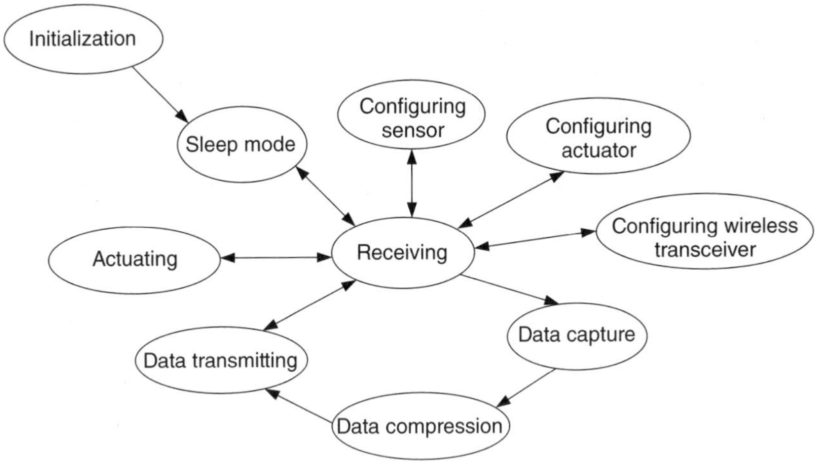

Figure 5.3. System-state machine in the implant unit.

with a high frequency in such a system. There are two major types of controllers in this architecture. The first type employs microcontrollers. It has better flexibility to change different functions by programming, but the size and power consumption is large, and the ROM has to be used for storage of program data. The second type is based on the ASIC design, which must follow certain design requirements. The final chip size, by contrast, is smaller, and is very suitable for low-power and small-area design. The critical drawback of this kind of chip is a lack of flexibility for various applications. Determining which type of controller should be selected involves a trade-off between low power and flexibility according to the requirements of the specific application.

Most of system-level strategies for low-power design of an implant unit are related to the system-state machine realized in the controller. For example, different states need different circuit modules to work in different clock frequencies. Shutting off the clock of an idle module and decreasing the clock frequency of the different modules as low as possible will save a great deal of power for the system, which is achieved through clock management in the controller. So, the design of the system-state machine should be analyzed very carefully. The system-state machine is shown in Figure 5.3.

1. Initialization: All of the values of the registers in the implant unit are initialized while the power is on.
2. Sleep mode: The main clock is stopped. Only the low clock works.
3. Receiving: The command or actuating data from outside are received in this state. If no command or actuating data can be received during the specified interval time, the system will enter into sleep mode. This mechanism guarantees that the system can come back to sleep mode automatically to save power in case the wireless communication channel is so bad that the command cannot be received.

4. Configuring the sensor, wireless transceiver, and actuator: The values of registers in the sensor, wireless transceiver, and actuator are configured, respectively, in this state. The clock frequency of the configuring operation is usually low. For example, only a 400-kHz clock is needed for inter-integrated circuits bus (I2C) serial communication [71].
5. Data capture: One frame data with specified size is captured by the sensor in this state.
6. Data compression: The captured data are compressed with high quality for the specified medical requirements.
7. Data transmitting: The compressed data are transmitted outside by the wireless transceiver.
8. Actuating: In this state, the actuator interface provides the operational data required for the actuator, and according to the data, the actuator decides how to act on the living tissue.

5.3.5 Low Clock Controller

In most implantable medical applications, the implant unit does not have to work continuously. So, a sleep mode is usually used to save power. This mode is usually realized by a low clock controller whose work clock frequency is very low, at 32 kHz or even lower. When the system enters into sleep mode, only the low clock controller is working, and the system's main clock is shut off by the low clock controller. It is important that the asynchronous circuit design be carefully considered because two different clock domains exist between the low clock controller and the controller.

5.3.6 Power Management

Minimum average power is not always the objective when designing battery-based implantable medical applications. Rather, what really matters for this kind of system is ensuring long battery lifespan. There are two power management methods. One is a closed-loop power management method, in which the decision rules used to control the system operation state are based on the observation of a battery's output voltage. The other is an open-loop power management method that reaches decisions about component shutdown independently from battery voltage measurement. The open-loop method is normally simpler but less effective than closed-loop ones [72].

Power management in the implant unit must be able to inform the controller to control the system to operate in different modes, which trades off performance for power consumption. When one module in the system is idle, the power management can force a transition to sleep mode or power down mode. The consequences for such a transition are latency and power. Generally, there is a delay and energy cost for activating an implantable system in power down mode. Even if the system is not idle, in some cases, it can be forced into a mode where it operates with lower performance and with reduced power. One important function of the power management unit is to try to avoid the concurrent actions of the different parts of the circuits considering the nature of the battery-driven system [73].

5.4 A DESIGN CASE: LOW-POWER DIGITAL INTEGRATED CIRCUIT DESIGN FOR WIRELESS CAPSULE ENDOSCOPY

5.4.1 Introduction to Wireless Capsule Endoscopy

Compared with conventional endoscopic techniques, wireless capsule endoscopy allows us to directly study the entire small intestine and does not require any sedation, anesthesia, or insufflation of the bowels. The noninvasiveness of the procedure improves patient comfort and eliminates complications associated with endoscopic procedures. So, more and more people have started researching in the field of wireless endoscopy, and some preliminary progress has been achieved [2, 74–77]. The small bowel capsule designed by Given Imaging works for less than 8 hours (generally, it takes capsules about 10–48 hours, typically 24 hours, to move from mouth to evacuation [2]), the image frame rate is slow (2 frames per second [fps]), and image size is small (256×256), which results in the fact that there is not enough time for the capsule to check the large intestine and some regions of the small intestine in which doctors are interested in examining but which are often left out. The device made by the Korean [74] is an analog system in which the anti-interference ability is less effective than that of the digital system, which seriously affects the quality of the digestive tract images. Its capsule size is also large (the diameter of the capsule is 11 mm). In the wireless endoscopic system made by the Japanese [75], the speed of the capsule cannot be controlled. Olympus is using a CCD image sensor in their small-bowel capsule to get better image quality. Its capsule works at a 2 fps image frame rate only for less than 8 hours [77]. A Chinese wireless small-intestinal capsule (OMOM) has gone through three technical design changes, with improved image quality, and currently costs less than the other systems [78]. A Korean small-intestinal capsule (MIRO) includes some novel technological features, including two external electrodes and a single skin electrode to facilitate transmission of image data using the body as an electrical conductor, thus eliminating the need for RF (wireless) data transmission, and as a result, saving power [78]. The capsule can work for about 11 hours with a 3 fps image frame rate [76].

In order to lower the power consumption inside the capsule, this chapter proposes a new system of architecture for wireless endoscopy and an analog–digital (A/D) mix-mode, low-power IC architecture inside the wireless endoscopic capsule for diagnoses of the whole human digestive tract and real-time endoscopic image monitoring. Furthermore, the design of an analog power management unit (APMU) and the key part of the proposed system, a digital IC inside the capsule, are both discussed in detail later. This chapter proposes a low-power digital IC design inside the capsule for 8 fps with 320 × 288 image size, with adjustable capsule speed and whole digestive tract check. Analysis of power consumption in existing capsules shows that the power needed to transmit image data occupies about 90% of the total power in the capsules. So, a new, low-complexity and high-quality image compression algorithm is used here to reduce the power needed for transmitting image data and the communication bandwidth, and to provide a high image frame rate (8 fps with 320 × 288 image size) and a long, continuous work time. Furthermore, a novel digital VLSI architecture, a bus encoding technique, and a power management method are also applied for low-power design, and a microcurrent generator is used to stimulate the muscles of the digestive tract to contract and to push the capsule forward, which will enable a doctor to move the capsule to those parts of the digestive tract requiring examination.

5.4.2 System Architecture

The architecture of the proposed wireless endoscopy system is given in Figure 5.4. It consists of a capsule, a portable device, and a computer with a wireless transceiver. The capsule, which will be swallowed by the patient, is hermetically packaged. Inside the capsule, a CMOS image sensor is applied to capture digestive tract images, and an IC chip is used for signal processing and wireless communication. There are four light-emitting diodes (LEDs) inside the capsule for illumination: two white LEDs for normal two-dimensional (2D) images and two infrared LEDs for three-dimensional (3D) range images. Two oxide batteries are installed in the capsule. Outside the patient's body, there is a transceiver connected to a computer or a portable recording device. The image data can be stored in the recording device, mounted on the waist of the patient, and will be used by the doctor afterward. The patient can move and wander freely during this time. This mode is called "offline diagnosis." Another mode is called "online diagnosis," in

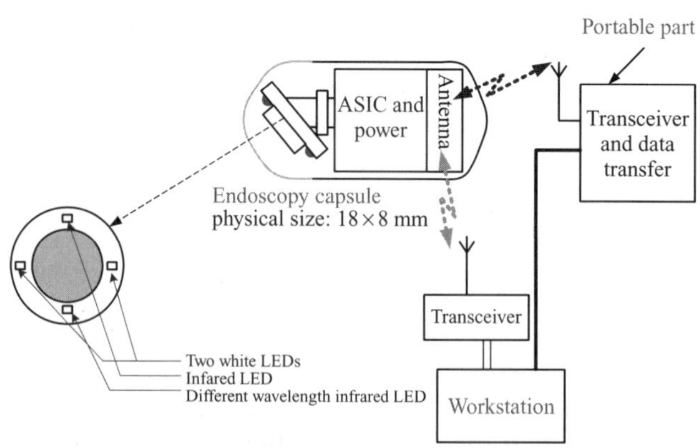

Figure 5.4. The architecture of wireless endoscopy system.

which the capsule is controlled by the computer and the image data are displayed on or stored directly in the computer. In the online diagnosis mode, doctors can justify the state of the capsule, such as the image size, the visual angle, the speed, and the luminance. The proposed system can provide a high-resolution image (320 × 288) and a high image frame rate (8 fps). Furthermore, bidirectional communication makes it possible for the outside of the capsule to control the behavior of the circuits inside the capsule. In order to realize a diagnosis of the whole human digestive tract, three key strategies are applied here. The first strategy is to design the low-power IC inside the capsule. The second strategy is to make the circuits inside the capsule enter into sleep mode to save power when the capsule passes the areas of the digestive tract that are not of interest. The third strategy is to generate current to stimulate the muscles of the digestive tract to contract and to push the capsule forward when the doctor wishes to bypass such areas.

5.4.3 Analog–Digital Mix-Mode Low-Power Integrated Circuit Architecture and Analog Power Management Unit Design Inside the Capsule

5.4.3.1 Analog–Digital Mix-Mode Low-Power Integrated Circuit Architecture. The A/D mix-mode low-power IC architecture inside the capsule is proposed as illustrated in Figure 5.5. It consists of two parts. One is the digital circuits, which include static random access memory (SRAM), image compression, baseband communication processing, a control unit, an interface with the CMOS image sensor, digital power management, LEDs, and current drivers. The other is a wireless RF transceiver that transmits the compressed image data and receives the control command from outside the capsule. The wireless transceiver [79] designed by our RF team provides ASK modulation with 2 Mb/s for image data transmitting and on–off keying (OOK) demodulation with 128 Kb/s for control command receiving. Its power consumption is about 8.3 mW at 1.8 V.

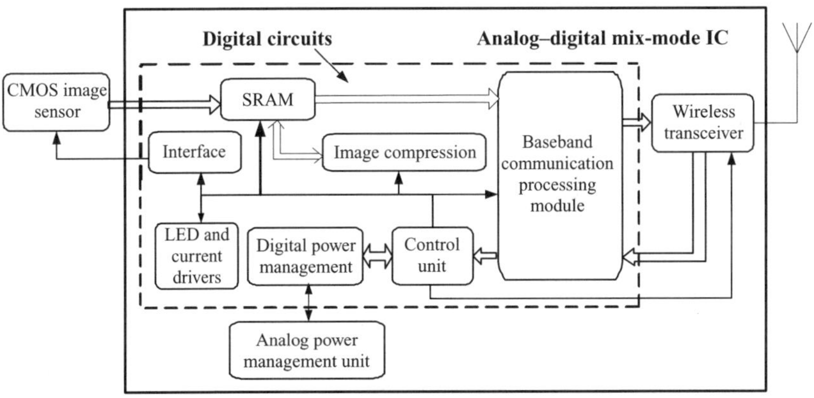

Figure 5.5. Analog–digital mix-mode ASIC architecture inside the capsule.

A low-power CMOS image sensor is a commercial chip, which is used to implement the capture of raw image data with a Bayer color filter array (CFA). The digital circuits inside the capsule are in the dashed frame of Figure 5.5. In the A/D mix-mode single chip, the low-power digital circuit design is vital for the diagnosis of the entire human digestive tract.

5.4.3.2 Analog Power Management Unit Design Inside the Capsule. The APMU inside the capsule not only has to supply on-chip functional blocks but also has to deliver the power for additional components connected to the chip, for example, the CMOS image sensor, the RF transceiver, and the LEDs for illumination. The miniature size requirement of a wireless endoscopic system makes a very severe demand on the area of the printed circuit board (PCB) as well as on the number of off-chip components inside the capsule. The overall analog power management architecture, which allows one to reduce the on-board components, is explored by considering the chip architecture and the capsule architecture.

To make a better compromise among power conversion efficiency, on-chip area, off-chip components, and PCB routing requirements, two possible power supply architectures that both meet the integrated requirements have been considered and compared, as shown in Figure 5.6. The difference between the two architectures is the selection of power supply conversion using either linear or switched regulator types. The first architecture is built without a DC–DC converter and is purely based on the low dropout (LCO) regulator, as shown in Figure 5.6. The second one uses a DC–DC converter to generate the voltage supply for the digital core. The selection of the types of regulator is, in general, a trade-off between power conversion efficiency and system cost. A linear regulator results in lower voltage ripple, fewer off-chip components, and less chip area;

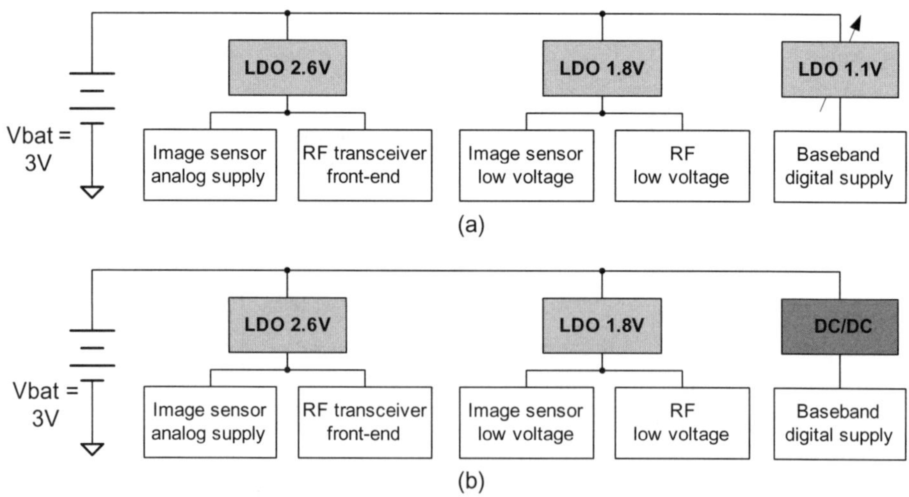

Figure 5.6. Two different power supply architectures.

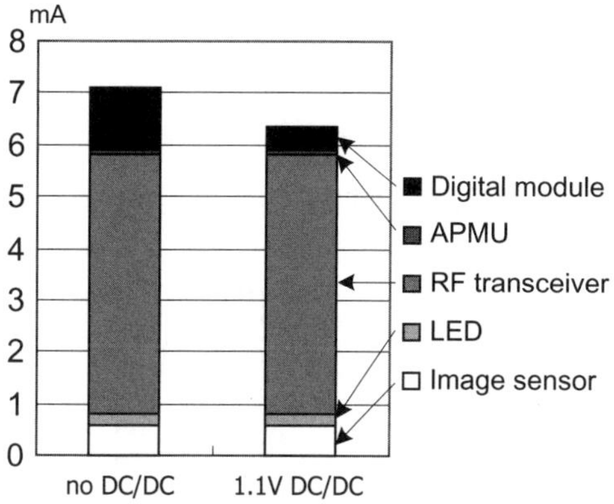

Figure 5.7. Current consumption for different blocks.

however, it results in lower efficiency if the voltage difference between the input and output is large. On the other hand, switched regulators exhibit higher efficiency; however, more on-chip area for the more complex switching stage, one more off-chip power inductor, higher voltage ripple, and electromagnetic interference (EMI) can cause problems. Moreover, additional noise-reduction filtering has to be utilized to suppress the spurious switching in the receive and transmit band for the RF part [80].

In order to assess the two architectures, the achievable current consumption and implementation costs are evaluated. The average current consumption of all blocks with a 2 fps frame rate is summarized in Figure 5.7. It is shown that the voltage of analog and RF parts is too high to justify a switching regulator. Therefore, they are supplied directly by LDOs. Only the current consumed by the digital core is relevant with the use of switching DC–DC converters. It is notable that, for the purpose of ultralow power, the supply voltage of the digital core can be scaled down from 1.8 to 1.1 V without degrading the performance. This results in a more than 60% power reduction in the digital core. With voltage scaling, the application of a DC–DC converter saves only about 10.5% of the battery current compared with the purely LDO architecture. Therefore, considering off-chip components, chip area, and other aspects, the pure LDO architecture is employed in the design.

The 3.3-V maximum operating voltage of the 0.18-μm CMOS technology facilitates the design of the LDO, which regulates the battery voltage 3V down to a desired output voltage. The LDO can be directly powered by battery without overvoltage problems. Although it is possible to design an LDO with small on-chip load capacitance to reduce the off-chip components, this has to be done with the need of a large die area; a high, steady current [81]; or gate over driven [82, 83], which is not tolerant for the low-cost and low-power target. Thus, the load capacitance is designed as the only off-chip component.

Figure 5.8 illustrates the LDO circuitry design. It is composed of an error amplifier, a unit–gain buffer, a PMOS pass device, a feedback network, and a loading capacitance. The PMOS pass device MP exhibits minimum channel length

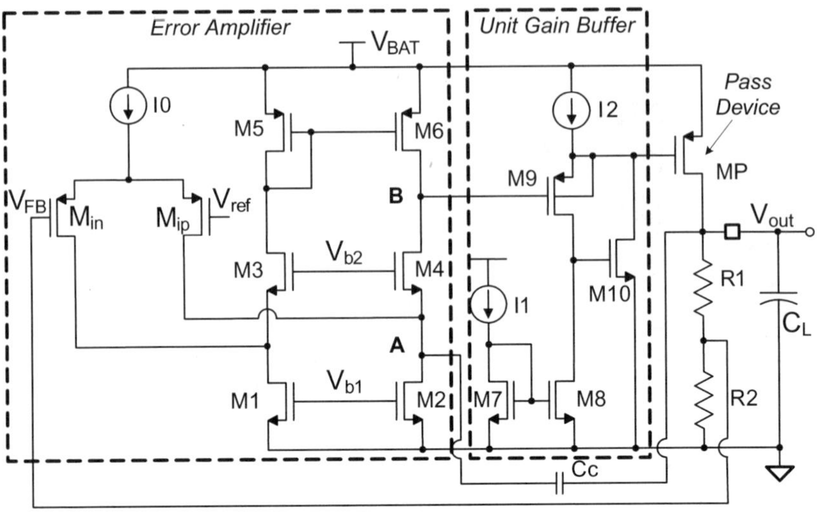

Figure 5.8. Schematic of the LDO.

and very large width since it is designed to source 30-mA current. This, in turn, makes the parasitic capacitances associated with this device very large. This affects the circuit's frequency performance by reducing the value of the parasitic pole present at the gate of pass device [82]. Therefore, phase–margin degrades and stability may be compromised unless the output impedance of the error amplifier is reduced accordingly. As a result, a unit–gain buffer stage derived from Fiocchi and Gatti [84] is designed. The unity gain buffer isolates the large capacitance from the sensitive error amplifier and pushes the pole at the gate of the pass transistor to high frequency.

The Ahuja compensation method [85], rather than the Miller compensation method, is used. The cascade transistor M4 in the error amplifier is used as the transistor in the feedback path. The transconductance of the cascade transistor M4 is designed to be large enough to make the source of M4 virtual ground. Therefore, any current flowing through Cc goes through the source of M4 and comes out of the drain. Due to the low impedance present at node A, the pole at node A is pushed to higher frequency, which improves the phase margin of the feedback loop. Compared with the Miller compensation, which connects Cc to high impedance node B, the Ahuja compensation exhibits excellent phase margin across all loads. The Ahuja compensation also provides higher power supply rejection (PSR) than the classical Miller compensation. The Miller topology has less PSR because the Miller capacitance is feeding any ripple from the supply to node B directly to the output of the LDO. The Ahuja topology has excellent PSR because of the absence of such a path.

The wireless endoscopic system is equipped with white LEDs as illumination devices, which need a constant voltage higher than the battery voltage to provide constant light density. This is achieved by the integrated charge pump. One major

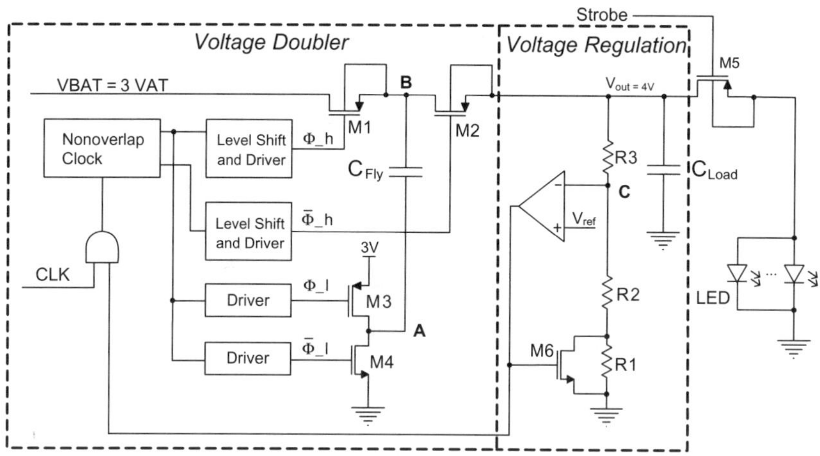

Figure 5.9. Schematic of the charge pump.

design challenge of integrated charge pump is the high output voltage compared with the 3.3-V maximum operating voltage of 0.18-μm CMOS technology. Any transistor that withstands a voltage difference across their terminals of up to 4V or more would face lifetime reduction. However, a long lifespan is not required in our application since the endoscopic capsule only needs to work for tens of hours. In addition, LEDs are not conducted continuously. Therefore, short-time overvoltage is allowed in this specific application.

The charge pump is mainly composed of a voltage doubler and voltage regulation as shown in Figure 5.9. The circuit works as follows. During the clock phase when M1 and M4 are on and M2 and M3 are off, the CFly is charged to the VBAT. During the clock phase, when M1 and M3 are off and M2 and M3 are on, the voltage at node A is driven to VBAT, and the voltage over CFly cannot be changed, which makes the voltage at node B twice that of the supply voltage. The charge on the CFly will be transferred to CLoad to accommodate the load current. A comparator U2 is used for regulating the output voltage to a desired value. The output of the comparator would change to ground if the output voltage on CLoad is higher than the specific level. Then the clock is gated and the voltage doubler is turned off. When the output voltage is lower than the specific level as CLoad is discharged by ILoad, the clock will be opened and CLoad will be charged again. As a result, a constant voltage is generated. Resistors R2 and R3 are used as voltage divider to generate feedback voltage. A small resistor, R1, and M6 are used to prevent unnecessary switches of the output of the comparator if the voltage at node C is very near V_{ref}. M5, controlled by strobe signal, acts as a switch to charge the LED.

For this application, the charge pump is discharged only at the moment of capturing images. To save energy, two different operating modes are implemented. If the strobe signal is active, the charge pump works in high-speed mode with a 1-MHz clock frequency and high-bias current of the comparator, which

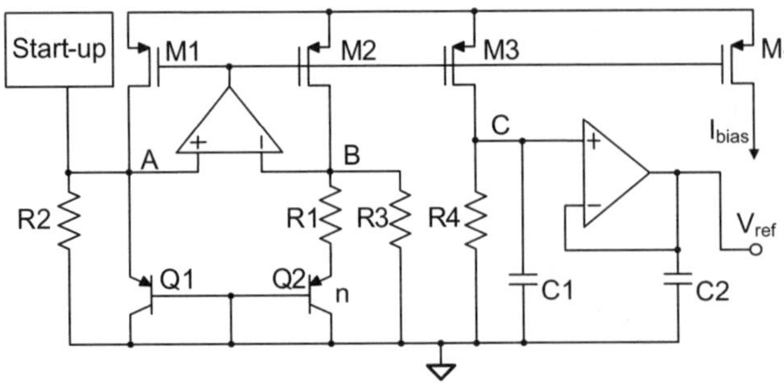

Figure 5.10. Schematic of the bandgap reference.

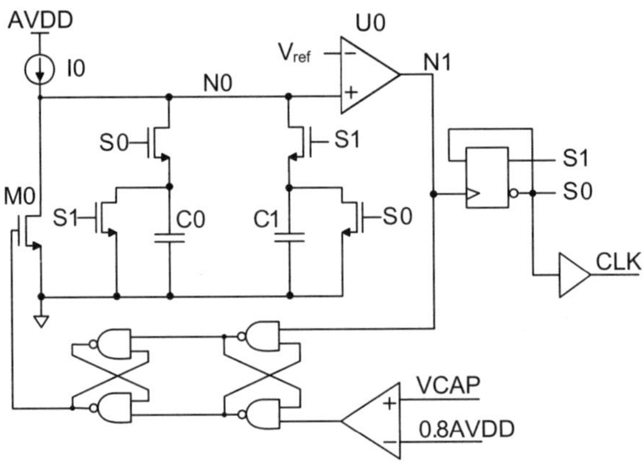

Figure 5.11. Schematic of the oscillator.

results in large output current sourcing ability and low-voltage ripple. Otherwise, 20-KHz clock frequency and low-bias current are employed to sustain the output voltage in order to ensure a prompt reaction of the strobe signal. With this approach, the power consumed by the charge pump is greatly reduced when the LED is off.

A fully integrated bandgap reference voltage generator provides a reference voltage and bias currents for the other circuit. It adopts a classical, self-biased cascade structure to eliminate the dependence of supply voltage and temperature, as shown in Figure 5.10. OPA2 forms a unit-gain amplifier to enhance the driving capability of the V_{ref}. The current consumed by the bandgap reference is 12 μA.

The on-chip oscillator is designed not only to provide a clock for the charge pump but also to provide a system clock for the digital core and CMOS image sensor. Thus, an off-chip crystal oscillator can be eliminated to reduce the PCB area.

The circuit of the oscillator is shown in Figure 5.11. Capacitors C0 and C1 are charged and discharged by turns. Once one of the capacitors is charged above V_{ref}, the output of comparator U0 would change, which causes the output of D flip-flop to invert. Then the capacitor would be discharged and the other capacitor starts to repeat the same process. The frequency can be adjusted by controlling I0. A sawtooth wave, which is composed of voltages of C0 and C1, will be generated at node V_{CAP}. The output of comparator U0 is a square wave with certain duty cycle. The D flip-flop acts as a frequency divider that generates a square wave with a 50% duty cycle. In addition to comparator U0, a second comparator, U1, is used to prevent the oscillator from stopping, if the frequency is too high to be handled by comparator U0. The reference voltage of comparator U1 is higher than that of U0. Once the voltage of node V_{CAP} is higher than the reference, the M0 will discharge the capacitors and reset the oscillator. In order to make the frequency independent of the variation of battery voltage, the oscillator is powered by the output of the LDO.

Due to a limited ability of sourcing current for batteries, a large start-up peak current may cause an apparent voltage drop in the batteries, which will result in a failure of start-up. In order to minimize the peak current, soft-start ability should be included in the APMU. This is achieved by generating a different start signal at a different time slot for each block. The time intervals between two start signals are specified by the RC delay. Therefore, blocks in the APMU will be enabled in a predefined sequence and a large start-up current can be avoided, which results in a more stable start-up process. Furthermore, the reset signal of the digital core is also generated at the end of the start-up process when the supply voltage is stable.

In order to save PCB area, only large capacitances are used as off-chip components, and all the other capacitances and resistors are fully integrated. The process variation makes it very difficult to promise precise values for on-chip capacitances and resistors, which will cause variations in the circuit performance, such as in the voltage level of bandgap reference or in the frequency of the oscillator. Therefore, resistors and capacitances with trim ability are widely used. The trim options are controlled by the configuration register files, which can be accessed by the digital core. The configuration process can be started by wireless communication. This facilitates the configuration even though the endoscopic system is hermetically encapsulated within the package.

Furthermore, the regulators for analog and digital power supply lines are separated locally with adequate space between one another. It helps to reduce the inductive coupling of the digital power supply, where mainly the high-frequency switching current generates spuriously in the analog domains while the average DC causes no trouble. Each block is surrounded by p-plus and n-well guard rings to achieve further isolation. Neither triple-well nor deep n-well was utilized to reduce technology cost.

5.4.4 Low-Power Digital Integrated Circuit Design Inside the Capsule

In this section, the low-power digital circuit design mainly focuses on system level and register transfer level (RTL).

5.4.4.1 *System-Level Low-Power Design Strategies.* In the low-power circuit design, the most efficient stage is the system-level low-power design because the largest possible power consumption can be saved at this level [86, 87]. The following system-level strategies are applied.

Half-duplex transmission and a simple ASK modulation [88] are used for low-power consumption. The LEDs are controlled to flash only at the moment of image capture. With respect to the battery-driven system [72], one function of the digital power management unit is to avoid the concurrent actions of the different parts of the circuits while maintaining the data transmitting rate.

Raw Bayer CFA image data are used in the CMOS image sensor in that the quantity of the Bayer CFA data is only one-third of the full color image data. So, at the algorithm level, a new near-lossless image compression algorithm with low computational complexity is proposed to decrease communication bandwidth and save the transmitting power.

The image sensor has two catalogs: a CCD image sensor and a CMOS image sensor. Due to one-chip circuit design and the low-power characteristics of the CMOS image sensor, these are the ideal choices.

An on-chip, 10-kHz clock and a 40-MHz clock are used to support the sleep mode and work mode of the system. When the capsule passes those areas of the digestive tract that are not of interest, the sleep mode can be started to save power. A large number of clock-gating techniques are used in the circuit design.

5.4.4.2 *Digital IC Block Diagram Inside the Capsule.* Based on the above system-level, low-power design strategies, a low-power digital IC block diagram inside the capsule is proposed as illustrated in Figure 5.12. It can be divided into five parts, namely, the control part, the interface with the image sensor, image compression, baseband communication processing, and drivers. The low-power circuit design of the five parts will be discussed next.

5.4.4.3 *System-State Analysis Inside the Capsule.* With the given arrangement of circuit tasks inside the capsule, the system-state machine can be obtained as shown in Figure 5.13.

1. Initialization: All of the values of the registers in the capsule are initialized while the power is on.
2. Sleep mode: The main clock is stopped. Only the low clock works.
3. Receiving: Commands from outside, the capsule are received in this state. If no command can be received during the specified interval time, the system will enter into sleep mode. This mechanism guarantees that the system can come back to the sleep mode automatically to save power in case the wireless communication channel is so bad that a command cannot be received. Because a command is very short, the data rate of commands is set to 128 Kb/s to reduce the design complexity of the RF receivers.
4. CMOS image sensor setting and wireless transceiver setting: The values of registers in the CMOS image sensor and the wireless transceiver are set,

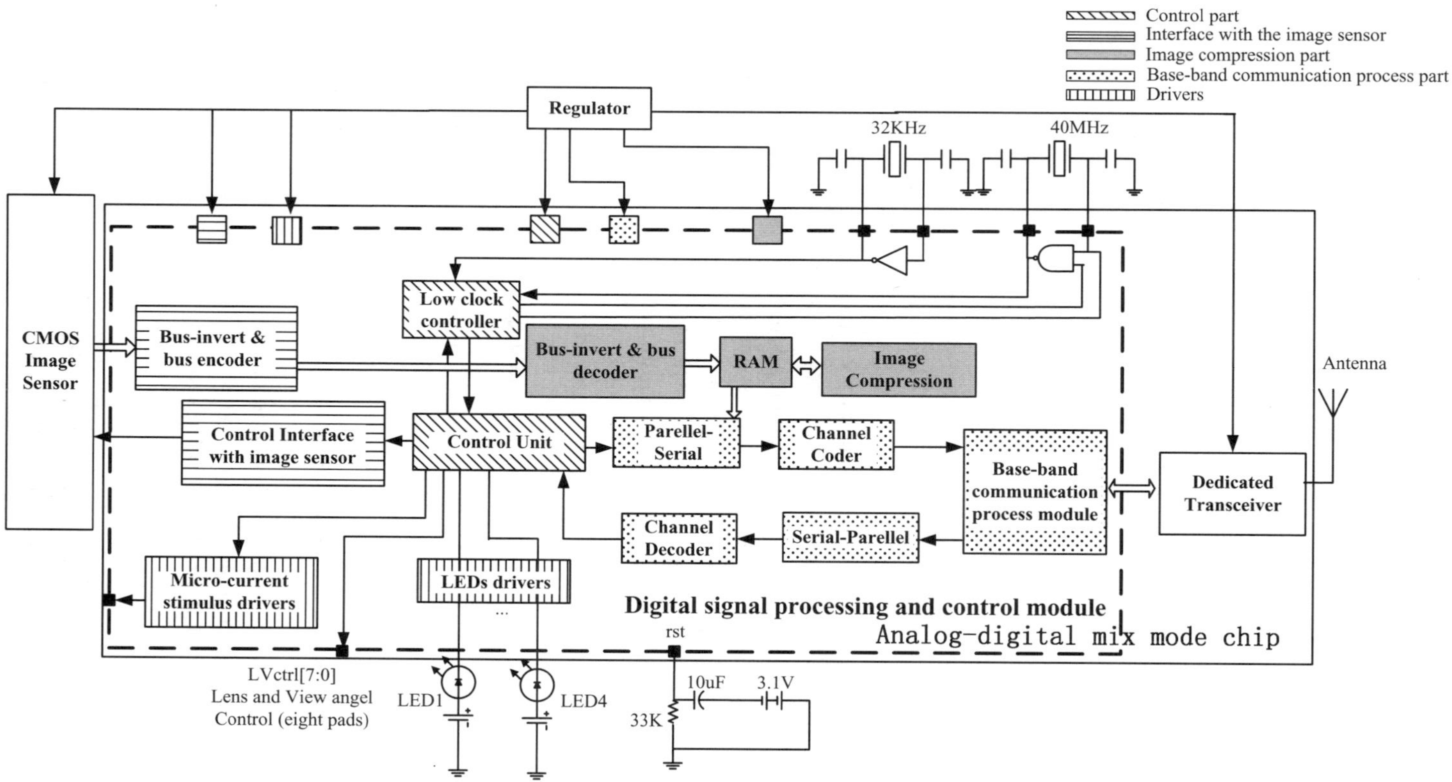

Figure 5.12. VLSI architecture of the digital IC inside the capsule.

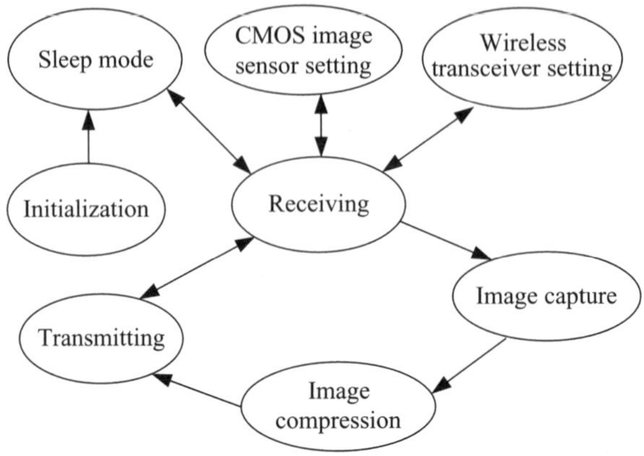

Figure 5.13. System-state machine inside the capsule.

respectively, in this state. Only a 400-kHz clock is needed for I2C serial communication [71].

5. Image capture: One-frame image data with a specified size is captured and saved in the RAM. The clock frequency of the image sensor is 20 MHz.
6. Image compression: The image data readout from the RAM is compressed by proposed near-lossless and lossless image compression with low computational complexity, which is similar to the work by Xie et al. [89]. The compressed image data are saved back into the RAM, so the automatic retransmission request (ARQ) communication scheme can be used for high-image-quality wireless communication [90].
7. Transmitting: The compressed data stored in the RAM are transmitted to outside the capsule by the dedicated wireless transceiver. Only a 2-MHz clock is needed in this state because the transmitted data rate is 2 Mb/s.

Different states only need different circuit modules to work in different clock frequencies. So, shutting off the clock of an idle module and decreasing the clock frequency of the different modules as low as possible will save a great deal of power for the system.

Image compression plays a very important role in the low-power design inside the capsule. By power simulation analysis, it was found that the power consumed in the transmission state occupies above 90% of the total power in the capsule without image compression. With our image compression schedule, approximately 48% of the power can be saved.

5.4.4.4 Low-Power Circuit Design. Control Part. Three stages of clock management architecture are proposed for low-power design as illustrated in Figure 5.14. Every stage has a different priority. The control unit is the control core inside the capsule. The clocks used in the low clock controller and the control unit are 32 kHz and 40 MHz, respectively.

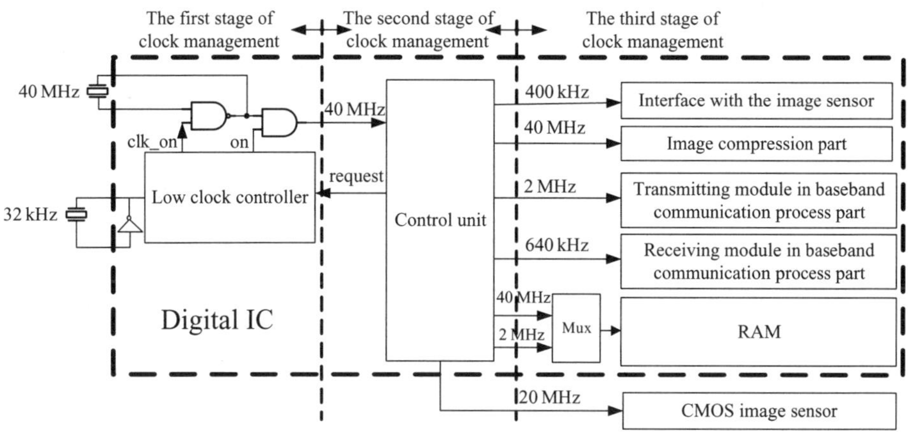

Figure 5.14. The clock management architecture in the wireless endoscopy capsule.

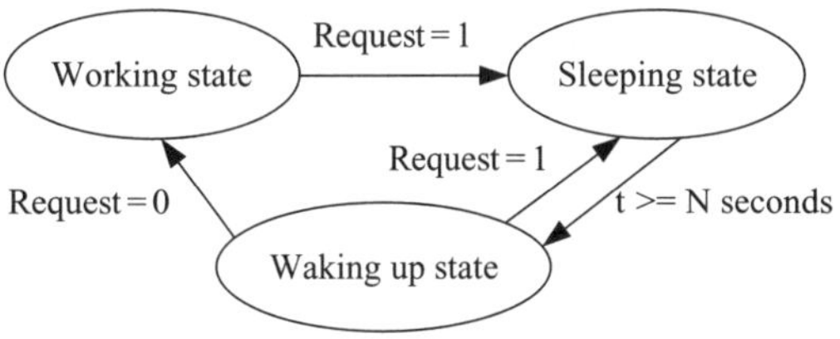

Figure 5.15. The state machine in a low clock controller.

The first stage, which has the highest priority, is the low clock controller, which controls the 40-MHz clock input of the control unit. The state machine in the low clock controller is shown in Figure 5.15. To decrease power consumption in sleep mode, a new stoppable ring crystal oscillator is applied as illustrated in Figure 5.16. When the system enters sleep mode, the low clock controller will stop the 40-MHz crystal from oscillating by controlling the AND gate. When the signal "request" from the control unit is one, the low clock controller will set the clk_en to shut off the 40-MHz crystal. The 40-MHz clock output is delayed L msec by the delay module, "delay L milliseconds," which assures that the 40-MHz clock is outputted stably to the control unit when the circuit inside the capsule is woken up. The latch prevents glitches on the "on" signal from propagating to the register's clock pin. The two-stage flip-flops are used as a synchronizer to avoid metastability problems between the low clock controller and the control unit. Although this scheme cannot eliminate the metastability problem, it minimizes the probability of synchronous failure. Because the United Macroelectronics Corporation (UMC) 0.18-m CMOS process is used, we can obtain the estimated mean time between failure $MTBF \approx 10^{260}$ years based on the equation in Shear [91] and in Brown and Feher [92]. The synchronous failure can be neglected here.

Thus, the 40-MHz crystal is stopped from oscillating and only the low clock controller works at 32-kHz frequency when the system is in sleep mode. By using

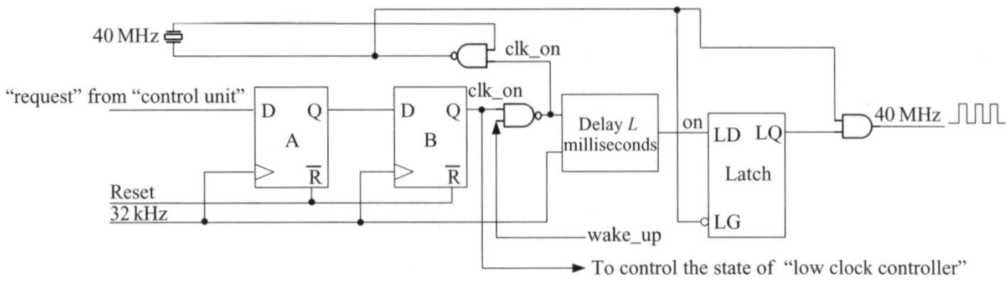

Figure 5.16. The control circuit of a 40-MHz crystal.

the PrimePower tool of the SYNOPSYS Corporation, the simulation results show that the power of the digital circuits dissipates in sleep mode by about 40 μW.

In the second stage of clock management, the control unit controls the clock input of all other modules except the low clock controller. Each module in the third stage has its own clock management.

This architecture of clock management decreases the clock frequency of the different modules as low as possible to reduce power dissipation. The power simulation with the PrimePower tool shows that this design can save 46% power inside the capsule compared with design without clock management. This architecture also supports the module-based design for low power.

Interface with the Image Sensor. One interface is the control interface with image sensors between digital IC and CMOS image sensors, and this interface supports I2C serial communication standards [71]. The other is the bus inverter and bus encoder and decoder that are applied as the data bus interface between the RAM and the CMOS image sensor. Each input–output (I/O) signal transition dissipates a substantial amount of power [93]. Bus coding methods can reduce I/O bus-switching activity. There is an 8-bit-width data bus between the RAM and a CMOS image sensor in this system, and the neighboring pixels' output from the image sensor have strong correlation, so the Hamming-distance-based bus-invert method [94, 95] is applied. This method can reduce the bus-switching activity and thus decrease the I/O peak power dissipation by 50% and the I/O average power dissipation by 25%. The block diagram is shown in Figure 5.17.

The procedure of the bus-invert method for even bus width is given here [96]:

1. Compute H_d, that is, the Hamming distance between the next data value and the present bus value (not including the invert line).
2. If $H_d > n/2$ or if $H_d = n/2$ and the present value on the invert line is one, set the next value on the invert line to one and make the next bus value equal to the inverted data value.
3. Otherwise, set the next bus value equal to the next data value, and set the invert line to zero.
4. At the receiver side, the receiver decodes the bus value according to the value on the invert line.

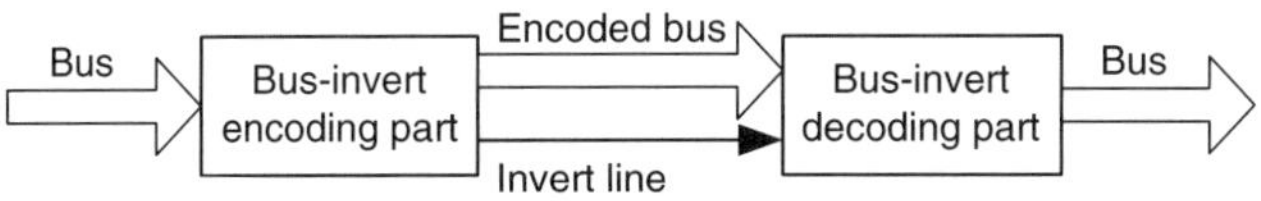

Figure 5.17. The block diagram of the bus-invert method.

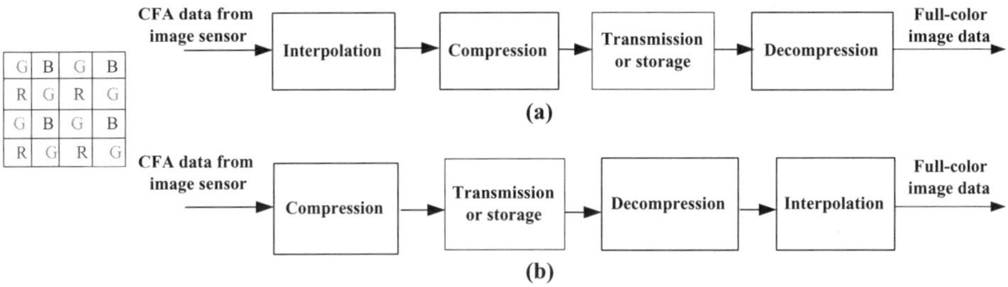

Figure 5.18. Block diagrams of two schemes for CFA image data compression: (a) conventional interpolation-first scheme and (b) new compression-first scheme.

Image Compression Part. *Image Compression Algorithm.* In most conventional applications of image sensors, the captured data with a Bayer CFA pattern [97] are interpolated into a full-color image and compressed before the data are transmitted or stored [98–100]. The conventional algorithm is shown in Figure 5.18a. Works by Toi and Ohita, Lee and Ortega, and Koh et al. [101–103] have demonstrated that by using a compression-first scheme as illustrated in Figure 5.18b, more pertinent information is retained, so a lower compression rate and higher image quality can be achieved.

The CMOS image sensor used in the wireless endoscopy system can output image data with a Bayer CFA format. In order to to meet the requirements of low-power design as well as high-compression performance, a novel near-lossless and lossless compression-first algorithm for the images with Bayer CFA is proposed for this system [104]. The architecture of the compression algorithm is illustrated in Figure 5.19. The CFA data are low-pass filtered directly in the RGB space. The G, R, and B components of the Bayer data are filtered by the low-pass filters I and II, respectively. The quincunx G component is then transformed into a rectangular array. Finally, three components are compressed using the JPEG-LS image file compression algorithm [105]. Although the introduction of the low-pass filters leads to a small loss of high frequency data, a good compression ratio comes out with high fidelity. The reconstructed image quality can be adjusted by changing the quality control factor. The specified region of interest (ROI) can be coded losslessly when adjusting the ROI parameters. The quality control factor and the ROI parameters are used as the input parameters of the low-pass filters. The corresponding decompression algorithm is shown in Figure 5.19b. It is a simple reverse procedure of the compression. The method can provide a lower bit rate (bits/pixel) and lower complexity than any conventional

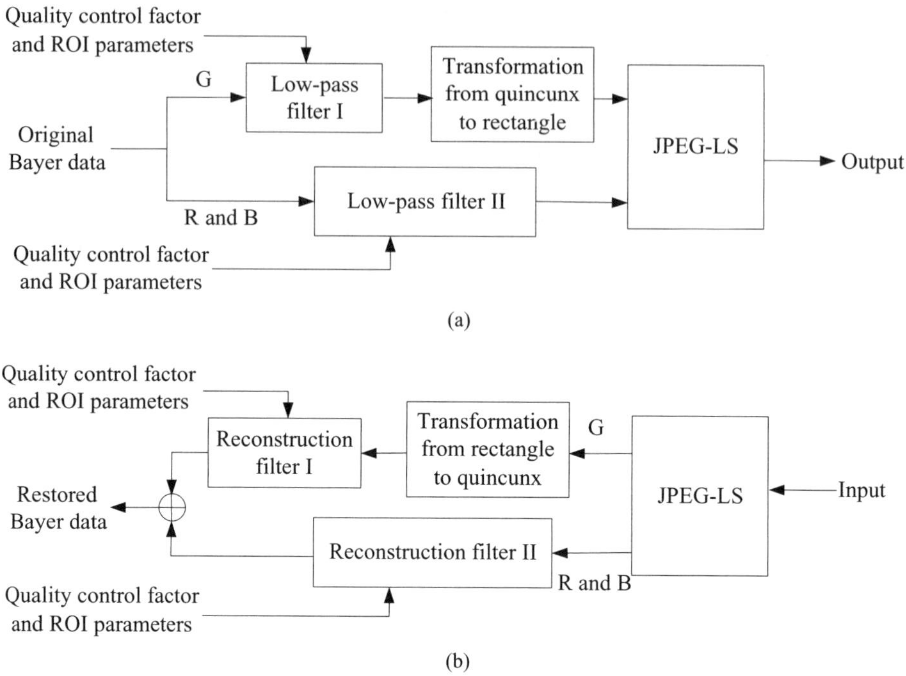

Figure 5.19. Block diagrams: (a) the proposed compression algorithm and (b) corresponding decompression algorithm.

interpolation-first methods [98–100] and other existing similar first algorithm presented [103] with high image quality.

The low-pass filters and the rounding operations in the near-lossless image compression algorithm are described by Xie et al. [89]. The lossless compression of the ROI is realized this way in that the pixels in the ROI are not filtered by the low-pass filters according to the ROI parameters. The ROI parameters contain information about the location and shape of the ROI in an image.

Compression performance. Table 5.1 illustrates a comparison of the following compression methods:

1. Proposed algorithm: It is proposed in this chapter.
2. JPEG-LS for CFA data: The CFA raw data are directly compressed via JPEG-LS. In this method, G, B, and R components are compressed by JPEG-LS separately.
3. JPEG-LS with near-parameter equal to two: Bayer data are compressed by a JPEG-LS near-lossless compression algorithm in which the near-parameter equals to two. In this algorithm, no pixel has an error of

TABLE 5.1. Compression Results for the Six Digestive Tract Images

	Image (256 × 256)	A	B	C	D	E	F
Proposed algorithm ($q = 1$)	PSNR[a]	46.891	46.882	46.921	46. 875	46.904	46.913
	CR[b]	1.962	2.176	2.223	2.256	2.081	2.411
Proposed algorithm ($q = 0.25$)	PSNR	52.815	52.798	53.006	52.962	52.991	53.017
	CR	3.095	3.331	3.581	3.636	3.209	3.752
JPEG-LS with near-parameter equal to two	PSNR	45.387	45.734	45.736	45.219	45.581	45.592
	CR	2.168	2.412	2.228	2.6311	2.174	2.415
Structure conversion method [103]	PSNR	51.731	51.698	51.812	51.296	51.799	51.825
	CR	3.815	4.022	4.206	4.318	4.031	4.527

[a] PSNR values in dB.
[b] CR values in bits per pixel.
CR: compression rate.

more than two intensity levels, which is the same as for the proposed algorithm.

4. Structure conversion method: This is realized according to the first algorithm presented by Koh et al. [103]. Note that in this method, JPEG-LS is used instead of JPEG.

Table 5.1 shows that the proposed compression algorithm has the best compression performance when six typical digestive tract images with 256 × 256 size are compressed. The average compression ratio can reach 2.18 bits/pixels with an approximate peak signal-to-noise ratio (PSNR) of 46.897 dB for digestive tract images. The proposed algorithm has the same complexity as the structure conversion method. In addition, the method provides lossless compression for the ROI. Note that no pixel has an error of more than two intensity levels in this compression algorithm.

VLSI Architecture of the Compression Algorithm. Preprocessor. A low-complexity VLSI architecture of the preprocessor is proposed as illustrated in Figure 5.20. The synchronous signals from the CMOS image sensor include vertical and horizontal synchronous outputs and pixel clock output. The synchronous signals are used to locate the current pixel, that is, the column and row information. According to this information, the control unit identifies which type (G or B/R) the current pixel is, and then controls the working states of filters I and II, as well as the read and write operations of the two-line SRAM.

The hardware overhead of the two filters is low. There is only one RAM read operation and four add operations for each pixel, and each pixel needs three 8-bit registers to store the value of its three neighbors. By this structure, the filters can

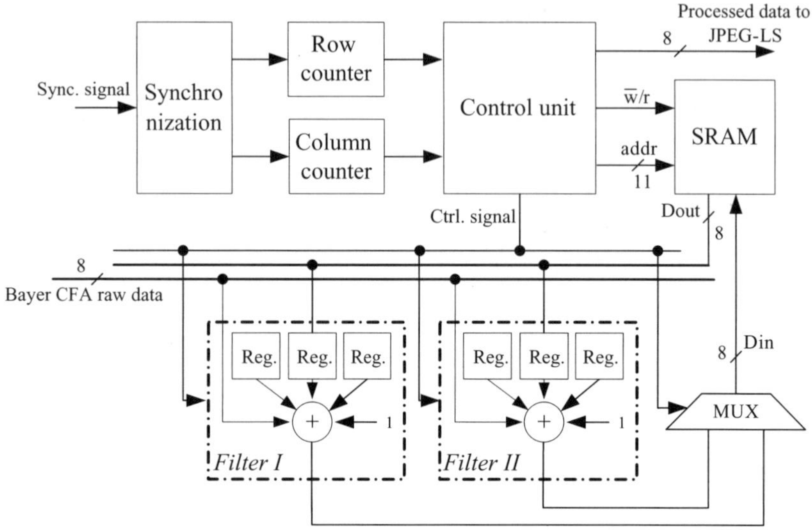

Figure 5.20. Architecture of the preprocessor.

access real-time data processing. The transformation and combination can be realized by controlling the SRAM write operation. Finally, the processed data are sent to the JPEG-LS engine for compression.

JPEG-LS engine. Although the JPEG-LS algorithm is easier for hardware implementation compared to many other lossless algorithms, its implementation for high-speed applications is difficult due to poor parallelizability. During context modeling, once two continuous pixels have the same context, the computation for the second pixel must wait for the completion of the first one, for the updating of the context history. In a work by DeMichele et al. [8], a limited parallelism is obtained only when the computations do not depend on previous ones. In Savakis and Pioriun, and Ferretti and Boffadossi [106, 107], more on-chip memory and logic gates are sacrificed in exchange for higher processing rates. However, none of these three methods can realize real-time data processing, and the methods in both Savakis and Pioriun, and Ferretti and Boffadossi [106, 107] suffer from degradation of compression ratios.

For the purpose of real-time data processing and low-power application, a fully pipelined architecture with a clock management scheme is proposed for the JPEG-LS engine, as shown in Figure 5.21. It consists of four parts: (1) a mode decision module; (2) a clock controller; (3) three parallel pipelines, including a regular pipeline, a run pipeline, and an interrupt pipeline; and (4) a two-tier data packer. These modules work in four different interlaced clock domains, the generations of which are under the control of the dedicated clock controller. The clock management scheme ensures the performance of bottleneck calculations,, helps reduce the clock frequency on noncritical paths, and controls the shutting

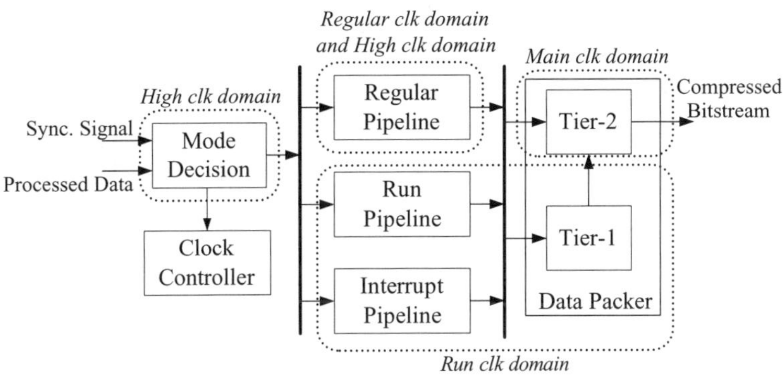

Figure 5.21. Proposed architecture for JPEG-LS encoder.

down of the working clock for each module so as to reduce the overall power consumption. With this pipelined architecture, real-time data processing, that is, 1 pixel/clock, can be achieved.

Mode decision module. The mode decision module works in a high frequency clock domain, which is four times that of the main working clock. It consists of a single port SRAM for buffering one line of a frame, and a finite-state machine (FSM) including four states. The data in the current line will be used at the same time as the neighboring pixels of data in the next line. A four-state FSM is applied for read/write control of the SRAM, while it selects a mode (regular, run, or interrupt) the current pixel should enter by the values of the neighboring pixels and informs the clock controller.

Clock controller. This JPEG-LS engine contains four different clock domains in all, which are indicated by dashed rectangles in Figure 5.21. The task of the clock controller is to control the generation and working states of the three clocks, that is, the main clk, regular clk, and run clk. The architecture of the clock controller is depicted in Figure 5.22.

The main clk is generated by four divisions of high clk through two T flip-flops, and then sent to the regular clk/run clk generation circuit for processing, respectively. The regular clk generation circuit is shown in the dashed block diagram. If a pixel in the regular mode occurs, the port "regular mode" valid would appear to be "1;" otherwise "0" would appear. Reg 1 to Reg n compose an n-bit right shift register. The information about the validness of regular mode is passed down through the right shift register and remains valid for n clock cycles because of the OR gate. The latch and the AND gate compose a typical clock gated circuit. The number of the stages in a regular pipeline is denoted by n. Thus, the clock generation circuit can provide adequate clocks for all the computations in a regular pipeline, as well as shut down the clock immediately unless another pixel in regular mode occurs. The run clk generation circuit works in the same way as the regular clk generation circuit, but it provides a clock for the run

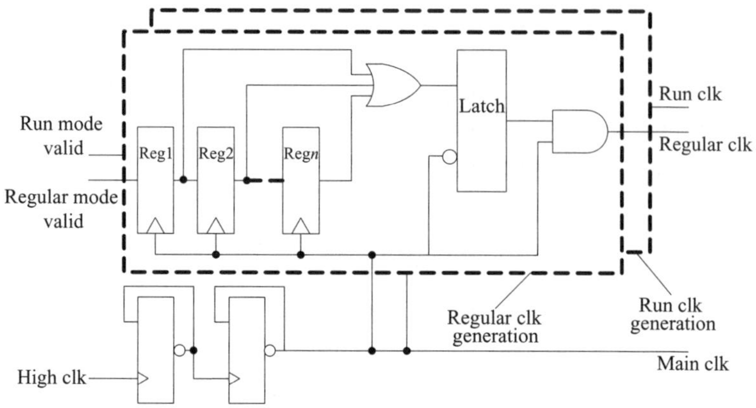

Figure 5.22. Architecture of the clock controller.

pipeline, the interrupt pipeline, and tier-1 of the data packer. The value of n is the overall latency of the three modules.

Parallel pipelines. In this design, the regular pipeline, the run pipeline, and the interrupt pipeline, which are parallel to each other, can work simultaneously to acheive a high data processing speed. The regular pipeline works in the regular clk domain, the other two in the run clk domain. Therefore, the idle pipeline can be shut down immediately through the clock management scheme previously mentioned while not affecting the normal working states of the other pipelines.

The regular pipeline is used for encoding the pixels in regular mode. It is arranged with three stages and has a latency of three main clk cycles (Fig. 5.23). In the first stage, the fixed prediction value and context of the current pixel are calculated. The second stage is divided into four small intervals working in the high clk domain, in which an FSM with four states is used to look up the context table, calculate the parameters for Golomb–Rice encoder [108], and update the context table. The context table, including variables A, B, C, and N, is built out of a single-port SRAM. A Golomb–Rice encoder is implemented in the third stage with two outputs generated, indicating the code and its valid length, respectively.

The latency of the run pipeline is one clock cycle, during which run length is encoded with two outputs generated, indicating the variable length code and the length of valid bits, respectively. The interrupt pipeline has a latency of three clock cycles, the tasks in each stage are similar to that in the regular pipeline. Since there exist only two contexts, registers are used instead of the SRAM in the second stage, and thus the high clk is unnecessary. The interrupt mode always follows the run mode, so both of the two pipelines work in the run clk domain.

Two-tier data packer. The data packer is used to convert variable length compressed data into a fixed-length compressed data stream, 32 bits in this design. Since an interrupt pixel always follows run length coding, tier-1 of the data packer is used to combine the run-length coding results with the Golomb coding results

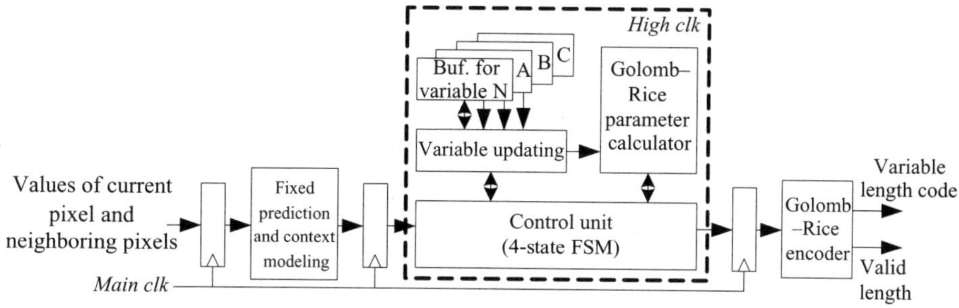

Figure 5.23. Block diagram of a regular pipeline.

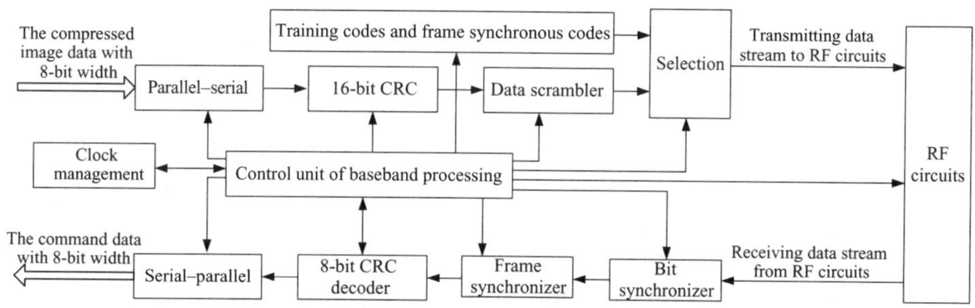

Figure 5.24. VLSI architecture of the baseband communication processing part.

of the interrupt pipeline. Tier-2 of the data packer, which works in the main clk domain, is used to combine the results of tier-1 with the Golomb coding results of the regular pipeline, and output 32-bit fixed-length code to the output bitstream.

Baseband Communication Processing Part. VLSI architecture of baseband communication processing is shown in Figure 5.24. The compressed Bayer CFA data are read out from the SRAM. Then, the data are protected by the International Telephone and Telegraph Consultative Committee (CCITT) CRC-16 frame checked sequence (FCS), of which undetected error probability can be neglected when the BER is smaller than 10^{-} [109]. To avoid long, continuous zeros or ones in the transmitted bits, the polynomial $G(x) = Z^{-7} + Z^{-4} + 1$ is applied here to scramble all bits transmitted. Before the image data are transmitted, the training codes and synchronous codes are transmitted first by the control unit of the baseband processing. The transmitted data rate is 2 MHz, and the work clock frequency of the transmitting circuits is also 2 MHz.

When the system inside the capsule is in the receiving command state, the data received by the RF circuits will be synchronized by a two-stage synchronizer, that is, a bit synchronizer and a frame synchronizer. The data are then sent into an 8-bit CRC decoder and a serial–parallel converter. Finally, the command is output to the control unit to be parsed. The received data rate is only 128 Kb/s. The work clock frequency of the receiving circuits can be decreased to 640 kHz to save power. The clock manage module will shut off the clocks of those modules in idle state.

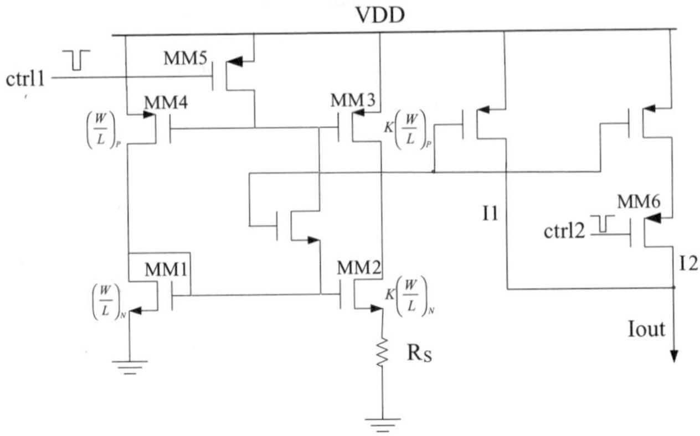

Figure 5.25. A microcurrent generator of stimulating muscle of digestive tract.

Drivers. Each LED is driven by an output buffer with a 24-mA CMOS output. The current stimulus drivers can generate current to stimulate the muscles of the digestive tract to contract, thereby pushing the capsule forward when the physician wants to bypass a region of the tract that is not of interest. Considering that the temperature inside the human body is almost invariable, the simple driver circuit as illustrated in Figure 5.25 is used. A control signal, "ctrl1," is applied to the gate of the pMOS transistor MM5 when the control signal is high and the output current is zero. Otherwise, the microcurrent "I_{out}" is output to stimulate the muscle. The pMOS transistor MM6 is used to control whether current "I1" is output or not. When control signal "ctrl2" is low, the output stimulus current is $I_{out} = I1 + I2$. Otherwise, the output current is I1, which is given by

$$I1 = I2 = \frac{2}{\mu_n C_{ox}(W/L)_N} \bullet \frac{1}{R_S^2}\left(1 - \frac{1}{\sqrt{K}}\right)^2, \tag{5.1}$$

where μ_n is the mobility of charge carriers of NMOS transistors and W, L, and C_{ox} represent the transistor's width, length, and gate capacitance per unit area, respectively. The W/L of the NM2 and NM3 are K multiples of the NM1 and NM4. In order to guarantee that microcurrent will not harm the patient, I1 and I2 are set at 10 mA by setting the parameters in Equation 5.1. So, the maximum stimulus current is 20 mA. The doctor can control the stimulus microcurrent intensity by external command.

5.4.5 Implementation Results

5.4.5.1 Field-Programmable Gate Array Verification System Design and Testing Results. Before designing the ASIC for the capsule, a field-programmable gate array (FPGA) verification system is built to test the whole

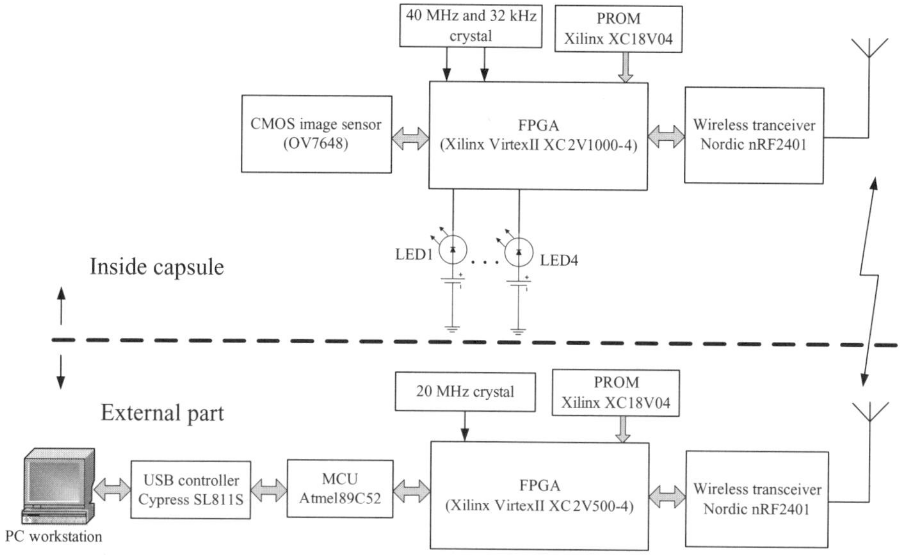

Figure 5.26. FGPA verification system.

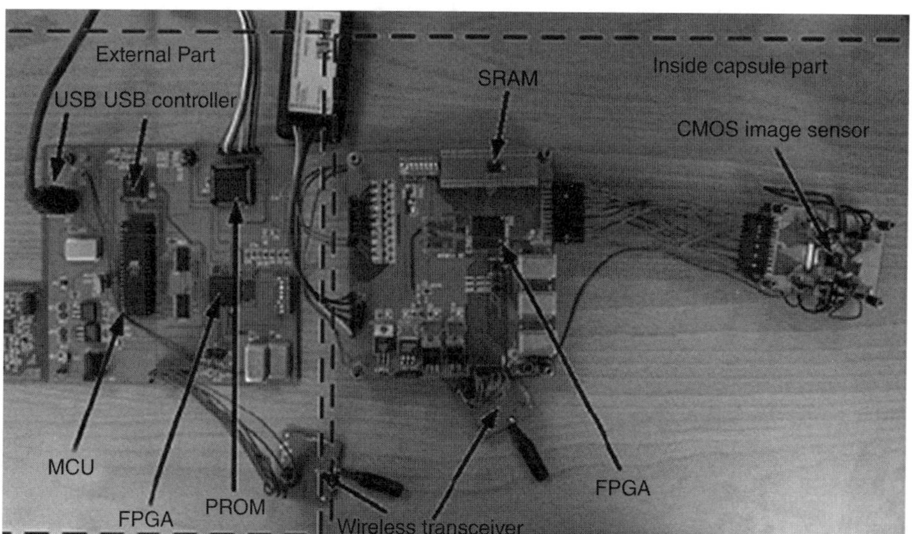

Figure 5.27. The photo of the FPGA testing system.

scheme of the wireless endoscopy system as illustrated in Figure 5.26. The photograph of the real PCB board is shown in Figure 5.27.

The testing system also verifies the low-power design and our power management strategies. Figure 5.28 shows the comparison results of the power consumption inside the capsule. The x-axis denotes time, and the y-axis denotes the voltage

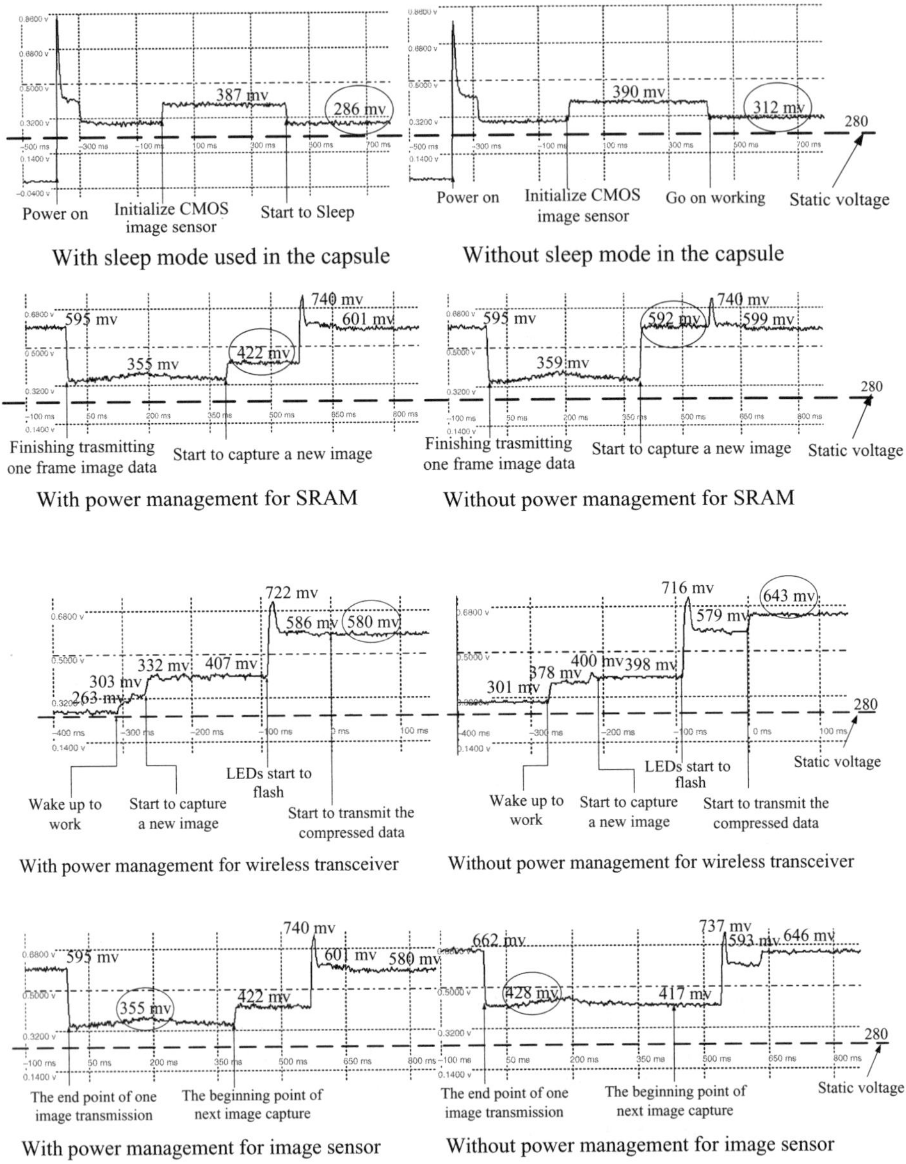

Figure 5.28. The comparison results of the power consumption inside the capsule.

of the sampling resistance. The voltage value shows the corresponding power consumed inside the capsule. Static voltage "280" in this figure denotes the power consumption of the FPGA system without our logical circuits. The measurement results show that the power management can save a great deal of power inside the capsule. Although the measurement cannot illustrate the power consumption inside the capsule accurately, it shows the obvious efficiency of the power man-

TABLE 5.2. Digital Chip Characteristics

Process	0.18-μm, Six-Layer CMOS
Sleep clock frequency	32 kHz
Work clock frequency	40 MHz
Die size	3.0 × 4.2 mm
Gate size	90k gates + 93.81 Kb SRAM
Power consumption in sleep mode	60 μW
Power consumption in work mode	6.2 mW (8 fps, 320 × 288)
Power supply	1.8 V (core)/3.3 V (I/O)

agement used inside the capsule. The simulation results also show that the low-power design of the digital circuits can save 46% of power in the capsule compared with a non-low-power design by using the power analysis software PrimePower.

5.4.5.2 Application-Specific Integrated Circuit Design and Results. The 0.18-μm CMOS process technology is used for the digital IC inside the capsule. The characteristics of the digital chip in the capsule are summarized in Table 5.2. The die area is 3 × 4.2 mm. Seven on-chip SRAMs are used to store one frame image with a maximum size of 320 × 288 = 92.16 Kb and JPEG-LS parameters of about 1.65 Kb. The amount of on-chip memory required is about 103 Kb. This occupies about 70% of the area of the chip. The power consumption in active mode is 6.2 mW with an 8 fps frame rate and 320 × 288 image size. The chip microphotograph is shown in Figure 5.29. A wireless endoscopic capsule prototype has been developed, as shown in Figure 5.30. Two button batteries, LEDs, the ASIC, image sensor, RF transceiver, and antenna are hermetically encapsulated within the package with a diameter of 11.3 mm and a length of 26.7 mm. Here, the button battery size limits the diameter of capsule. The prototype system performance has been fully characterized.

5.4.5.3 Estimation of the Continuous Working Hours. The average power consumption inside the capsule is less than that of a digital IC (6.2 mW) + RF IC (8.3 mW), which is 14.5 mW, and the power dissipation in the sleep mode, including the CMOS image sensor, is about 100 W. Assuming that the average capacity of one battery that can be easily bought in any market is approximately 70 mAh at 1.55 V, the continuous working time of the capsule is approximately 70 × 1.55 2/14.5 ≈ 15 hours. Note that the sleep mode in the capsule makes it possible for the whole digestive tract to be checked by the designed wireless endoscopic capsule. The wireless transceiver we have previously designed [79] has a small size (<2.8 × 2.5 mm) and the size of the digital IC is 3.0 × 4.2 mm, so all of the circuits including the CMOS image sensor can be easily put into such a small-size capsule (9 × 20 mm, if the full-custom battery could be obtained).

Compared with previous methods [2, 77], the proposed system architecture has many advantages, which are summarized in Table 5.3. The proposed digital

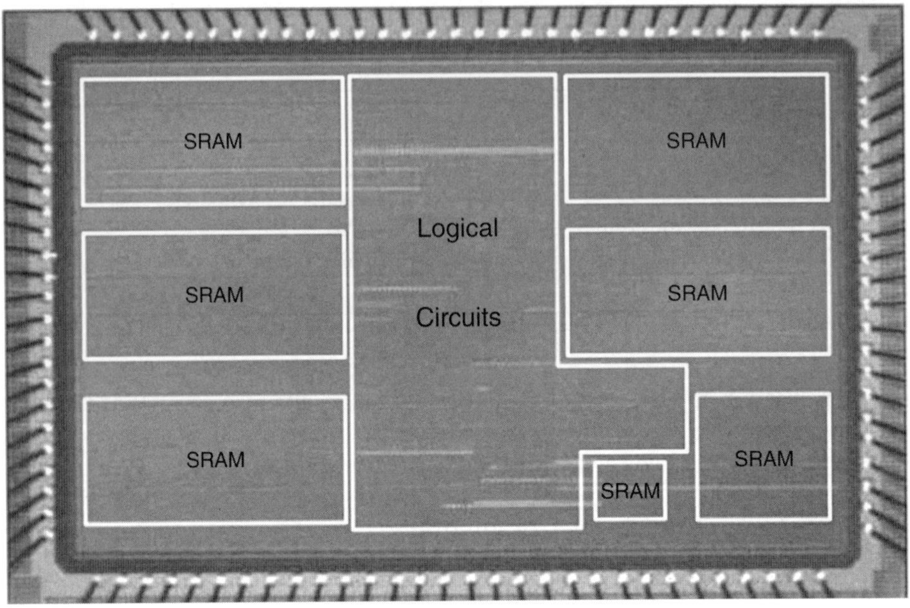

Figure 5.29. Die microphotograph of the digital IC inside the capsule.

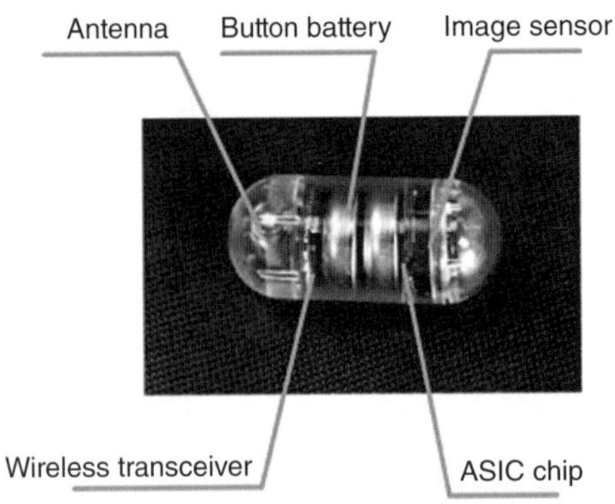

Figure 5.30. Photo of the wireless endoscopic capsule.

VLSI architecture brings more flexible, useful functions to our system than those by Park et al. [74], such as controllable image frame rate, image compression, controllable speed of the capsule movement, and two working modes. Moreover, because the ARQ communication scheme can be used in the proposed system, it provides higher quality image communication than that of the other two systems when the bit error probability of communication channel is larger than 10^{-4}. The proposed image compression algorithm gains about four times the image frame rate of that by Iddan et al. and Olympus [2, 77]. The proposed generator of the current can push the capsule forward when the doctor wants

TABLE 5.3. Comparison of Three Different Wireless Endoscopy Systems

Features	Given Imaging [2] PillCam SB	Olympus [77] Endocapsule	Proposed System
Image signal	Digital	Digital	Digital
Resolution	65,536 pixels	<65,536 pixels	92,160 pixels
Communication way	Unidirectional	Unidirectiona;	Bidirectional
ARQ scheme	No	No	Yes
Image frame rate	Fixed 2 fps	Fixed 2 fps	0~8 f/s, controlled
Image data	2D	2D	2D and 3D
Image compression	No	No	Lossless/near-lossless
Checked parts	Small intestine	Small intestine	All parts of the digestive tract
Continuous working hours	<8 hours	<8 hours	>14 hours
Speed of the capsule movement	No	No	Yes
Working mode [109]	Offline diagnosis	Offline, online diagnoses	Offline, online diagnoses

to neglect any parts not of interest, which gives our system more energy to check the areas of interest of the digestive tract than the other two systems have.

5.5 CONCLUSION

With the development of microelectronics, compact and high-energy batteries, hermetic packages, and wireless communication technologies, the implantable electronic device is playing a more and more important role in medical applications. Through the elaborate analyses of all kinds of typical implantable devices, this chapter extracted a general model of implantable electronic devices and discussed each module consisted of the implant unit in this model as well its low-power design. This chapter also concluded with a generic architecture of the digital circuit of the implant unit. Low-power design considerations of digital circuits in this architecture were discussed in detail.

Furthermore, a low-power digital IC design for wireless capsule endoscopy, a typical implantable medical device, was presented. In this low-power design case, a new architecture of the wireless endoscopic capsule system and an A/D mix-mode low-power IC architecture inside the endoscopic capsule for the diagnosis of the entire human digestive tract and real-time endoscopic image monitoring were proposed. Considering the characteristics of the wireless endoscopy system, the system-level low-power design strategies were given first. With these low-power strategies, the digital IC design for the wireless endoscopic capsule was discussed in detail. A novel architecture of three-stage clock management for low power was given. It can save 46% of the power in the capsule compared to a non-low-power design. To decrease the power consumption during sleep mode, a stoppable ring crystal oscillator with minimal overhead was applied in the designed system. A new image-compression algorithm based on the Bayer CFA format and its corresponding VLSI architecture were both proposed for low-power and low-communication data-rate application. The power simulation results showed that the system with the proposed image compression can save approximately 48% power compared with the system without image compression. All of these low-power designs in the digital IC have been verified in the FPGA system. A digital chip was implemented in the 0.18-m CMOS process, and its power consumption was 6.2 mW, which assures that the continuous work time of the capsule is at least 15 hours. The low-power and small-size chip make it possible for the whole digestive tract to be checked, and the capsule size will be smaller than 9×20 mm if a full-custom battery can be obtained.

REFERENCES

[1] S. Reichelt, J. Fiala, A. Werber, K. Förster, C. Heilmann, R. Klemm, and H. Zappe, "Development of an implantable pulse oximeter," *IEEE Trans. Biomed. Eng.*, 55(2), pp. 581–588, 2008.

[2] G. Iddan, G. Meron, A. Glukhovsky, et al., "Wireless capsule endoscopy," *Nature*, 405, pp. 417–418, 2000.

[3] S. Sarkar, D. Ritscher, and R. Mehra, "A detector for a chronic implantable atrial tachyarrhythmia monitor," *IEEE Trans. Biomed. Eng.*, 55(3), pp. 1219–1224, 2008.

[4] P. Cong, D. J. Young, B. Hoit, and W. H. Ko, "Novel long-term implantable blood pressure monitoring system with reduced baseline drift," *Proceedings of the 28th IEEE EMBS Annual International Conference, USA*, pp. 1854–1857, August 2006.

[5] J. Ohta, A. Higuchi, A. Tagawa, et al., "An implantable CMOS image sensor for monitoring deep brain activities of a freely moving mouse," *IEEE Biomedical Circuits and Systems Conference, BioCAS*, pp. 269–272, 2008.

[6] W. F. Agnew and D. B. McCreery, Eds., *Neural Prostheses—Fundamental Studies*. Englewood Cliffs, NJ: Prentice-Hall, 1990.

[7] K. D. Wise, D. J. Anderson, J. F. Hetke, D. R. Kipke, and K. Najafi, "Wireless implantable microsystems: High-density electronic interfaces to the nervous system," *Proc. IEEE*, 92(1), pp. 76–97, 2004.

[8] G. A. DeMichele, P. R. Troyk, D. A. Kerns, and Weir R., "An implantable myoelectric sensor based prosthesis control system, engineering in medicine and biology society, 2006," *EMBS '06. 28th Annual International Conference of the IEEE*, pp. 2970–2973, August 30–September 3, 2006.

[9] C. Son and B. Ziaie, "Wireless implantable passive microdosimeter for radiation oncology," *IEEE Trans. Neural Syst. Rehabil. Eng.*, 55(6), pp. 1772–1775, 2008.

[10] S. R. Duggan, "Drug administration device," U.S. Patent: 4 692 147, September 8, 1987.

[11] D. A. LaVan, T. McGuire, and R. Langer, "Small-scale systems for *in vivo* drug delivery," *Nat. Biotechnol.*, 21(10), pp. 1184–1191, 2003.

[12] M. Ghovanloo and K. Najafi, "A wireless implantable multichannel microstimulating system-on-a-chip with modular architecture," *IEEE Trans. Neural Syst. Rehabil. Eng.*, 15(3), pp. 449–457, 2007.

[13] M. Mazza, P. Renaud, D. C. Bertrand, and A. M. Ionescu, "CMOS pixels for subretinal implantable prosthesis," *IEEE Sens. J.*, 5(1), pp. 32–37, 2005.

[14] R. E. Isaacs, D. J. Weber, and A. B. Schwartz, "Work toward real-time control of a cortical neural prosthesis," *IEEE Trans. Rehabil. Eng.*, 8(2), pp. 196–198, 2000.

[15] R. S. Sanders and M. T. Lee, "Implantable pacemakers," *Proc. IEEE*, 84(8), pp. 480–486, 1996.

[16] F. A. Spelman, "The past, present, and future of cochlear prostheses," *IEEE Eng. Med. Biol. Mag.*, 18(3), pp. 27–33, 1999.

[17] Freedonia Group, *Implantable Medical Devices to 2011*. Freedonia Group, 2007.

[18] L. S. Y. Wong, S. Hossain, A. Ta, J. Edvinsson, D. H. Rivas, and H. Naas, "A very low-power CMOS mixed-signal IC for implantable pacemaker applications," *IEEE J. Solid-State Circuits*, 39(12), pp. 2446–2456, 2004.

[19] C.-J. Cheng, C.-J. Wu, and S.-Y. Lee, "Programmable pacing channel with a fully on-chip LDO regulator for cardiac pacemaker," *IEEE Asian Solid-State Circuits Conference*, pp. 285–288, November 2008.

[20] H. Makino, Y. Saitoh, Y. Mitamura, and T. Mikami, "Implantable defibrillator with high-output pacing function after defibrillation," *Proc. IEEE*, 76(9), pp. 1187–1193, 1988.

[21] J. A. Warren, R. D. Dreher, R. V. Jaworski, J. J. Putzke, and R. J. Russie, "Implantable cardioverter defibrillators," *Proc. IEEE*, 84(3), pp. 468–479, 1996.

[22] T. J. Gale, "Modeling defibrillation electrode performance," *27th Annual International Conference of the Engineering in Medicine and Biology Society, 2005*, pp. 3885–3888, January 2006.

[23] F.-G. Zeng, S. Rebscher, W. Harrison, X. Sun, and H. Feng, "Cochlear implants: System design, integration, and evaluation," *IEEE Rev. Biomed. Eng.*, 1, pp. 115–142, 2008.

[24] J. N. Fayad, S. R. Otto, R. V. Shannon, and D. E. Brackmann, "Cochlear and brainstem auditory prostheses 'neural interface for hearing restoration: Cochlear and brain stem implants'," *Proc. IEEE*, 96(7), pp. 1085–1095, 2008.

[25] E. Zrenner, "Will retinal implants restore vision?" *Science*, 295, pp. 1022–1025, 2002.

[26] J. D. Weiland and M. S. Humayun, "Visual prosthesis," *Proc. IEEE*, 96(7), pp. 1076–1084, 2008.

[27] C.-C. Wang, T.-J. Lee, Y.-T. Hsiao, U. F. Chio, C.-C. Huang, J.-J. J. Chin, and Y.-H. Hsueh, "A multiparameter implantable microstimulator SOC," *IEEE Trans. Very Large Scale Integration (VLSI) Syst.*, 13(12), pp. 1399–1402, 2005.

[28] J. Mou'ine, K. A. Ammar, and Z. Chtourou, "A completely programmable and very flexible implantable pain controller," *Proceedings of the 22nd Annual International Conference of the IEEE Engineering in Medicine and Biology Society, 2000*, 2, pp. 1104–1107, July 2000.

[29] B. J. Wenzel, J. W. Boggs, K. J. Gustafson, and W. M. Grill, "Detecting the onset of hyper-reflexive bladder contractions from the electrical activity of the pudendal nerve," *IEEE Trans. Neural Syst. Rehabil. Eng.*, 13(3), pp. 428–435, 2005.

[30] J. Mou'ine, D. Brunner, and Z. Chtourou, "Design and implementation of a multichannel urinary incontinence prosthesis," *Proceedings of the 22nd Annual International Conference of the IEEE Engineering in Medicine and Biology Society, 2000*, 2, pp. 1100–1103, July 2000.

[31] R. Sarpeshkar, W. Wattanapanitch, S. K. Arfin, B. I. Rapoport, S. Mandal, M. W. Baker, M. S. Fee, S. Musallam, and R. A. Andersen, "Low-power circuits for brain–machine interfaces," *IEEE Trans. Biomed. Circuits Syst.*, 2(3), pp. 173–183, 2008.

[32] R. H. Olsson and K. D. Wise, "A three-dimensional neural recording microsystem with implantable data compression circuitry," *IEEE J. Solid-State Circuits*, 40(12), pp. 2796–2804, 2005.

[33] J. N. Y. Aziz, K. Abdelhalim, R. Shulyzki, et al., "256-channel neural recording and delta compression microsystem with 3D electrodes," *IEEE J. Solid-State Circuits*, 44(3), pp. 995–1005, 2009.

[34] M. Min and T. Parve, "Improvement of lock-in electrical bio-impedance analyzer for implantable medical devices," *IEEE Trans. Instrum. Meas.*, 56(3) 2007, pp. 968–974.

[35] B. Palan, K. Roubik, M. Husak, and B. Courtois, "CMOS ISFET-based structures for biomedical applications," *1st Annual International Conference Microtechnologies in Medicine and Biology*, Lyon, France, pp. 502–506, October 2000.

[36] A. Errachid, P. Godignon, A. Ivorra, J. Millan, J. Aguilo, and J. Baucells, "Implementation of multisensor silicon needles for cardiac application," *International Semiconductor Conference, CAS 2001 Proceedings*, 1, pp. 273–276, October 2001.

[37] D. Maillefer, H. van Lintel, G. Rey-Mermet, and R. Hirschi, "A high-performance silicon micropump for an implantable drug delivery system," *Proceedings of the 12'h IEEE MEMS 1999 Technical Digest*, Orlando, FL, 1/17–21/99, pp. 541–546, 1999.

[38] X. Xie, C. Zhang, and Z., "A review of the implantable electronic devices in biology and medicine," *Chin. J. Electron.*, 32(3), pp. 462–467, 2004.

[39] K. Goto, T. Nakagawa, and S. Kawata, "An implantable power supply with an optically rechargeable lithium battery," *IEEE Trans. Biomed. Eng.*, 48(7), pp. 830–833, 2001.

[40] S.-Y. Lee and S.-Y. Lee, "An implantable wireless bidirectional communication microstimulator for neuromuscular stimulation," *IEEE Trans. Circuits Syst. I Regular Pap.*, 52(12), pp. 2526–2538, 2005.

[41] Z. Yang, W. Liu, and E. Basham, "Inductor modeling in wireless links for implantable electronics," *IEEE Trans. Magn.*, 43(10), pp. 3851–3860, 2007.

[42] S. J. Rebscher, M. Heilmann, W. Bruszewski, N. H. Talbot, R. L. Snyder, and M. M. Merzenich, "Strategies to improve electrode positioning and safety in cochlear implants," *IEEE Trans. Biomed. Eng.*, 46(3), pp. 340–352, 1999.

[43] D. K. Peterson, M. L. Nochomovitz, T. A. Stellato, and J. T. Mortimer, "Long-term intramuscular electrical activation of the phrenic nerve: Safety and reliability," *IEEE Trans. Biomed. Eng.*, 41(12), pp. 1115–1126, 1994.

[44] S. Itoh, S. Kawahito, and S. Terakawa, "A CMOS one-chip wireless camera with digital image transmission function for capsule endoscopes," *IEEJ Trans. SM*, 126(7), pp. 318–324, 2006.

[45] Y. Hu and M. Sawan, "CMOS front-end amplifier dedicated to monitor very low amplitude signal from implantable sensors, circuits and systems," *Proceedings of the 43rd IEEE Midwest Symposium*, Lansing MI, pp. 298–301, January 2000.

[46] R. R. Harrison and C. Charles, "A low-power low-noise CMOS amplifier for neural recording applications," *IEEE J. Solid-State Circuits*, 38(6), pp. 958–965, 2003.

[47] K. Papathanasiou and T. L. Ehmann, "An implantable CMOS signal conditioning system for recording nerve signals with cuff electrodes, circuits and systems," *Proceedings ISCAS 2000, The 2000 IEEE International Symposium*, Geneva, Switzerland, pp. 281–284, 2000.

[48] T. M. Seese, H. Harasaki, G. M. Saidel, and C. R. Davies, "Characterization of tissue morphology, angiogenesis, and temperature in the adaptive response of muscle tissue to chronic heating," *Lab. Invest.*, 78(12), pp. 1553–1562, 1998.

[49] P. Mohseni and K. Najafi, "A fully integrated neural recording amplifier with DC input stabilization," *IEEE Trans. Biomed. Eng.*, 51(5), pp. 832–837, 2004.

[50] G. Bonfini, A. S. Brogna, C. Garbossa, L. Colombini, M. Bacci, S. Chicca, F. Bigongiari, N. C. Guerrini, and G. Ferri, "An ultralow-power switched opamp-based 10-B integrated ADC for implantable biomedical applications," *IEEE Trans. Circuits Syst. I Regular Pap.*, 51(1), pp. 174–177, 2004.

[51] P.-Y. Robert, B. Gosselin, A. E. Ayoub, and M. Sawan, "An ultra-low-power successive-approximation-based ADC for implantable sensing devices" *49th IEEE International Midwest Symposium on Circuits and Systems, MWSCAS '06*, 1, pp. 7–11, August 2006.

[52] K. Abdelhalim, L. MacEachern, and S. Mahmoud, "A nanowatt successive approximation ADC with offset correction for implantable sensor applications," IEEE International Symposium on Circuits and Systems, ISCAS 2007, pp. 2351–2354, May 2007.

[53] Y.-H. Liu, C.-J. Tung, and T.-H. Lin, "A low-power asymmetrical MICS wireless interface and transceiver design for medical imaging," *IEEE Biomedical Circuits and Systems Conference, BioCAS 2006*, pp. 162–165, November 29–December 1, 2006.

[54] D. Panescu, "Emerging technologies [wireless communication systems for implantable medical devices]," *IEEE Eng. Med. Biol. Mag.*, 27(2), pp. 96–101, 2008.

[55] A. M. Sodagar and K. Najafi, *Wireless interfaces for implantable biomedical microsystems*, 49th IEEE International Midwest Symposium on Circuits and Systems, MWSCAS '06, 2, pp. 265–269, August 2006.

[56] P. D. Bradley, "An ultra low power, high performance medical implant communication system (MICS) transceiver for implantable devices," *IEEE Biomedical Circuits and Systems Conference, BioCAS 2006*, pp. 158–161, November 29–December 1, 2006.

[57] D. W.-Y. Chung, A. C. Albason, A. S.-L. Lou, and A. Hu, *A 2-MHz Bi-directional wireless transceiver system for implantable glucose sensors*, IEEE International Symposium on Circuits and Systems, pp. 645–648, May 2009.

[58] S. Mandal and R. Sarpeshkar, "Power-efficient impedance-modulation wireless data links for biomedical implants," *IEEE Trans. Biomed. Circuits Syst.*, 2(4), pp. 301–315, 2008.

[59] D. Halperin, T. Kohno, T. Kohno, K. Fu, and W. H. Maisel, "Security and privacy for implantable medical devices," *IEEE Pervasive Comput.*, 7(1), pp. 30–39, 2008.

[60] K. Y. Yazdandoost and R. Kohno, "An antenna for medical implant communications system," *European Microwave Conference*, pp. 968–971, Oct. 2007.

[61] T. Karacolak, R. Cooper, and E. Topsakal, "Electrical properties of rat skin and design of implantable antennas for medical wireless telemetry," *IEEE Trans. Antennas Propagation*, 57(9), pp. 2806–2812, 2009.

[62] S. Soora, K. Gosalia, M. S. Humayun, and G. Lazzi, "A comparison of two and three dimensional dipole antennas for an implantable retinal prosthesis," *IEEE Trans. Antennas Propagation*, 56(3), pp. 622–629, 2008.

[63] P. Gerrish, E. Herrmann, L. Tyler, and K. Walsh, "Challenges and constraints in designing implantable medical ICs," *IEEE Trans. Device Mater. Reliability*, 5(3), pp. 435–444, 2005.

[64] "IEEE standard for safety levels with respect to human exposure to radio frequency electromagnetic fields, 3 kHz to 300 GHz," IEEE C95.1-1991, pp. P13, P20, 2006.

[65] D. C. Suresh, Agrawal B., Najjar W., and Yang J., "A tunable bus encoder for off-chip data buses," *Proceedings of the 2005 International Symposium on Low Power Electronics and Design, ISLPED '05*, pp. 319–322, August 2005.

[66] A. M. Kamboh, M. Raetz, K. G. Oweiss, and A. Mason, "Area-power efficient VLSI implementation of multichannel DWT for data compression in implantable neuroprosthetics," *IEEE Trans. Biomed. Circuits Syst.*, 1(2), pp. 128–135, 2007.

[67] L. A. Koyrakh, "Data compression for implantable medical devices," *Comput. Cardiol.*, 35, pp. 417–420, 2008.

[68] D. L. Donoho, "Compressed sensing," *IEEE Trans. Inf. Theory*, 52(4), pp. 1289–1306, 2006.

[69] M. Aghagolzadeh and K. Oweiss, "Compressed and distributed sensing of neuronal activity for real time spike train decoding," *IEEE Trans. Neural Syst. Rehabil. Eng.*, 17(2), pp. 116–127, 2009.

[70] M. A. Lebedev and M. A. L. Nicolelis, "Brain–machine interfaces: Past, present and future," *Trends Neurosci.*, 29(9), pp. 536–546, 2006.

[71] Philips Semiconductors, "The I2C-Bus specification V-2.0, Philips semiconductors," 1998.

[72] L. Benini, A. Bogliolo, and G. D. Michelli, "Battery-driven dynamic power management," *IEEE Des. Test Comput.*, 18(4), pp. 53–60, 2001.

[73] L. Benini, A. Bogliolo, and G. De Micheli, "A survey of design techniques for system-level dynamic power management," *IEEE Trans. Very Large Scale Integration (VLSI) Syst.*, 8(3), pp. 299–316, 2000.

[74] H. J. Park, H. W. Nam, B. S. Song, et al., "Design of bi-directional and multi-channel miniaturized telemetry module for wireless endoscopy," *Microtechnologies in Medicine & Biology 2nd Annual International IEEE-EMB Special Topic Conference*, Madison, WI, pp. 273–276, 2002.

[75] Rf.Co., Ltd., "RF—The next generation capsule endoscope Sayaka," http://www.rfamerica.com/sayaka/index.html.

[76] Intromedic Co., Ltd., "Capsule endoscopy," http://www.intromedic.com/.

[77] Olympus Co., Ltd., "Capsule endoscopy system," http://www.olympuschina.com/productContent/index.aspx?nodeid=75.

[78] P. Swain, "The future of wireless capsule endoscopy," *World J. Gastroenterol.*, 14(26), pp. 4142–4145, 2008.

[79] X. Xie, G. Li, X. K. Chen, et al., "A novel low power IC design for bi-directional digital wireless endoscopy capsule system," 2004 IEEE International Workshop on Biomedical Circuits and Systems (BioCAS04), Singapore, pp. S1.8.5–S1.8.8, 2004.

[80] M. Hammes, C. Kranz, D. Seippel, et al., "Evolution on SoC integration: GSM baseband-radio in 0.13 μm CMOS extended by fully integrated power management unit," *IEEE J. Solid-State Circuits*, 43(1), pp. 236–244, 2008.

[81] P. Hazucha, T. Karnik, B. A. Bloechel, et al., "Area-efficient linear regulator with ultra-fast load regulation," *IEEE J. Solid-State Circuits*, 40(4), pp. 933–939, 2006.
[82] G. A. Rincon-Mora and P. A. Allen, "A low-voltage, low quiescent current, low drop-out regulator," *IEEE J. Solid-State Circuits*, 33(1), pp. 36–44, 1998.
[83] K. N. Leung and P. K. T. Mok, "A capacitor-free CMOS low-dropout regulator with damping-factor-control frequency compensation," *IEEE J. Solid-State Circuits*, 38(10), pp. 1691–1702, 2003.
[84] C. Fiocchi and U. Gatti, "A very flexible BiCMOS low-voltage high-performance source follower," *Proceedings of the 1999 IEEE International Symposium on Circuits and Systems, ISCAS '99*, 2, pp. 212–215, June 1999.
[85] B. K. Ahuja, "An improved frequency compensation technique for CMOS operational amplifiers," *IEEE J. Solid-State Circuits*, 18(6), pp. 629–633, 1983.
[86] A. Stammermann, L. Kruse, W. Nebel, et al., "System level optimization and design space exploration for low power," *Proceedings of the 14th International Symposium on Systems Synthesis*, pp. 142–146, 2001.
[87] A. Raghunathan, N. K. Niraj, and S. Dey, *High-Level Power Analysis and Optimization*. Boston: Kluwer, 1998.
[88] Z. Cao and Y. Qian, *Modern Communication Theory*. Beijing, China: Tsinghua University Press, 1992.
[89] X. Xie, G. L. Li, X. W. Li, et al., "A new high quality image compression method for digital image sensors with Bayer color filter arrays," 2004 IEEE International Workshop on Biomedical Circuits and Systems (BioCAS04), Singapore, pp. S3.3-13–S3.3-16, 2004.
[90] X. Xie, G. Li, and Z. Wang, "ARQ scheme with adaptive block size for bidirectional wireless endoscopy system," *IEEE Proc. Commun.*, 153(5), pp. 611–618, 2006.
[91] D. Shear, "*Exorcise metastability from your design*," Elec. Design News, December 10, 1992.
[92] C. Brown and K. Feher, "Measuring metastability and its effect on communication signal processing systems," *IEEE Trans. Instrum. Meas.*, 46(1), pp. 61–64, 1997.
[93] M.-J. E. Lee, W. J. Dally, P. Chiang, et al., "Low-power area-efficient high-speed I/O circuit techniques," *IEEE J. Solid-State Circuits*, 35(11), pp. 1591–1599, 2000.
[94] R. J. Fletcher, "Integrated circuit having outputs configured for reduced state changes," U.S. Patent 4 667 337, May 1987.
[95] M. R. Stan and W. P. Burleson, "Bus-invert coding for low power I/O," *IEEE Trans. VLSI Syst.*, 3, pp. 49–58, 1995.
[96] J. M. Rabaey, "Exploring the power dimension," *Proceedings of the Custom Integrated Circuit Conference*, pp. 215–220, May 1996.
[97] B. E. Bayer, "Color imaging array," U.S. Patent 3,971,065, 1976.
[98] C. Herley, "A post-processing algorithm for compressed digital camera images, image processing, 1998. ICIP 98," *Proceedings*, 1, pp. 396–400, 1998.
[99] K. Yoon, C. Kim, B. Lee, et al., "Single-chip CMOS image sensor for mobile applications," *IEEE J. Solid-State Circuits*, 37(12), pp. 1839–1845, 2002.
[100] Y. H. Jung, J. S. Kim, et al., "Design of real-time image enhancement preprocessor for CMOS image sensor," *IEEE Trans. Consum. Electron.*, 46(1), pp. 68–75, 2000.
[101] T. Toi and M. Ohita, "A subband coding technique for image compression in single CCD cameras with Bayer color filter arrays," *IEEE Trans. Consum. Electron.*, 45(1), pp. 176–180, 1999.

[102] S. Y. Lee and A. Ortega, "A novel approach of image compression in digital cameras with a Bayer color filter array," *IEEE Int. Conf. Image Process.*, 3, pp. 482–485, 2001.

[103] C. C. Koh, J. Mukherjee, and S. K. Mitra, "New efficient methods of image compression in digital cameras with color filter array," *IEEE Trans. Consum. Electron.*, 49(4), pp. 1448–1456, 2003.

[104] X. Xie, G. L. Li, and Z. H. Wang, "Low-complexity and high-efficiency image compression algorithm for wireless endoscopy system," *J. Electron. Imaging*, 15(2), pp. 023017-1-15, 2006.

[105] M. Weinberger, G. Seroussi, and G. Sapiro, "The LOCO-I lossless image compression algorithm: Principles and standardization into JPEG-LS," *IEEE Trans. Image Process.*, 9(8), pp. 1309–1324, 2000.

[106] A. Savakis and M. Pioriun, "Benchmarking and hardware implementation of JPEG-LS, ICIP'02," Rochester, NY, 2, pp. II-949–II-952, September 2002.

[107] M. Ferretti and M. Boffadossi, "A parallel pipelined implementation of LOCO-I for JPEG-LS, ICPR'04," 1, pp. 769–772, 2004.

[108] S. W. Golomb, "Run-length encodings," *IEEE Trans. Inf. Theory*, IT-12, pp. 399–401, 1966.

[109] T. Fujiwara, T. Kasami, A. Kitai, et al., "On the undetected error probability for shortened hamming codes," *IEEE Trans. Commun.*, 33(6), pp. 570–574, 1985.

PART II

BIOSENSORS AND CIRCUITS

6

AFFINITY-BASED BIOSENSORS: STOCHASTIC MODELING AND FIGURES OF MERIT

Shreepriya Das, Haris Vikalo, and Arjang Hassibi

Biosensors, broadly defined, are sensors specifically designed to identify and estimate biological analytes. Along with the advancement of the biotechnology industry, biosensor systems have also witnessed a significant growth not only in terms of the number of applications but also in robustness, accuracy, and cost-efficiency. The most prominent biosensor systems include metabolite testers (e.g., blood glucose monitoring sensors), DNA and protein microarrays, and a plethora of environmental sensors. In this chapter, we focus on affinity-based biosensors and provide a detailed description of the biophysical processes that occur within these platforms. Our primary focus is to examine the statistical and computational methods to model and simulate these processes. Based on our formulations, we will also propose figures of merits that can be used to evaluate and compare the performance of biosensors. One important motivation behind this approach is to connect the models and figures of merits to the front-end electronic interfaces that most biosensor platforms have.

6.1 MODELING BIOSENSORS: INTRODUCTION

Affinity-based biosensors use selective binding and interaction between certain biomolecules (recognition probes) and specific target analytes to determine the presence of the latter in the biological samples. The essential role of the biosensor

CMOS Biomicrosystems: Where Electronics Meet Biology, First Edition. Edited by Krzysztof Iniewski.

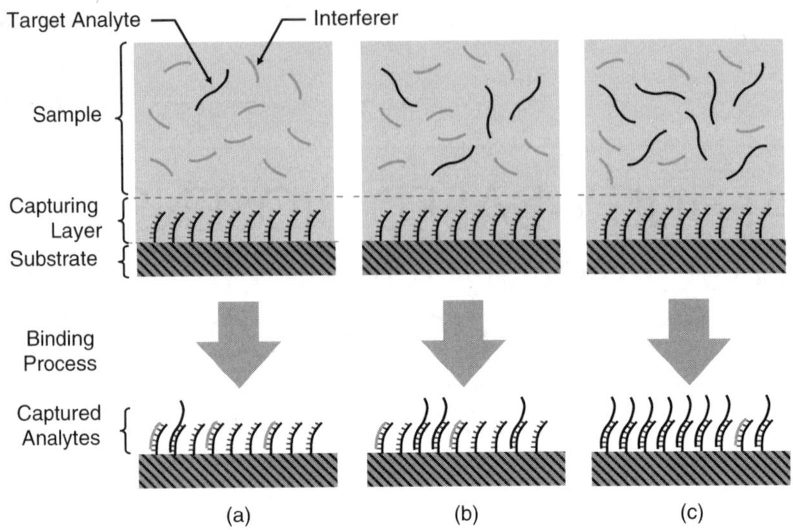

Figure 6.1. Capturing of analytes by the probes in biosensors can be in the (A) interference-limited, (B) target analyte-limited, or (C) saturation region.

platform (e.g., microarrays [1,2] and immunoassays [3]) is to facilitate the binding of the probe–target complexes to produce a detectable signal, which correlates with the presence of the target and, conceivably, its abundance. The minimal components required for affinity-based detection include a molecular recognition layer that includes the capturing probes, integrated within or intimately associated with a signal-generating physicochemical transducer [4], and a readout device.

To generate a target-specific signal, the target analytes in the sample volume first need to collide with the recognition layer, interact with the probes, and ultimately take part in a transduction process (Fig. 6.1). The analyte motion in typical biosensor settings (e.g., aqueous biological mediums) is dominated by diffusive spreading, which from a microscopic point of view is a probabilistic mass transfer process [5]. Accordingly, the analyte collisions with the probes become stochastic processes. Moreover, because of the quantum mechanical nature of chemical bond forming [6–9], interactions between the probes and the analyte molecules are also probabilistic, adding more uncertainty to the biosensing procedure. On top of these two processes, there is also the transducer and readout circuitry, which adds more noise to the already uncertain process.

Besides the inevitable uncertainty associated with the binding, transduction, and readout in all practical biosensors, binding of other species to the probes (nonspecific binding) is also possible. Nonspecific binding, for example, cross-hybridization in DNA microarrays, limits the minimum detectable level (MDL) [10–14] of the biosensor system. It is generally less probable than the specific binding when the target analyte and the interfering species have the same abundance; nevertheless, if the concentration of the nonspecific analyte becomes much higher than the target analyte, nonspecific bindings may dominate the measured signal and can even lead to false positives. It is fair to say that even the best biosensors today have significant amounts of "background" signal that can be associated to nonspecific bindings and interactions. Hence, reduction or accurate estimation of this background is one of the foremost challenges in this field.

The analysis of signals in biosensors has been traditionally focused on computing the ensemble (expected) behavior of the capturing process. These approaches have utilized computational fluid dynamic (CFD) techniques, molecular dynamic methods, and stoichiometric reaction rates to formulate and predict the kinetics of bindings and the evolution of the analytes' spatial concentration during the biosensing process [15–24]. While these approaches and models provide great insight into the binding kinetics, response time, and mass transport processes within biosensors systems, they do not offer quantitative analysis related to the uncertainty, noise, experiment-to-experiment variations, and ultimately, signal-to-noise ratio (SNR) of biosensors. The main reason behind this limitation is the fact that such models are constructed based on macroscopic formulations (e.g., diffusive spreading) that use continuous approximations for inherently discrete parameters such as analyte concentrations. Furthermore, the stochastic characteristics of mass transport and binding that are the main sources of uncertainty, are *averaged out* by utilizing deterministic formulations. While the analysis of noise in the context of biosensors has been a relatively recent effort, there has been extensive work on the stochastic nature of biochemical and molecular reactions in biophysics and systems biology [25–34]. Merging of these two is a promising approach toward creating a comprehensive model for biosensor systems that not only predicts the expected behavior but also the statistics of the signals.

6.2 BIOSENSOR MODEL: DETERMINISTIC AND STOCHASTIC

6.2.1 Analyte Motion: Brownian Motion

Molecules, cells, and many other analytes immersed in the aqueous mediums of biosensor platforms are subject to thermal fluctuations; and, in particular, sensory platforms such as electrophoresis systems are subject to electromagnetic or mechanical forces. While mechanical movement (e.g., convection) and electromagnetic forces (e.g., electrophoresis) are (in most cases) deterministic, thermal fluctuations are random in nature. These thermal fluctuations of a particle from a microscopic point of view is nothing but Brownian motion, and this results in a diffusive spreading phenomena in macroscopic systems.

The deterministic formulation for the diffusion equation of analyte particles with spatial concentration of $C(X,t)$, where X is a three-dimensional vector representing the three spatial coordinates, follows Fick's law of diffusion. This formula is given by

$$\frac{\partial C(X,t)}{\partial t} = D\nabla^2 C(X,t), \tag{6.1}$$

where D is the diffusion constant and is assumed to be constant in all directions here.

The actual random sample paths of individual molecules can be generated using a random walk model that essentially simulates the particle stochastic motion. The distance traveled by an analyte molecule in a given time interval Δt in each coordinate can be found by picking three random numbers from a normal distribution with mean zero and variance Δt, multiplying them with half the diffusion constant and adding it to the present coordinates. (This essentially follows from the Langevin equations that describe diffusion [35].) This process can be repeated for a large number of Δt's until equilibrium is reached. Multiple realizations of this method (i.e., Monte Carlo simulation) allow the calculation of the probability density function (pdf) of large number of analyte particles. Clearly, this is computationally very exhaustive and is practical when the number of analytes is small.

Another approach to deal with stochastic motion of the molecules is to use Markov chain models [10] and associate a probability (i.e., transition probability) for analyte molecules moving from one coordinate to another in time Δt. Such an approach can provide closed-form formulations for the pdf. In addition, using such an approach, we can form stochastic differential equations (SDE) for the pdf of molecules within the solutions using mathematical formulations such as Fokker–Planck (FP) equations [36]. The complexities of all of these methods are comparable, and generally, the applications denote which one is more suitable.

It is important to realize that diffusive spreading is not the only means of mass transport in biosensors. It is quite possible to have particle drift generated by electric fields and other *deterministic* sources. Incorporation of all this into the above models is straightforward and is done by adding a drift component into the formulations.

Now that a statistical model of analyte motion within the solution has been developed, it is necessary to incorporate the boundary of the biosensor system. The boundary conditions for conventional biosensors are either purely reflective (e.g., inert chamber walls and solution surfaces) or selectively absorbing (e.g., surfaces where the capturing probes exist). Incorporation of reflective boundaries into the model is carried out by assigning zero probability for particles to move beyond the boundaries, which effectively means that the probability flux at the reflecting boundaries goes to zero. However, accommodation of selective absorption requires a probabilistic model for the collision and interaction processes that we will look into in the next section.

6.2.2 Analyte Capture: Reaction

The capturing process in biosensors is a dynamic process that occurs over time and is a function of analyte diffusion, concentration, temperature, density of

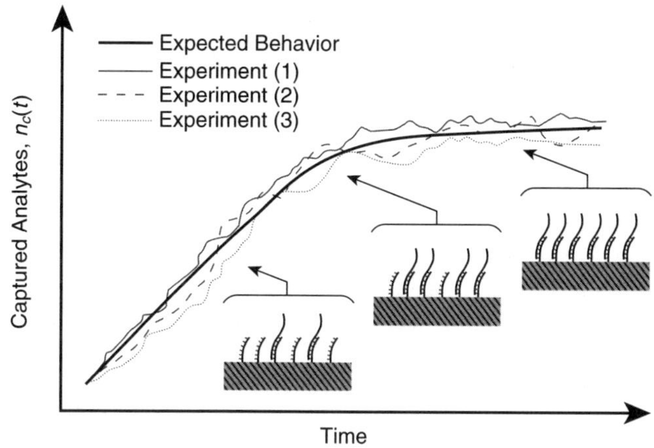

Figure 6.2. Typical binding kinetics for an affinity-based biosensor.

capturing probes, and the reaction surface-to-volume ratio. For simplicity, diffusion is initially ignored in this section and just the chemical kinetics is examined. Figure 6.2 illustrates a typical dynamical process of an individual biosensor capturing, where the total number of captured analytes is denoted by $n_c(t)$. It essentially starts from $n_c(0)=0$, and increases until it reaches the biochemical steady state, that is, the state where the capturing and release processes have equal rates, thus making the ensemble average of $n_c(\infty)$ constant.

The set of probes (denoted by P) is assumed to be confined (immobilized) in a certain coordinate. Any meaningful interaction between the set of analytes or targets (denoted by T) and P at time t occurs only if a target molecule is in intimate proximity of$P$. If the bulk phase reaction between the analyte $T$and the capturing probes P has an association rate k_1, and disassociation rate k_{-1}, then

$$\begin{cases} T+P \xleftrightarrow{k_1,k_{-1}} TP \\ \dfrac{d[TP]}{dt} = k_1[T][P]-k_{-1}[TP], \end{cases} \tag{6.2}$$

where the symbol [] indicates concentration of the species. From Equation 6.2, the expected value (ensemble average) of $n_c(t)$ (which is the cardinality of TP), $\langle n_c(t)\rangle$, can be approximated by the biochemical rate equation

$$\frac{d\langle n_c(t)\rangle}{dt} = k_1(n_t-\langle n_c(t)\rangle)(n_p-\langle n_c(t)\rangle)-k_{-1}\langle n_c(t)\rangle, \tag{6.3}$$

where n_p is the total number of capturing probes immobilized at the surface that is the cardinality of P, and n_t is the total number of analytes (T) in the reaction chamber. Note that n_p is typically described in terms of concentration; however, here the number of analytes is used since n_p needs to have the same unit as $n_c(t)$ (which is the quantity that is to be computed). Clearly, this does not change the generality of the formulations.

Equation 6.4 describes the ensemble average behavior of the capturing process; however, to evaluate the uncertainty of the capturing of the system, a probabilistic description of $n_c(t)$ is needed. One approach to describe the stochastic nature of the capturing process is to form a discrete-time Markov chain model of the system similar to the previously described approach in modeling Brownian motion. In this model, the states correspond to the possible physical coordinates of the analytes. Hence, in each time epoch Δt, the transition probabilities of the Markov model represent how likely it is that an analyte particle moves from one state to another. A simpler model would be to consider just two states, the "free" state (F) and the "captured" state (C). In the following formulations, the effects of diffusion are then absorbed into the coefficient, yet this does not affect the generality of the approach. The transition probabilities can be calculated based on the parameters of the analyte random walk, and the reaction kinetics (k_1 and k_{-1}), that is

$$\begin{cases} p(F \to C) = \alpha k_1 \Delta t = k_1^* \Delta t \\ p(C \to F) = k_{-1} \Delta t \end{cases}. \tag{6.4}$$

The parameters $p(F \to C)$ and $p(C \to F)$ are also defined as the association and disassociation transition probabilities, respectively. The quantity α describes the effects of immobilization of probes and the finite reactive distance between the probe and target, which is the same as k_1^* in Equation 6.4.

The formulations in Equations 6.2 and 6.4 assume infinite capturing capacity for probes, which is not always realistic. If there are n_p immobilized capturing sites, then the association and disassociation probabilities for the immobilized P species become functions of $n_c(t)$, the number of captured analytes, such that

$$\begin{cases} p(F \to C) = k_1^* \left(\dfrac{n_p - n_c(t)}{n_p} \right) \Delta t \\ p(C \to F) = k_{-1} \Delta t \end{cases}, \tag{6.5}$$

where the term $(n_p - n_c(t))/n_p$ is the probability of finding a free probe at the capturing site. The incorporation of Equation 6.5 clearly results in an inhomogeneous Markov process since the transition probabilities have become a function of target analyte distribution (see Fig. 6.3). To take this constraint into account, Equation 6.3 is modified as

$$\frac{d\langle n_c(t)\rangle}{dt} = k_1^* \left(\frac{n_p - \langle n_c(t)\rangle}{n_p} \right) (n_t - \langle n_c(t)\rangle) - k_{-1} \langle n_c(t)\rangle. \tag{6.6}$$

In Equation 6.6, k_1^* is the association rate assuming an unlimited abundance of fixed capturing probes, and the term $(n_p - \langle n_c(t) \rangle)/n_p$ represents the availability of the probes (i.e., the probability of finding an unoccupied probe). Although the coefficient k_{-1} in Equation 6.6 is independent of the reaction volume dimensions and only a function of the reaction energetics, k_1^* is a function of the reaction volume dimensions, mass transport processes, and the surface-to-volume ratio.

To derive the distribution of $n_c(t)$, here the FP equation (also known as the Kolmogorov forward equation) [36–40] is used to describe the statistics of the capturing and release process, and ultimately calculate the time evolution of the pdf using its available solution. Note that the master equation can be used to calculate this exactly and may be the only resort when dealing with a very small number (e.g., less than 20) of analytes and probes [11]. Nevertheless, for a sufficiently large number of analytes and probes in a solution, which is a common scenario in most biosensor assays, the FP equation provides a good approximation.

To evaluate the time evolution of the number of the captured analytes $n_c(t)$, it is necessary to formulate its differential, dn_c, as a function of dt, $d[W]$ (where W denotes the Wiener process), and functions $\mu(n_c,t)$ and $\sigma(n_c,t)$ in the form of

$$dn_c = \mu(n_c,t)dt + \sigma(n_c,t)d[W]. \tag{6.7}$$

The pdf of $n_c(t)$, denoted by $f_n(n_t,t)$, is given by the following partial differential equation (generally referred to as the FP equation [36]),

$$\frac{\partial f_n(n_t,t)}{\partial t} = -\frac{\partial}{\partial n_c}[\mu(n_c,t)f_n(n_t,t)] + \frac{1}{2}\frac{\partial^2}{\partial n_c^2}\left[|\sigma(n_c,t)|^2 f_n(n_t,t)\right]. \tag{6.8}$$

For a Wiener process W, $d[W]/dt = N(0,dt^{-1/2}) = dt^{-1/2}N(0,1)$, where $N(0,1)$ is a zero-mean normal distribution. Therefore, Equation 6.7 can be rewritten as

$$\frac{dn_c}{dt} = \mu(n_c,t) + \sigma(n_c,t)dt^{-1/2}N(0,1), \tag{6.9}$$

which points out that in order to compute $f_n(n_t,t)$, it is necessary to describe dn_c/dt as a function of a normal distribution. If the Markov model of the system in Figure 6.3 is examined and the assumption that there are a large number of particles participating in the capturing and release processes is valid, $n_c(t+\Delta t)$ can be defined as a function of $n_c(t)$ by

$$\begin{aligned} n_c(t+\Delta t) - n_c(t) = {} & p_C(n_t - n_c(t)) - p_R n_c(t) \\ & + \left[p_C(1-p_C)(n_t - n_c(t))\right]^{1/2} N_1(0,1) \\ & + \left[p_R(1-p_R)(n_c(t))\right]^{1/2} N_2(0,1), \end{aligned} \tag{6.10}$$

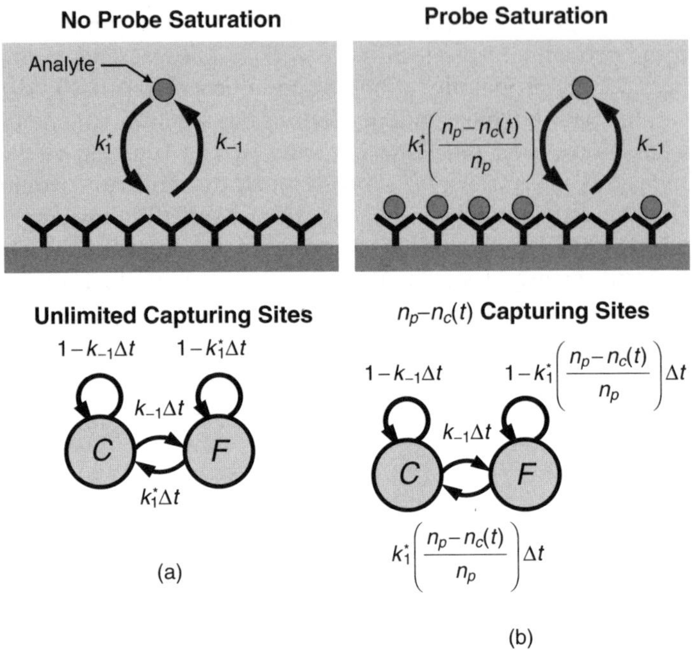

Figure 6.3. Capturing rate and Markov chain model for (A) biosensor with no probe saturation, and (B) with probe saturation.

where p_C is the probability of capturing from the free state and p_R is the probability of release into the free state. The underlying assumption is that the binding process in the time epoch Δt is a sum of a large number of independent Bernoulli random trials with the probabilities p_C and p_R of the capturing and release of $n_t - n_c(t)$ and $n_c(t)$ analyte particles, respectively. This assumption is what leads to the two normal distributions in Equation 6.10. If p_C and p_R are expressed in terms of the parameters of the Markov model, and by calculating the limit when $\Delta t \to dt$, the differential equation describing dn_c/dt as a function of a single normal distribution, $N(0,1)$, can be rewritten as

$$\begin{aligned}\frac{dn_c}{dt} &= k_1^*\left(\frac{n_p-n_c}{n_p}\right)(n_t-n_c)-k_{-1}n_c \\ &+\left[k_1^*\left(\frac{n_p-n_c}{n_p}\right)(n_p-n_c)+k_{-1}n_c\right]^{1/2} dt^{-1/2}N(0,1).\end{aligned} \tag{6.11}$$

Note that in the derivation of Equation 6.11, $1-k_1^*(n_p-n_c)/n_p dt \approx 1$, and $1-k_{-1}dt \approx 1$ has been assumed. From Equation 6.11, $\mu(n_c,t)$ and $\sigma(n_c,t)$ that appear in Equation 6.9 turn out to be

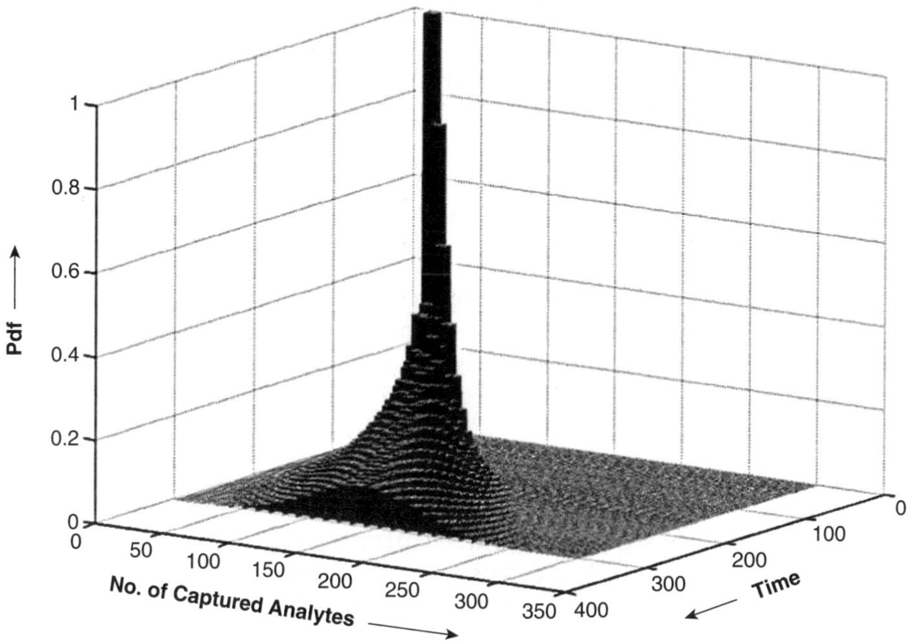

Figure 6.4. Evolution of pdf in time at the outset simulated using the Fokker–Planck equation.

$$\mu(n_c,t)) = k_1^*\left(\frac{n_p - n_c}{n_p}\right)(n_t - n_c) - k_{-1}n_c \tag{6.12}$$

and

$$\sigma(n_c,t) = \left[k_1^*\left(\frac{n_p - n_c}{n_p}\right)(n_t - n_c) + k_{-1}n_c\right]^{1/2}. \tag{6.13}$$

One important consideration is that in order to simulate the FP equation, additional boundary conditions need to be imposed. In this case, the probability flux must be zero at the boundaries. The initial conditions are that the pdf is equal to 1 at $n_c = 0$ and $t = 0$. Figure 6.4 shows the time evolution of the pdf obtained from the FP equation.

6.2.3 Coupling Diffusion and Reaction

Having explained both diffusion and reaction in sufficient detail, it is apparent that the actual process in biosensors involves both of them. Assuming that the probes are immobilized in a defined region (e.g., the bottom surface, typically), it is possible to solve the coupled equation involving Equations 6.1 and 6.8 to find the pdf. Also, while no distinction has been made between different probe molecules earlier, it is possible to define a matrix C with n_p states such that any element in this matrix represents the probability of an analyte being attached to the corresponding probe. A vector of such matrices can then be used to represent different probe spots in a given biosensor such as a DNA microarray. For brevity, the details of this are omitted [10].

6.2.4 Transduction

The transduction process and the output signal in biosensors are defined as a function of $n_c(t)$. For simplicity, the transduction process is represented by a transduction factor T, and the output signal $s(t)$ is defined by

$$s(t) = T \cdot n_c(t) + u_T(t). \tag{6.14}$$

$u_T(t)$ is a random process describing the additive noise of the transducer. The transduction factor T in Equation 6.14 signifies the process wherein binding results in a detectable signal or measurable parameter. For example, the unit of T in a chemiluminescence-based biosensor is photons/second/target, while in impedimetric biosensors it is $\Delta\Omega^{-1}$/target. As implied by Equation 6.14, the transduction process itself may introduce some uncertainty in the form of an additive noise component represented by $u_T(t)$. As discussed earlier, $n_c(t)$ is a random process and its expected behavior (ensemble average) is denoted by $\langle n_c(t)\rangle$; hence, Equation 6.14 can be rewritten as

$$s(t) = T \cdot (\langle n_c(t)\rangle + u_c(t)) + u_T(t) = T \cdot \langle n_c(t)\rangle + T \cdot u_c(t) + u_T(t). \tag{6.15}$$

In Equation 6.15, the random process $u_c(t)$ describes the deviation of the actual analyte concentration (perturbation) from the expected value $\langle n_c(t)\rangle$, where $\langle u_c(t)\rangle = 0$ for all t. The function $T \cdot u_c(t)$ therefore describes the observed *biochemical noise* of the system.

Since the transduction mechanism varies significantly among biosensor platforms, one needs to reexamine Equation 6.15 and derive a formulation that makes their performances comparable. One approach is to refer the biosensor output signal to the input of the biosensor system that is essentially the number of analytes in the sample. In this approach, the biosensor equivalent signal $s_b(t)$ represents the number of *perceived* target analytes in the sample plus its uncertainty. To carry out the procedure, Equation 6.15 is divided by T, and hence $s_b(t)$ can be formulated by

$$s_b(t) = \langle n_c(t)\rangle + u_c(t) + \frac{u_T(t)}{T}. \tag{6.16}$$

If there are K stages after the transducer (Fig. 6.5), each with gain G_i and additive noise $u_i(t)(i = 1, \cdots, K)$, one can incorporate them in Equation 6.16 such that

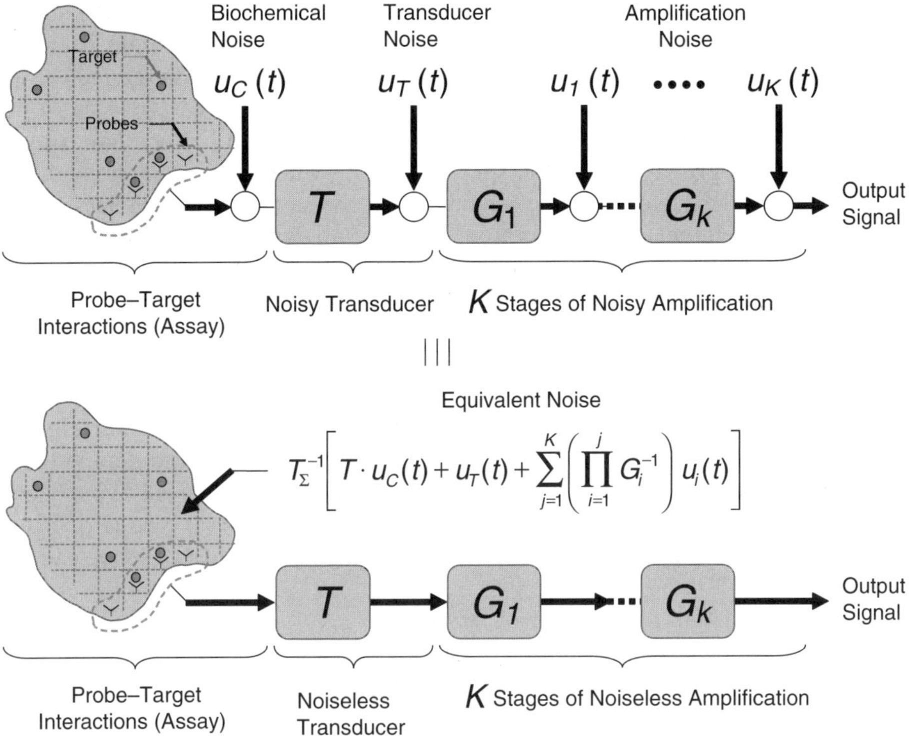

Figure 6.5. Block diagram of the different stages of a general biosensor platform.

$$s_b(t) = \langle n_c(t) \rangle + u_c(t) + \frac{u_T(t) + \sum_{i=1}^{K}\left(\prod_{j=1}^{i} G_j^{-1}\right) \cdot u_i(t)}{T}. \tag{6.17}$$

Equations 6.16 and 6.17 provide the number of observed (or existing) target analytes in the biosensor system. In the next section, these stochastic formulations are applied to calculate the noise characteristics of biosensor platforms.

Note that the effective number of analytes so obtained is input referred with respect to the biosensor platform. However, this still does not account for the fact that we are primarily interested in an accurate quantification of n_t, not of $n_c(t)$; hence, computing the uncertainty with respect to n_t. A more detailed explanation on this is provided in Section 6.4.

6.2.5 Effects of Interference, Nonspecific Binding, and Background

The expressions for noise in the previous sections all assumed that there is only one analyte species (the target) that can bind to the capturing probes. However, in all practical biosensors, other species coexist with the target in the sample, and they may or may not bind to the probes. These species are the interferers and their capturing incidents are called nonspecific bindings. Accordingly, the signal generated by the transducer that originates from nonspecific bindings is called nonspecific signal, or in certain biological detection platforms, it is referred to as the background signal.

A nonspecific binding event (e.g., cross-hybridization in DNA microarrays) is generally less probable than the specific binding when target analyte and the interfering species have the same abundance in the sample. Nevertheless, when the ratio of the target analyte to the nonspecific analyte is small, contributions of nonspecific binding may dominate the measured signal. To incorporate the effects of nonspecific binding in the previously discussed formulations, the uncertainty associated with the concentration of interferers needs to be assessed. This defines whether one should consider nonspecific binding incidents coming from a deterministic or a random concentration.

To examine the general case of detection in the presence of interference, assume that m interferers coexist with the target in the sample, and denote their captured amounts by $n^{(1)}, n^{(2)}, \cdots$, and $n^{(m)}$, where for the ith species the expected number of analytes is $\langle n^{(i)} \rangle$, with variance of $\sigma_{(i)}^2$. In the general case (Eq. 6.16), the equilibrium equation for the input-referred signal with interference can be rewritten as

$$s_b(t) = \langle n_c(t) \rangle + \sum_{i=1}^{m} \langle n^{(i)}(t) \rangle + \left(u_c(t) + \sum_{i=1}^{m} u_c^{(i)}(t) \right) + \frac{u_T(t)}{T}, \tag{6.18}$$

where $\langle n_c^{(i)}(t) \rangle$ and $u_c^{(i)}(t)$ are the ensemble average and biochemical noise of the ith interferer species, respectively.

6.3 SIGNAL-TO-NOISE RATIO AND NOISE FIGURE DEFINITIONS

The SNR of a biosensor platform is defined as the ratio of the square of the detected target analyte (i.e., signal power) over its noise variance. The general expression for the SNR at time t is given by

$$\mathrm{SNR}(t) = \frac{\langle s_b(t) \rangle^2}{\langle [s_b(t) - \langle s_b(t) \rangle]^2 \rangle}, \tag{6.19}$$

where $s_b(t)$ is the number of perceived target analytes at measured time t. By employing Equation 6.16, Equation 6.19 can be rewritten as

$$\mathrm{SNR}(t) = \frac{(\langle n_c(t) \rangle)^2}{\langle u_c(t) \rangle^2 + \dfrac{\langle u_T(t) \rangle^2}{T^2}} \tag{6.20}$$

or

$$\mathrm{SNR}(t) = \frac{(\langle n_c(t) \rangle)^2}{\sigma_c^2(t) + \dfrac{\sigma_T^2}{T^2}}. \tag{6.21}$$

The first term in the denominator of Equation 6.21 describes the biochemical shot noise (Poisson noise) of the biosensor, while the second term is the added noise of the transducer. Hence, the SNR of the biosensor excluding any transducer noise (i.e., ideal detector) has an upper bound defined by this shot noise that is generally referred to as the quantum-limited (QL) SNR [10].

Equation 6.21 can be further generalized for the case where the biosensor comprises of other amplification stages as in Equation 6.17. In this case, the SNR becomes

$$\mathrm{SNR} = \frac{(\langle n_c(t) \rangle T)^2}{T^2 \sigma_c^2(t) + \sigma_T^2 + \sum_{i=1}^{K} \left(\prod_{j=1}^{i} G_j^{-1} \right) \sigma_i^2}, \tag{6.22}$$

where σ_i^2 represents the variance of the noise added by the ith amplification stage.

To evaluate the range of detection of biosensors, similar to all sensing systems, one needs to find the MDL as well as the highest detectable level (HDL). On the lower side of sensing, the MDL or n_{MDL} is defined as the analyte level where the biosensor SNR is equal to some minimum acceptable value, $\mathrm{SNR}_{\min}$. However, for the same $\mathrm{SNR}_{\min}$, there will be another analyte level that is larger than n_{MDL}. This value is the HDL, and the analyte concentration level associated with it is n_{HDL}.

The detection dynamic range (DR) is another important figure of merit in biosensors. This parameter is generally defined as the range of n_t, within which $\mathrm{SNR}_t(t)$ is higher than $\mathrm{SNR}_{\min}$, the minimum acceptable SNR. The dynamic range of the biosensor, DR, is then defined as

$$\mathrm{DR} = \frac{n_{\mathrm{HDL}}}{n_{\mathrm{MDL}}}.$$

The closed-form for DR is generally complicated and is derived by solving Equation 6.22 for $\mathrm{SNR}_{\min}$ and is commonly a concave function. It is important also to realize that without having nonlinearity in binding (e.g., probe saturation) n_{HDL} does not exist and hence DR will be infinite. Obviously, in all practical biosensors there will be some kind of saturation and therefore DR will be finite.

In addition to SNR, one can define the noise figure (NF) of the biosensor as another figure of merit for the biosensing process. This figure of merit describes the ratio of the added noise to the input noise that is the unavoidable shot noise. The NF is defined by

$$\mathrm{NF} \equiv \frac{\text{Total output noise power}}{\text{Output noise power due to input source}}. \tag{6.23}$$

From Equation 6.17 and for simplicity referring everything to the transducer output,

$$\mathrm{NF} = 1 + \frac{\sigma_T^2}{T^2\sigma_c^2(t)} + \frac{\sum_{i=1}^{K}\left(\prod_{j=1}^{i} G_j^{-1}\right)\sigma_i^2}{T^2 \cdot \sigma_C^2(t)}. \tag{6.24}$$

Note that when n_c becomes large, Equation 6.22 becomes proportional to n_c, while Equation 6.24 is inversely proportional to n_c, which indicates that by increasing the concentration (or quantity) of the analytes, the system generates more noise. However, SNR increases and NF decreases. This phenomenon, which one expects from the inherent Poisson noise characteristics of the binding events, sets no upper bound for the concentration levels that the biosensor is able to detect. However, from Equation 6.6, one knows that probe saturation will eventually occur as one keeps increasing the amount of analytes and this will lead to a decrease in the corresponding SNR.

6.4 TRANSIENT SIGNAL-TO-NOISE RATIO ANALYIS

At this point, based on the models for diffusion, chemical binding and transduction that occur concurrently, one can simulate various characteristics of the biosensor and assess the output signal of the system. In this section, the focus will be on transient analysis of noise in biosensors and on some of the metrics that might be useful in comparing the performance of biosensors. The previous section provides a comprehensive outline of the most important definitions such as SNR and NF. This section looks at a simplified version of transient SNR analysis in the absence of diffusion.

The pdf $f_n(n_t,t)$, which is calculated using Equation 6.8, provides ground for the statistical analysis of biosensor systems. It describes not only the kinetics of the capturing and the expected rate of bindings but also the uncertainty associated with them. This section characterizes certain properties of biosensors as figures-of-merits based on $f_n(n_t,t)$. The approach is to utilize $\langle f_n(n_t,t)\rangle$ and $\langle f_n^2(n_t,t)\rangle$ in the definition of these parameters. $\langle f_n(n_t,t)\rangle$ and $\langle f_n^2(n_t,t)\rangle$ are the first- and second-order moments of $n_c(t)$ and can be used to evaluate how fast the biosensor, on an average sense, captures analytes and what the standard deviation of $n_c(t)$ is as time progresses.

As in all sensors, the most important characteristic of biosensor systems is their SNR. The QL SNR of $n_c(t)$ (i.e., the best possible SNR of a biosensor) $\mathrm{SNR}_c(t)$, can be defined as the ratio of $\langle f_n(n_t,t)\rangle$ (signal) and its variation $\langle f_n^2(n_t,t)\rangle-\langle f_n(n_t,t)\rangle^2$ (noise), that is

$$\mathrm{SNR}_c(t)=10\log\frac{\langle f_n(n_t,t)\rangle^2}{\langle f_n^2(n_t,t)\rangle-\langle f_n(n_t,t)\rangle^2}. \tag{6.25}$$

Previously, it was shown that an appropriate way of defining the SNR is to refer the output signal (i.e., $n_c(t)$) to the input (i.e., n_t). We revisit our SNR and NF definitions based on this requirement.

To calculateSNR$_t(t)$, first write $n_c(t)$ as a function of $\langle f_n(n_t,t)\rangle$ and $\tilde{n}_c(t)$, the temporal fluctuation of the system,

$$n_c(t)=\langle f_n(n_t,t)\rangle+\tilde{n}_c(t), \tag{6.26}$$

where $\langle\tilde{n}_c(t)\rangle=0$ and $\langle\tilde{n}_c^2(t)\rangle=\langle f_n^2(n_t,t)\rangle-\langle f_n(n_t,t)\rangle^2$. If we assume that the sensor has a transducer with a gain of T that adds a zero-mean additive noise of $u_T(t)$, we can formulate the output of the biosensor system, $s(t)$, as

$$s(t)=T\cdot n_c(t)+u_T(t) \tag{6.27}$$

or

$$s(t)=T[\langle f_n(n_t,t)\rangle+\tilde{n}_c(t)]+u_T(t). \tag{6.28}$$

In order to evaluate the effects of both transducer and biochemical noise in the detection, we can refer the $s(t)$ to the input of the system that is n_t. This is essentially an inverse mapping such that

$$n_t=g_t^{-1}\left[\frac{s(t)-u_T(t)}{T}-\tilde{n}_c(t)\right] \tag{6.29}$$

or

$$n_t=g_t^{-1}[n_c(t)-\tilde{n}_c(t)] \tag{6.30}$$

where g_t^{-1} represents a generally nonlinear inverse function and $g_t(n_t)=\langle f_n(n_t,t)\rangle$. By making a small-signal noise assumption (i.e., $\tilde{n}_c(t)<<T^{-1}s(t)$ and $T^{-1}u_T(t)<<T^{-1}s(t)$) and truncating the Taylor's series expansion of g_t^{-1} after the second term, Equation 6.30 becomes

$$n_t = g_t^{-1}\left(\frac{s(t)}{T}\right) - \left(\left.\frac{\partial g_t^{-1}(x)}{\partial x}\right|_{x=s(t)}\right)\left(\tilde{n}_c(t) + \frac{u_T}{T}\right). \tag{6.31}$$

Subsequently, with a little algebraic manipulation and assuming that $\tilde{n}_c(t)$ and $u_T(t)$ are uncorrelated, the variance of n_t, denoted by $\langle \tilde{n}_t^2 \rangle$, can be derived as

$$\langle \tilde{n}_t^2 \rangle = \left(\left.\frac{\partial g_t^{-1}(x)}{\partial x}\right|_{x=s(t)}\right)^2 \left(\langle \tilde{n}_c^2(t) \rangle + T^{-2}\sigma_T^2\right). \tag{6.32}$$

By definition, the input-referred $\text{SNR}_t(t)$ is then

$$\text{SNR}_t(t) = 10\log \frac{n_t^2}{\langle \tilde{n}_t^2 \rangle}. \tag{6.33}$$

Now, if we use Equation 6.32, we can rewrite Equation 6.33 as

$$\text{SNR}_t(t) = 10\log \frac{n_t^2}{\left(\left.\frac{\partial g_t^{-1}(x)}{\partial x}\right|_{x=s(t)}\right)^2 \left(\langle \tilde{n}_c^2(t) \rangle + T^{-2}\sigma_T^2\right)}. \tag{6.34}$$

If g_t^{-1} is a one-to-one function (i.e., a bijection, and hence invertible), we have

$$\left.\frac{\partial g_t^{-1}(x)}{\partial x}\right|_{x=s(t)} = 1\Bigg/\left(\left.\frac{\partial g_t(x)}{\partial x}\right|_{x=n_t}\right) = 1\Bigg/\left(\left.\frac{\partial \langle f_n(n_t,t)\rangle}{\partial x}\right|_{x=n_t}\right). \tag{6.35}$$

We know that $g_t(n_t) = n_c(t) - \tilde{n}_c(t) = \langle f_n(n_t,t)\rangle$. Hence, we have

$$\text{SNR}_t(t) = 10\log \frac{n_t^2}{\langle f_n^2(n_t,t)\rangle - \langle f_n(n_t,t)\rangle^2}\left(\left.\frac{\partial \langle f_n(x,t)\rangle}{\partial x}\right|_{x=n_t}\right)^2. \tag{6.36}$$

Note that for the case with no saturation, n_c and n_t are linearly related and g_t is easily seen to be invertible. If we assume that the rate equations provide us with this expected binding kinetics,

$$\langle n_c(t)\rangle = \frac{k^* n_t}{k^* + k_{-1}}\left(1 - \exp\left(-\left(k^* + k_{-1}\right)t\right)\right), \tag{6.37}$$

and with saturation, it can be shown that

$$\langle n_c(t)\rangle = \frac{n_t n_p}{\frac{n_t+n_p}{2}+\frac{k_{-1}n_p}{2k^*}+\alpha\coth\left(\alpha\frac{k^*}{n_p}t\right)}, \tag{6.38}$$

where $\alpha=\sqrt{\left(\frac{n_t+n_p}{2}+\frac{k_{-1}n_p}{2k^*}\right)^2-n_t n_p}$.

Differentiating Equation 6.38 with respect to n_t, one gets a quantity that is always positive. This implies that the function g_t is one-to-one, and hence invertible. Equations 6.26–6.38 provide a detailed treatment of SNR derivation in the transient case.

It is also possible to add nonspecific hybridization as interference to define the signal-to-interference-and-noise ratio (SINR) as well. For an infinite number of probes (and neglecting any variation of the interfering species), this is of the form

$$\mathrm{SINR}_t(t)=10\log\frac{n_t^2}{\left(\left.\frac{\partial g_t^{-1}(x)}{\partial x}\right|_{x=s(t)}\right)^2\left(\langle\tilde{n}_c^2(t)\rangle+\left(\sum_{i=1}^{m}\langle n^{(i)}(t)\rangle\right)^2+T^{-2}\sigma_T^2\right)}. \tag{6.39}$$

For a finite number of probes, such expressions are more cumbersome to obtain because the signal and the interfering species are no longer independent, although the previous formulation may still be applied in the initial stages of the reaction where the effect of finite probes is negligible.

For most practical cases, neglecting the variation of the interfering species is a valid assumption. If the background variation were to be as large as the signal variation, it would effectively mean that the background is as large as the signal. In typical platforms, this makes the SNR too low for proper detection.

To find the DR of the system with interference one needs to find $n_{\min}$ and $n_{\max}$ in the presence of interference. Finding $n_{\min}$ requires solving Equation 6.39. Deriving $n_{\max}$ is more complicated. In most biosensors, the concentration levels of analytes exposed to capturing probes are kept low enough to ensure that the background and nonspecific binding do not drive the system into saturation. As a result, the upper limit of detection $n_{\max}$ can still be dominated by the captured target analyte rather than the interferers. Thus, to find $n_{\max}$in biosensors, one generally does not need to take into account the interferer analytes, which makes the calculation less complicated.

Based on the definition of NF, we have

$$\mathrm{NF}=1+\frac{\sigma_T^2}{T^2\sigma_c^2(t)}+\frac{\left(\sum_{i=1}^{m}\langle n^{(i)}(t)\rangle\right)^2}{\sigma_c^2(t)}+\frac{\sum_{i=1}^{K}\left(\prod_{j=1}^{i}G_j^{-1}\right)\sigma_i^2}{T^2\sigma_C^2(t)}. \tag{6.40}$$

For simplicity, the NF has not been referred to the input. Also, we have used the assumptions made while deriving the SNR in the presence of noise and interference. Note that for a sensor that has a large background and noise variation, improving the noise performance of the biosensor and the additional amplification stages itself will not make a significant improvement to the performance of the system. This has profound implications for sensor designers and needs to be carefully interpreted while designing specifications for such biosensors.

The above statement however, needs to be read with caution. It is possible that in biosensor platforms that typically read probe signals from single spots, the background is so large that it will effectively "kill" the signal even if the transducer is noiseless. However, with careful design of probes and by using smart signal processing techniques where the signal from multiple probes is interpreted jointly, it is possible to overcome the serious limitations that background noise imposes. A more detailed discussion of this, however, is beyond the scope of the present text.

6.5 SIMULATIONS

The formulations provided above intuitively tell us how things get worse as the background and sensor noise become large. In this section, we address an important question in the biosensor community—the effect of scaling. We illustrate how scaling adversely affects the SNR.

Transduction noise and background has been neglected in our simulations and do not affect the basic results. Most of the simulations carried out focus on the effect of saturation on these figures of merit. In particular, the effect of isomorphic scaling of the biosensor will be analyzed, that is, the effect of shrinking the biosensor (and hence reducing the number of probes).

Figure 6.6 shows an example of what the aforementioned formulations can provide for a biosensor when $n_t >> n_p$ (i.e., when saturation is apparent). In Figure 6.6A, 10 Monte Carlo simulations of a binding process based on the Gaussian approximation of the Markov chain model is shown. The stochastic differential equation is simulated by a Milstein scheme. As expected, the signal (captured analytes) shows significant fluctuation during the transient and saturation phases. The results of the proposed method, based on FP equations, are illustrated in Figure 6.6B, which shows that is it possible to predict not only the ensemble average of the capturing but also its variation (in fact, the entire pdf can be generated; in some sense therefore, the FP equation is an overkill—the first two moments are sufficient to completely determine the figures of merit that are mostly used here). One important observation based on these particular results is that the variance of the captured analytes (i.e., biochemical noise) does not monotonically increase as the capturing process progresses. This shows that the biosensor noise cannot be simply modeled by Poisson noise when saturation (nonlinearity) occurs.

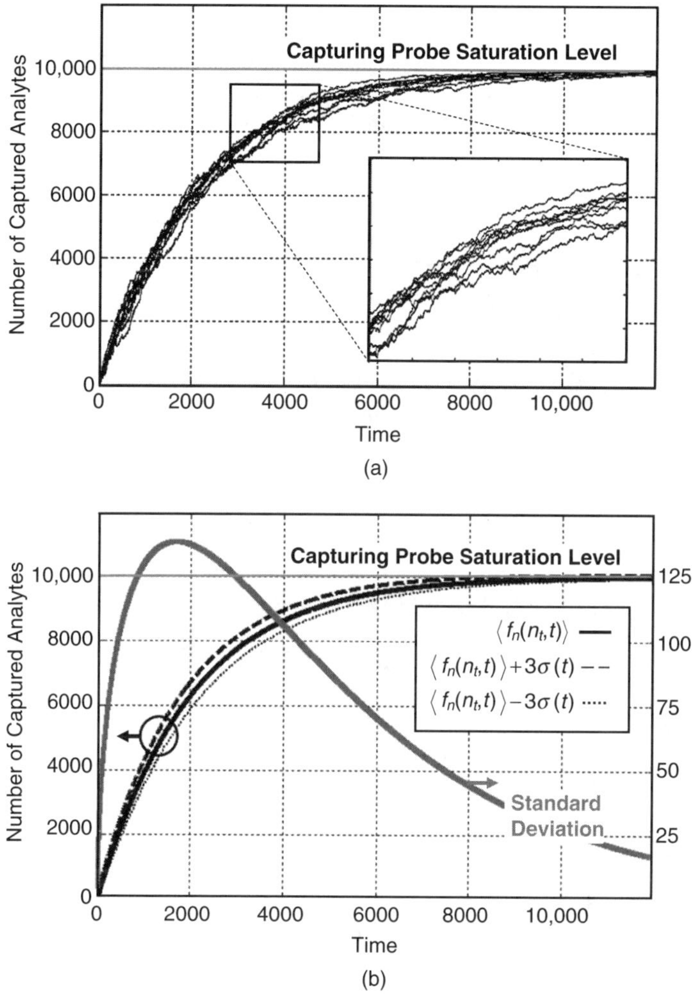

Figure 6.6. (A) Monte Carlo simulation using stochastic differential equations. (B) Ensemble average and variation of the capturing process obtained using Fokker–Planck (FP) equations.

We look at the response curves in terms of the settling time (defined as the time taken for the signal to reach 90% of its steady-state value) or t_r. In Figure 6.7, $\langle f_n(n_t,t)\rangle$ has been computed numerically with the biosensor system subject to different surface-to-volume scaling factors, meaning that the reaction volume is kept constant while the sensing area is reduced by a scaling factor of α. n_t is kept constant in both cases. Observe that for the $n_t >> n_p$ regimes, there is a negligible change in t_r, while in $n_t \approx n_p$, the settling decreases as α increases. This can be explained by the fact that when the number of analytes decreases as the capturing progresses, the likelihood of finding an analyte near the capturing probe gets smaller, and hence the capturing processes becomes slower.

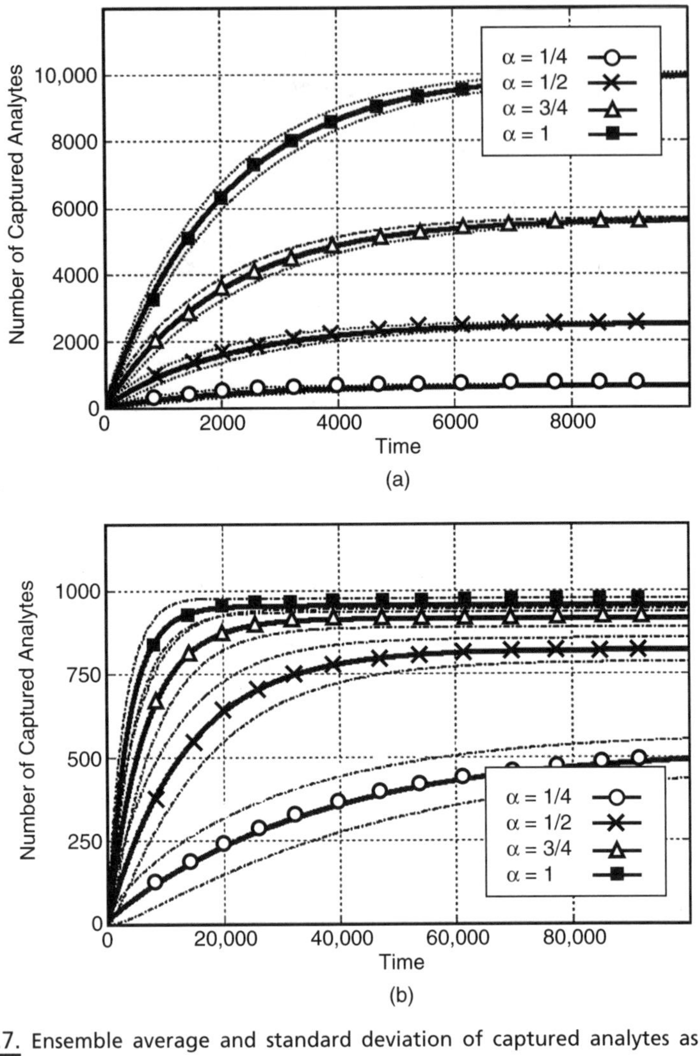

Figure 6.7. Ensemble average and standard deviation of captured analytes as a surface-to-volume scaling factor (α) for (A) excess analyte (i.e., $n_t >> n_p$), and (B) depleting analyte (i.e., $n_t \approx n_p$) regions.

To study the effects of scaling on noise and SNR, both $\langle f_n(n_t,t)\rangle$ and $\langle f_n^2(n_t,t)\rangle$ are computed numerically as shown in Figure 6.7. In Figure 6.8, $\text{SNR}_t(t_0)$ of the system when $n_t \approx n_p$ for a fixed time, t_0, using Equation 6.36 as a function of input n_t is plotted. These results show that scaling reduces the maximum achievable SNR. As expected, SNR is a concave function and as the surface-to-volume ratio becomes smaller, both MDL and HDL become smaller. The reduction in MDL can be traced back to biochemical noise; however, the reduction in HDL is due to the reduction in n_p as α decreases. It is important to realize that by changing

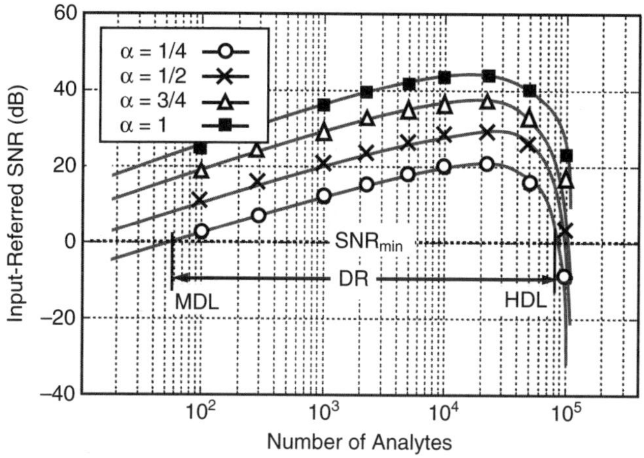

Figure 6.8. Effect of surface-to-volume scaling factors (α) on $SNR_t(t)$, DR, MDL, and HDL for a typical biosensor.

t_0, the relative position and concave shape of the graphs will stay the same, but the maximum achievable SNR will change.

These formulations and results prove that in certain cases scaling reduces the quality of the biosensor by making it more susceptible to biochemical noise. It also reduces the HDL of the sensor system by lowering the capturing probes' saturation level. Both of these impediments decrease the achievable DR. Therefore, biosensor designers should take all these issues into account when designing biosensors, in addition to the practical considerations related to background noise, transducer, readout circuitry, and data processing design.

6.6 CONCLUSION

A comprehensive stochastic model for biosensing has been provided, starting from the actual physical processes that occur. These models naturally lend themselves to be used for system-level statistical simulations of biosensors. Useful sensor specifications can then be developed on top of this, with the ultimate goal of improved sensor design (including the electronic reading apparatus) as well as detection algorithms. We provide representative simulations to highlight the general stochastic nature of the signal and the effect of saturation. In particular, we addressed the important issue of scaling through simulations. While these simulations do contain certain simplifying assumptions, they can provide important insight for the biosensor designer. By incorporating the different processes enumerated earlier, one can include them into computer-aided design (CAD) tools and provide means to compute the kinetics and noise behavior of these systems. With sufficient computational power, the dream of doing large-scale *in silico* biological experiments might even turn out to be true.

REFERENCES

[1] M. Schena, *Microarray Analysis*. New York: Wiley, 2003.

[2] M. Schena, *Protein Microarrays*. Sudbury, MA: Johns and Bartlett, 2003.

[3] E. P. Diamandis and T. K. Christopoulos, Eds., *Immunoassays*, 1st ed. San Diego, CA: Academic, 1996.

[4] G. T. A. Kovacs, *Micromachined Transducers*. New York: McGraw-Hill, 1998.

[5] H. C. Berg, *Random Walks in Biology*. Princeton, NJ: Princeton University Press, 1993.

[6] N. Levine, *Quantum Chemistry*, 5th ed. New York: Prentice-Hall, 1999.

[7] N. G. Van Kampen, *Stochastic Processes in Physics and Chemistry*. Amsterdam: North-Holland, 1981.

[8] J. I. Steinfield, J. S. Fransisco, and W. L. Hase, *Chemical Kinetics and Dynamics*, 2nd ed. New York: Prentice-Hall, 1998.

[9] J. Hubble, "Monte-Carlo simulation of biospecific interactions," *Biotechnol. Lett.*, 22(18), pp. 1483–1486, 2000.

[10] A. Hassibi, H. Vikalo, and A. Hajimiri, "On noise processes and limits of performance in biosensors," *J. Appl. Phys.*, 102, pp. 1–12, 2007.

[11] S. Das, H. Vikalo, and A. Hassibi, "On scaling laws of biosensors: A stochastic approach," *J. Appl. Phys.*, 105(10), pp. 102021–102027, 2009.

[12] Y. Tu, G. Stolovitzky, and U. Klein, "Quantitative noise analysis for gene expression microarray experiments," *Proc. Natl. Acad. Sci.*, 99, pp. 14031–14036, 2002.

[13] A. Hassibi, S. Zahedi, R. Navid, R. W. Dutton, and T. H. Lee, "Biological shot-noise and quantum-limited signal-to-noise ratio in affinity-based biosensors," *J. Appl. Phys.*, 97, pp. 1–10, 2005.

[14] K. Linneti and M. Kondratovich, "Partly nonparametric approach for determining the limit of detection," *Clin. Chem.*, 50(4), pp. 732–740, 2004.

[15] P. Schuck, "Kinetics of ligand binding to receptor immobilized in a polymer matrix, as detected with an evanescent wave biosensor. I. A computer simulation of the influence of mass transport," *Biophys. J.*, 70, pp. 1230–1249, 1996.

[16] V. Chan, D. J. Graves, and S. E. McKenzie, "The biophysics of DNA hybridization with immobilized oligonucleotide probes," *Biophys. J.*, 69, pp. 2243–2255, 1995.

[17] M. A. Livshits and A. D. Mirzabekov, "Theoretical analysis of the kinetics of DNA hybridization with gel-immobilized oligonucleotides," *Biophys. J.*, 71, pp. 2795–2801, 1996.

[18] D. Erickson, D. Li, and U. Krull, "Modeling of DNA hybridization kinetics for spatially resolved biochips," *Anal. Biochem.*, 317, pp. 186–200, 2003.

[19] T. M. Squires, R. J. Messinger, and S. R. Manalis, "Making it stick: Convection, reaction, and diffusion in surface-based biosensors," *Nat. Biotechnol.*, 26, pp. 417, 2008.

[20] J. A. Benn, J. Hu, B. J. Hogan, R. C. Fry, L D. Samson, and T. Thorsen, "Comparative modeling and analysis of microfluidic and conventional DNA microarrays," *Anal. Biochem.*, 348, pp. 284–293, 2006.

[21] C. Gadgil, A. Yeckel, J. J. Derby, and W. S. Hu, "A diffusion–reaction model for DNA microarray assays," *J. Biotechnol.*, 114, pp. 31–45, 2004.

[22] J. Bishop, A. M. Chagovetz, and S. Blair, "Kinetics of multiplex hybridization: Mechanisms and implications," *Biophys. J.*, 94, pp. 1726–1734, 2008.

[23] Y. Zhang, D. A. Hammer, and D. J. Graves, "Competitive hybridization kinetics reveals unexpected behavior patterns," *Biophys. J.*, 89, pp. 2950–2959, 2005.

[24] A. W. Peterson, L. K. Wolf, and R. M. Georgiadis, "Hybridization of mismatched or partially matched DNA at surfaces," *J. Am. Chem. Soc.*, 124, pp. 14601–14607, 2002.

[25] D. T. Gillespie, "Exact stochastic simulation of coupled chemical reactions," *J. Phys. Chem.*, 81, pp. 2340–2361, 1977.

[26] D. Gillespie, "General method for numerically simulating stochastic time evolution of coupled chemical-reactions," *J. Comput. Phys.*, 22, pp. 403–434, 1976.

[27] M. A. Gibson and J. Bruck, "Efficient exact stochastic simulation of chemical systems with many species and many channels," *J. Phys. Chem.*, 104, pp. 1876–1889, 2000.

[28] A. Kierzek, "Stocks: Stochastic kinetic simulations of biochemical systems with Gillespie algorithm," *Bioinformatics*, pp. 18–13, 2003.

[29] Z. Wu and R. Irizarry, "Stochastic models inspired by hybridization theory for short oligonucleotide arrays," *J. Comput. Biol.*, 12(6), pp. 882–893, 2005.

[30] R. Nadon and J. Shoemaker, "Statistical issues with microarrays: Processing and analysis," *Trends Genet.*, 18(5), pp. 265–271, 2002.

[31] K. Chen, T. Wang, H. Tseng, C. Ying, F. Huang, and C. Kao, "A stochastic differential equation model for quantifying transcriptional regulatory network in saccharomyces cerevisiae," *Bioinformatics*, 21-12, pp. 2883–2890, 2005.

[32] T. Lu, D. Volfson, L. Tsimring, and J. Hasty, "Cellular growth and division in the Gillespie algorithm," *Syst. Biol.*, 1(1), pp. 121–128, 2004.

[33] A. Becskei and L. Serrano, "Engineering stability in gene networks by autoregulation," *Nature*, 405, p. 590, 2000.

[34] W. J. Blake, M. Kaern, C. R. Cantor, and J. J. Collins, "Noise in eukaryotic gene expression," *Nature*, 422, pp. 633–637, 2003.

[35] D. Bedeaux, K. Lindenberg, and K. Shuler, "On the relation between master equations and random walks and their solutions," *J. Math. Phys.*, 12, p. 2116–2123, 1971.

[36] H. Risken, *The Fokker–Planck Equation: Methods of Solutions and Applications*, 2nd ed. Springer Series in Synergetics. Berlin: Springer, 1989.

[37] B. Oksendal, *Stochastic Differential Equations: An Introduction with Applications*. Berlin and New York: Springer, 2003.

[38] P. Kloeden and E. Platen, *Numerical Solution of Stochastic Differential Equations*. Berlin and New York: Springer-Verlag, 1992.

[39] E. Allen, *Modeling with Ito Stochastic Differential Equations*. Dordrecht: Springer, 2007.

[40] F. Moss and P. McClintock, *Noise in Nonlinear Dynamical Systems: Volume I—Theory of Continuous Fokker-Planck Systems*. Cambridge, U.K.: Cambridge University Press, 1989.

7

FABRICATION EXAMPLES BASED ON STANDARD CMOS AND MEMS PROCESSES

Bernard Courtois

7.1 THE NEED FOR INFRASTRUCTURES

In microelectronics in general, infrastructures that offer custom integrated hardware manufacturing services are important for several reasons:

- They allow students and researchers to access professional facilities at a reasonable cost.
- They allow companies to access small-volume production, which would otherwise be difficult to obtain directly from manufacturers.

The needs of universities, research laboratories, and companies can be summarized as follows:

- Universities need to have access to technology for teaching their students. Those students will be in the industry in the future; as such, they have to be trained in state-of-the-art technology processes.
- Research laboratories usually need to have high-performance technologies to test new concepts. The quality of research results depends mostly on the quality of the technologies used. Access to up-to-date technologies is therefore a necessity.
- Industrial users also need to access state-of-the-art technologies. This is vital for industrial users. The development of a product is usually long

CMOS Biomicrosystems: Where Electronics Meet Biology, First Edition. Edited by Krzysztof Iniewski.

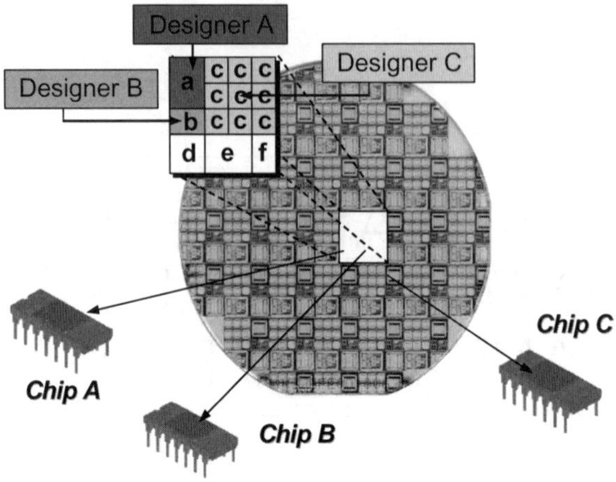

Figure 7.1. MPC/MPW techniques.

(more than 1 or 2 years). It is necessary that industrial users have access to up-to-date processes to ensure the quality of the product.

Infrastructures are also important because of the leverage effect they allow. They are a means to make the development of many projects easier by reducing time and expense when funding for individual projects is less efficient (in terms of the number of projects). The major issue is to obtain an affordable cost.

A large number of complex technological operations are required for integrated circuit (IC) fabrication, but circuits are inexpensive owing to the fact that most of those operations are repetitive. Each processed wafer of silicon is cut into hundreds of dies. For some of the slowest and costliest operations, batches of hundreds of wafers are processed together. That means that tens of thousands of circuits are fabricated simultaneously.

For noncollective operations, such as testing and packaging, operations are highly automated, using mass-production techniques. These very expensive techniques seem out of reach for research and educational centers for IC design. The design of a circuit by students, however, must be pursued to its conclusion, which means fabrication, but given that a student will only need to produce a few chips, mass production is not necessary. The basic idea of a multiproject chip (MPC) is to collectively process different circuits. High fabrication costs can then be shared. To do so, a great number of elementary circuits are put side by side to be reproduced on a wafer. The fabrication yield must be excellent, at least constant, since circuits cannot be tested before being sent back to the designer. This good yield is obtained through industrial production processes, depicted here in Figure 7.1. The production approach is known in general as the MPC/multiproject wafer (MPW) technique. Manufacture of prototypes or low-volume production is inexpensive because of a shared wafer cost.

In addition to low cost, an affordable turnaround time should be available. This is pictured in Figure 7.2. A total turnaround time of about 12 weeks can be obtained by services like CMP.

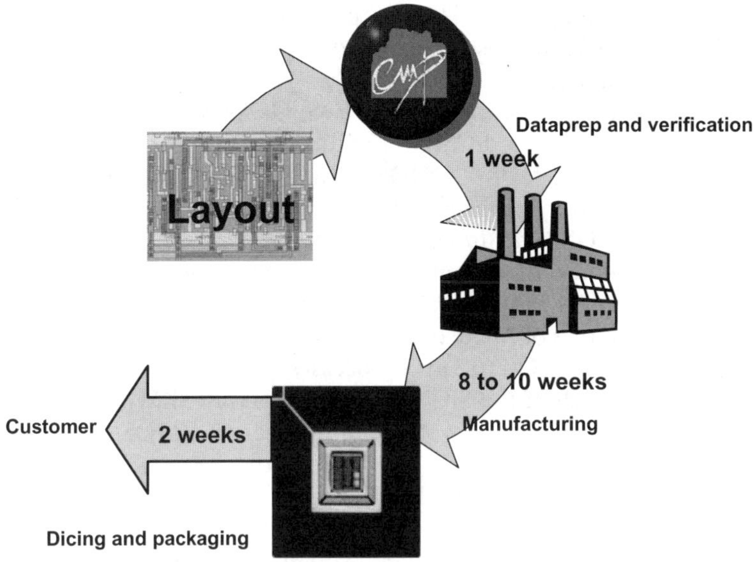

Figure 7.2. From layout to packaged chips.

Using such industrial processes as described above, CMP opened its services for prototyping and low-volume production to industries as early as 1990. Low-volume production service is aimed at helping small- and medium-size enterprises (SMEs) to produce relatively small numbers of circuits (say a few hundred or a few thousand), which they would be unable to obtain directly from manufacturers. A center like CMP, then, interfaces the IC manufacturers and the SMEs.

Such an infrastructure allows the design of custom hardware from standard processes; that is, no custom process development is required. Targeting an appropriate application to produce the custom hardware only requires design specifications.

7.2 INTEGRATED CIRCUIT MANUFACTURING AT CMP

CMP is a nonprofit service, reporting to the French National Council for Research (CNRS) and to universities in Grenoble. A review of early efforts can be found in Courtois [1].

7.2.1 Development at CMP

Several periods may be distinguished in CMP's development:

1981–1982: CMP launched the NMOS
1983–1984: Development of NMOS process and launch of the complementary metal-oxide semiconductor (CMOS) process

1984–1986: Development of CMOS process
1987–1989: NMOS process abandoned; the frequency of CMOS runs was increased
1990–1994: Launch of Bipolar, BiCMOS, GaAs MESFET, GaAs HEMT, and advanced CMOS (0.5 μm TLM) processes
1995–1997: Launch of CMOS- and GaAs-compatible microelectromechanical system (MEMS), DOEs, and deep-submicron CMOS (0.25 μm 6LM) processes
1998: Launch of silicon surface micromachining; GaAs MESFET process abandoned
1999: Launch of SiGe, deep-submicron CMOS (0.18 μm 6LM), SOI/SOS CMOS (0.5 μm) processes
2000: Launch of the SiGe BiCMOS (0.35 μm 5LM) process
2001: Launch of the very-deep-submicron CMOS (0.12 μm 6LM) process
2002: Launch of the Inp HBT process
2003: Launch of the 0.35 μm CMOS-Opto process
2004: Launch of the very-deep-submicron CMOS (90 nm, 7LM) and HBT Sige:C BiCMOS 0.25 μm processes
2006: Launch of the CMOS (65 nm, 7LM) process
2008: Launch of the CMOS 45 nm process

7.2.2 Processes Available

Presently, the processes available for IC manufacturing are depicted in Table 7.1.

TABLE 7.1. IC Processes Available

Austriamicrosystems	0.35 μ CMOS C35B4C3
	0.35 μ CMOS C35B4M3
	0.35 μ CMOS-Opto C35B4O1
	0.35 μ CMOS Flash C35B4E3
	0.35 μ SiGe BiCMOS S35D4M5
	0.35 μ HV-CMOS H35B4D3
STMicroelectronics	45 nm CMOS CMOS045
	65 nm SOI
	65 nm CMOS CMOS065
	90 nm CMOS CMOS090
	130 nm CMOS HCMOS9GP
	130 nm SOI
	0.25 μ SiGe:C BiCMOS7RF
OMMIC	0.2 μ HEMT GaAs ED02AH

7.2.3 Integrated Circuit Design Kits and CAD Software

Design kits and libraries are distributed by CMP for most of the processes and most commonly used CAD tools. CMP sometimes develops design kits in cooperation with the manufacturers and CAD vendors. CMP also offers special CAD software conditions from a few CAD vendors. As a focal point, CMP also distributes information on configuration files, converters, and so on. About 40 design kits are available for each process and the main CAD tools.

7.2.4 Test and Packaging

Packaging and testing services are also offered. Various types of packages are supported, including DIL, SOIC, CQFP, JLCC, and PGA. Testing of prototypes is usually done by the final user. Upon request, especially for low-volume production, CMP may take over the testing together with manufacturing.

7.2.5 Key Figures

Since 1981, CMP has served more than 1000 institutions from 66 countries in various processes. Support to industries started in 1990. CMP was ISO 9002-1994 certified from 2000 to 2003. It is working on certification for the revised standard, ISO 9002-2000.

7.2.6 Recent Developments

Recent developments include the move to very-deep-submicron processes: 130 nm CMOS, 90 nm CMOS, 65 nm CMOS and 65 nm SOI, 45 nm CMOS, 0.35 μ HBT SiGe BiCMOS, and 0.25 μ Sige:C HTB BiCMOS from STMicroelectronics and the exploration of new MEMS fabrication offerings.

7.2.7 The Move to Very-Deep-Submicron Processes

CMP introduced the 130 nm CMOS process as early as 2001. A total of 250 circuits were fabricated. It introduced the 90 nm CMOS process in 2004, and 242 circuits have been fabricated since then. In 2006, the 65-nm CMOS process was launched and 57 circuits have been fabricated so far. This means a total of more than 500 circuits manufactured for about 50 research laboratories and industrial companies. These processes have been very well received. Let us discuss in detail what happened with the 90-nm CMOS processs. The 90-nm CMOS process was announced in 2004, and the first design rules manuals (DRMs) and design kits were shipped to designers that same year. The list of institutions that have used the 90-nm CMOS process to date is depicted in Table 7.2. One can notice a number of top-level universities in Europe and North America mostly. All Canadian universities are using the 90-nm CMOS process. The move to 65 nm has started. The 65 nm CMOS process was announced in 2006. Table 7.3 depicts

TABLE 7.2. Institutions Having Submitted Circuits 90-nm CMOS

Institution	Town	Country
University of Calgary	Calgary	Canada
University of Waterloo	Waterloo	Canada
Carleton University	Carleton	Canada
Dalhousie University	Halifax	Canada
University of Guelph	Guelph	Canada
University of Saskatchewan	Saskatoon	Canada
University of British Columbia	Vancouver	Canada
University of Toronto	Toronto	Canada
Ecole Polytechnique de Montréal	Montréal	Canada
McGill University	Montréal	Canada
CMC Microsystems	Kingston	Canada
University of Alberta	Edmonton	Canada
University of Macau	Macau	China
Technical University of Denmark	Lyngby	Denmark
VTT Information Technology	Espoo	Finland
University of Turku	Turun Yliopisto	Finland
ISEN	Lille	France
IMEP	Grenoble	France
THALES	Palaiseau	France
University of Stuttgart	Stuttgart	Germany
University of Paderborn	Paderborn	Germany
RWTH Aachen	Aachen	Germany
U. degli studi di Pavia	Pavia	Italy
Politecnico di Milano	Milano	Italy
U. degli studi di Pisa	Pisa	Italy
University of Perugia	Perugia	Italy
University of Parma	Parma	Italy
University of Modena	Modena	Italy
Istituto Nazionale di Fisica Nucleare	Pavia	Italy
University of Oslo	Oslo	Norway
Norwegian University of Science and Technology	Trondheim	Norway
Novelda as	Kviteseid	Norway
University of the Philippines	Quezon City	Philippines
Universidad Politechnica de Catalunya	Barcelona	Spain
Instituto Microelectronica Sevilla	Seville	Spain
Linköping U.—ISY	Linkoping	Sweden
ETH Zentrum IIS	Zurich	Switzerland
University of Neuchatel	Neuchatel	Switzerland
CERN	Geneva	Switzerland
Imperial College London	London	United Kingdom
SUN Microsystems	Montain View	United States
UC Berkeley—BWRC	Berkeley	United States
University of Michigan	Ann Arbor	United States
Stanford University	Stanford	United States
Massachusetts Inst. of Technology	Cambridge	United States

TABLE 7.2. (*Continued*)

Institution	Town	Country
UCLA	Los Angeles	United States
Achronix Semiconductor LLC	Ithaca	United States
University of Texas	Dallas	United States
Georgia Institute of Technology	Atlanta	United States
University of Virginia	Charlottesville	United States
University of Washington	Seattle	United States
Total	51 institutions from 14 countries	

TABLE 7.3. Institutions Having Submitted Circuits 65-nm CMOS

Institution	Town	Country
Katholieke Universiteit	Leuven	Belgium
McGill University	Montréal	Canada
University of Alberta	Edmonton	Canada
University of British Columbia	Vancouver	Canada
University of Calgary	Calgary	Canada
University of Toronto	Toronto	Canada
University of Waterloo	Waterloo	Canada
University of Stuttgart	Stuttgart	Germany
IMS	Bordeaux	France
LAAS	Toulouse	France
ISEN	Lille	France
LETI/CEA	Grenoble	France
ENST	Paris	France
Politecnico di Milano	Milan	Italy
Nanyang Technological University	Singapore	Singapore
Universitat Politechnica de Catalunya	Barcelona	Spain
UC Berkeley—BWRC	Berkeley	United States
University of California	Davis	United States
University of Michigan	Ann Arbor	United States
Georgia Institute of Technology	Atlanta	United States
University of Virginia	Charlottesville	United States
University of Minnesota	Minneapolis	United States
University of Pretoria	Pretoria	South Africa
Total	23 institutions from nine countries	

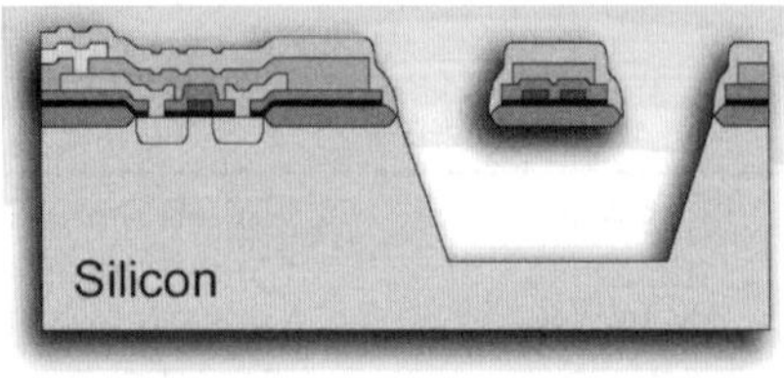

Figure 7.3. Bulk micromachining cross section.

the list of institutions that have had circuits manufactured in 65 nm. As this table shows, many top-level universities in Europe and in North America moved to the 65 nm CMOS process.

7.3 MICRO-ELECTROMECHANICAL SYSTEMS MANUFACTURING AT CMP

To address many real-life applications, ICs are necessary, but additional, basically mechanical features are also often necessary. All these features are usually provided by MEMS.

There are two families of MEMS. The first is bulk micromachining, which is also called volume micromachining. With bulk micromachining MEMS the substrate is etched in depth, with either a wet or a dry method, on either the front or back sides. This kind of MEMS is mostly used for beams, bridges, and thin structures. The second is surface micromachining. This method uses sacrificial layers, grown during the fabrication process and then removed during the post-process to create movable structures. This kind of MEMS is often used for capacitive devices.

Several types of MEMS are available from CMP, and these are classified into two categories. The first category consists of two bulk micromachining MEMS, one based on a standard CMOS process and the other on a standard BiCMOS process, which release structures with a post-process step, without any additional mask. These two processes/post processes offered by CMP allow the integration of both electronics and mechanical structures on the same circuit. The second category consists of the specific MEMS processes such as the Multi User MEMS Processes (MUMPs)® family from MEMSCAP and Sandia Ultra-planar Multi-level MEMS Technology V (SUMMiT V)™ from Sandia, which provide either surface or volume micromachining. In these processes, very advanced systems can be created on moveable platforms.

7.3.1 Bulk Micromachining Micro-ElectroMechanical Systems

CMP offers the possibility to fabricate MEMS using a low-cost 0.35 μm CMOS process from Austriamicrosystems. Structures are released after circuit fabrication with a humid tetramethylammonium hydroxide (TMAH) solution to etch the silicon as shown in Figure 7.3. Systems like bridges, micromirrors, comb drives, or sensors can be made.

The second bulk micromachining possibility is based on the 0.25 μm BiCMOS process from STMicroelectronics. This BiCMOS7RF process has five metal layers, the top one being thick metal, vertical NPN with Ft = 55 GHz, and is con-

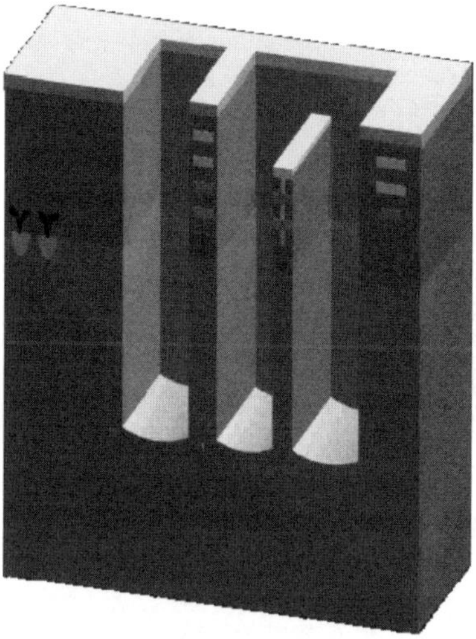

Figure 7.4. ASIMPS overview.

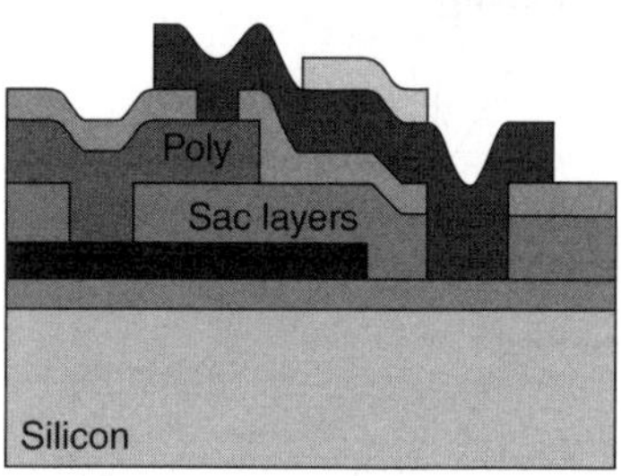

Figure 7.5. PolyMUMPs cross section.

venient for radio frequency (RF) designs. It includes MIM capacitors, inductors, and bipolar components. The associated post-process is called application-specific integrated MEMS process service (ASIMPS) and is made at the Carnegie Mellon University (CMU). Mechanical structures are released by reactive ion etch (RIE) and then by deep reactive ion etch (DRIE, as illustrated in Fig. 7.4). Potential devices to be designed and fabricated in the process include accelerometers, gyroscopes, RF MEMS communication systems (with resonator oscillators, RF filter, and high-Q inductors), infrared sensors and imagers, electrothermal converters, and force sensors. The technology enables the integration of multiple devices on the same chip. For example, high-Q inductors and micromechanical resonators can be combined for CMOS RF applications. In another example, multiple accelerometers are integrated on the chip to create a three-axis inertial measurement system. Furthermore, both the communications and accelerometer systems can be combined to form a wireless microsensor system.

Two complementary design kits are provided by CMP: one from STMicroelectronics for the electronic part and one from CMU for the MEMS part.

7.3.2 Micro-Electromechanical Systems Processes

CMP offers processes dedicated for MEMS. The MUMPs family includes PolyMUMPs, which is a polysilicon/gold surface micromachining, using sacrificial

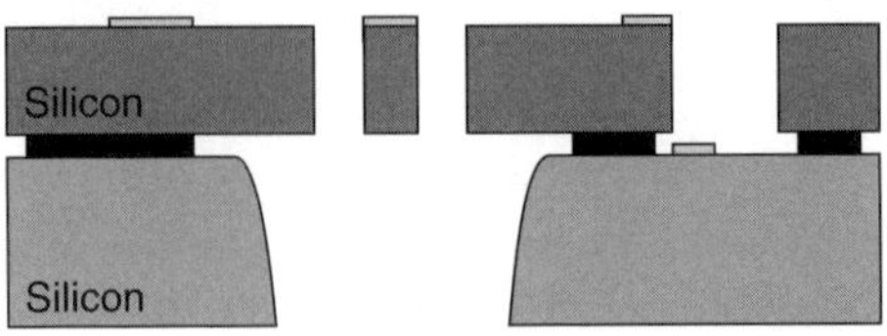

Figure 7.6. SOIMUMPs cross section.

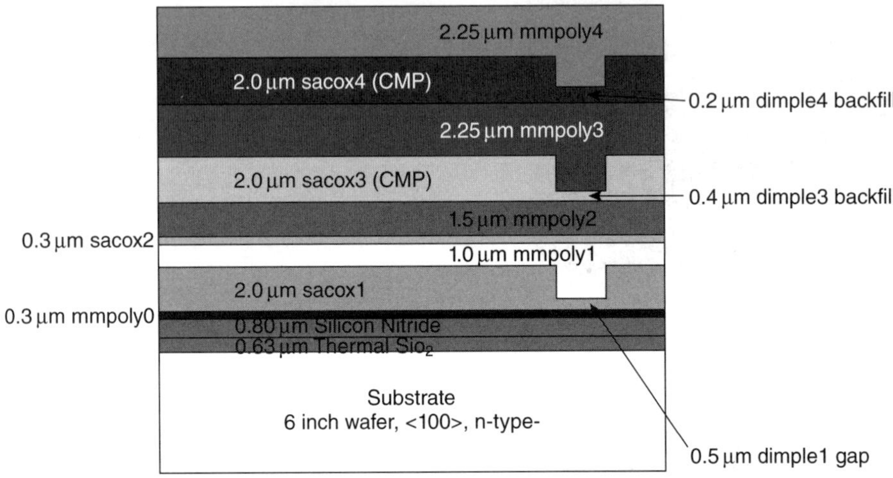

Figure 7.7. SUMMit V cross section.

layers to suspend structures for which Figure 7.5 shows the process cross section; SOIMUMPs, which uses DRIE on silicon on insulator (SOI) technology. This process enables etching of the front and back sides of a wafer to completely suspend the structures (Fig. 7.6); and MetalMUMPs, which uses a thick nickel electroplated layer. With these last two processes, the substrate is etched. For all MUMPs processes, CMP provides Cadence and Tanner design kits, with technology files and design rule checking (DRC) implementation.

The SUMMiT V from Sandia is also available. This process uses five polysilicon layers, all planarized (Fig. 7.7), offering flexibility and producing a mechanical robustness in the devices.

Systems like comb actuators, meshing gears and transmission dynamometers, laminated support springs, steam engines, microengines and micromachines, motors, mirrors and optical encoders, microsensors, RF MEMS, and linear racks can be fabricated. Two design kits are available through CMP. MEMS Pro from SoftMEMS and Autocad 2000. Both enable DRC verification, and 2D and 3D visualizations. Table 7.4 summarizes the MEMS processes available from CMP. With the wide variety of processes in the portfolio, a designer can imagine very complex mechanical structures and find a solution for fabrication. The designs can be complex, either with the electronics management and control of the MEMS or in the movable structures. CMP also provides some support on MEMS design and provides design kits upon request.

TABLE 7.4. MEMS Processes Available

Integrated micromachining	Base Austriamicrosystems 0.35 μ Base STMicroelectronics 0.25 μ BiCMOS Post-process ASIMPS from CMU
Specific MEMS	PolyMUMPs from MEMSCAP MetalMUMPs from MEMSCAP SOIMUMPs from MEMSCAP SUMMiT V from SANDIA

7.4 OTHER MAJOR INFRASTRUCTURES

Many countries made pioneering efforts during the late 1970s to the early 1980s to establish infrastructures. These efforts are documented in Courtois [1]. The first cooperative initiative in Europe was EUROMOS in 1985, undertaken by CMP in France, Darmstadt in Germany, Norchip in Denmark, and NIHE in Ireland. Next came the time of EUROCHIP and CHIPSHOP. Details can be found in Courtois [2].

Today, there are seven major national services in the world: CIC in Taiwan, CMC in Canada, CMP in France, ICC in China, IDEC in Korea, MOSIS in the United States, and VDEC in Japan. They are described in the CMP Annual Reports. Three of them, CMC, CMP, and MOSIS, decided to cooperate in 2002. It might happen later on that further cooperative arrangements will be developed.

7.5 ICs AND MICRO-ELECTROMECHANICAL SYSTEMS FOR BIOMEDICAL APPLICATIONS

Biomedical applications of electronics and MEMS in general include implant devices and biosensors, DNA-based systems, analytical protein arrays, and cell-based systems [3]. Two basic technological prerequisites are microfluidic platforms and separation-based tools on chips. The goal of this section is not to present an exhaustive panorama of all the types of biomedical applications that can be devised with some types of electronics and MEMS but only to address a few examples of what can be achieved with standard processes, for example, those available from CMP. Users do not need to call for specific custom process developments nor worry about the manufacturing; they only need to design.

7.5.1 Complementary Metal-Oxide Semiconductors for Neurosciences

CMOS ICs can be used for interfacing with cells and biological objects. Both ICs and neurons (and more generally, electroactive cells) work electrically. Electrons and holes in semiconductors and ions in cells are the information carriers. Neurons transmit information along nerves through the action potential, that is, the depolarization of their membranes. Due to differences in ion concentration between sides of the cell membranes, neurons present a negative potential inside the membrane. When membrane proteins open ion channels, a depolarization occurs and propagates along the nerve; this is the propagation of the action potential. By placing a metallic or insulator/semiconductor structure in the vicinity of a neuron membrane, it is possible to measure the depolarization and thus to access the electrical activity of cells.

The idea of trying to build an electrical connection between a living cell and an electronic circuit was developed in the 1970s. This idea was based on the measurement of the extracellular, instead of the intracellular, potential, and thus presented the possibility that a noninvasive method for accessing the electrical activity of cells could be devised.

Microelectrode arrays (MEAs) were then developed, and these succeeded in not only measuring the electrical activity of neurons and tissues (the spikes) but also in interfering with and initiating action potentials in neurons. Needle-shaped microelectrodes have also been developed in order to be implanted *in vivo* in cerebral tissue and then record its electrical activity. Despite these remarkable results, MEAs suffered from limitations in terms of signal/noise ratio and integration possibilities.

In 1991, Peter Fromherz [4] worked on a silicon/neuron junction and then developed the first real connection between a neuron and an IC. Such efforts have been pursued toward greater integration, and soon real communication between an IC and a neuron [5, 6] was achieved. By communication, we mean initiation of an action potential in a neuron, propagation to other neurons, and then reading of the signals in these other neurons through other microelectrodes in the IC.

The integration of real neuron networks with ICs is a very promising technique for neuroscience. However, some specific care must be taken for the coupling. For biocompatibility reasons, it is not possible to directly connect a culture medium to the surface of an IC. Aluminum, as an example of metal present in IC connection pads, is not compatible with neurons. Several techniques have been developed to overcome this problem, including the use of capacitive electrodes (i.e., silicon dioxide is biocomptabile) or the covering of metal electrodes with noble metals such as platinum. It is shown as an example in Figure 7.8 where a square platinum plate covers the top metal opening.

Apart from an electrical interface with electrically active cells, ICs have been used in several other biological applications, such as the measure of ion concentration in the vicinity of cells. This has been done with the purpose of studying the ionic activity of cells (through membrane proteins) in the presence of drugs in the culture medium. Several studies report the use of ion-sensitive field effect transistors to measure ionic concentration. Another application in cell biology has been to use an IC for localization and immobilization of cells. Manaresi et al. [7], have created an array of photodiodes/electrodes included in a microfluidic system. Once a cell is detected through the array of photodetectors, it is

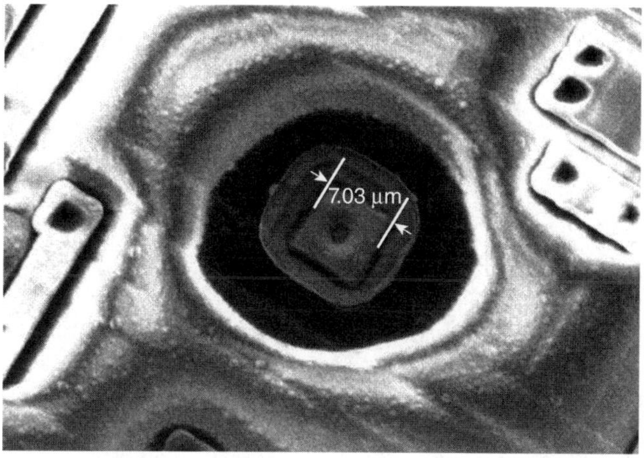

Figure 7.8. Scanning electron microscope (SEM) picture of the grid electrode of an ion-sensitive field effect transistor (ISFET) covered with a platinum layer. This electrode is a part of an ISFET sensor matrix implemented on CMOS.

kept trapped by means of a vertical dielectrophoretic well. This system allows the control of cell population on top of an IC.

All previously described applications have been developed to measure the electrical or ionic activity of living neurons. Meanwhile, there is intense research activity in the field of mimicking the behavior of neurons and synapses, with the goal of building artificial analog neuron networks. Neuron networks have been intensively studied and modeled using computers. In the case of neuromorphic ICs [8], a physical implementation of a neuron is made on silicon. This has the advantage of being real time and could be used both for the study of computing techniques and also toward the goal of hybridation with a real neuron network. Figure 7.8 shows an example of such an analog neuron network; it was devised by researchers at the University of Bordeaux [8]. This chip emulates the electrical activity of neurons using a biophysical model (Hodgkin–Huxley formalism). Five neurons have been integrated and are fully tunable. Their model cards are stored in an analog memory cell array. Such application-specific integrated circuits (ASICs), as shown in Figure 7.9, form the computation core of a complete simulation system dedicated to the investigation of the dynamics of biomimetic neural networks.

7.5.2 Bulk Micromachining for Biomedical Applications

Bulk micromachining allows the fabrication of various types of sensors for biomedical applications. In the following section, an acoustic sensor for otorhinolaryngology (ORL) surgery is briefly described.

7.5.2.1 Acoustic Sensor for Otorhinolaryngology Surgery. This project is under development jointly by the TIMA Laboratory in Grenoble and the

Figure 7.9. A neuromimetic and modular ASIC: integration of biomimetic neurons.

Hopital Nord in Grenoble. In the field of ORL, middle-ear surgery aims at correcting certain types of hearing loss or at treating certain diseases. Among different kinds of techniques, the ossiculoplasty attempts to re-establish a connection between the tympanic membrane and the oval window. This surgery involves ossicular chain reparation or reconstruction with appropriate replacement prosthesis. The three elements of the ossicular chain (the stapes, incus, and malleus), the smallest bones in the human body, provide the sound energy transfer between the tympanic membrane and the inner ear. Successful surgery can lead to the correction of hearing loss owing to tympanic membrane anomalies or a discontinuity or fracture of the ear bones. There exist a number of different techniques to reconstruct the ossicular chain using either biomaterials or various other materials such as titanium, gold, or ceramics. In spite of this progress, surgical operation of the middle ear remains difficult because of a large number of factors. Moreover, there are no available means of perioperative monitoring that can give necessary feedback to the surgeon.

The project is aimed at the development of a micromachined vibration sensor that works in the audible frequency range from 1 to 5 kHz, which is required by ORL surgeons. Such a sensor, used during a surgery, will make it easier for a surgeon to make a decision as to whether or not the realized ossiculoplasty provides an optimal transfer of the acoustic signal from the tympanic membrane to the inner ear. The simplified picture showing the main parts of the human ear and the procedure using the vibration sensor is shown in Figure 7.10. A sound source located in front of the patient's outer ear generates a test signal that

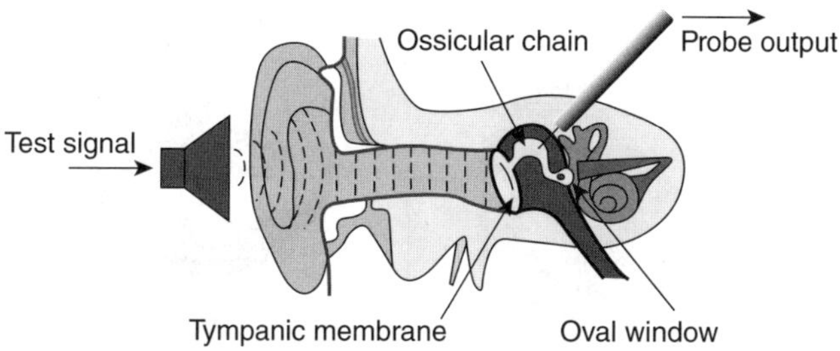

Figure 7.10. Illustration of the basic parts of the human ear and the use of the sensor.

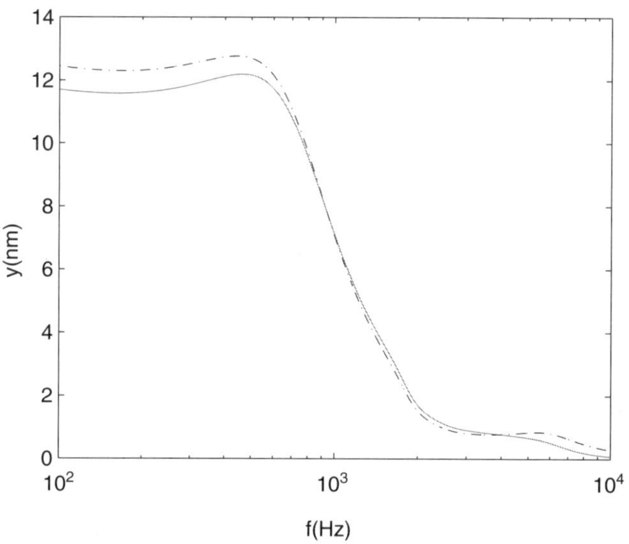

Figure 7.11. Middle-ear bone displacements as a result of the behavioral model of the ear (sound level 80 dB).

propagates through the external ear to the tympanic membrane. The movement through the tympanic membrane is transferred via the ossicular chain to the input of the inner ear represented by the oval window. The vibration sensor put in contact with any part of the ossicular chain will thus provide real-time information about its degree of mobility and about the quality of the propagated sound signal.

The MEMS-based approach to the sensor design is motivated by the small size and low mechanical impedance of the ossicular chain. A micromachined sensor tip will provide the possibility of vibration measurement by physical contact with no side effects to ear function. A careful design of the sensor is required in order to overcome the ultralow level of vibrations (see Fig. 7.11). The curve in Figure 7.11 shows middle-ear displacement values generated by the

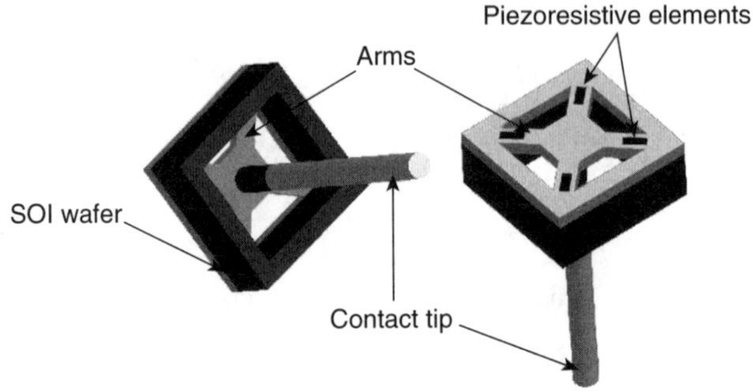

Figure 7.12. Principle of the sensor structure.

sound pressure level of 80 dB on the tympanic membrane as obtained from a behavioral model of the ear.

Different possible arrangements of the sensor are investigated. A sensor with a contact tip placed perpendicularly to the sensitive element and composed of four arms equipped with piezoresistive gauges is shown in Figure 7.12.

The sensitive element of the sensor is made from an SOI wafer. This kind of substrate facilitates the fabrication of arms with uniform thickness. The silicon arms are made using front-side micromachining. The whole sensitive structure is suspended on the cavity obtained with DRIE from the back side of the wafer. The contact tip is formed by a glass fiber attached with a central stem obtained after the patterning and the etching of the bulk silicon layer. Attention must be paid to the resulting characteristics of the mechanical structure. In particular, the mechanical impedance at the end of the tip must match that of the middle ear's ossicular chain. Too high values of the mechanical impedance may affect the function or even damage the structure of the ear; too low impedance value would not ensure the optimal transfer of the tip movement toward the piezoresistive gauges. One of the results of the sensor structure finite element (FE) modeling is shown in Figure 7.13. The zones of maximal stress on the arms as a result of force load at the end of the tip can be identified here.

Another important issue is in piezoresistive gauge optimization. Extremely low displacement values require a high signal-to-noise ratio achieved by optimal geometry and placement of the gauges, proper doping of the silicon layer, and low-noise electronics applied at the front end.

7.5.3 Multi User MEMS Processes for Biomedical Applications

MUMPs allow the manufacturing of devices for various biomedical applications. The following section successively addresses research applications and commercial applications.

7.5.3.1 Research Applications. The following examples come from Canadian universities. The projects have been collected by Canadian

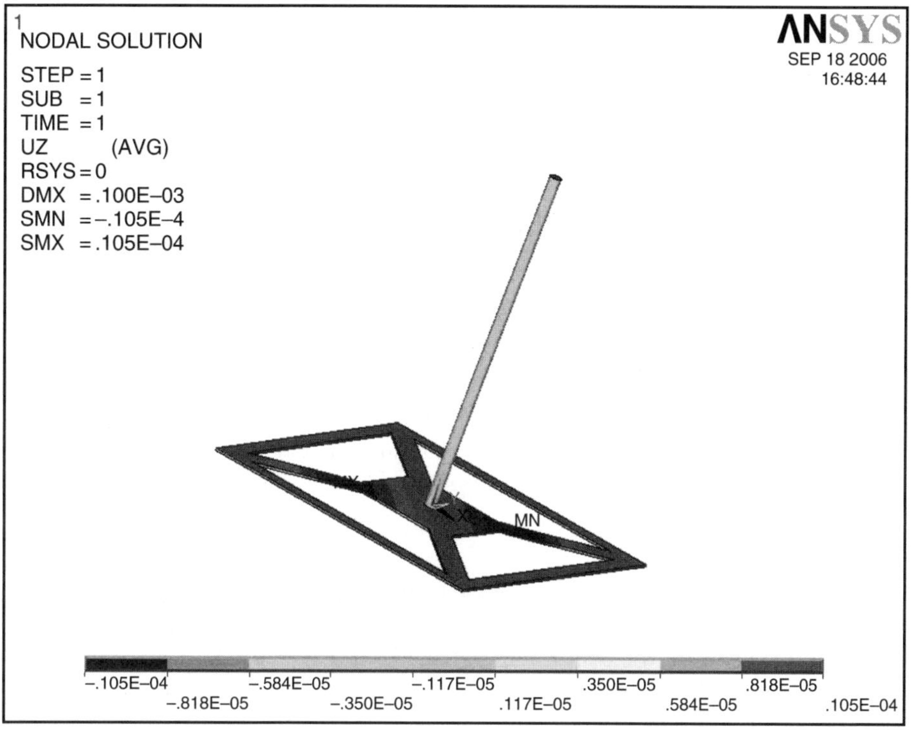

Figure 7.13. Piezoresistive sensor structure FE modeling.

Microelectronics Corporation (CMC), a service similar to CMP, which services Canadian universities. The first example comes from Kaler and Mintchev [9]. The device would extract blood like a mosquito would, electronically analyze the sample, and then transmit the data to a wireless device to monitor and control the insulin infusion pump so that the glucose balance in the body of a patient with diabetes is maintained throughout the day. Figure 7.14 depicts the device in detail. The following is taken from the designers.

The very small volume of blood (<1 mL) delivered by the sampling process is stored in a miniature blood compartment, where the microsensor converts the blood element of interest (e.g., the glucose level) into an electrical signal. This automated and self-calibrated procedure is performed by a microsensor integrated inside the blood compartment.

The major building blocks consist of (1) a microneedle; (2) a microactuator integrated with the microneedle; (3) a microsensor implemented inside a blood-collecting compartment; (4) a microelectronic stage including analog signal conditioning, an analog-to-digital converter, controlling electronic circuitry, and a digital radio-frequency transceiver; (5) a microbattery providing the energy to operate the MEMS device; (6) an associated packaging to protect the delicate microsystem inside; (7) an adhesive and antiseptic layer placed between the skin

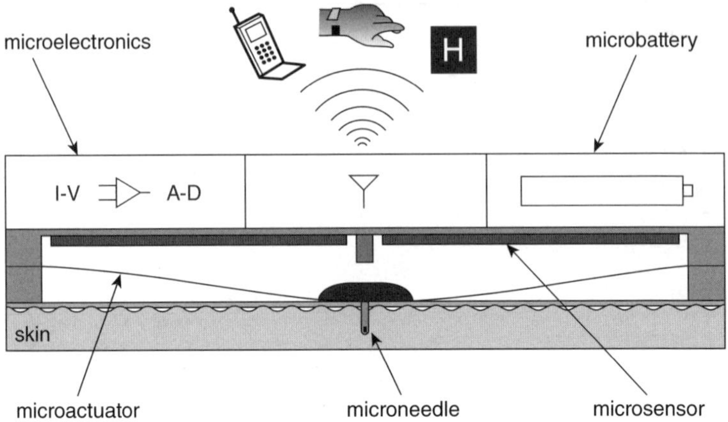

Figure 7.14. The e-Mosquito cell: The device's building blocks include microneedle, microactuator, microsensor, microelectronics, and microbattery (courtesy of the University of Calgary).

surface and the device; and (8) an enclosing band-aid to attach the e-Mosquito™ patch safely onto the skin. An array of single-use and individually actuated e-Mosquito cells form the disposable patch, and a matrix of 180 e-Mosquito cells can provide periodic blood sampling for up to 1 week, assuming that blood monitoring is required every hour.

Another example comes from Dalhousie University, Mechanical Engineering Department. Ted Hubbard et al. developed a microgripper to be used in mechanical testing of cells and bacteria, cell manipulation, and medical screening. Initial designs were made on MUMPs; they next moved to Micragem, an SOI MEMS technology from Micralyne that is available from CMC. The following is taken from Fraser, Hubbard, and Kujath [10]. An electrothermal microgripper is used. Typical displacements for chevron actuators are in the range of a few micrometers, so mechanical amplifiers are needed to increase the motion.

7.5.3.2 Commercial Applications. The following examples come from MEMSCAP. The products did not come directly from MUMPs, but MUMPs have served as test vehicles for various components. The first example is a pressure sensor that MEMSCAP is manufacturing for CardioMEMS. The wireless pressure sensor is inserted during the minimally invasive repair of abdominal aortic aneurysms (AAAs) or thoracic aortic aneurysms (TAAs), via a catheter into a patient's aneurysm sac. The small size, durability, and lack of wires and batteries enable the system to last for and transmit data over the lifetime of the patient without requiring repeated procedures (Fig. 7.15).

The second example is a wireless imaging system made for Given Imaging with endoscopy in mind. The tiny camera contained in a capsule captures images of the gastrointestinal tract as it travels through the body and transmits the

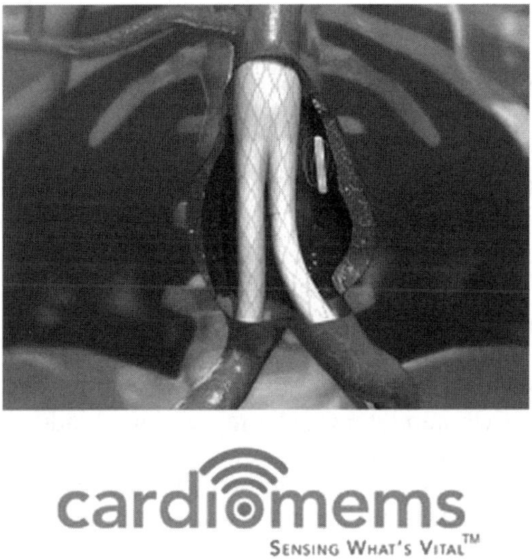

Figure 7.15. Wireless AAA pressure sensor from CardioMEMS.

Figure 7.16. PillCam Capsule from Given Imaging.

images to a computer so a physician can view them and make a diagnosis (Fig. 7.16).

7.5.3.3 Application-Specific Integrated Micro-Electromechanical Systems Process Service for Biomedical Applications. ASIMPS from CMU supports the design of biomedical applications. Here, an example is a bone implantable stress sensor. The following is taken from Alfaro et al. [11]. The clinical management of skeletal trauma and disease relies on radiographic imaging to infer bone quality. However, bone strength does not necessarily correlate well with image intensity. There is a need for a safe and convenient way to measure bone strength *in situ*. The goal is to present a new technique to directly measure bone strength *in situ* at a microlevel scale through a MEMS sensor. The proposed

MEMS stress imager comprises an array of piezoresistive sensor "pixels" to detect stress across the interfacial area between the MEMS chip and bone with resolution to 100 Pa, in 1 second averaging. The sensors are integrated within a textured surface to accommodate sensor integration into bone. From initial research, surface topography with 30- to 60-μm features was found to be conducive to guiding new cell growth. Finite element analysis (FEA) has led to a sensor design for normal and shear stress detection.

In contrast to the local nature of the validation in single-sensor experiments, an array of piezoresistive elements offers the possibility of global data over an entire surface on the order of 1–4 mm^2.

7.5.3.4 A System View: Biomedical Science and Beauty. This last section gives a detailed example of a whole system including various kinds of sensors, ASICs for signal processing, data acquisition, and expert systems. It comes from IntuiSkin, a wholly owned subsidiary of the MEMSCAP Group. IntuiSkin is focused on innovative, technology-based skin-care solutions. These solutions provide an answer to the new and strong consumer demand for technology-based cosmetology. They allow the characterization of the skin in general, in order to recommend suitable cosmetic treatments.

The Visio probe uses its sensors to capture skin images with extreme precision. The system enables many measurements, including measurements of wrinkles, sebum, hairiness, dark spots, and clogged pores or bacterial infection. The Physio probe contains sensors and extracts *in vivo* the key characteristics of the skin. This probe measures, among other parameters, hydration, transepidermal water loss (TEWL), and skin temperature. Several pieces of equipment and products have been devised by IntuiSkin to address various needs. Skin Evidence addresses the medical market. It is an answer to the practicians' needs in detection and cure, as well as in choosing specific treatments (peeling, injection, and/or fillers), and to the clinicians' needs in pre- and postsurgical support. It is expected that the system will be used by skin specialists, dermatologists, dermatocosmeticians, and plastic surgeons in their clinics or in their offices.

The other market addressed by the general concept of IntuiSkin is the beauty market. The IOMA Beauty Diag™ measures seven main dysfunctions of the skin: hydration and UV damage, fine lines, wrinkles and elasticity, redness, bacterial infection, sebum, and dark spots. It is expected that aestheticians will use this equipment to recommend the best treatments.

The last exemplified systems, the Sensicards™, are portable systems for consumer use. Sensicards enable the optimization of cosmetic product use, to watch and check skin evolution, so as to enjoy the sun's benefits without constraints. Sensicards are systems equipped with sensors, microelectronics, data displays, batteries, a communications interface, and data analysis and management. They are custom made and integrated in a light and resistant packaging, about the size of a credit card. The UV-Sensicard is expected to be used for protection of the skin against the sun. Once the sun protection factor (SPF) of the available sun cream is selected, this card enables the measurement and display of the intensity of the ambient UV, and recommends a sun cream reapplication time for each skin type. Other Sensicards have been developed along the same general concept.

7.6 CONCLUSIONS

Several conclusive aspects are addressed. The first one deals with biomedical applications [13]. Next, directions for service organizations like CMP are discussed.

7.6.1 What Is Important for the Biomedical Community?

Many kinds of biomedical applications have been addressed in this chapter, ranging from neuroscience to surgical aids, to endoscopy, to skin treatment. Many other kinds of applications may be devised in the future. For example, going further from building upon applications for dermatology, hardware devices might be designed in view of the emerging market in dermonutrition or nutricosmetics. Danone is offering yogurts "nourishing the skin from inside," which is being marketed as nutritional food that fights skin aging, and L'Oreal, together with Nestlé, is offering nutraceuticals with cosmetic benefits (the so-called beauty pills), In both cases, the efficiency of the products can be scientifically measured by specific devices.

What is important for the biomedical community is that education and research should take advantage of these infrastructures in the same way as education and research in microelectronics took advantage of these infrastructures in the 1980s. At that time, these infrastructures offered to electrical engineering and computer science students, teachers, and researchers the possibility of focusing on the design of complex circuits, hence focusing on applications as these infrastructures freed them from the burden of the manufacturing processes, the cost of their projects, or both. Today, various CMOS and MEMS processes allow students, teachers, and researchers to focus on biomedical applications. Not all possible applications can be realized using standard processes offered by service organizations like CMP, but these service organizations are continuously expanding their portfolios of available technologies. CMC from Canada recently introduced, for example, access to a microfluidics platform. In addition to providing fixtures for custom fluidic microchips, the platform advantageously uses multiple technologies—photonics, electronics, and embedded software, which can expand the set of biomedical applications currently targeted by teachers, researchers, and students.

7.6.2 Where Should Service Organizations Like CMP Go?

Remarks on directions for today and tomorrow are addressed in the following section.

7.6.2.1 Today: Cooperation Is a Must. Key issues at CMP in 2007 include the following:

- more and more circuits: +25% from 2005 to 2006, +22% from 2006 to 2007;
- industrial circuits are maintained to about 20% of the total number of circuits and low-volume production is provided up to tens of wafers;
- a large portfolio of technologies (17 different processes from low-cost processes to very advanced ones) for ICs and MEMS.

CMP is doing well, as with other service organizations. But no single service organization can offer a very large portfolio on its own. Cooperation between service organizations is a must.

In 2004, CMP cooperated with CIC in Taiwan. Since 2002, a major cooperative effort is being undertaken by CMC, CMP, and MOSIS. These three infrastructure services announced their combined effort in June 2002. Since then, the cooperative effort has steadily expanded [14]. CMP has also set up cooperative endeavors with ICC in China and IDEC in Korea. Separately, CMP has also set up a number of bilateral cooperative programs with special groups in various countries and has established distributors in several parts of the world.

7.6.2.2 Tomorrow. Several comments are addressed in the following, along two broad lines: more Moore, and more than Moore; and two considerations: going global and excelling.

More Moore.

It has been recognized that students, researchers, and SME designers must be provided with the possibility of having their circuits fabricated. From its inception in 1981, CMP has successfully pursued this goal and has experienced significant growth to reach and to keep its present level. Our success is partly a result of our basic service principles, which compel us to operate along industrial and advanced process lines. Advanced processes are more and more necessary because of the need for and needs of very skilled designers, and because CAD industrial software (which is replacing university CAD software) is more widely available to universities. Since new versions of CAD software are targeted for industrial use, there is no choice but to use advanced processes. Industry also makes more use of CMP services. During the 1980s, the available CMP processes were not very advanced, but they approached more and more those of state-of -the-art industries during the 1990s owing to CAD software and in response to increasing industry use of CMP. Since then, CMP has always offered state-of-the-art processes.

But cost and acceptance are slowing down the use of ever-advancing processes. Practical difficulties are popping up, including power density, temperature, variability, leakage power, and analog design. The costs rise if circuits are not manufactured in very large volumes. A way to overcome the cost issue is to choose larger die sizes for current-generation geometrics. Nevertheless, fundamental limits originating in thermodynamics, quantum mechanics, electromagnetics, and so on, will stage show stoppers sooner or later.

MORE THAN MOORE. The quest for ever larger densities may also be satisfied with three-dimensional (3D) integration though possibly not including very advanced-process dies that could reduce costs. System-in-package (SiP) integration allows heterogeneous integration, such as sensors, and electronics, as opposed to systems on chip (SoC) in which all parts are manufactured in the same process, which is not always optimal.

Three-dimensional processes using through silicon vias (TSVs) allow going one step farther; interconnections are shorter, hence leading to power savings and better performance, and there are fewer inputs and outputs (I/Os), so less power is consumed. Heterogeneous integration is also possible for SiP. There are still concerns to be addressed such as thermal issues (increased density may lead to hot spots, electromigration, etc.), power issues (no power savings if current-generation geometrics are used), and CAD software. CMP has recently introduced 3D processes using TSVs.

A step further may be required for some applications where the substrate is required to be flexible, as for many biomedical applications. Organic electronics also allow 3D integration with a low cost for large areas, and for heterogeneous integration. Biocompatibility is also a plus for biomedical applications. Chuo and Kaminska [15] have introduced a good example of such systems. A multilayer polymer microsensor system allows skin-tissue conformity since the substrate can bend up to 30°. It includes various sensors, signal conditioning and processing, RF communication, and power from disposable or rechargeable batteries. The prototype includes electrodes for electrocardiogram (ECG) and a MEMS accelerometer for body positioning calculation.

It is also recognized that complementary developments must be addressed in order to address more diversified needs. In this respect, CMP has been a pioneer. In 1995, we were the first service in the world to offer MEMS processes. Farther along the way, fields beyond mechanics–electronics, such as photonics, optics, fluidics, and so on, are to be addressed. CMP will be actively promoting these developments in the future.

GOING GLOBAL. Cooperation between service organizations has already been mentioned. On the side of the users of CMP and of other similar services, the design way is also getting more and more global in the sense that more and more intellectual property blocks may come from various sources. These blocks are a consequence of the ever-increasing complexity of designs, including parts coming from various teams, companies, and countries. An initiative is being developed to address this unintended obstacle created by our cooperative progress: the Global Education for Microelectronic Systems (GEMS) [16].

EXCELLING. Globalization also requires that we excel in order to stay ahead of others. This is important in a global market environment where some countries can offer low production costs, and high production costs in other countries may make their production services less attractive to a consumer. Some countries that were low cost a few years ago are already experiencing competition from other countries offering even lower costs, forcing some to outsource their own outsourcing [17]. The way to combat this is to stay ahead of the others. The way to stay ahead is to educate and to research using top-level electronic processes available from top-level services like CMP.

REFERENCES

[1] B. Courtois, "MPC services available worldwide," *APCCAS '94 IEEE Asia-Pacific Conference on Circuits and Systems*, Taipei, Taiwan, December 5–8, 1994.

[2] B. Courtois, "Infrastructures for education and research: From national initiatives to worldwide development," Technical University of Darmstadt, M. Glesner 60th birthday ceremony, August 29, 2003.

[3] G. Urban, Ed., *BioMEMS*. Springer, 2006.

[4] P. Fromherz et al., "A neuron silicon junction: A retzius cell of the leech on an insulated gate field effect transistor," *Science*, 252(1952), pp. 1290–1293, 1991.

[5] P. Fromherz and A. Stett, "Silicon-neuron junction: Capacitive stimulation of an individual neuron on a silicon chip," *Phys. Rev. Lett.*, pp. 1670–1673, 1995.

[6] P. Fromherz et al., "A 128 × 128 bio-sensor array for extracellular recording of neural activity," *International Solid State Circuits Conference (ISSCC)*, 2003.

[7] N. Manaresi et al., "A CMOS chip for individual cell manipulation and detection," *IEEE J. Solid-State Circuits*, 38(12), pp. 2297–2305, 2003.

[8] S. Renaud et al., "Neuromimetic ICs with analog cores: An alternative for simulating spiking neural networks," *Proceedings of the IEEE 2007 International Symposium on Circuits and Systems, (ISCAS '07)*, New Orleans, LA, pp. 3355–3358, 2007.

[9] K. Kaler and M. P. Mintchev. "Electronic Mosquito: A Semi-Invasive MEMS for Blood Sampling and Analysis," University of Calgary, Electrical and Computer Engineering Department, 2003.

[10] J. Fraser, T. Hubbard, and M. Kujath, "Theoretical and experimental analysis of an off-chip microgripper," *Can. J. Elect. Comput. Eng.*, 31(2), pp. 77–85, 2006.

[11] F. Alfaro et al., "BioImplantable bone stress sensor," *Proceedings of the International Conference of IEEE Engineering in Medicine and Biology Society (EMBS)*, Shanghai, China, September 1–4, 2005.

[12] J. M. Karam, "Panel on success stories in MEMS-based systems," *Design, Test, Integration and Packaging of MEMS and MOEMS Conference*, Nice, France, April 9–11, 2008.

[13] B. Courtois, B. Charlot, G. Di Pendina, and L. Rufer, "Infrastructures for education, research and industry: CMOS and MEMS for BioMed," Invited paper at the *12th World Multi-Conference or Systemics, Cybernetics and Informatics (WMSCI 2008)*, Orlando, FL, June29–July 2, 2008.

[14] C. Pina, B. Courtois, and D. Gale, "CMC, CMP and MOSIS 'The Scale of Cooperation Increases as the Dimensions of Microchips Decrease'," *3rd International Conference on Microelectronic Systems Education*, Anaheim, CA, June 1–2, 2003.

[15] Y. Chuo and B. Kaminska, "*Multiparameter single locus integrated multilayer polymer microsensor system*", to appear in IEEE Trans. on BIOCaS, 2008.

[16] A. Rucinski, "Global education for microelectronic systems." Available: http://www.ieee-gems.edu.

[17] "India outsources its own outsourcing–Pioneer nation fends off new rivals," *International Herald Tribune*, September 25, 2007.

8

CMOS CAPACITIVE BIOINTERFACES FOR LAB-ON-CHIP APPLICATIONS

Ebrahim Ghafar-Zadeh

8.1 INTRODUCTION

Laboratory-on-chip (LoC) is a multidisciplinary approach used for the miniaturization, integration, and automation of biological assays [1, 2]. Biology and chemistry are experimental sciences that continue to evolve and develop new protocols. Each protocol contains step-by-step laboratory instructions, lists of necessary equipment, and required biological and/or chemical supplies [3]. A biological laboratory contains various pieces of equipment used for performing different types of biological protocols. The engineering aspect of LoC design aims to embed all these components into a single chip for specific applications. Several clear advantages of this technology over conventional approaches, including portability, full automation, ease of operation, low sample consumption, and fast assays time, make LoCs suitable for highly throughput screening, early detection of disease, point-of-care testing, and environmental assessment [4, 5].

Standard complementary metal-oxide semiconductors (CMOS) technology, by offering a distinct cost, integrated circuits (ICs), and embedded sensors and/or actuators, as such, is a good fit for implementation of some of the essential functions of LoCs [6, 7]. Indeed, microelectronic CMOS technology allows one to fabricate an active substrate with different types of sensing or actuating sites that can be addressed in a random access mode by means of decoders. Photodiodes [8], ion-selected field effect transistors (ISFETs) [9], lateral/substrate bipolar transistors [10], and microelectrodes, realized in the topmost metal layer, are the front-end devices in the CMOS process, which can be employed for optical

CMOS Biomicrosystems: Where Electronics Meet Biology, First Edition. Edited by Krzysztof Iniewski.

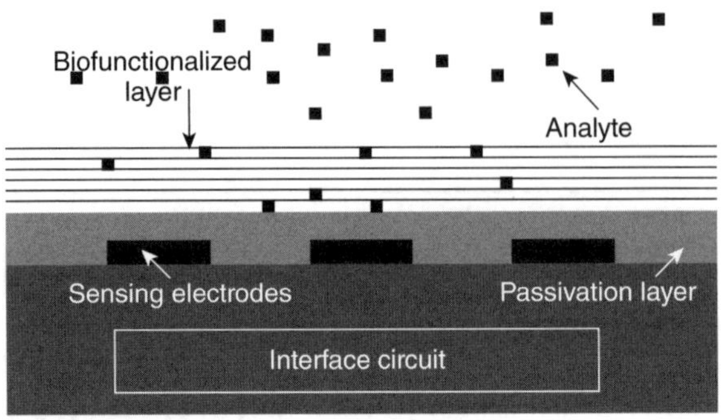

Figure 8.1. Illustration of CMOS capacitive sensor consisting of integrated circuit, sensing electrodes and functionalized sensing layer.

sensing, electrical charge detection, temperature measurement, and electrochemical sensing, respectively. Additionally, the topmost metal layer can be patterned so as to realize dielectrophoresis microelectrodes or magnetic microcoils for cellular manipulation [11, 12]. Among various microelectronic devices employed for LoCs, CMOS capacitive biointerfaces have received significant interest for several applications, including DNA detection, antibody–antigen recognition, and bacteria sensing [13–15]. As shown in Figure 8.1, the sensing electrodes on the top of CMOS chip serve as an interface between the microelectronic readout system and the biofunctionalized sensing layer. The presence of the analyte results in changing the physiochemical properties of the sensing layer. Such minute physiochemical changes can be detected using high-precision capacitive interface circuits.

The main components of CMOS capacitive biosensors, which are sensing electrodes, biofunctionalized sensing layers, and interface circuitries, are described in Sections 2, 3, and 4 respectively. In Section 5, the microfluidic packaging of CMOS sensors will be put forward. This section is followed by a conclusion in Section 6.

8.2 ON-CHIP SENSING ELECTRODES

The capacitive sensing electrodes can be placed on top of the CMOS chip using standard design rules. The performance of sensing electrodes is influenced by several parameters, including the shape and dimension of electrodes. Additionally, the compatibility of the sensing layer or bioparticles with the material of the metallic electrode and/or the insulation layer above the electrodes is a very important issue in capacitive biosensors. To date, several techniques to create capacitive sensing electrodes above the CMOS chip for biological and chemical sensors have been reported. Herein, a review of these on-chip microelectrode techniques is presented for biological applications.

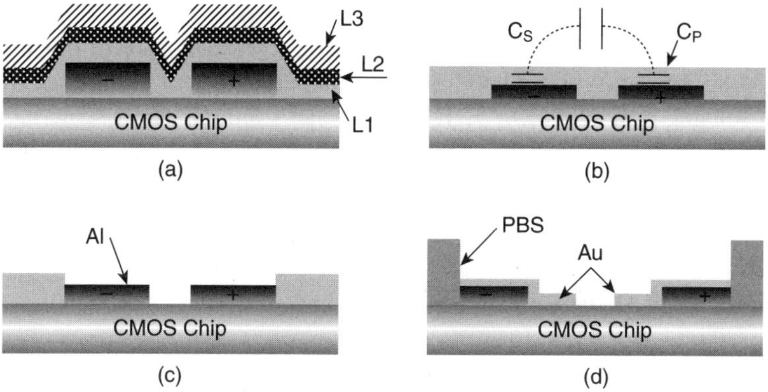

Figure 8.2. On-chip capacitive electrodes: (a) passivated electrodes with three layers (L1, L2, and L3), (b) passivated electrode with one layer (reported for cell monitoring), (c) uncovered aluminum (Al) electrode, and (d) on-chip gold electrode using micromachining procedure upon the fabrication of chip using standard CMOS processes.

8.2.1 Passivated Electrodes

In standard CMOS technology, the passivation layers (e.g., 0.18 μm CMOS process includes silicon oxide, silicon nitride, and polyamide) are stacked in the last step of the process (Fig. 8.2a). A thin passivation layer with uniform thickness is used to isolate the sensing electrodes from the conductive electrolyte. As reported by Prakash et al., the passivated electrode can be employed for monitoring the cell activity [16]. In this case, a large parasitic capacitance C_P (e.g., 0.5-μm CMOS process, 0.05 fF/μm^2 [17, 18]) across the passivation layer is approximately discounted in series with small sensing capacitance C_S (see Fig. 8.2b). The passivated sensing electrode atop the CMOS chip can also be used for bacterial growth monitoring as reported by Ghafar-Zadeh et al. [19]. In this work, the passivation layer in between the electrodes was removed in order to increase the sensitivity and dynamic range of sensing electrodes. The passivated electrode method can also be employed for other applications, including polyelectrolyte layer detection and organic solvent monitoring [20–22].

8.2.2 Aluminum Electrodes

Aluminum is widely used for electrical contacts and interconnections in microelectronic technologies. The topmost metal layer in standard CMOS processes is also made of aluminum, and a small impurity concentration of silicon (e.g., 0.18 μm CMOS, Al/1%Si, Al with 1% silicon) [23]. The passivation layers can be removed from the top of the electrodes in order to make a direct contact with the aluminum electrode (Fig. 8.2c). The "pad" mask layer in this technology can be selected by the designer if needed for passivation layer removal.

Aluminum is not widely used for biosensing purposes as opposed to gold and platinum. However, a native Al_2O_3 layer (≈10 nm) on the surface of the aluminum increases the durability in typical biosensors, as reported for DNA detection and bacteria-sensing purposes [15, 23]. It is worth mentioning that by selecting a thick topmost metal layer in the CMOS process and also by selecting a pad–etch mask, a larger space in between the electrodes can be made; therefore, a strong electrical field is created in this space.

8.2.3 On-Chip Gold Electrodes

Gold is a stable noble metal and a far better electron conductor than aluminum, copper, or even silver. This highly defined and conductive surface of gold may be ideal for several biosensing applications, including bacterial growth monitoring, virus detection, and DNA detection [24–27]. The gold layer can be readily fabricated using commercially available lithographic technologies on a CMOS chip using CMOS-compatible micromachining procedures with a low temperature [28–32] (Fig. 8.2d). These procedures are performed to connect the deposited gold to the underlying circuitry deep in the CMOS chip. By removing the silicon oxide and silicon nitride layers above the via, titanium (Ti) and titanium nitride (TiN) are thereafter sputtered as an excellent diffusion barrier material. Tungsten is used to fill the opening. The double layer of Ti/Pt is also deposited above the created via, followed by appropriate thicknesses (e.g., Ti/Pt/Au 50 nm/50 nm/500 nm [32]) of gold using a lift-off process. A resist layer, such as polybenzoxazole (PBS), is coated on top of the gold electrode and other parts of the chip. This layer is thereafter developed and baked at a certain temperature.

8.3 CAPACITIVE BIOCHEMICAL METHODS

In this section, we discuss the most important capacitive affinity biosensors, including DNA hybridization detection, antibody–antigen recognition, organic solvent detection, chemical gas detection, polyelectrolyte layer recognition, bacterial growth monitoring, and protein-based sensing techniques.

8.3.1 DNA Detection

DNA molecules consist of sequences of four building blocks (A, G, C, T). The hydrogen bonding of two complementary single strands of DNA, such as AGCATA and TCGTAT, is called hybridization. This process is used to detect the unknown strands that wander up and down in a DNA sample while the known strands are immobilized on a substrate such as a gold layer (in Fig. 8.3a). The hybridization of complementary strands can be detected optically by fluorescence tags, which are already labeled onto the input DNA molecules [33–35]. A decrease in the electrochemical capacitance of an electrode–solution interface after being exposed to unknown fragments can also be detected by a

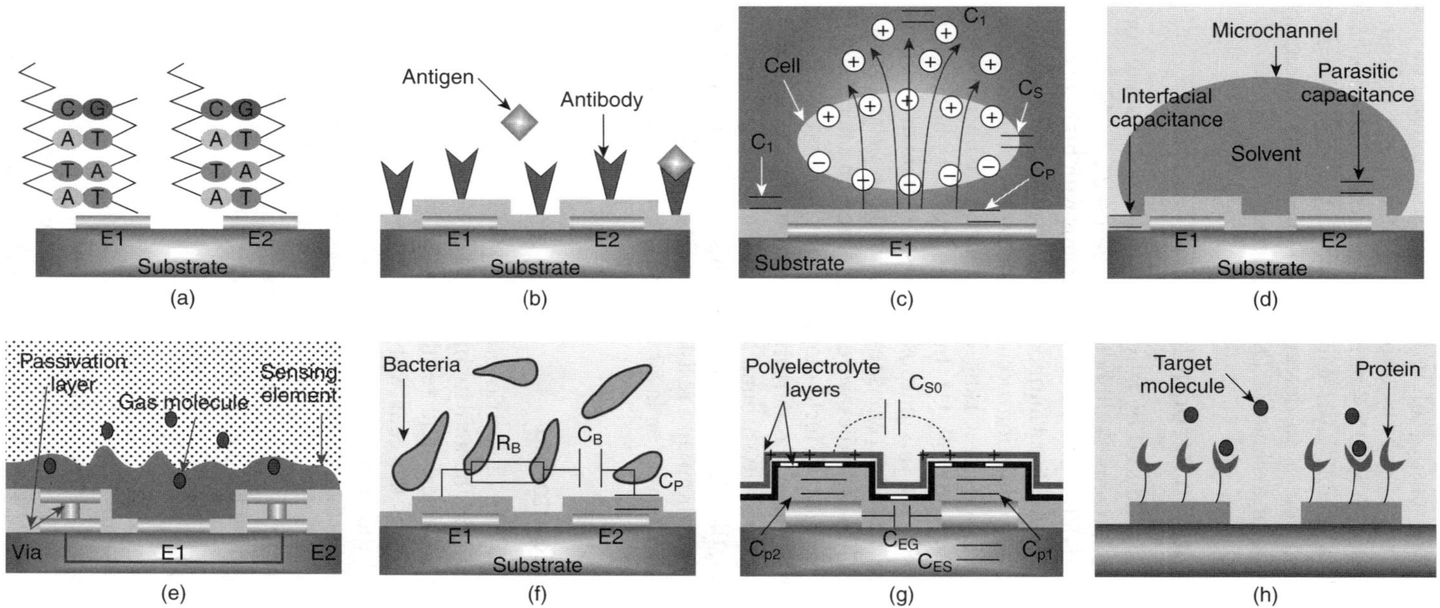

Figure 8.3. Capacitive biointerfaces methods for (a) DNA detection, (b) antibody–antigen recognition, (c) cell monitoring, (d) organic solvent detection, (e) gaseous chemical sensor, (f) bacteria sensor, (g) polyelectrolyte layer detection, and (h) detection of conformational change of proteins.

low-complexity capacitive sensor [36]. In this direction, Guiducci et al. reported the realization of an array of gold electrodes and subsequent readout circuitry using a CMOS process [37]. The DNA molecules on gold electrodes can be electrically modeled with a double-layer capacitance in parallel with resistors as reported [37]. The main goal is to detect this double-layer capacitance, which depends on the presence or absence of DNA molecule.

8.3.2 Antibody–Antigen Recognition

Antibodies are proteins produced by a living creature to inactivate foreign particles, which are called antigens. Antibodies can be used as a biofunctionalized sensing layer in order to detect the unknown antigens with high specificity. When an antigen binds to an immobilized antibody layer, such as antihuman transferrin [38] at the substrate, a modulation occurs in thickness or dielectric constant parameters. This modulation in sensing capacitance can be detected through a capacitive sensor (Fig. 8.3b) [39–41]. In one of the few published reports of using capacitive methods for this purpose, Balasubramanian et al. reported a virus detection through a standard enzyme-linked immunosorbent assay (ELISA) [42] on a CMOS chip. The antibodies were immobilized on silicon dioxide, where a thick metal layer in the CMOS technology was used to increase the sensitivity. Also, a temperature treatment was performed prior to measurement in order to decrease the ionic conductivity around the sensing sites. After applying the temperature treatment, the electrical model of this sensor can be shown purely with two capacitances (without any resistance), which are the parasitic capacitance created across the passivation layer and the sensing capacitance above the chip (Fig. 8.3b).

8.3.3 Living Cell Monitoring

The surface charge [43] or dielectric property [44] of a living cell can play essential roles in measuring the activity of living cells. These charges are moved upon exposure to an electrical field resulting in a dipole moment as shown in Figure 8.3c. The contact of the cells with the passivation layers on top of the sensing electrodes and the created dipole moment are significantly detected by a capacitive sensor. In this figure, the sensing capacitance, interfacial capacitance, and parasitic capacitance across the passivation layer are denoted as C_S, C_I, and C_P, respectively. In order to detect such a sensing capacitance (C_S), a floating capacitor technique can be used as described in Section 4.

8.3.4 Organic Solvent Sensors

The detection of organic solvents such as benzene, toluene, trichloroethylene, and tetrachloroethylene, which cause many health problems, including breast cancer, is a very critical task for environmental monitoring applications [45]. The dielectric constants of solvents can be used as the selectivity and/or sensitivity factor for their detection purposes. The electrical model of the sensing electrodes exposed to the organic solvents is shown in Figure 8.3d, while a low electrical conductivity is expected for most polar solvents. This figure shows the sensing capacitance (C_S), the interfacial capacitance (C_I) created on the surface of the passivation layer, and the parasitic capacitance (C_P) created across the passivation layer. The large value of C_I and C_P can be discounted in the series with C_S. For this purpose, a CMOS capacitive sensor has already been reported using the 0.18 μm CMOS process [46]. This sensor chip features an array of interdigitated electrodes and a readout circuitry as reported [47]. In another effort, a capacitive sensor was reported for the detection of gaseous organic solvents [48] by realizing quasi-interdigitated electrodes above a CMOS chip. As shown in Figure 8.3e, the electrodes were implemented through the two topmost metal layers using an etching process to remove the passivation layers in between the electrodes. A polymeric recognition element, such as polyetherurethane (PEUT), is also used to coat the electrode. This polymer layer allows the absorption of target chemical molecules, depending on their concentration. Consequently, the output of capacitive sensor varies as a function of the concentration of organic solvent with specific dielectric constants.

8.3.5 Bacterial Growth Monitoring

Growth of bacteria occurs via the division of one bacterium into two bacteria. The growth and/or settling of bacteria create a biofilm on the sensing electrode that consequently varies the dielectric constant and conductivity of this biofilm over time [49]. A medium solution is used to grow bacteria. For example, Luria-Bertani (LB) is the medium of *Escherichia coli*. Figure 8.3e shows the electrical equivalent models of the sensing electrodes exposed to the bacteria solution. C_B and R_B are the corresponding capacitance and electrolyte resistance of the bacteria solution, respectively. In this simplified model, the effect of the double-layer capacitances (C_{db}) is discounted in series with other small capacitances. This capacitive model is the principle of on-chip capacitive bacteria growth technique reported by Ghafar-Zadeh et al. using *E. coli* bacteria with two different concentrations [50].

8.3.6 Polyelectrolyte Monolayer

By injecting the dilute aqueous solutions of positively and negatively charged molecules or the so-called polyelectrolyte onto the substrate, an arbitrary number of times, the ultrathin polyelectrolyte layers (PLs) can be built up as a result of the electrostatic attraction between them. The thickness of ultrathin PLs is on the order of a few nanometers [51]. This low-temperature process with no destructive chemical solutions is CMOS compatible, and the ultrathin PLs such as the polycation chitosan and the polyanion alginate polyelectrolytes can be detected using active sensors in a CMOS chip [52]. Figure 8.3g shows a simple method of creating such multilayers of charged molecules on a CMOS chip. The first layer is always a positively charged polyethyleneimine (PEI) layer, which is formed on the surface of the chip to initiate the sequential adsorption of the weak polyelectrolytes. This layer can be followed by alginate or chitosan, layers subsequently as reported [53]. In this figure, C_{p1} and C_{p2} represent the parasitic capacitances between the electrodes and the analyte, and C_{EG} and C_{ES} are the parasitic capacitances between the electrodes and the substrate. Based on this model, the formation of the first PL results in a large capacitance, $C_S = C_{PL1}$ (see Fig. 8.3a). In this case, the total capacitance standing on the right electrode is approximately equal to ($Cp1 \| Cp2 + C_{EG} + C_{ES}$).

8.3.7 Detection of Protein Conformation

Proteins, with their biologically active macromolecules that can be easily introduced to living environments and their unique binding properties, are good candidates for the creation of high-specificity biosensors. There are many proteins that can be used for various biomolecules [54]. Proteins can be functionalized through genetic engineering procedures and thereafter immobilized on a gold substrate using the self-assembly monolayer (SAM) technique as a linking layer [55, 56]. SAMs have been widely used for studying biological and chemical processes, molecular interactions, and electrochemical and electronic applications [57]. The binding of target molecules with proteins results in a conformational changes accordingly. Conformational changes in proteins upon binding with the corresponding molecules can be detected through various devices, including capacitive sensors (Fig. 8.3h).

8.4 CAPACITIVE INTERFACE CIRCUITS

The design criteria for capacitive interface circuits for LoC applications differ from the criteria for conventional micro-electromechanical systems (MEMS)-based applications, such as acceleration, vibration, or pressure [58]. A movable sensing electrode should be implemented through MEMS procedures and then connected to an interface circuit for measurement purposes, but the surface electrodes can be directly realized on an IC chip fabricated through standard CMOS technology. This is why a large parasitic capacitance (C_p) on the order of a few picofarads (pF) is created under the pad, and consequently, this capacitance should be canceled using further circuitries (e.g., op-amp with negative feedback). Additionally, a MEMS-based capacitive sensor is often operated over a continuous and lengthy measurement time. For example, a sensory device fixed to a piece of electromachinery should be able to monitor vibration for weeks or months without error. A built-in self-calibration module is often incorporated in interface capacitive circuits to correct the accumulating errors in the MEMS-based capacitive sensors [59].

A capacitive sensor for LoC applications does not suffer from such a problem. This is because the subtraction of sensing capacitances in the presence, rather than the absence, of analyte is measured in an LoC system for only a short period of time. On the other hand, a MEMS-based capacitive sensor is prevented from having direct contact with fluids, dust, or even air by an appropriate vacuum chamber in order to obtain reliable measurement results, but an LoC chip is always exposed to aqueous biological or chemical samples. For this, the presence of nonbinded bioparticles or other remnants in the aqueous samples may interfere with the functioning of the sensor chip over time. It is necessary to clean LoC chips in order to maintain their function and output dynamic range. Based on the above discussion, the design strategy of MEMS-based capacitive sensors differs from the design and implementation of capacitive sensors for LoCs. In the remainder of this section, a brief description of the most important interface circuitries used for LoC applications are put forward.

8.4.1 Capacitive Methods for Laboratory-on-Chip Applications

8.4.1.1 Switch Capacitor Technique. Among the few capacitive techniques for LoC applications, switched capacitor interface circuitries have been reported for chemical toxic detection [60, 61]. Figure 8.4 shows a second-order switch capacitor ΣΔ modulator that was characterized as a fully differential capacitive interface sensor. By applying the reference voltages (V_{ref}) and clock pulses (Φ's), and by using an integrating capacitor (C_{int}) and other capacitors, the subtraction of sensing (C_S) and reference (C_R) capacitances is converted to digital output (y).

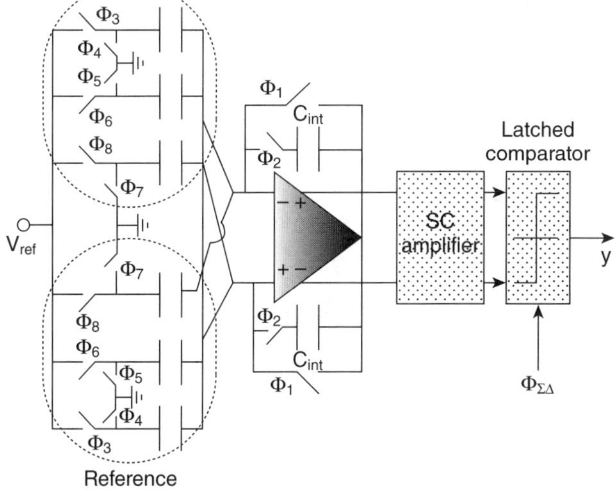

Figure 8.4. Capacitive interface circuit through conventional switched capacitor topology.

8.4.1.2 Time Constant Technique. Another effort was made by Stagni to detect DNA using a CMOS capacitive sensor [62]. The principle of this technique is shown in Figure 8.5a. In fact, the measurement of capacitance in series with an unknown resistance (R) can be performed using a resistor-capacitor (RC) technique. As shown in this figure, the capacitance is charged and discharged when the applied pulses with the certain amplitude of V_P and period of T become positive and negative, respectively. A voltage comparator is also used to detect the voltage value (V_c) dropped on sensing capacitor. Once this voltage becomes higher than a reference voltage (V_{ref}), the directions of reference current (I_{ref}) and voltage (I_{ref}) are reverted simultaneously through SW1-SW4. As the time constant of sensing electrodes is influenced by the capacitive electrode–solution interface, the frequency of the circuitry shown in Figure 8.5a depends on the sensing capacitance.

8.4.1.3 Capacitive Integrating Technique. The sensing capacitance C_S can also be measured using a capacitive inverter amplifier as shown in Figure 8.5b [63]. By applying a voltage pulse V_p, the output voltage (V_{out}) of the op-amp becomes proportional to sensing capacitance C_S. In this figure, C_R is a reference capacitance and node "A" is virtually grounded. As the parasitic capacitance C_{p2}, standing in between nodes "A" and "O," is in parallel with C_{in}, the dominator of its function (V_{out}/V_P) can be expressed simply as $C_{in} + C_{p2}$; consequently, this results in a lower amplification factor as compared to amplification in the ideal

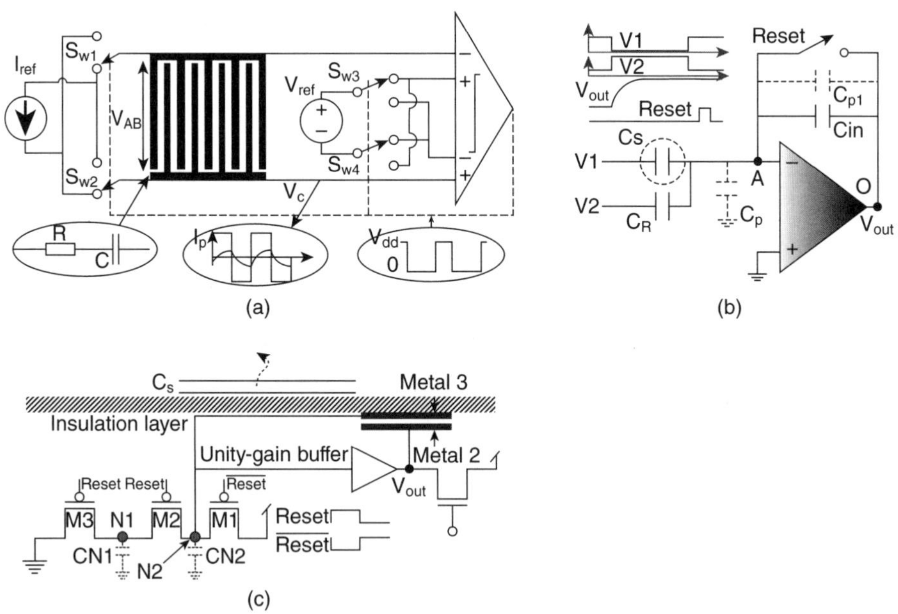

Figure 8.5. Schematic of capacitive readout circuits based on (a) time constant method, (b) capacitive integrator, and (c) a floating capacitor technique for LoC applications.

case with no parasitic capacitance C_{p2}. The dynamic range of this circuit depends on C_S, which is equal to the summation of a constant value C_{S0} and the variable sensing capacitance ΔC_S. The dynamic range of output voltage of this design is increased by an appropriate value of C_R ($\approx C_{S0}$).

8.4.1.4 Float Capacitor Technique. A floating capacitor technique can be used to detect sensing capacitance. As can be seen in Figure 8.5c, three transistors, M1–M3, charge and discharge the nodal parasitic capacitances C_{N1} and C_{N2}. By minimizing the nodal parasitic capacitances, the variation of output voltage can be increased significantly. For this purpose, the fringe capacitances between the sensing electrode and the substrate are cancelled by shielding the sensor circuitry using metal 2 with a potential that tracks the sensing electrode potential using a unity-gain buffer. This circuit topology can be employed for cell monitoring. The passivated electrodes are exposed to living cells in order to monitor their viability/mortality and to track cancer-cell proliferation as described by Prakash and Abshire [18].

8.4.1.5 Charge-Based Capacitance Measurement Techniques. Recently, core-charge-based capacitance measurement (core-CBCM) techniques for designing on-chip capacitive interface circuits for LoC applications have been reported. These methods with low complexity over their circuit topology and high precision emerge as a good candidate for CMOS capacitive biosensors. In the next section, the basics of this method are described, and its circuit topology design is evaluated.

8.4.2 Core-Charge-Based Capacitance Measurement Capacitive Sensor for Laboratory-on-Chip Applications

The CBCM method was invented as an accurate technique for the characterization of interconnect capacitance in deep-submicron CMOS ICs [64, 65]. In a CBCM structure, two signal pulses ($\Phi 1$ and $\Phi 2$) are applied to two pairs of n/pMOS transistors in order to frequently charge and discharge the interconnect capacitance C_L. Based on this method, the subtraction of the charging/discharging currents of two branches is equal to the multiplication of power supply V_{dd} and frequency f, respectively (see Fig. 8.6a). Even though CBCMs were originally proposed for the measurement of static capacitance in deep CMOS chips, this measurement technique has recently attracted attention for the measurement of dynamic capacitance on the upper surface of the CMOS chip. CBCM offers low-complexity and high-precision methods to detect the capacitance variation of the sensing electrode realized on the topmost metal layer of the CMOS chip. The new advantage of CBCM was successfully exploited for biological and chemical applications such as detection of DNA, cellular activity, and bacterial growth, and organic solvent monitoring [14, 19, 66, 67]. In the remainder of this section, the modifications that have been applied over the circuitry of core-CMOS capacitive sensor for LoC application are described.

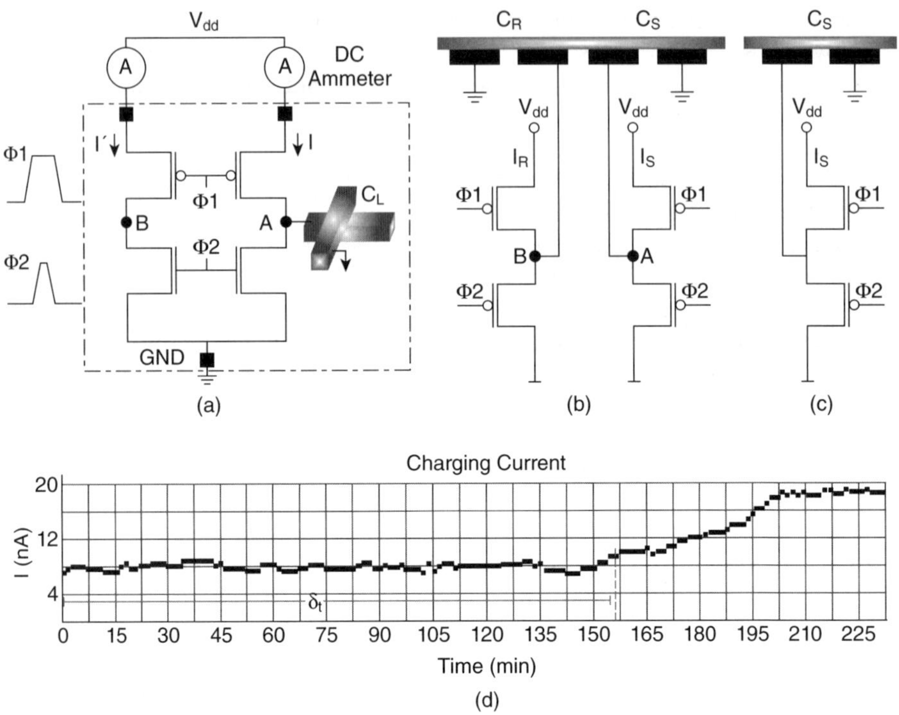

Figure 8.6. CBCM structures for the measurement of (a) interconnect capacitance, and the measurement of sensing capacitance through (b) four-transistor and (c) two-transistor structures, along with (d) experimental results of a capacitive bacterial growth monitoring using a two-transistor structure. GND, ground.

8.4.2.1 Charge-Based Capacitance Measurement Capacitive Sensor

Four-Transistor Method. As shown in Figure 8.6b, two similar capacitive electrodes, C_R and C_S, are connected to nodes A and B. C_R and C_S play the role of reference and sensing electrodes, respectively. Based on CBCM method, the subtraction of charging currents, I_R and I_S, is proportional to the subtractions of the capacitances standing on nodes A and B. Therefore, by assuming that similar sensing electrodes and transistors are present on both sides of the CBCM structure, the sensing capacitance (ΔC) can be obtained from the following equation:

$$\Delta I = f \cdot V_{dd} \cdot \Delta C, \tag{8.1}$$

where $\Delta C = C_S - C_R$ and $\Delta I = I_S - I_R$.

Two-Transistor Method. The capacitance changes of a single electrode can be measured in two steps using only two p/n transistors instead of using the four-transistor technique (see Fig. 8.6c). In the first step, the charging current I_1 is measured using a DC ammeter (A) while the channel is empty. The combination of all parasitic capacitances related to MOS transistors and electrodes are included in the current measured in the first step. At the second step, the solution is injected into the microchannel and the charging current (I_2) is measured. The differential current $I_1 - I_2$ is proportional to sensing capacitance ΔC. This simple method was reported by Ghafar-Zadeh et al. using the 0.18 μm CMOS process for organic solvent detection [22]. In another effort, they reported on bacterial growth monitoring through a similar CBCM structure. Figure 8.6d shows the variation of charging current over time. As shown in this figure, the bacteria starts growing after a sufficient time (δt) as expected from basic microbiology. δt depends on many factors, including the temperature and the medium solution. In this experiment, the ambient temperature was used to demonstrate the viability of this measurement system for portable bacteria sensors. Incorporating a temperature control system might increase the temperature of the sensing site and, consequently, decrease the time δt.

8.4.2.2 Charge-to-Voltage Converter for Charge-Based Capacitance Measurement Structure. The first step toward a core-CBCM integrated sensor is the design of an interface circuit for the measurement of charging current. An alternative method is the conventional capacitive integrating method as shown Figure 8.7a [37]. In this figure, the discharging current is conducted to the integrating capacitor (C_{in}) through node A. By applying clock pulses, the average current in the discharging half-period results in an output voltage. The value of voltage can be calculated from the following relation [37]:

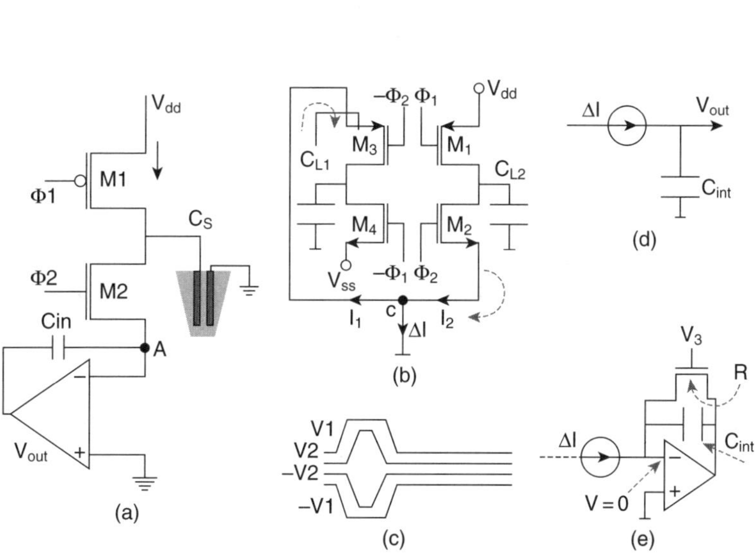

Figure 8.7. A core-CBCM capacitive interface circuit: (a) op-amp-based current integrator (b) current differential technique by applying (c) clock pulses along with (d) simple RC and (e) op-amp-based integrators.

$$V_{out} = \left(-V_{dd} + \frac{T}{C_{in}} \cdot I_{dc}/2\right) - \frac{C_S}{C_{in}} \cdot V_P, \tag{8.2}$$

where C_{in}, V_P, and $T = 1/f$ are the integrating capacitor, the amplitude, and the period of the applied clock pulses, respectively. I_{dc}, in the relationship, is also the quiescent current. Owing to negative feedback established by feedback capacitance C_{in}, the op-amp keeps a constant voltage on the source of the nMOS transistor (M2). This method has been successfully applied for DNA detection by Guiducci et al. using discrete devices [37].

By changing the CBCM circuitry and the applied clock pulses, the discharging currents can simply be subtracted in a circuit topology like the one shown in Figure 8.7b without any op-amp [68]. This differential CBCM structure operates in two phases by nonoverlapping clocks $-\Phi 1$, $-\Phi 2$, $\Phi 1$, and $\Phi 2$, as shown in Figure 8.7c. In the first phase, M2 and M3 are turned on while M1 and M4 are turned off. The discharging currents of C_{l1} and C_{l2} are subtracted in node C. The subtracted current can be integrated through either simple RC circuitry or an op-amp-based integrator as shown in Figure 8.7d,e. The main problem of this circuit topology is an error associated with ΔI. In fact, I_1 and I_2 are not the same even if similar devices are used in this topology.

8.4.2.3 Modified Current-to-Voltage Converter for Charge-Based Capacitance Measurement Structure. As shown in Figure 8.8a, the current mirror can sense and amplify the CBCM charging current (I_S) with a current gain equal to A_I. In this figure, once $\Phi 1$ is low and the voltage on C_S starts rising rapidly, the following relation is established among the charging current through Cs, the drain current of M_3, and the charging current through C_{in} to this capacitance-to-voltage converter (CVC) unit:

$$V_{out} = \frac{C_S}{C_{in}} \cdot \frac{W_4}{W_3} \cdot \frac{L_3}{L_4} (V_{dd} - V_{TP}), \tag{8.3}$$

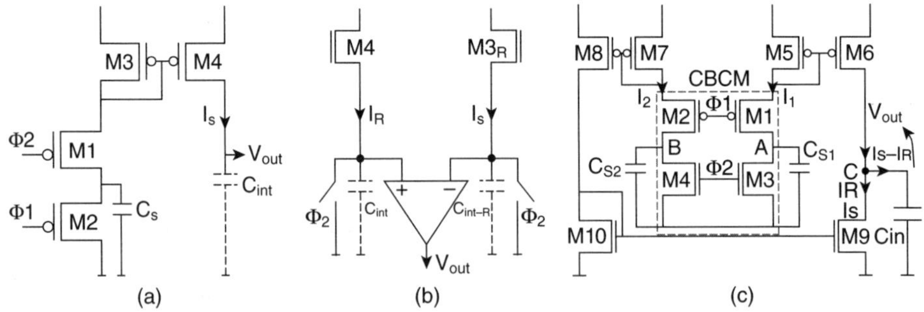

Figure 8.8. Current-based core-CBCM structures: (a) current-to-voltage converter (CVC) unit, (b) voltage differential, and (c) current differential methods.

where W_4/L_4 and W_3/L_3 are aspect ratios of M4 and M3, respectively. As $C_S = \Delta C + C_0$, it is obvious that the dynamic range of V_{out} remains limited because $\Delta C >> C_0$. In order to extract the sensing capacitance ΔC from C, two different solutions have already been reported [69, 70].

VOLTAGE-DIFFERENTIAL METHOD. The first solution is to generate a reference voltage, V_R, by employing a replica of the circuit shown in Figure 8.8b with C_R instead of C_S. A differential voltage amplifier can be used to subtract the V_S from V_R as shown in Figure 8.10b [70]. Therefore, by assuming a symmetrical circuitry with similar transistors ($M1 \equiv M1_1$, $M2 \equiv M2_1$, $M3 \equiv M3_1$, $M4 \equiv M4_1$) and the same capacitance value for the integrating capacitor ($C_{int} \equiv C_{int\text{-}1}$), the output voltage can be derived from Equation 8.2. A problem arises from the voltage-differential amplifier. The higher sensitivity and, consequently, higher output voltage of each CVC unit pushes the voltage-differential amplifier into the nonlinear region, thereby derogating the resolution of the capacitive sensor. A simple solution to avoid both the above-mentioned problems is to subtract the charging currents resulting from the sensing and reference capacitances prior to injection in the integrating capacitors. We will progress forward on this topic by describing a differential current method.

CURRENT-DIFFERENTIAL METHOD. A solution for this problem is to generate a reference current, I_R, by employing a replica of the CVC unit with C_R instead of C_S. The differential current $I_S - I_R$ is injected into C_{in} (Fig. 8.10c). Owing to its symmetry and differential operation, V_{out} can be proportional to the differential current and, subsequently, the differential capacitance ($C_S - C_R$) using the circuitry shown in Figure 8.8c. In this figure, two current mirrors (M5–M6 and M7–M8) amplify (A_I) the charging currents I_1 and I_2, and the third current mirror (M9–M10) is employed to transfer I_2 to node C. The difference between the DC components of I_1 and I_2 is amplified and then injected into C_{in}. The output voltage of the amplifier should be buffered in the output stage of the circuit, and the integrating capacitor C_{in} should be reset in each trigger of the clock pulse $\Phi 2$ by adding more transistors to the circuit.

8.5 MICROFLUIDIC PACKAGING

A microfluidic packaging should be integrated with the microelectronic chip in order to protect the bonding wires and other circuitries from direct contact with fluids. Leakages of biofluids, which are most often ionically conductive, create parasitic capacitances/resistances and, consequently, degrade the performance of integrated sensor. For this, several techniques have been reported to create microfluidic channels for biological applications. Among these, only a few CMOS-compatible techniques can be used for hermetic and reliable microfluidic packaging above CMOS chips. These methods, including on-chip micromachining and adhesive, are briefly described in the next section.

8.5.1 Complementary Metal-Oxide Semiconductor-Compatible Microfluidic Packaging Methods

8.5.1.1 In-Chip Microfluidic Channel. A shallow microchannel can be realized using the standard metal layer (aluminum) inside the CMOS chip. In fact, by using traditional CAD tools, a metal layer is selected and patterned at the top of the sensing site, as shown schematically in Figure 8.9a [71]. This conductor (and vias) played the role of the sacrificial layer, which can be etched using 80% phosphoric acid, 5% nitric acid, 5% acetic acid, and 10% water. This procedure was successfully employed to create a monolithic integration into a microelectronic interface circuit for sensing the flow rate of liquids [72].

8.5.1.2 On-Chip Microfluidic Channel. A spin-coated and patterned polyimide can be formed on the SiGe IC as the sidewalls of a microfluidic channel. Lee et al. proposed a CMOS/microfluidic hybrid microsystem for two-dimensional (2D) magnetic [73] manipulation system based on this technique. A glass cover slip is also sealed on top of the sidewalls and connected to the inlet and outlet fluidic tubes in order to circulate the biological solutions in the microfluidic system. This low-temperature method could similarly be applied to CMOS sensors for other LoC applications (Fig. 8.9b).

A microfluidic structure can be fabricated using a variety of polymeric techniques (e.g., hot embossing) and adhesively attached onto CMOS chips using glue or a low-temperature, plasma-bonding method (see Fig. 8.9c) [12]. The hot

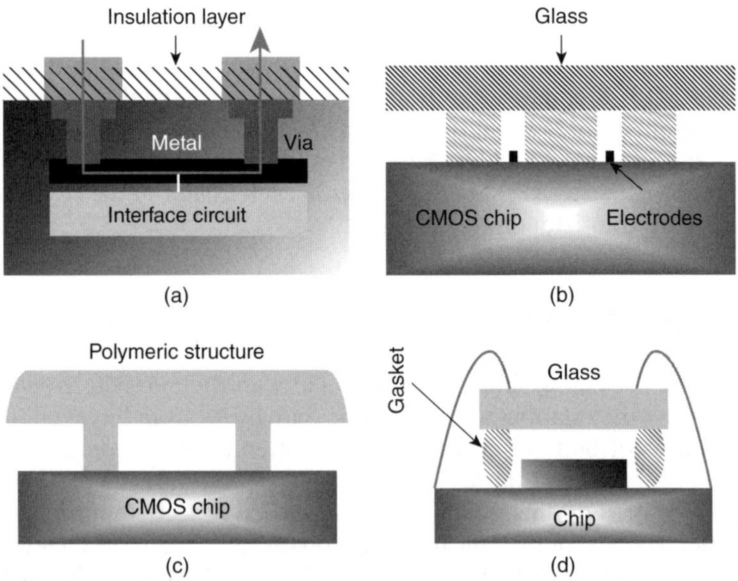

Figure 8.9. Microfluidic packaging methods: (a) in-chip, (b) on-chip surface machining, (c) on-chip adhesive bonding, and (d) rapid prototyping methods.

embossing technique allows a high-precision replication of features from a mold onto thermoplastic materials. Chartier et al. successfully demonstrated the fabrication of a polymer-based microfluidic structure using hot embossing for this purpose [74]. The implemented microfluidic structure was integrated onto a CMOS-based LoC for bioparticle detection and manipulation. In addition, a follow-up paper describes the fabrication of microfluidic networks on the same CMOS-based LoC using a dry-film resist [75].

Besides these methods, many other rapid prototyping methods can be performed to cover the chip with a window through which the analyte can be inserted onto the sensor chip. Among these techniques, Medoro et al. reported a nonconventional method using simple laboratory devices (see Fig. 8.9d) to create a well on top of the dielectrophoresis electrodes for cell manipulation [76].

In addition to the above-mentioned methods, recently, a robotic-based microfluidic packaging has been reported for CMOS-based LoC applications. This technique is called direct-write assembly, and it is described in the next section.

8.5.2 Direct-Write Microfluidic Packaging Procedure

The direct-write microfluidic fabrication process is a low-complexity method to create microfluidic structure above microelectronic devices [77]. By applying pressurized air, functional materials, or so-called inks, are deposited in a trajectory that is preloaded into the robotic system. This ink plays the role of a sacrificial layer that is encapsulated with liquid epoxy. The ink is finally extracted after hardening of the epoxy.

Several parameters, such as the speed of the robot, the inner diameter of the nozzle, the distance between the substrate and nozzle, air pressure, and ambient conditions, influence the performance of this process. Using custom software , a three-axis robot is programmed to carry a barrel of ink in a desired trajectory [78]. Ghafar-Zadeh et al. reported a six-step procedure (Fig. 8.10a–f) to create microfluidic packaging above a CMOS chip [79] using the direct-write assembly method. This procedure is described in the next subsection.

8.5.2.1 A Procedure for Microelectronic Sensors. Before starting the three-step direct-write microfluidic fabrication process, the conductors should be covered so as to avoid direct contact with fluids in the channels. For this, (1) a partially cured epoxy resin (Epon 828, Momentive, Columbus, OH) is dispensed (Champion 8200 dispenser, Creative Automation Co., Sun Valley, CA) on the packaged chip in order to encapsulate the bonding wires (Fig. 8.10a) and (2) a paste-like organic ink (a mixture of petroleum jelly and a microcrystalline wax) is extruded (Ultra® 2400, EFD Inc., East Providence, RI) through a micronozzle and deposited on the substrate. During the extrusion, a micropositioning robot (Model I&J 2200, Production Automation Co., Eden Prairie, MN) moves the nozzle across a desired trajectory (Fig. 8.10b). This sacrificial ink structure preserves its shape during epoxy encapsulation. Following the ink deposition process,

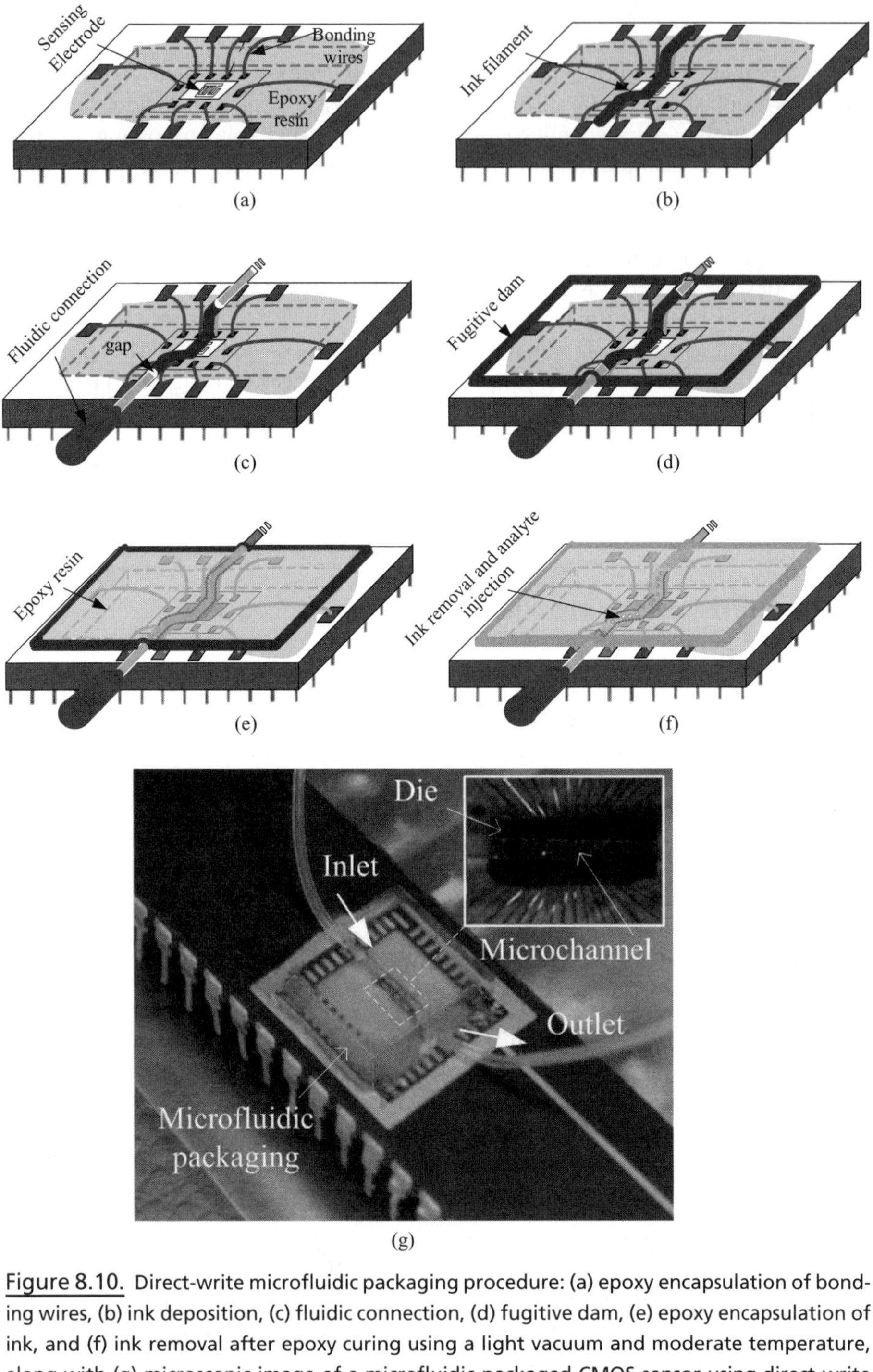

Figure 8.10. Direct-write microfluidic packaging procedure: (a) epoxy encapsulation of bonding wires, (b) ink deposition, (c) fluidic connection, (d) fugitive dam, (e) epoxy encapsulation of ink, and (f) ink removal after epoxy curing using a light vacuum and moderate temperature, along with (g) microscopic image of a microfluidic packaged CMOS sensor using direct-write technique.

(3) the microscale fluidic fittings (nozzle or tube) are placed and secured at the desired locations close to the deposited ink on the chip using a few drops of hot glue (Fig. 8.10c); (4) a fugitive dam is created using another ink deposition (Fig. 8.10d); (5) a low-viscosity epoxy resin is dispensed on the deposited ink within the encapsulation boundary (Fig. 8.10e). Curing of the resin occurs at room temperature over 24 hours. This epoxy encapsulation process creates a strong and hermetic bond on the uneven surface of the loose die. It is obvious that an open-top channel can be performed by using less volume of epoxy. (6) The fugitive ink is melted at ~75°C and expelled under a light vacuum or air pressure (Fig. 8.10f). Hot water is injected through the channel to remove the ink remnants. Just after this step, an analyte solution can be directly injected into the fabricated microchannel on the microelectronic chip for sensing purposes. Figure 8.10g shows the capacitive sensor incorporated with microfluidic packaging using the above-mentioned direct-write microfluidic packaging procedure [22].

8.6 CONCLUSION

Capacitive biointerfaces emerge as important techniques for biosensing purposes such as DNA detection, virus detection, and bacterial growth monitoring. The design and implementation of CMOS capacitive biointerfaces featuring interdigitated electrodes, intermediate layers, interface circuitries, and microfluidic packaging were described in this chapter. The standard CMOS technology serves as the chassis on which various integrated circuitries are built for a variety of applications, such as image sensors and accelerometers. However, several more steps, including microfluidic packaging, should be taken to create a hybrid-technology platform for biological LoC applications. Several tasks should still be completed in the future in order to design and implement fully integrated capacitive biosensors through standard CMOS process. For instance, the provided CAD tools for capacitance characterization are applicable for microelectronic devices in deep CMOS chips, but the parasitic capacitance between the topmost metal layer and the substrate cannot be measured precisely. The minimum requirement in the design of a capacitive sensor is the estimation of the parasitic capacitance created by the sensing electrode above the CMOS chip. Additionally, the CAD tools used for the design of microelectronic circuitries are implemented based on very accurate models provided for CMOS transistors and other microelectronic devices. The design of capacitive biosensors on CMOS chips would not be possible without exact models of biological samples and/or recognition elements formed at the top of the chips.

The design of a generic capacitive sensor LoC is another important challenge for microelectronic designers. An optimized generic system should feature a large array of capacitive sensors, offset cancellation and calibration modules, and a high-resolution low-speed analog to digital converter (ADC), preferably using a sigma–delta modulation technique. Many techniques have been developed to design and implement capacitive biosensors on CMOS chips; however, further studies are required to solve all associated practical problems, such as compatibility with a wide range of biological materials, cleaning procedures, and the implementation of a large array of capacitive sensors for multisensing purposes.

REFERENCES

[1] C. D. Chin, V. Linder, and S. K. Sia, "Lab-on-a-chip devices for global health: Past studies and future opportunities," *Lab Chip*, 5(1), pp. 41–57, 2007.

[2] H. Craighead, "Future lab-on-a-chip technologies for interrogating individual molecules," *Nature*, 442(7101), pp. 387–393, 2006.

[3] R. S. Tuan and C. W. Lo, *Developmental Biology Protocols: Methods in Molecular Biology*, Vol. I. Totowa, NJ: Human Press, 2000.

[4] R. P. Hertzberg and A. J. Pope, "High-throughput screening: New technology for the 21st century," *Curr. Opin. Chem. Biol.*, 4(4), pp. 445–451, 2000.

[5] P. V. Lode, "Point-of-care immunotesting: Approaching the analytical performance of central laboratory methods," *Clin. Biochem.*, 38(7), pp. 591–606, 2005.

[6] J. Byungchul and A. Hassibi, "Biosensor systems in standard CMOS processes: Fact or fiction?" *IEEE Trans. Ind. Electron.*, 56(4), pp. 979–985, 2009.

[7] E. Ghafar-Zadeh, M. Sawan, and D. Therriault, "CMOS-based capacitive sensor lab-on-chip: A multidisciplinary approach," *Analog Integr. Circuits Signal Process.*, 59(1), pp. 1–12, 2009.

[8] L. Pancheri, M. Scandiuzzo, D. Stoppa, and G. F. D. Betta, "Low-noise avalanche photodiode in standard 0.35-μm CMOS technology," *IEEE Trans. Electron. Devices*, 55(1), pp. 457–461, 2008.

[9] C. G. Jakobson, U. Dinnar, M. Feinsod, and Y. Nemirovsky, "Ion-sensitive field-effect transistors in standard CMOS fabricated by post processing," *IEEE J. Sens.*, 2(4), pp. 279–287, 2002.

[10] M. A. P. Pertijs, G. C. M. Meijer, and J. H. Huijsing, "Precision temperature measurement using CMOS substrate PNP transistors," *IEEE J. Sens.*, 4(3), pp. 294–300, 2004.

[11] G. Medoro, C. Nastruzzi, C. R. Guerrieri, R. Gambari, and N. Manaresi, "Lab on a chip for live-cell manipulation," *Des. Test Comput. IEEE*, 24(1), pp. 26–36, 2007.

[12] L. Hakho, L. Yong, R. M. Westervelt, and D. Ham, "IC/microfluidic hybrid system for magnetic manipulation of biological cells," *IEEE J. Solid-State Circuits*, 41(6), pp. 1471–1480, 2006.

[13] L. Yue, C. Vancura, K. U. Kirstein, J. Lichtenberg, and A. Hierlemann, "Monolithic resonant-cantilever-based CMOS microsystem for biochemical sensing," *IEEE Trans. Circuits Syst. I*, 55(9), pp. 2551–2560, 2008.

[14] C. Stagni, C. Guiducci, L. Benini, B. Ricco, S. Carrara, C. Paulus, M. Schienle, and R. Thewes, "A fully electronic label-free DNA sensor chip," *IEEE J. Sens.*, 7(4)2007, pp. 577–585.

[15] N. Nikkhoo, C. Man, K. Maxwell, and P. G. Gulak, "A 0.18 μm CMOS integrated sensor for the rapid identification of bacteria," *IEEE International Solid-State Circuits Conference (ISSCC) Digest of Technical Papers*, pp. 336–617, February 3–7, 2008.

[16] S. B. Prakash, P. Abshire, M. Urdaneta, and E. Smela, "A CMOS capacitance sensor for cell adhesion characterization," IEEE *International Symposium on Circuits and Systems (ISCAS)*, Kobe, Japan, May 2005.

[17] S. B. Prakash and P. Abshire, "A CMOS capacitance sensor that monitors cell viability," *IEEE Conference on Sensors*, Irvin, CA, October 2005.

[18] S. B. Prakash and P. Abshire, "Tracking cancer cell proliferation on a CMOS capacitance sensor chip," *Biosens. Bioelectron.*, 23(10), pp. 1449–1457, 2008.

[19] E. Ghafer-Zadeh, M. Sawan, A. Shabani, M. Zourob, and V. Chodavarapu, "Bacteria growth monitoring through a CMOS based capacitive sensor." 14th IEEE International Mixed-Signal, Sensors, and Systems Test Workshop (IMS3TW), Vancouver, BC, Canada, 2008.

[20] E. Ghafar-Zadeh and M. Sawan, "A charge based sigma delta capacitive sensor for ultrathin polyelectrolyte layer detection." Northeast Workshop on Circuits and Systems (NEWCAS), Montreal, QC, Canada, June 2008.

[21] E. Ghafar-Zadeh and M. Sawan, "Toward fully integrated CMOS based capacitive sensor for lab-on-chip applications." IEEE International Workshop on Medical Measurements and Applications (MeMeA) 2008, Ottawa, ON, Canada, pp.77–80, May 2008.

[22] E. Ghafar-Zadeh and M. Sawan, "A hybrid microfluidic/CMOS capacitive sensor dedicated to lab-on-chip applications," *IEEE Trans. Biomed. Circuits Syst.*, 1(4), pp. 270–277, 2007.

[23] A. Hassibi, H. Vikalo, and A. Hajimiri, "On noise processes and limits of performance in biosensors," *J. Appl. Phys.*, 102, pp. 1–12, 2007.

[24] S. M. Radke and E. C. Alocilja, "Design and fabrication of a microimpedance biosensor for bacterial detection," *Sens. J. IEEE*, 4(4), pp. 434–440, 2004.

[25] M. Hnaiena, M. F. Diouanib, S. Helalia, I. Hafaida, W. M. Hassena, N. J. Renaultc, A. Ghramd, and A. Abdelghani, "Immobilization of specific antibody on SAM functionalized gold electrode for rabies virus detection by electrochemical impedance spectroscopy," *J. Biochem. Eng.*, 39(3), pp. 443–449, 2008.

[26] Y. Y. Chena, A. C. Shua, H. Y. Hsub, H. Y. Changb, T. R. Yewa, Y. C. Lua, and Y. S. Chuanga, "Bacteria detection utilizing electrical conductivity," *Biosens. Bioelectron.*, 23(12), pp. 1856–1861, 2008.

[27] A. Sassolas, B. D. Leca-Bouvier, and L. J. Blum, "DNA biosensors and microarrays," *Chem. Rev.*, 108(1), pp. 109–139, 2008.

[28] M. Datta, S. A. Merritt, and M. Dagenais, "Electroless remetallization of aluminum bond pads on CMOS driverchip for flip-chip attachment to vertical cavity surface emitting lasers(VCSEL's)," *IEEE Trans. Components Packaging Technol.*, 22(2), pp. 299–306, 1999.

[29] L. Berdondini, P. D. Van der Wal, N. F. De-Rooij, and M. Koudelka-Hep, "Development of an electroless post-processing technique for depositing gold as electrode material on CMOS devices," *Sens. Actuators B Chem.*, 99(2), pp. 505–510, 2004.

[30] F. Hofmann, A. R. Frey, B. Holzapfl, M. Schienle, C. Paulus, P. Schindler-Bauer, R. Thewes, R. Hintsche, E. Nebling, J. Albers, and W. Gumbrecht, "Passive DNA sensor with gold electrodes fabricated in a CMOS backend process," *32nd European Conference on Solid-State Device Research Conference*, September 2002.

[31] F. Hofmann, A. Frey, B. Holzapfl, M. Schienle, C. Paulus, P. Schindler-Bauer, D. Kuhlmeier, J. Krause, R. Hintsche, E. Nebling, J. Albers, W. Gumbrecht, K. Plehnert, G. Eckstein, and R. Thewes, "Fully electronic DNA detection on a CMOS chip: Device and process issues." Electron Devices Meeting (IEDM '02), pp.488–491, December 2002.

[32] C. Guiducci, C. Stagni, A. Fischetti, U. Mastromatteo, L. Benini, and B. Riccò, "Microelectrodes on a silicon chip for label-free capacitive DNA sensing," *IEEE J. Sens.*, 6(5), pp. 1084–1093, 2006.

[33] X. Chen, L. Jiong, W. Yijin, C. Lu, L. Zuhong, and M. Chan, "A CMOS-compatible DNA microarray using optical detection together with a highly sensitive nanometallic particle protocol," *IEEE Electron. Device Lett.*, 26(4), pp. 240–242, 2005.

[34] T. D. Huang, S. Sorgenfrei, P. Gong, R. Levicky, and K. L. Shepard, "A 0.18-μm CMOS array sensor for integrated time-resolved fluorescence detection," *IEEE J. Solid-State Circuits*, 44(5), pp. 1644–1654, 2009.

[35] G. Sigalov, J. Comer, G. Timp, and A. Aksimentiev, "Detection of DNA sequences using an alternating electric field in a nanopore capacitor," *Nano Lett.*, 8(1), pp. 56–63, 2008.

[36] C. Berggren, P. Staêlhandske, J. Brundell, and G. Johansson, "A feasibility study of a capacitive biosensor for direct detection of DNA hybridization," *Electroanalysis*, 11(3), pp. 156–160, 1999.

[37] C. Guiducci, C. Stagni, G. Zuccheri, A. Bogliolo, L. Benini, B. Samor`ı b, and B. Riccò, "DNA detection by integrable electronics," *Biosens. Bioelectron.*, 19, pp. 781–787, 2004.

[38] S. Q. Hua, Z. Y. Wua, Y. M. Zhoua, Z. X. Caoa, G. L. Shen, and R. Q. Yu, "Capacitive immunosensor for transferrin based on an o-aminobenzenthiol oligomer layer," *Anal. Chim. Acta*, 458(2), pp. 297–304, 2002.

[39] H. Berneya, J. Aldermana, W. Lanea, and J. K. Collins, "A differential capacitive biosensor using polyethylene glycol to overlay the biolayer," *Sens. Actuators B Chem.*, 44(1), pp. 578–584, 1997.

[40] S. Satyanarayanaa, D. T. McCormickb, and A. Majumdar, "Parylene micro membrane capacitive sensor array for chemical and biological sensing," *Sens. Actuators B Chem.*, 115(1, 23), pp. 494–502, 2006.

[41] J. W. Chunga, S. D. Kima, R. Bernhardtb, and J. C. Pyun, "Application of SPR biosensor for medical diagnostics of human hepatitis B virus (hHBV)," *Sens. Actuators B Chem.*, 111(11), pp. 416–422, 2005.

[42] A. Balasubramanian, B. Bhuva, R. Mernaugh, and F. R. Haselton, "Si-based sensor for virus detection," *IEEE J. Sens.*, 5(3), pp. 340–344, 2005.

[43] S. B. Prakash and P. Abshire, "On-chip capacitance sensing for cell monitoring applications," *IEEE J. Sens.*, 7(3), pp. 440–447, 2007.

[44] K. Asami, E. Gheorghiu, and T. Yonezawa, "Real-time monitoring of yeast cell division by dielectric spectroscopy," *Biophys. J.*, 76(6), pp. 3345–3348, 1999.

[45] E. Lynge, A. Anttila, and K. Hemminki, "Organic solvents and cancer," *Cancer Causes Control*, 8(3), pp. 406–419, 1997.

[46] E. Ghafar-Zadeh, M. Sawan, and D. Therriault, "A 0.18-μm CMOS capacitive sensor lab-on-chip," *Sens. Actuators A Phys.*, 141(2), pp. 454–462, 2008.

[47] E. Ghafar-Zadeh and M. Sawan, "CMOS based capacitive detection lab-on-chip: Design and experimental results," *IEEE Custom Integrated Circuits Conference (CICC)*, San Jose, CA, 2007.

[48] C. Hagleitner, D. Lange, A. Hierlemann, O. Brand, and H. Baltes, "CMOS single-chip gas detection system comprising capacitive, calorimetric and mass-sensitive microsensors," *IEEE J. Solid-State Circuits*, 37(12), pp. 1867–1878, 2002.

[49] P. Antoniou, J. Hamilton, R. Jain, B. Holloway, B. Koopman, G. Lyberatos, and S. A. Svoronos, "Effect of temperature and pH on the effective maximum specific growth rate of nitrifying bacteria," *Water Res.*, 24(1), pp. 97–101, 1990.

[50] E. Ghafar-Zadeh, M. Sawan, and V. P. Chodavarapu, "Micro-organism-on-chip: Emerging direct-write CMOS-based platform for biological applications," *IEEE Trans. Biomed. Circuits Syst.*, 3, pp. 212–219, 2009.

[51] G. Decher and J. B. Schlenoff, "Chapter I. Multilayer thin films," in *Polyelectrolyte Multilayers: An Overview*, G. Decher and J. B. Schlenoff, Eds. Weinheim: Wiley-VCH Verlag GmbH, 2002.

[52] E. Ghafar-Zadeh and M. Sawan, "Towards fully integrated lab-on-chip: Design, assembly and experimental results," *Int. J. Adv. Media Commun.*, 3(1), pp. 154–166, 2009.

[53] E. Ghafar-Zadeh and M. Sawan, "A core-CBCM sigma delta capacitive sensor array dedicated to lab-on-chip applications," *Sens. Actuators A Phys.*, 144(2), pp. 304–313, 2008.

[54] J. M. Cooper and A. E. G. Cass, *Biosensors*. Oxford: Oxford University Press, 2003.

[55] A. Ulman, "Formation and structure of self-assembled monolayers," *Chem. Rev.*, 96(4), pp. 1533–1554, 1996.

[56] F. Tao and S. L. Bernasek, "Understanding odd–even effects in organic self-assembled monolayers," *Chem. Rev.*, 107(5), pp. 1408–1453, 2007.

[57] J. H. Fendler, "Chemical self-assembly for electronic applications," *Chem. Mater.*, 13(2), pp. 3196–3210, 2001.

[58] W. Bracke, R. Puers, and C. C. Van-Hoof, *Ultra Low Power Capacitive Sensor Interfaces*. Springer, 2007.

[59] N. Yazdi, H. Kulah, and K. Najafi, "Precision readout circuits for capacitive microaccelerometer," *IEEE Proceedings of Sensors*, 2004.

[60] A. Hierlemann, "CMOS-based chemical sensors," in *CMOS-MEMS*, H. Fujita and D. Liepmann, Eds. Berlin: Wiley-VCH Verberg GmbH & KGaA, Advanced Micro and Nanosystems series, 2008.

[61] C. Hagleitner, A. Hierlemann, O. Brand, and H. Baltes, "Sensor technology CMOS single chip gas detection systems: Part I," *Sens. Update*, 11(1), pp. 101–155.

[62] C. Stagni, C. Guiducci, L. Benini, B. Ricco, S. Carrara, C. Paulus, M. Schienle, and R. Thewes, "CMOS DNA sensor array with integrated A/D conversion based on label-free capacitance measurement," *IEEE J. Solid-States Circuits*, 41(12), pp. 2956–2964, 2006.

[63] A. Romani, N. Manaresi, L. Marzocchi, G. Medoro, A. Leonardi, L. Altomare, M. Tartagni, and R. Guerrieri, "Capacitive sensor array for localization of bioparticles in CMOS lab-on-a-chip," *Digest of Technical Papers, IEEE ISSCC Conference*, pp. 224–225, 2004.

[64] D. Sylvester, J. C. Chen, and H. Chenming, "Investigation of interconnect capacitance characterization using charge-based capacitance measurement (CBCM) technique and three-dimensional simulation," *IEEE J. Solid-State Circuits*, 33(3), pp. 449–453, 1998.

[65] D. Sylvester and W. Chenming, "Analytical modeling and characterization of deep-submicrometer interconnect," *Proc. IEEE*, 89(5), pp.634–664, 2001.

[66] S. B. Prakash, P. Abshire, M. Urdaneta, and E. Smela, "A fully differential CMOS capacitance sensor design, testing and array architectur." 2005. IEEE International Symposium on Circuits and Systems (ISCAS), Kobe, Japan, May 2005.

[67] E. Ghafar-Zadeh and M. Sawan, "Charge-based capacitive sensor array for CMOS-based laboratory-on-chip," *IEEE International Conference on Sensors*, Daegu, South Korea, 2006.

[68] E. Ghafar-Zadeh and M. Sawan, "A highly linear capacitive sensor circuit based on a novel CBCM method." 3th International IEEE Northeast Workshop on Circuits and Systems (NEWCAS), Quebec City, 2005.

[69] I. Evans and T. York, "Microelectronic capacitance transducer for particle detection," *IEEE Sens. J.*, 4(3), pp. 364–372, 2004.

[70] E. Ghafar-Zadeh and M. Sawan, "A charge based capacitive sensor array for lab-on-chip applications," *IEEE J. Sens.*, 8(4), pp. 325–332, 2008.

[71] A. Rasmussen, M. Gaitan, L. E. Locascio, and M. E. Zaghloul, "Fabrication techniques to realize CMOS-compatible microfluidicmicrochannels," *J. Microelectromech. Syst.*, 10(2), pp. 286–297, 2001.

[72] A. Rasmussen, Implementation and modeling of microfluidic components realized using CMOS technology. Angela, D.Sc., George Washington University, 2002.

[73] H. Lee, D. Ham, and R. M. Westervelt, "Chapter III. CMOS/microfluidic hybrid systems," in *CMOS Biotechnology*, H. Lee, R. M. Westervelt, and D. Ham, Eds. New York: Springer, 2008.

[74] I. Chartier, C. Bory, A. Fuchs, D. Freida, N. Manaresi, M. Ruty, J. Bablet, and L. Fulbert, "Fabrication of hybrid plastic-silicon micro-fluidic devices for individual cell manipulation by dielectrophoresis," *Proc. SPIE*, 5345, 2004.

[75] P. Vulto, N. Glade, L. Altomare, J. Bablet, L. Tin, G. Del-Medoro, I. Chartier, N. Manaresi, M. Tartagni, and R. Guerrieri, "Microfluidic channel fabrication in dry film resist for production and prototyping of hybrid chips," *J. Lab Chip*, 5, pp. 158–162, 2005.

[76] M. Tartagni, L. Altomare, R. Guerrieri, A. Fuchs, N. Manaresi, and G. Medoro, "Microelectronic chips for molecular and cell biology," *Sens. Update*, 13(1), pp. 155–200, 2003.

[77] D. Therriault, S. R. White, and J. A. Lewis, "Chaotic mixing in three-dimensional microvascular networks," *Nat. Mater.*, 2(4), pp. 265–271, 2003.

[78] D. Therriault, R. F. Shepherd, S. R. White, and J. A. Lewis, "Fugitive inks for direct-write assembly of 3-D microvascular networks," *Adv. Mater.*, 17(4), pp. 395–399, 2005.

[79] E. Ghafar-Zadeh, M. Sawan, and D. Therriault, "A microfluidic packaging technique for lab-on-chip applications," *EEE Trans. Adv. Packaging*, 32(2), pp. 410–416, 2009.

9

LENSFREE IMAGING CYTOMETRY AND DIAGNOSTICS FOR POINT-OF-CARE AND TELEMEDICINE APPLICATIONS

Sungkyu Seo, Ting-Wei Su, Anthony Erlinger, and Aydogan Ozcan

9.1 INTRODUCTION

The cost of wireless cell phone technology has significantly decreased over the last decade. As a result, today, cell phones are in use even in the developing world as highlighted in Figure 9.1 [1]. Furthermore, the technical capabilities of existing cell phones are rapidly improving, which permits various different applications to run on today's cell phones, together with an enormous amount of computational power that is readily available within a very compact platform. This impressive advancement is one of the central building blocks of the emerging fields of "telemedicine" and "wireless health." Specifically, utilizing this advanced, state-of-the-art cell phone technology toward point-of-care diagnostics and/or microscopic imaging applications can offer numerous opportunities to improve health care, especially in the developing world where medical facilities and infrastructure are extremely limited or do not even exist. For example, in such resource-scarce settings, it is rather difficult to build a fully functional medical laboratory.

CMOS Biomicrosystems: Where Electronics Meet Biology, First Edition. Edited by Krzysztof Iniewski.

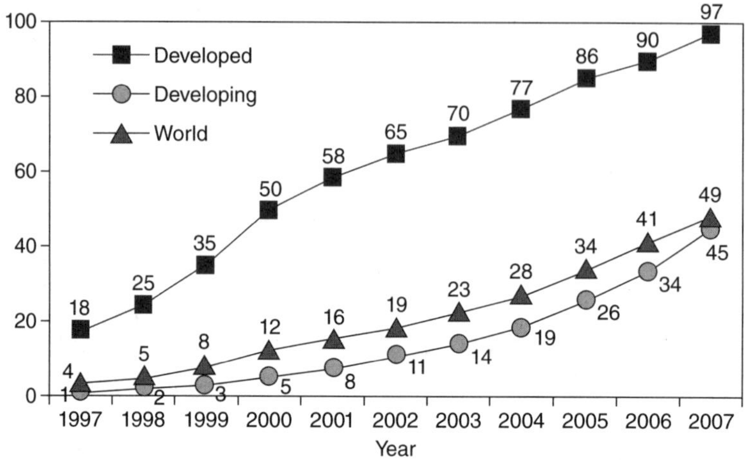

Figure 9.1. Mobile phone subscribers per 100 inhabitants for developing and developed countries, between 1997 and 2007 [1]. Copyright © ITU.

Not only are most medical test equipment expensive to start with, but the running costs of such high-tech laboratory devices are also beyond the budget limits of the developing countries. In the meantime, most cell phones today are already equipped with advanced sensing and imaging systems that can be utilized for various health monitoring applications, taking over some of the functions of a medical laboratory. With the fast pace of technological and manufacturing advances, our cell phones will soon be equipped with quite advanced systems that will all be custom designed for a specific patient, becoming a valuable tool for telehealth and the soon-to-come era of "personalized medicine."

For this revolution in health care to occur, on-chip systems that can potentially be miniaturized to a self-contained unit equipped with a wireless transmitter are urgently needed. From a business point of view, in an ideal setting, such systems should better be fully compatible with the existing designs of our cell phones, which will make their use in the developing world more cost-effective. The topic of this chapter is geared toward shedding some light to this direction, and therefore, in this chapter, we investigate and summarize the use of on-chip cytometry and diagnostics for point-of-care and telemedicine applications. We will also review the basics of a new, lensfree, on-chip cytometry and diagnostics platform that our group recently demonstrated as a promising technology for wireless health applications, especially toward global health-related problems.

9.2 CLINICAL NEED FOR CYTOMETRY AND ITS SIGNIFICANCE FOR BIOMEDICAL DIAGNOSTICS

For medical diagnostics, blood (together with saliva, sputum, and urine) is one of the most important subjects of interest. In vertebrata, blood is composed of three major cell types, including red blood cells (RBCs), white blood cells (WBCs), and platelets suspended in plasma, which also contains many other components, for example, hormones, proteins, and ions. Diagnosis of various diseases, such as acquired immunodeficiency syndrome (AIDS) and malaria, can be performed by analyzing the collected blood sample of the patient, which is primarily conducted through popular laboratory tools such as flow cytometers. There are numerous other diseases that can be accurately diagnosed based on the increase and/or decrease of the count of specific blood cells. Table 9.1 lists some of these diseases that can be diagnosed from a complete blood analysis in comparison to the reference range of healthy people [2–4]. For human immunodeficiency virus (HIV)/AIDS, examples of essential biological indicators would include the counting of CD4+ T lymphocytes, RNA testing, and antibody testing [3]. Malaria, another widespread disease in developing countries, also requires optical imaging techniques involving microscopic examination of blood samples of the patient. Since there are four major types of parasites known to cause malaria, and each has unique spatial characteristics, a microscopic analysis of blood smears can provide reliable diagnostic results [5].

Among existing cytometry approaches, flow cytometry is a technological breakthrough especially for the diagnosis of infectious diseases. It meets all the key requirements of diagnostic assays by providing rapid, specific, and highly accurate statistical information on the state of the patient. Additionally, the use of flow cytometry-based, multiplexed immunoassays enables more enhanced diagnostics for global health-related infectious diseases, allowing simultaneous detection of more than one analyte in a clinical sample [6]. For instance, overall specificity and accuracy of lymphoma diagnosis combined with flow cytometry and immunocytochemistry were considerably higher than those diagnosed without them [7]. Depending on the density, uniformity, shape, and color of the cells/bacteria to be screened, the spectrum of this powerful technology includes a boundless variety of applications in biomedicine. As a result, today, flow cytometry defines the state of the art especially in hematological diagnostics for carrying out various tasks such as CD4+ T cell counting for HIV, differential leukocyte counts, the detection of malaria parasites in whole blood [8], or the analysis of parasitemia during *Plasmodium falciparum* infection [9]. Other significant clinical applications of cytometry in biomedicine include, but are not limited to, detection and quantification of circulating tumor cells, and detection of bacteria in food and drinking water [10–13]. Starting on the next section, we will discuss technical details of modern cytometry tools.

9.3 MODERN CYTOMETRY TECHNOLOGIES

Counting cells using a microscope is one of the most widespread methods in clinical settings. This is especially true in the developing world, where more advanced medical devices are lacking. For manual counting of a heterogeneous cell solution, hemacytometers are frequently used to control the sample volume

TABLE 9.1. A Summary of Diseases That Can Be Diagnosed through Complete Blood Count [2–4]

Type of Tests		Reference Range	Unit	Increase	Decrease
RBC	RBC count	4.2E6–6.9E6	cells/μL	Polycythemia vera, myeloproliferative syndrome, low oxygen level, malignancy, renal cell carcinoma, dehydration, cardiovascular disease	Anemia, hemolytic anemia, leukemia, bone marrow failure, malnutrition, chronic renal failure, hemorrhage, erythropoietin deficiency, hemolysis, multiple myeloma, over dehydration
	Hgb	Male: 13–18 Female: 12–16	g/dL		
	Hct	Male: 45–62 Female: 37–48	%		
	MCV	76–100	μm^3	B_{12} and folate deficiency, alcohol abuse, liver disease marrow aplasia, myelofibrosis, reticulocytosis, hypothyroidism	Iron deficiency, thalassemia, hemoglobinopathy, anemia of chronic disease, sideroblastic anemia, chronic renal failure, lead poisoning
	MCH	27–32	pg/cell	Macrocytic anemia	Abnormal hemoglobin synthesis, microcytic anemia
	MCHC	32–36	%	Spherocytosis,unstable hemoglobin	Iron-deficiency anemia
	RDW	11.5–14.5	%	Mixed population of RBCs, immature RBCs	Iron deficiency, thalassemia, anemia

WBC	WBC count	4.3E3–10.8E3	cells/μL	Leukocytosis may indicate infection, inflammation, cancer, leukemia	Autoimmune conditions, severe infection, bone marrow failure, congenital marrow aplasia
	Neutrophil	38–80	%	Bacterial infection	Leukopenia
	Lymphocyte	15–40	%	Epstein–Barr virus, hepatitis, pertussis, toxoplasmosis, tuberculosis, brucellosis, acute lymphoblastic leukemia	HIV or other viral, bacterial, or fungal infection; malnutrition; systemic lupus erythematosus; rheumatoid arthritis
	Monocyte	0–13	%	Chronic inflammation, hyperadrenocorticism, immune-mediated disease, pyogranulomatous disease, necrosis	High risk of bacterial infection
	Eosinophil	0–8	%	Hypereosinophilic syndrome, allergic disorders, cholesterol embolization, Churg–Strauss syndrome, myeloid leukemia, Hodgkin's disease, Gleich's syndrome, Addison's disease	Use of steroids and bacterial infection
	Basophil	0–2	%	Myeloproliferative disorders, chronic granulocytic leukemia	–
Platelet	Platelet count	1.5E4–3.5E4	cells/mL	Myeloproliferative disease, myelofibrosis, chronic myelogenous leukemia, polycythemia vera, hyposplenism, hemorrhage, iron deficiency	Leukemia, myelodysplastic syndrome, liver failure, sepsis, systemic viral or bacterial infection, dengue fever, hereditary syndromes
	MPV	7.5–11.5	μm^3	Idiopathic thrombocytopenic purpura, Bernard–Soulier disease, May–Hegglin anomaly	Aplastic anemia, Wiskott–Aldrich syndrome, thrombocytopenia-absent radii, storage pool disease

Hgb: hemoglobin; Hct: hematocrit; MCV: mean corpuscular volume; MCH: mean corpuscular hemoglobin; MCHC: mean corpuscular hemoglobin concentration; RDW: red cell distribution width; MPV: mean platelet volume.

of interest [14]. Even though the use of a hemacytometer is a practical and cost-effective method in measurement of the cell count, the accuracy of a manual count is degraded by several error sources such as the inhomogeneous distribution of the cells throughout the sample volume. A further drawback of this simpler method is that it cannot distinguish living or dead cells, unless stains such as Trypan blue are applied to the sample. We should also note that while a hematologist can manually count around three to four cells within a second using a hemacytometer, a modern flow cytometer can monitor cells faster, typically at speeds reaching >10–50K cells/s [15]. Starting at the next subsection, we will review several modern cytometry approaches that are either widely used today in clinical settings or that hold significant potential.

9.3.1 Flow-Based Optical Approaches for Cytometry

9.3.1.1 Conventional Benchtop Optical Flow Cytometry. As we briefly touched earlier, flow cytometry is a powerful technology that enables the counting, analysis, and sorting of cells flowing through a microconcentrated liquid channel by collecting information of fluorescence and/or light scattering from the cells [15]. At its core, it utilizes the mechanism of light scattering (forward and side scattering) and fluorescence emission to record the unique statistical signature of a cell type of interest using photomultiplier tubes (PMTs) and/or other photodiodes.

Modern flow cytometers consist of several key components such as a light source, flow channel, light detectors, color filters, and a data analysis system (see Fig. 9.2) [16]. Light sources that are often used in flow cytometry systems are

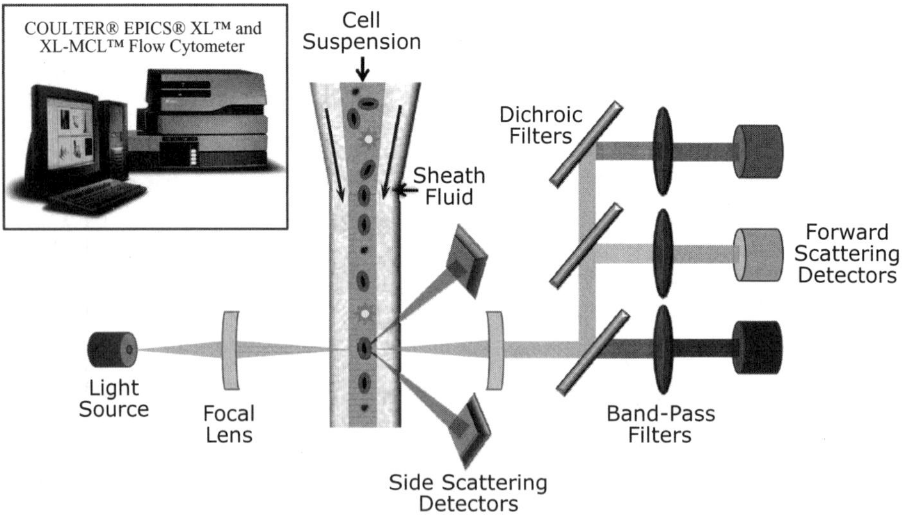

Figure 9.2. Schematic illustrations of the principles of conventional flow cytometry along with an example of commercialized product [16]. Copyright © Beckman Coulter.

lamps (mercury, xenon), lasers (argon, krypton, helium–neon, helium–cadmium, etc.), and diode lasers, each varying between 350 and 660 nm in wavelength. The liquid flow, which contains the target cells of interest, is hydrodynamically focused into a tiny central area of the fluidic channel by adopting sheath flow. The turbulent free flow provided by the sheath fluid permits the cells to be accelerated to form a fine focused column surrounded by a fluid of the same refractive index [16, 17]. For the detector end, most commercial flow cytometers use PMTs or avalanche photodiodes (APDs) to detect forward-scattered (FSC) and side-scattered (SSC) light, together with intrinsic/extrinsic fluorescence emission. The FSC light that is detected by a sensitive in-line detector provides the size information of the target cells. On the other hand, the SSC light, which is measured at tilted positions from the light source axis, can provide complementary information on the shape of the cells of interest. Fluorescence detection at various wavelengths of interest usually requires a more involved optical system that employs several dichroic filters as well as low-pass, high-pass, and/or band-pass filters, to block the excitation light while simultaneously permitting the detection of the weak fluorescent light (with power levels in the order of nanowatts) [18].

In order to make a final characterization decision, flow cytometry uses various analysis tools such as the histogram, dot plot, contour plot, and three-dimensional (3D) plot of the collected data. Since FSC and SSC contain size- and shape-relevant information, characterization of a heterogeneous cell solution containing, for example, RBCs, lymphocytes, monocytes, and granulocytes, can be achieved by an analysis of the two-dimensional (2D) map of FSC and SSC data [19].

9.3.1.2 On-Chip Optical Flow Cytometry. For point-of-care and telemedicine-related applications, there is an urgent need for miniaturized cytometry systems that can be used for diagnostics even in settings with limited resources. For this end, there have been a variety of studies to miniaturize conventional benchtop flow cytometers into portable on-chip microsystems. Most of these prior on-chip approaches have been primarily focused on the development of methodologies that can provide higher detection efficiencies, smaller optical components, and smaller sample volumes by integrating on a chip all the required components, such as microfluidic channels, light sources, optics, and detectors (see Fig. 9.3) [20].

To miniaturize the light source in microflow cytometry, researchers adopted commercially available waveguides coupled to an integrated source [21–23], where the cleaved ends of the waveguides are precisely aligned to the flow channel for detection of the scattered light from the cells. Recently, a complete lab-on-a-chip device with an integrated microfluidic dye laser, optical waveguides, a microflow channel, and a photodiode with fluorescent detection capability has been successfully demonstrated [24]. In the meantime, integration of multiple light sources on the same chip, while still feasible, can be a challenging task for this approach.

In order to miniaturize the flow channel, considerable effort was made to achieve reliable and rapid cell transport. Conventional flow cytometers use

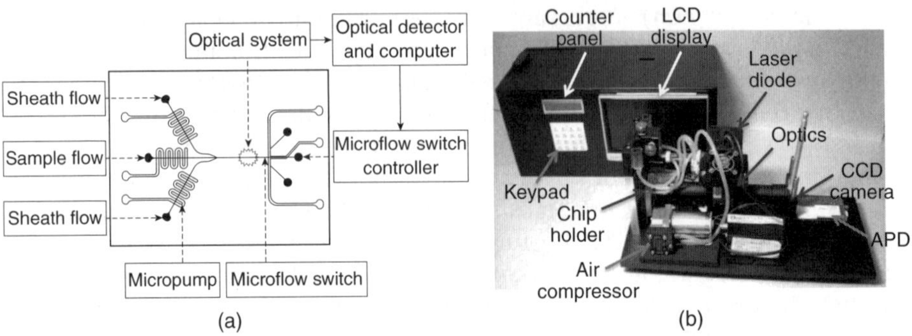

Figure 9.3. A cell counting/sorting system incorporated with a microflow cytometer chip. (a) Design of the chip for the microflow cytometer and (b) system configuration [20]. Copyright © IOP.

sheathed flow capillary, which has a main liquid inlet and sheath flow tubes [17]. To integrate this bulky hydrodynamic channel on a chip, researchers utilized the initial idea of Jacobson and Ramsey [25] that utilizes a "cross" structure to have electrokinetic focusing on microfabricated channels, especially at the intersection between the cells and the sheath streams. This has been one of the most widely adopted techniques to achieve 2D, and even 3D, hydrodynamic cell focusing [26–34]. To further minimize sample and reagent use, a sheath-free transport system has also been demonstrated, where a microchannel was fabricated on a silicon substrate to handle blood cells [35] such as RBCs, platelets, lymphocytes, monocytes, and granulocytes [36], or to detect albumin level in serum. Despite its simple structure, this sheath-free approach faced an important limitation that was caused by clotting of the tiny flow channel. This clotting issue, however, is not observed in conventional sheath flow-based channels since the cells are not in contact with the walls of the fluidic channel.

Dielectrophoresis (DP) is another mechanism that has been utilized to enable sheath-free flow in microcytometer systems. By using an electric field gradient, cells in a microchannel can be forced to move forward (positive DP) or backward (negative DP). The size of the cells, electric field strength, and the distance between each cell and the electrodes should all be taken into account to enable delicate control of the cell flow. Employing two sets of electrodes can, in principle, provide 3D focusing of the cell fluid, where one set of electrodes controls the horizontal focusing and the other set controls the vertical [34]. However, one caveat of this electrical fluid control method is that it can potentially cause cell damage through direct contact with the electrodes. This approach also suffers from lower throughput when compared with state-of-the-art hydrodynamic pressure flow.

For the detection end, a variety of optical devices such as PMTs [37], PIN photodiodes [38], single-photon avalanche diodes (SPADs) [39], charge-coupled devices (CCDs), and complementary metal-oxide semiconductor (CMOS) devices have been employed in existing flow cytometers. Due to their high sensitivity, PMTs and SPADs can easily measure less than 1 nW of fluorescence light with a very low background noise. PMTs have a relatively high internal gain, while their bulky size significantly limits their use in microflow cytometers. SPADs can be a good alternative to PMTs due to their compactness, and thus they are more suitable to be utilized as miniaturized detectors in portable cytometry systems. The relatively higher cost of PMTs and SPADs makes PIN diodes also attractive alternatives; however, since PIN diodes have no internal gain nor photon multiplication, they are not suitable for applications that require a highly sensitive detector. Apart from this disadvantage, PIN diodes are still significant for designing miniaturized cytometers since they permit on-chip integration through conventional microfabrication processes.

9.3.2 Electrical Approaches for Cytometry

Detection of the electrical properties of a heterogeneous solution is another analysis method utilized in cell counting. In this electrical counting scheme, which was originally invented by Wallace Coulter in 1953 [40], the cells of interest are guided through a narrow orifice or microfluidic channel with electrodes on both ends. The electrical properties, for example, impedance, of a suspension containing cells vary depending on the volume of the cells, creating corresponding pulses, which are then amplified to generate an electrical output signal corresponding to each cell. After its initial proof of concept, the Coulter counter principle has significantly evolved by utilizing some additional discrimination parameters, for example, conductivity in radio frequency (RF) and light scattering measurements, such that it can also characterize the intracellular structure and granularity, becoming a widely used tool in modern cytometry [16]. The cell count information can be extracted by analyzing the modulation in the electrical impedance and light scattering of the cell suspension due to the passage of the cells through the microchannel. Since the cell membrane passes only the high-frequency current but not the low frequencies, taking measurements in several different electrical frequencies gives further insight on the dielectric properties of the cell membrane and the cytoplasm. This process is the core element in differentiating various cell types from each other based on their unique dielectric properties [41]. By using this electrical approach, Cheung et al. demonstrated that impedance changes that originate from cell size, membrane capacitance, and cytoplasm conductivity over a wide range of frequencies can be used to identify and count blood cells (see Fig. 9.4) [42, 43].

Fabrication of a microchip electrical cell counter usually involves a planar silicon structure covered with glass followed by several steps starting with positioning of alignment marks on both sides of the silicon wafer with an infrared aligner. The wafers are then etched to create inlet and outlet channels, after which a thin layer of titanium is deposited over the channels to guarantee perfect sidewall coverage, permitting a continuum of conductivity throughout the channels. After fabrication of the electrodes, the cell channel is dry-etched with layers of

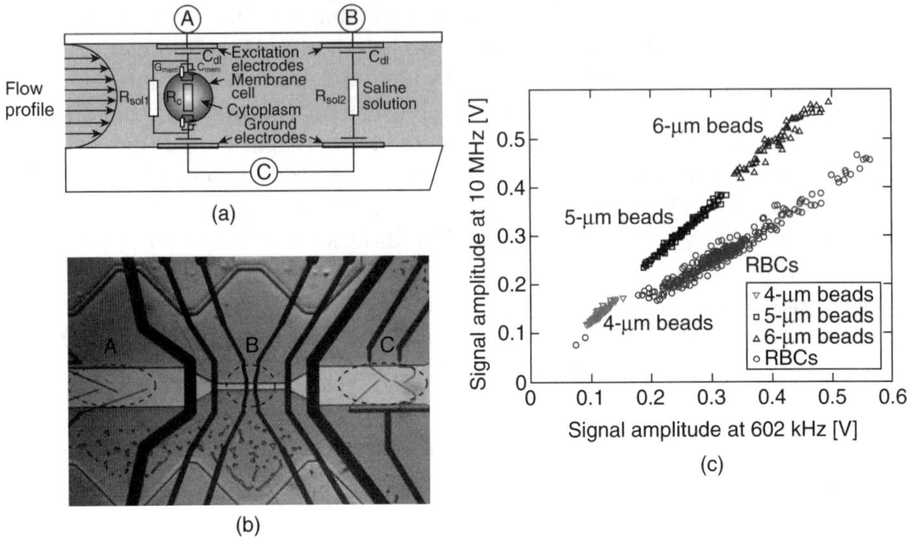

Figure 9.4. On-chip label-free flow cytometry using impedance spectroscopy. (a) Cell discrimination principle; (b) fabricated microfluidic chip composed of cell focusing electrodes (part A), measurement electrodes (part B), and sorting electrodes (part C); and (c) impedance correlation at a high and low frequency for RBC and polystyrene microbead differentiation [41]. Copyright © Wiley InterScience.

Si_3N_4 and SiO_2 and further etched again in KOH, such that the inlet and outlet channels become connected [43].

The precise control of the flow rate is a challenging task for both the orifice-based Coulter counter and its microfluidic counterpart. Too large of an aperture size will allow several cells to be probed at the same time, causing the cell pulses to overlap; on the other hand, small apertures exhibit a higher risk of clotting. As shown in Figure 9.5, several hydrodynamic-focusing techniques in 2D [44] or 3D [45], similar to the ones used in optics-based flow cytometers, were also applied to electrical impedance-based cytometry systems to keep the cell signatures separate from each other [46, 47].

Similar to optical flow cytometers, an important challenge for impedance-based cytometers is the specificity of characterization. Tagging cells with conductive or insulating particles has already been proposed but has not yet been successfully demonstrated [45]. As an alternative approach, immobilizing cells with antibody-coated surfaces and measuring the impedance change caused by the ions released after lysing the captured cells can provide highly specific characterization and counting of a target cell type of interest, as also shown in Figure 9.6 [48]. However, in such systems, the measurement results, that is, the cell count, strongly depend on the original ion concentration of each cell, which may have a large variation when analyzing cell samples from various patients.

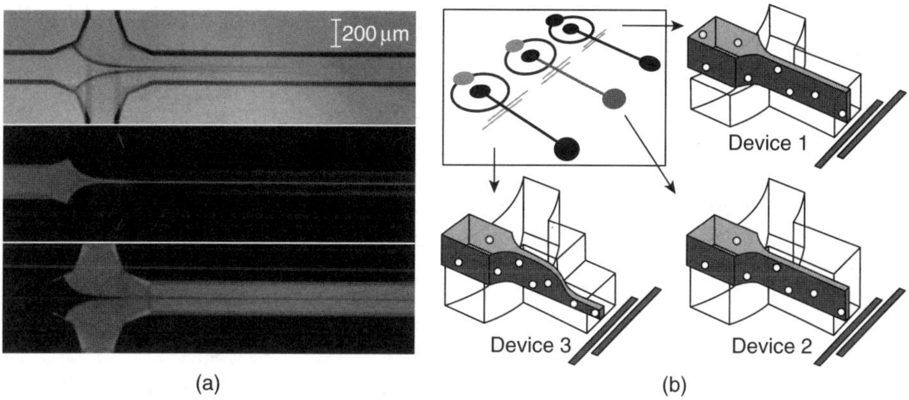

Figure 9.5. Hydrodynamic focusing techniques for microfluidic channels of on-chip flow cytometry. (a) Two-dimensional [43] and (b) three-dimensional hydrodynamic focusing [44]. Copyright © AIP.

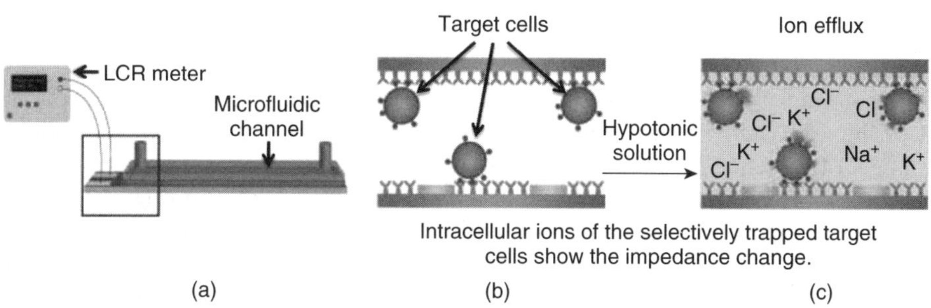

Figure 9.6. Cell detection and counting platform based on impedance spectroscopy [47]. Copyright © RSC Publishing.

This may present certain statistical challenges for the extension of this technique into clinical settings, which requires further validation steps.

9.4 AN EMERGING LENSLESS OPTICAL TECHNOLOGY FOR HIGH-THROUGHPUT ON-CHIP CYTOMETRY AND DIAGNOSTICS: LUCAS

As discussed in Section 9.3.1.2 to integrate the functionality of conventional benchtop flow cytometers onto a chip, there have been several systemized approaches [49–51]. Relatively recently, an alternative on-chip optical cytometry approach termed *lensless ultrawide field-of-view cell monitoring array platform based on shadow imaging* (LUCAS), which does not require any fluid flow during cell characterization and counting, was proposed [52, 53]. Compared with conventional cytometry approaches that are outlined in earlier sections, this lensfree, on-chip imaging platform, LUCAS, can rapidly characterize thousands of cells (based on their shadows or diffraction patterns), all in parallel, without the need for a mechanical scanning or any fluid flow [54]. Moreover, since LUCAS does not rely on any bulky and expensive optical components, such as microscope objectives or mechanical microstages, it holds significant promise to enable a compact, lightweight, and cost-effective point-of-care cell analysis platform that can detect and count major blood cell types, such as RBCs, WBCs, platelets, or even CD4+ and CD8+ cells when combined with surface chemistry-based selective approaches [55, 56].

9.4.1 Overview of the LUCAS Platform and Its Impact

The United Nations Programme on HIV/AIDS and World Health Organization (WHO) reported that, currently, more than 40 million people worldwide are suffering from HIV, particularly in the developing countries [57]. These HIV-positive patients need to be tested for their CD4+ T-cell counts, ideally every few months, to monitor the course of the disease. On the other hand, this is a rather demanding task, especially considering the limited resources of these developing countries. For this challenge and many other global health-related problems, the lensfree and compact platform of LUCAS may provide a promising tool for on-chip cytometry and telemedicine applications.

The LUCAS platform samples the diffraction signatures, that is, shadow images, of target cells that are illuminated by, for example, quasi-monochromatic light of various wavelengths or a simple light-emitting diode. To record the diffraction signatures of the cells all in parallel, the sample solution (either within a microfluidic channel or simply between two cover slips as in a hemacytometer) is placed on the top of an optoelectronic sensor array that has a controlled sample-to-sensor distance (Z). Therefore, the unique diffraction signatures/patterns of various cell types directly fall onto the sensor array, which forms the raw LUCAS image (see Fig. 9.7). This raw LUCAS image is then digitally processed to provide an improved signal-to-noise ratio (SNR) for each cell signature, which is an essential step for LUCAS-based automated characterization of a heterogeneous cell solution with the least possible error rate. In LUCAS, the characteristic diffraction signature of a specific cell type is affected by several key parameters such as cell size, shape, 3D morphology, refractive index, sample-to-sensor distance (Z), illumination wavelength, and illumination angle, as well as the coherence properties of the light source. The effects of these key parameters will be further discussed in the following sections.

In comparison to other existing cytometry techniques, LUCAS has several major advantages. First, it is a massively parallel on-chip imaging modality, where a complete reservoir of cells is immediately (i.e., within less than a second) imaged/monitored using an optoelectronic sensor array. Specifically, we have recently shown that the LUCAS platform can monitor a heterogeneous cell solution, all in parallel, over a field of view (FOV) of ~18 cm^2 and a depth of field (DOF) of ~5 mm [53–55]. In Figure 9.7, the extent of this ultrawide FOV is

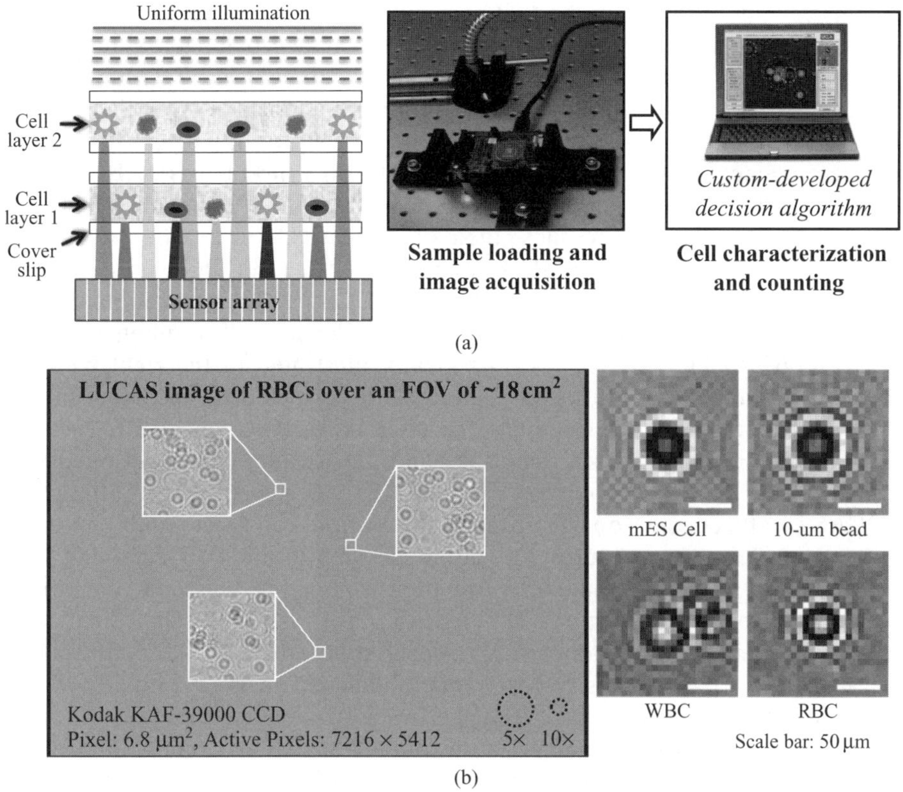

Figure 9.7. Overview of the LUCAS technique. (a) Cell detection and characterization procedures based on shadow images, and (b) various cell and micro-object signatures that can be rapidly characterized by ultrahigh throughput nature of LUCAS.

demonstrated and compared with the FOV of standard 5× and 10× microscope objective lenses. This comparison illustrates that LUCAS can monitor >2 orders of magnitude wider FOV than a regular optical microscope, without the need for any lenses, which makes it potentially quite compact and cost-effective. Furthermore, due to rapid progress in optoelectronics industry, LUCAS can be easily improved to have an even larger FOV (e.g., >20 cm^2) without significantly affecting its cost and complexity, translating to an imaging speed of >100,000 cells/s. This throughput simply does not exist in other on-chip cytometry systems and is quite comparable to that of the existing state-of-the-art, benchtop flow cytometers [16, 54].

Another significant aspect of LUCAS, which does not readily exist in other on-chip cytometer approaches, is that all the necessary basic components of LUCAS, including the sensor array and an illumination source, already exist in most commercially available cell phones. This feature is extremely important for telemedicine applications, as it implies that cost-effective modification of an existing cell phone device into a LUCAS device is feasible. When combined with simple sample preparation steps, this opportunity may provide a real revolution for combating infectious diseases, such as malaria, tuberculosis (TB), and HIV in resource-scarce settings, for example, even in rural districts of Africa.

For the initial demonstration of the LUCAS platform, both homogeneous and heterogeneous solutions of monocytes, NIH-3T3 (fibroblasts), RBCs, hepatocytes, yeast cells, bacteria, and so on, as well as polystyrene microbeads of various sizes, were tested to show their unique LUCAS diffraction signatures [53–55]. Since LUCAS is a lensless system and the sensor chips used in these studies had pixel sizes larger than 2 μm, these LUCAS cell signatures look highly pixelated. However, this is not an obstacle to hinder the cell counting and characterization capability of LUCAS. As clearly illustrated in the right panel of Figure 9.7b, individual diffraction signatures of various cells, including mouse embryonic stem (mES) cell, polystyrene microbead, RBC, and WBC (neutrophil), are quite unique in size, contrast, and 2D texture, which permit digital recognition of each cell type based on its LUCAS signature. The efficacy of LUCAS greatly depends on this property that various cell types exhibit uniquely different diffraction signatures for automatic identification of each cell type within a given FOV. We should also note that, as already discussed in our earlier work [54, 55], even if for certain cell types (e.g., CD4+ vs. CD8+ cells), the natural diffraction signatures are rather close to each other, the LUCAS platform can make use of surface chemistry-based microfluidic channels or antibody-coated microspheres to further increase its cell characterization specificity.

For automated cell characterization purposes, LUCAS relies on building statistical image libraries corresponding to each cell type of interest for a given illumination condition. A cell library's unique LUCAS signature, defined as L, is obtained by digitally averaging >30–50 arbitrary cells within the captured diffraction image of a homogenous cell solution of the same cell type. We can then define the LUCAS *deviation index* as the absolute difference between L and the light intensity of a region of interest (ROI) given by f. Quantitatively, at each image location (x, y), this deviation index can be calculated as

$$\mathrm{Dev}(x,y) = \sum_{(x',y')\in D_L} \left| f(x'+x, y'+y) - L(x',y') \right|,$$

where D_L refers to the domain of the image to be characterized. As expected, when there is a high degree of similarity between the cell library image L and the light intensity in the ROI (i.e., f), the deviation index will have a low value. Conversely, when there is a small amount of similarity, the deviation index will have a larger value. Since we have no a priori knowledge of each cell's location, we need to calculate the deviation index at each individual pixel within the entire FOV in order to find the location of each target cell type. For this purpose, we define a new variable, that is, the 2D correlation index, which can be defined as the normalized inverse of the deviation index, that is,

$$\text{Corr} = 1 - \frac{\text{Dev}(x,y) - \text{Dev}_{\text{min}}}{\text{Dev}_{\text{max}} - \text{Dev}_{\text{min}}},$$

where Dev_{max} and Dev_{min} are the maximum and minimum deviation values within a given LUCAS image, respectively. A correlation index of 1 indicates the minimum difference between L and the ROI, which is the most probable location for that particular cell type. On the other hand, a correlation value of 0 indicates the maximum deviation value in that particular ROI for the library image of L.

After the calculation of the 2D correlation index map, each cell type within the LUCAS FOV can be isolated by applying a desired threshold across that cell's corresponding correlation map. In practice, it has been found that it is best to first identify cells with a higher SNR. This ensures that each cell is counted more accurately when some cell types might have similar LUCAS signatures under certain illumination conditions. In other words, by giving a higher priority to the cells with higher SNR, we ensure that each cell is correlated properly to its corresponding signature. As each cell is counted, a binary mask is applied to prevent redundant characterization in subsequent correlation maps, such that once a decision is made on the cell type for a given ROI, subsequent analysis for the other cell types simply ignores that particular ROI. A minor drawback in the generation of a 2D correlation map is that it can be computationally cumbersome for large images. One method of mitigating this problem is subtracting the background noise via thresholding to further isolate the probable cell locations, thereby minimizing the number of locations where a correlation index would need to be computed for a given LUCAS image.

As further illustrated in the following sections, various digital image filters can also be applied before the count is performed to maximize contrast and SNR of the LUCAS images. Utilizing the above-discussed computational methods, Figure 9.8 successfully illustrates the automated LUCAS characterization results of a heterogeneous mixture of RBCs and 3-, 5-, 10-μm beads located at a plane of $Z = 625\,\mu\text{m}$ captured with a 2.2-μm square pixel-size CMOS image sensor (Micron Technology Inc., Boise, ID) under $\lambda = 550$-nm illumination with ~15-nm bandwidth.

9.4.2 Multicolor LUCAS: Tuning the LUCAS Diffraction Signatures by Varying the Wavelength

Another significant aspect of the LUCAS platform is that by varying the illumination wavelength, the characteristic diffraction signature of a given cell type can be significantly tuned. This feature is especially important to improve the performance of LUCAS-based automated cell characterization for high-throughput screening of a large sample volume. Through experimental results, we illustrated that the use of short wavelength illumination significantly enhances the digital SNR of the diffraction signatures, especially for cells that are located at long DOF values [55]. More specifically, Figure 9.9 illustrates the effect of the illumination

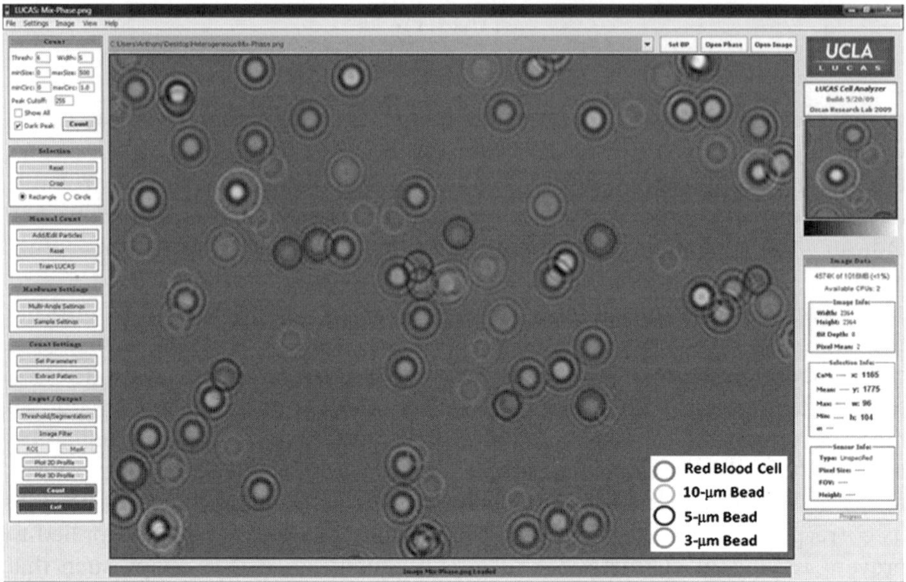

Figure 9.8. Screenshot of the automated LUCAS cell characterization algorithm for a heterogeneous solution of red blood cells and various sizes of polystyrene microbeads (3, 5, and 10 μm in diameter).

spectrum on LUCAS image quality for various micro-objects, such as 3-μm-diameter beads (a–c), RBCs (d–f), and alive (g–i) and fixed (j–l) yeast cells (*Schizosaccharomyces pombe*), all placed at $Z > 3$ mm above the sensor array. For comparison purposes, Figure 9.9a,d,g,j also shows regular transmission microscope images of the same FOV acquired with a 10× objective lens. In these figures, note that for shorter wavelength of illumination ($\lambda = 300$ nm with ~10-nm bandwidth), when compared with white light illumination, the texture of the shadow signatures of various micro-objects is now richer, with a significantly improved digital SNR by, for example, >10 dB; see Figure 9.9c,f,i,l. Furthermore, by employing a shorter, monochromatic wavelength at $\lambda = 300$ nm, the area of the sensor array occupied by the diffraction signature of each cell was narrowed to ~1/3 of the white light illumination case, which can potentially increase the throughput of imaging. On the other hand, at such short illumination wavelengths, for imaging cells located at a short DOF value, the diffraction signatures might exhibit spatial nonuniformities, and this effect would be more pronounced especially for a large pixel-size sensor array. One solution to this issue is to utilize a longer wavelength of illumination that can improve digital signature uniformity of cells located short DOF values by spreading the diffraction pattern. As a matter of fact, our earlier work already demonstrated the use of two different

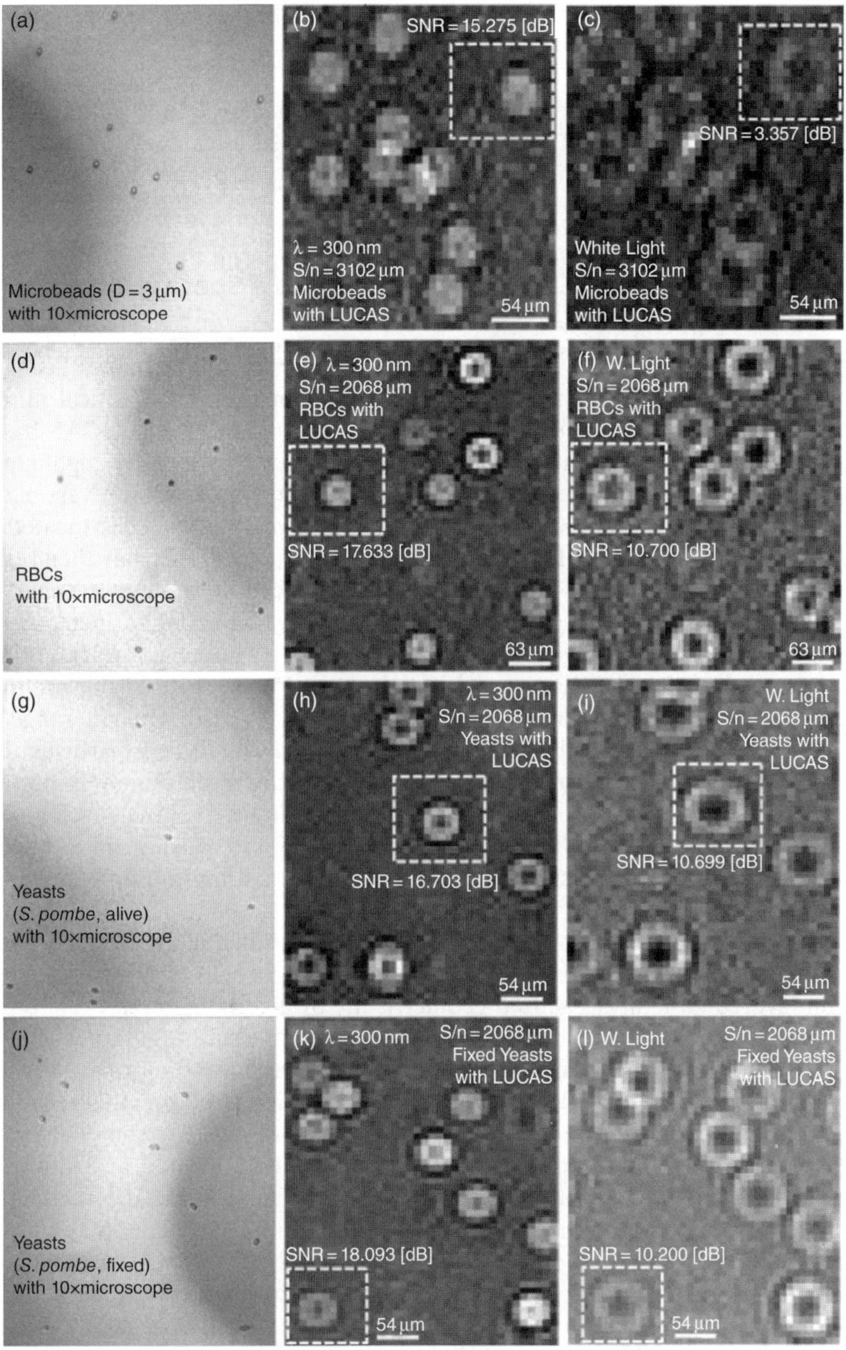

Figure 9.9. Effect of different illumination sources on the LUCAS signatures of various micro-objects such as 3-μm-diameter polystyrene beads (b,c), red blood cells (e,f), and alive (h,i) and fixed (k,l) yeast cells (*S. pombe*), all placed at >3 mm above from the sensor array. (a,d,g,j) Regular optical microscope images of 3-μm beads, RBCs, and alive and fixed yeasts cells taken with a 10× objective lens, respectively. Note that for the shorter illumination wavelength, that is, λ = 300 nm (b,e,h,k), when compared with the white light illumination (c,f,i,l), the texture and SNR of the LUCAS images are much improved, for example, by >10 dB.

wavelengths—one short and one long—to digitally increase the overall performance of LUCAS by combining the above-discussed advantages of both the short and the long illumination wavelengths [55] and will be further discussed in Section 9.4.4.

9.4.3 Multiangle LUCAS: Digital Zooming to a Specific Depth of Field

The LUCAS results that are examined so far were all acquired using a single illumination angle, that is, the vertical illumination. This vertical illumination angle provides various sources of information for LUCAS-based automated characterization of multiple layers of cells, as already highlighted in Figure 9.7. On the other hand, for different layers that have very close height values (i.e., $\Delta Z \leq 100\,\mu m$), the diffraction signatures of cells located at different microchannels could be quite similar to each other. This may then result in characterization errors in LUCAS, as some of the cells located at one channel can be misidentified to be at another one. Therefore, in order to increase the accuracy of the LUCAS platform along the vertical (Z) direction, we can utilize multiple angles of illumination, as highlighted in Figure 9.10. Using multiple angles of illumination, we can "digitally zoom" to a specific height range of interest and can numerically avoid all the other unwanted cells/particles that lie outside of the volume of interest [53]. This feature is extremely important for global health-related problems, since in resource-limited settings, each microfluidic device that needs to be tested can potentially have dust or other uncontrolled particles, all of which can then be avoided by using multiple angles of illumination.

The basic principle of multiangle LUCAS is illustrated in Figure 9.10. When the illumination angle is tilted, the shadow of the target cells or micro-objects in a multilayered structure is shifted geometrically by a certain amount relative to the vertical illumination angle. By analyzing the relative shift of a target shadow on the sensor array as a function of the illumination angle, the precise vertical position of the cell can be determined [53]. As mentioned before, this approach is quite useful especially when a target cell or micro-object type shows a similar shadow pattern when compared with another cell type located at a different layer. Based on the calibration results shown in Figure 9.10c, the multiangle LUCAS scheme can provide a vertical resolution of, for example, $<50\,\mu m$. Therefore, a cell type that is identified in a given LUCAS image can also be assigned its own precise 3D position. This, however, by no means, is the ultimate limit of Z resolution, and in fact, submicron depth localization accuracy over an ultralarge FOV can be readily achieved with the multiangle platform through advanced processing [53].

A proof of concept demonstration of multiangle LUCAS system is given in Figure 9.10e for a heterogeneous mixture of 5-, 10-, and 20-µm-diameter microbeads located at three different plane heights, that is, $Z = 1.2$, 1.3, and 1.4 mm. For this three-layered heterogeneous mixture, LUCAS images are captured at

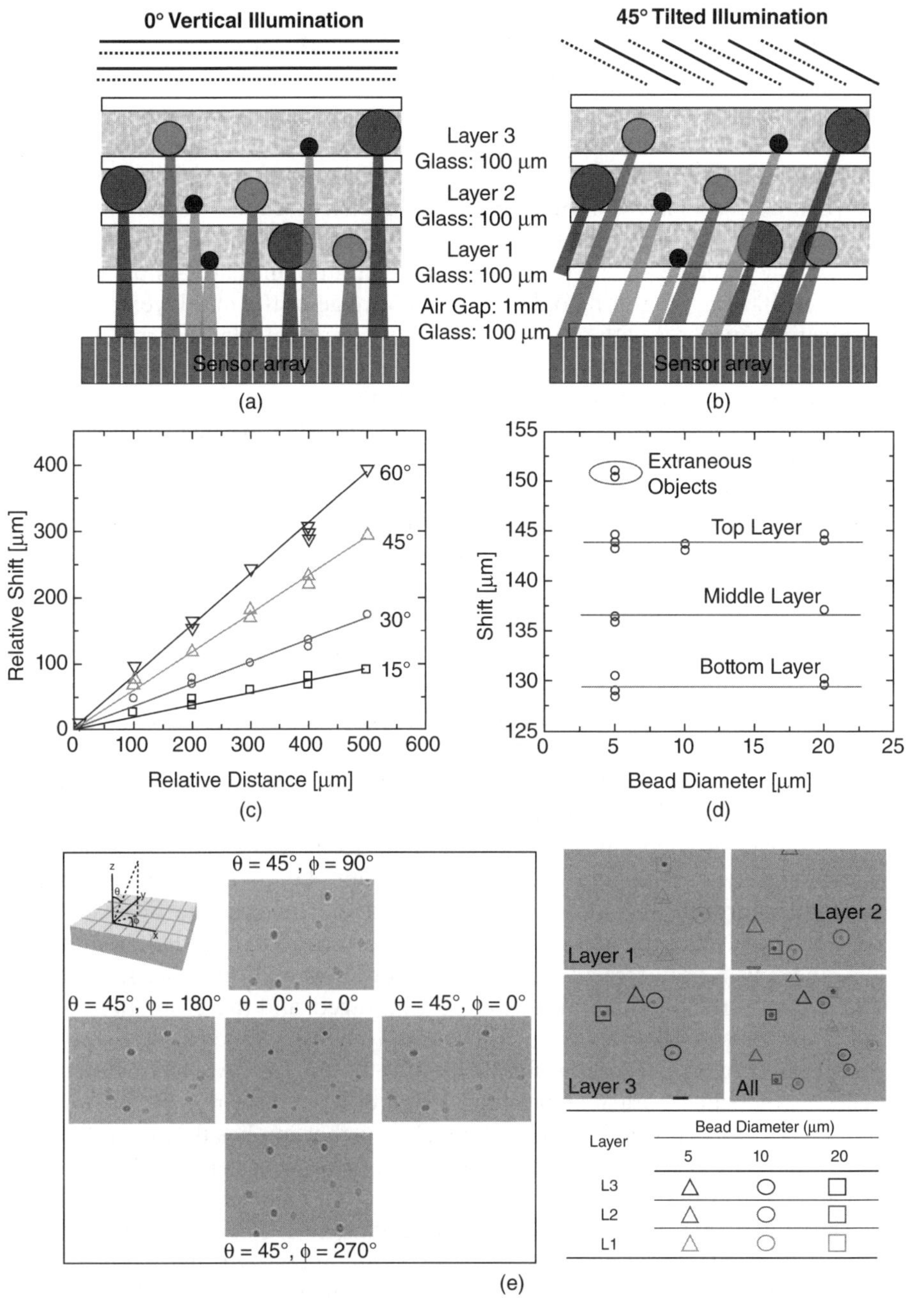

Figure 9.10. The principles of the multiangle LUCAS platform with (a) vertical illumination and (b) tilted illumination angles are illustrated. (c) The relationship between the relative heights of the objects and the lateral shift of their shadows at the sensor plane as a function of the illumination angle is given. (d) The principle of digital zooming in a multilayered structure is illustrated, showing that extraneous objects can be digitally excluded, while desired cell planes are uniquely recovered. (e) The characterization results for a three-layered microchannel are shown, where each channel includes a heterogeneous mixture of microbeads of three different sizes.

five different illumination angles, including one vertical illumination and four tilted illumination angles (each at 45°). For each one of the acquired images, the target micro-objects are identified based on their correlation index to the corresponding LUCAS library image. Here, we should note that the LUCAS library images of the same particle type located at different planes are almost the same since $\Delta Z = 100\,\mu m$. Therefore, only the library image of the mid layer is used for identification purposes within the 2D correlation map calculations, which significantly simplified the numerical calculations and reduced the computation time. To further "digitally" zoom to any one of these three layers separated by 0.1 mm, a backward projection from the position of the identified particle shadow to the position of the light source is calculated for each one of the illumination angles. Each line that connects the detected LUCAS shadow center to the light source contributes exactly one "count" to the "zoomed" plane of interest (located at a desired Z value), and for all the illumination angles, a summation of these counts are recorded for each cell shadow. The focus of all such lines for a given cell or particle shadow has a higher count number than other regions of the image and, therefore, marks the final vertical position of the micro-object corresponding to the shadow of interest. By utilizing this backward ray computation for each one of the identified diffraction patterns, not only the three different object planes (with $\Delta Z = 100\,\mu m$) are clearly separated from each other, but also extraneous particles, such as dust on the top of the cover glass, are accurately distinguished from the target particles as illustrated in Figure 9.10d.

The above-outlined multiangle LUCAS platform is especially important for imaging densely packed multiple layers of a sample solution. First, with this multiangle approach, the number of sample layers that can be probed is not limited by the SNR of the detected diffraction patterns. Since the height information of the cells is not extracted using the texture information of the shadow, the distance between individual layers can be minimized, and therefore, the SNR of the shadow image of the top layer can be kept to be very similar to the one from the bottom layer. This greatly simplifies automated LUCAS characterization process, as the same library image for a given cell type can now be utilized for all the layers of interest at the same time. This further implies that with the multiangle illumination scheme, the overall speed of characterization will be significantly improved since the number of 2D correlation map calculations that need to be computed is reduced. Therefore, in multiangle LUCAS, all the layers of interest can be identified and characterized simultaneously using the backward projection algorithm as outlined above. This reduced computational burden may lead to decreased power consumption, which is very important for point-of-care diagnostic applications utilizing a battery operation.

Second, by use of multiple angles of illumination, the system will be more tolerant to fabrication defects and extraneous particles such as dust. This benefit of digital zooming capability should make LUCAS characterization results more reliable and accurate, especially in resource-limited settings. And finally, the characterization errors caused by statistical overlap of cell shadows will be significantly reduced using multiple illumination angles. Because the same sample volume is probed multiple times at different angles, more information is gathered regarding the sample, which makes it more robust to such statistical errors caused by overlapping LUCAS signatures [53].

9.4.4 Noise-Reduction Techniques in LUCAS

In LUCAS images, there are several noise sources, such as unwanted microparticles on the sensor surface or within the sample volume, dead pixels of the optoelectronic sensor array, and detection noise. Furthermore, although a low DOF might enable a higher SNR in LUCAS signatures, it may also lead to characterization errors due to signature nonuniformity issues. The main cause for this reduced signature uniformity is pixelation error, caused by 2D sampling at the sensor plane. Therefore, especially at a lower DOF, corresponding to a lower Z value, the reduced width of the shadow pattern may result in nonuniform LUCAS signatures. As a result of this, the use of a shorter wavelength of illumination to increase SNR may not always guarantee a reliable LUCAS characterization. To address all of these issues, we recently devised several approaches that include background subtraction, use of adaptive noise-reduction filters, as well as the use of multiple illumination wavelengths (i.e., the hybrid approach) [55]. And in this subsection, we will briefly summarize these digital noise-reduction approaches.

As an initial step of the LUCAS characterization and to remove the undesired effects of dead pixels, defects, or other particles on the sensor array, a background image subtraction is implemented. A quantified example of the improvement of this background subtraction step is illustrated for LUCAS image of fixed yeast cells (*S. pombe*) located at a DOF of $Z = 5.1\,\text{mm}$, as shown in Figure 9.11a,b. Using the background subtraction scheme, several dead pixels and unwanted particles on the sensor array were digitally eliminated. This also resulted in a >3 dB increase in the digital SNR of the LUCAS image as shown in Figure 9.11b.

For further noise reduction in LUCAS images, we evaluated the performance of several widely used digital filters, including the enhanced Lee filter, Lee filter, Kuan filter, Yu filter, "A Trous" wavelet transform filter, hybrid median filter, symmetric nearest-neighbor filter, average filter, and adaptive Wiener filter [55]. These filters were applied to various LUCAS images following the above-described background subtraction step. Among these different types of digital filters, "A Trous" wavelet (Figure 9.11c) and enhanced Lee filters (Figure 9.11d) were particularly successful in improving the digital SNR of the LUCAS signature of the yeast cells (at Z ~5 mm above the sensor plane) by ~10 dB when compared with the original LUCAS image of Figure 9.11a. This significant SNR improvement enabled a noteworthy increase in the accuracy of the LUCAS decision algorithm as reported in our previous results [55].

To further improve the digital SNR of the diffraction signatures in LUCAS images, a shorter sample-to-sensor distance is required, which is especially important for weakly scattering objects or cells such as small bacteria. On the other

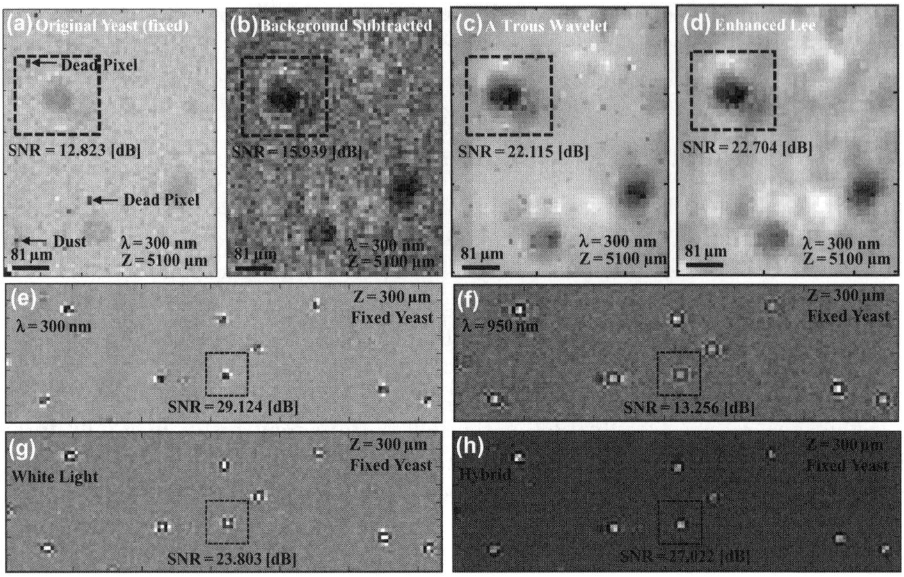

Figure 9.11. Noise-reduction techniques in LUCAS. Signal-to-noise ratio (SNR) of the LUCAS signatures of fixed yeasts (*S. pombe*) in (a) raw data format is improved by implementation of (b) background image subtraction or the use of various digital filters, that is, (c) A Trous wavelet filter and (d) enhanced Lee filter. LUCAS diffraction images captured under (e) $\lambda = 300$ nm, (f) $\lambda = 950$ nm, and (g) white light are shown. The figure in (h) illustrates a hybrid LUCAS image, which digitally combines (e) and (f).

hand, when the diffraction pattern of the cells or bacteria gets narrower by having small Z values, not only the uniformity of the LUCAS signatures may decrease, but also the 2D diffraction texture, which is used as the major input for LUCAS pattern recognition algorithm becomes less distinct. To resolve this potential issue, one possibility is to digitally combine multiple illumination wavelengths, as briefly pointed out in earlier discussions [55]. Figure 9.11e–g shows the diffraction images of fixed yeast cells located at $Z = 300\,\mu m$ for $\lambda = 300$ nm, 950 nm, and white light illumination, respectively. For the same sample solution, Figure 9.11h illustrates a hybrid LUCAS image that is reconstructed by the digital subtraction of the 950-nm LUCAS image from the 300-nm one. This hybrid approach exhibits a better SNR than the long wavelength of illumination, and it also improved the uniformity of the 2D texture of each cell signature when compared with the short wavelength of illumination such that the pattern recognition algorithm can now better identify the target image within the whole FOV and DOF. Thus, the digitally combined multicolor LUCAS images offer a higher SNR and better signature uniformity, as well as richer texture formation, which we further illustrated in our earlier work [55].

9.4.5 Holographic LUCAS with Spatially Incoherent Sources

So far, all the discussions above are based on spatially incoherent illumination (at the cell plane) where weak or nonuniform cell signatures can be improved by tuning illumination wavelengths, applying adaptive digital filters, or combining multiple images recorded at different wavelengths. In addition to these, by controlling the spatial coherence of the light source, for example, by using a variable pinhole in front of the incoherent light source, the digital SNR and the signature uniformity of the cell shadows can be significantly improved such that weakly scattering phase objects, for example, bacteria or blood cells, can be characterized with better sensitivity and specificity [58]. We refer to this improved technique as "holographic LUCAS," where the scattered light from the cell body and the unscattered reference light directly emanating from the pinhole source interferes, creating the "holographic shadow" of each cell at the sensor surface.

The use of spatially incoherent source, such as a light-emitting diode, is quite important for a significant reduction in cost, complexity, and dimensions of the instrument, all of which are requirements of telemedicine applications. Quite interestingly, for a completely incoherent source that is emitting from a finite aperture, free space propagation can introduce partial spatial coherence even at a distance of over a few centimeters. However, in conventional LUCAS discussed earlier, the diameter over which the cells face a spatially coherent beam is rather small when compared with the diameter of the cell; as a result of which, the phase information of the scattered waves cannot get embedded into the amplitude oscillations at the sensor array (i.e., self-interference of the scattered waves is recorded with conventional LUCAS platform). In holographic LUCAS, on the other hand, a variable pinhole in front of the source is used to enlarge coherence diameter at the cell plane to become larger than the cell diameter, as a result of which, the phase of the scattered fields can now be embedded into amplitude oscillations through holographic interference [58]. As a matter of fact, this embedded optical phase information, if processed appropriately, can bring much more than a mere signature or SNR improvement, which will be left as a topic to be discussed in a future publication.

To show the performance improvement of the holographic LUCAS system over the classical incoherent LUCAS platform, a heterogeneous solution containing RBCs, fixed yeast (*S. pombe*), and 10-μm polystyrene beads was imaged with and without a 100-μm-diameter pinhole, which controls the spatial coherence properties of the light at the cell plane. As shown in Figure 9.12, the diffraction signatures of various micro-objects captured with conventional incoherent LUCAS are significantly blurred when compared with the results of the holographic LUCAS platform. Figure 9.12c1,d1,e1 are 10× microscope images of a fixed yeast, an RBC, and a 10-μm bead, respectively, and their LUCAS images under incoherent illumination (Figure 9.12c2,d2,e2) are compared with those of the holographic LUCAS system (Figure 9.12c3,d3,e3). As illustrated in Figure 9.12, with the pinhole, owing to the increased spatial coherence diameter at the cell plane, the 2D shadow textures of the micro-objects become richer,

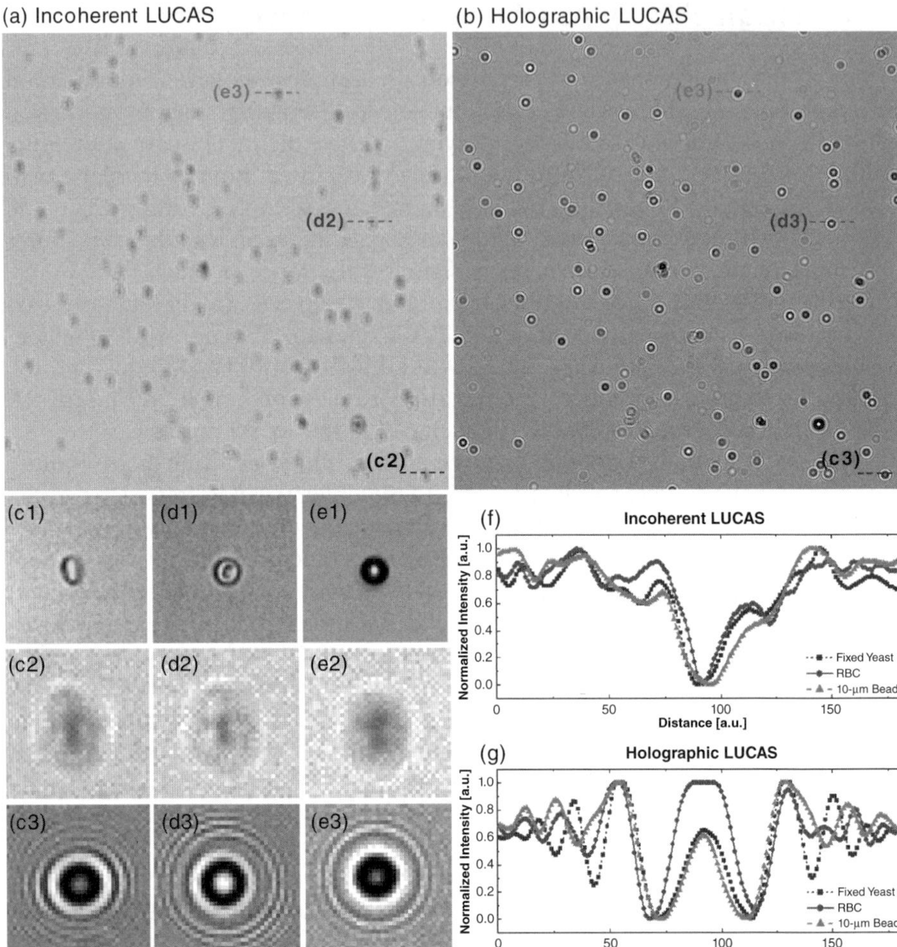

Figure 9.12. Holographic LUCAS imaging results are illustrated. For a heterogeneous solution of fixed yeast cells, red blood cells and 10-μm beads, diffraction signatures captured with (a) conventional incoherent LUCAS platform yield blurry images when compared with (b) the holographic LUCAS platform results. The figure (c1,d1,e1) shows 10× microscope images of fixed yeast, RBC, and 10-μm bead, respectively, and their LUCAS images under (c2,d2,e2) the conventional incoherent illumination were compared with those under (c3,d3,e3) the holographic illumination. Cross-sectional comparison of these images is also provided in (f) and (g), respectively.

which translates itself to a more robust cell characterization scheme as we already demonstrated in our earlier results [58].

To evaluate the usefulness of the holographic LUCAS technique in cytometry and diagnostics applications, automated counting results of RBCs for various concentration levels were compared with a hemacytometer as shown in

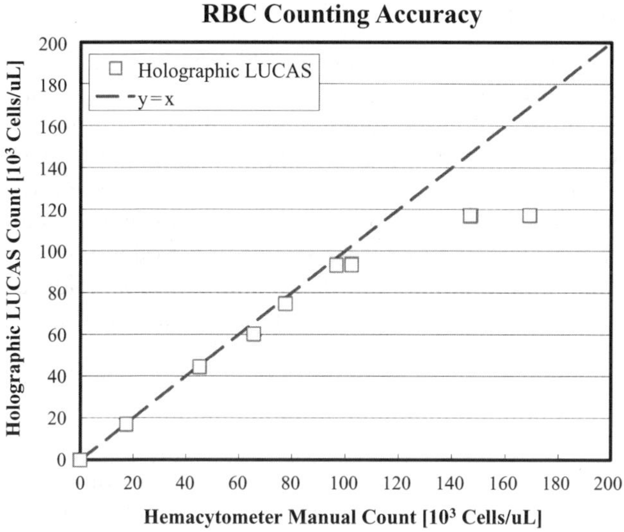

Figure 9.13. RBC counting performance of the holographic LUCAS platform. RBCs that are imaged under the holographic LUCAS platform were counted at various concentrations, up to ~200,000 cells/μL; the count results were compared against a commercially available hemacytometer. Holographic LUCAS platform successfully counted the red blood cells with an accuracy of ~95% up to a cell density of ~100,000 cells/μL. Beyond this level of cell density, the overlapping patterns at the sensor plane introduced errors in our automated characterization algorithm since the holographic patterns of the cells started to lose their unique signatures at such high cell densities.

Figure 9.13. In these experiments, holographic LUCAS automatically counted RBCs at various cell densities, up to ~200K cells/μL, by using its decision algorithm described in Figure 9.8, and the same samples were also counted through a hemacytometer (Cell-Vu CBC DRM-70, Millennium Sciences) to validate and compare our counting accuracy. Based on the experimental results shown in Figure 9.13, the holographic LUCAS platform successfully counted RBCs up to a cell density of ~100,000 RBCs/μL with an accuracy of >95%, compared with the results of the hemacytometer. Beyond a density of >100,000 cells/μL, because of the statistical overlap of the cell holograms with each other, our counting results started to get inaccurate.

In summary, by controlling the spatial coherence properties of the light source at the cell plane, the holographic LUCAS platform can achieve a significantly improved performance for automated characterization of a heterogeneous cell solution. Future LUCAS platforms will better control and make use of the information content of each cell hologram through more advanced processing algorithms, which we leave as a topic for an upcoming publication.

9.4.6 Future Directions in LUCAS

The overview of LUCAS-related results of this chapter presents just the initial steps of a very powerful approach to on-chip cytometry and medical diagnostics. Within the next couple of years, several new technological breakthroughs centered on LUCAS are expected. Among these, exploitation of the embedded phase information in LUCAS signatures in the form of digital reconstructions is a major advancement that will transform LUCAS into a new dimension. Furthermore, polarization and fluorescent imaging modalities, once made compatible with the LUCAS platform, will expand the application domain of the LUCAS platform into a complete new set of diagnostic needs.

9.5 CONCLUSION

In summary, we reviewed the recent progress in on-chip cytometry and diagnostic technologies that are relevant to point-of-care and wireless health applications. Specifically, we reviewed the basics of a recently introduced lensfree optical cytometry approach, which is termed LUCAS. This approach can rapidly characterize a heterogeneous cell solution of interest on a chip over an ultralarge FOV of ~18 cm^2 and a large DOF of ~5 mm. Unlike most existing cytometry technologies, the high-throughput platform of LUCAS does not utilize any fluid flow nor lenses and, therefore, may permit a cost-effective, lightweight, and compact platform for cytometry and diagnostics. Therefore, this emerging technology holds a significant potential for various point-of-care and wireless health applications, especially in developing countries where the resources are scarce.

REFERENCES

[1] International Telecommunication Union, "Market information and statistics," 2007. Available: http://www.itu.int/ITU-D/ict/statistics/maps.html (accessed August 31, 2008).

[2] Bloodbook.com, "Blood test results—normal ranges," 2007. Available: http://www.bloodbook.com/ranges.html#BLOOD (accessed August 15, 2008).

[3] National Library of Medicine, "Medical dictionary," 2008. Available: http://www.nlm.nih.gov/medlineplus/mplusdictionary.html (accessed August 13, 2008).

[4] Family Practice Notebook, "Mean corpuscular volume," 2008. Available: http://www.fpnotebook.com/Hemeonc/Lab/MnCrpsclrVlm.htm (accessed August 15, 2008).

[5] S. Mali, S. Steele, L. Slutsker, and P. M. Arguin, "Malaria surveillance—United States 2006," *MMWR Surveill. Summ.*, 57(SS05), pp. 24–39, 2008.

[6] C. D. Chin, V. Linder, and S. K. Sia, "Lab-on-a chip devices for global health: Past studies and future opportunities," *Lab Chip*, 7, pp. 41–57, 2007.

[7] M. Cristofanilli, "Circulating tumor cells, disease progression, and survival in metastatic breast cancer," *N. Engl. J. Med.*, 351, pp. 824–826, 2004.

[8] A. Ribeiro, A. E. Vazquez-Sequeiros, L. M. Wiersema, K. K. Wang, J. E. Clain, and M. J. Wiersema, "EUS-guided fine-needle aspiration combined with flow cytometry and immunocytochemistry in the diagnosis of lymphoma," *Gastrointest. Endosc.*, 53, pp. 485–491, 2001.

[9] American Cancer Society, "Detailed guide: HIV infection and AIDS: How is HIV/AIDS diagnosed?" 2008. Available: http://www.cancer.org/docroot/CRI/content/CRI_2_4_3x_How_Is_HIVAIDS_Diagnosed.asp (accessed July 9, 2008).

[10] T. J. Lynch, M. Toner, and D. A. Haber, "Detection of mutations in *EGFR* in circulating lung-cancer cells," *N. Engl. J. Med.*, 359, pp. 366–377, 2008.

[11] S. Maheswaran, L. V. Sequist, S. Nagrath, L. Ulkus, B. Brannigan, C. V. Collura, E. Inserra, S. Diederichs, A. J. Iafrate, D. W. Bell, S. Digumarthy, A. Muzikansky, D. Irimia, J. Settleman, and R. G. Tompkins, "Detection of mutations in EGFR in circulating lung-cancer cells," *N. Engl. J. Med.*, 359, pp. 366–377, 2008.

[12] J. Wang, N. Bao, L. L. Paris, H. Y. Wang, R. L. Geahlen, and C. Lu, 2008 "*American physical society meeting*." March 16–20, Pittsburgh, PA, 2008.

[13] I. V. Jani, G. Janossy, D. W. Brown, and F. Melody, "Multiplexed immunoassays by flow cytometry for diagnosis and surveillance of infectious diseases in resource-poor settings," *Lancet Infect. Dis.*, 2, pp. 243–250, 2002.

[14] R. Bowen, "Counting cells with a hemacytometer." Colorado State University, 2002. Available: http://www.vivo.colostate.edu/hbooks/pathphys/reprod/semeneval/hemacytometer.html (accessed August 9, 2008).

[15] R. Nunez, *Flow Cytometry for Research Scientists: Principle and Applications*. Wymondham, U.K.: Horizon Press, 2001.

[16] Beckman Coulter, "COULTER® EPICS® XL™ and XL-MCL™ flow," 2008. Available: Cytometerhttp://www.beckmancoulter.com/products/instrument/flowcytometry/epicsxl.asp (accessed August 30, 2008).

[17] P. J. Crosland-Taylor, "A device for counting small particles suspended in a fluid through a tube," *Nature*, 171, pp. 37–38, 1952.

[18] M. A. Owens and M. R. Loken, *Flow Cytometry: Principles for Clinical Laboratory Practice*. New York: Wiley-Liss, 1995.

[19] A. Givan, *Flow Cytometry: First Principles*. New York: Wiley-Liss, 2001.

[20] S.-Y. Yang, S.-K. Hsiung, Y.-C. Hung, C.-M. Chang, T.-L. Liao, and G.-B. Lee, "A cell counting/sorting system incorporated with a microfabricated flow cytometer chip," *Meas. Sci. Technol.*, 17, pp. 2001–2009, 2006.

[21] Q. Xiang, B. Xu, and D. Li, "Miniature real time PCR on chip with multi-channel fiber optical fluorescence detection module," *Biomed. Microdevices*, 9, pp. 443–449, 2007.

[22] N. Pamme, R. Koyama, and A. Manz, "Counting and sizing of particles and particle agglomerates in a microfluidic device using laser light scattering: Application to a particle-enhanced immunoassay," *Lab Chip*, 3, pp. 187–192, 2003.

[23] L. M. Fu, R. J. Yang, C. H. Lin, Y. J. Pen, and G. B. Lee, "Electrokinetically driven micro flow cytometers with integrated fiber optics for on-line cell/particle detection," *Anal. Chim. Acta*, 507, pp. 163–169, 2004.

[24] L. Novak, P. Neuzil, J. Pipper, Y. Zhang, and S. H. Lee, "An integrated fluorescence detection system for lab-on-a-chip applications," *Lab Chip*, 6, pp. 27–29, 2007.

[25] S. C. Jacobson and J. M. Ramsey, "Electrokinetic focusing in microfabricated channel structures," *Anal. Chem.*, 69, pp. 3212–3217, 1997.

[26] A. Wolff, P. Nielsen, U. D. Larsen, P. Friis, G. Goranovic, C. R. Poulsen, J. P. Kutter, and P. Telleman, "Integrating advanced functionality in a microfabricated high-throughput fluorescent-activated cell sorter," *Lab Chip*, 3, pp. 22–27, 2003.

[27] C. Lancaster, A. Kokoris, M. Nabavi, J. Clemmens, P. Maloney, J. Capadanno, J. Gerdes, and C. F. Battrell, "Rare cancer cell analyzer for whole blood applications: Microcytometer cell counting and sorting subcircuits," *Methods*, 37, pp. 120–127, 2005.

[28] C. Simonnet and A. Groisman, "Focusing microparticles in a microfluidic channel with standing surface acoustic waves (SSAW)," *Anal. Chem.*, 78, pp. 5653–5663, 2006.

[29] X. C. Xuan and D. Q. Li, "Focused electrophoretic motion and selected electrokinetic dispensing of particles and cells in cross-microchannels," *Electrophoresis*, 26, pp. 3552–3560, 2005.

[30] R. Yang, D. L. Feeback, and W. Wang, "Microfabrication and test of a three-dimensional polymer hydro-focusing unit for flow cytometry applications," *Sens. Actuators A*, 118, pp. 259–267, 2005.

[31] G. B. Lee, C. C. Chang, S. B. Huang, and R. J. Yang, "An optimal three-dimensional focusing technique for micro-flow cytometers," *J. Micromech. Microeng.*, 16, pp. 1024–1032, 2006.

[32] M. A. McClain, C. T. Culbertson, S. C. Jacobson, and J. M. Ramsey, "Flow cytometry of *Escherichia coli* on microfluidic devices," *Anal. Chem.*, 73, pp. 5334–5338, 2001.

[33] R. J. Yang, C. C. Chang, S. B. Huang, and G. B. Lee, "A new focusing model and switching approach for electrokinetic flow inside microchannels," *J. Micromech. Microeng.*, 16, pp. 341–348, 2006.

[34] C. H. Lin, G. B. Lee, L. M. Fu, and B. H. Hwey, "Vertical focusing device utilizing dielectrophoretic force and its application on micro flow cytometer," *J. Microelectromech. Syst.*, 13, pp. 923–932, 2004.

[35] E. Altendorf, D. Zebert, M. Holl, and P. Yager, *Transducers*. Chicago, IL: Elsevier, 1997.

[36] L. K. Chau, T. Osborn, C. C. Wu, and P. Yager, "Microfabricated silicon flow-cell for optical monitoring of biological fluids," *Anal. Sci.*, 15, pp. 721–724, 1999.

[37] C. Gruden, S. Skerlos, and P. Adriaens, "FEMS microbiology," *Ecology*, 49, pp. 37–49, 2004.

[38] T. Kamei, B. M. Paegel, J. R. Scherer, A. M. Skelley, R. A. Street, and R. A. Mathies, "Integrated hydrogenated amorphous Si photodiode detector for microfluidic bioanalytical devices," *Anal. Chem.*, 75, pp. 5300–5305, 2003.

[39] S. Niehren, W. Kinzelbach, S. Seeger, and J. Wolfrum, "An all-solid-state flow cytometer for counting fluorescent microspheres," *Anal. Chem.*, 67, pp. 2666–2671, 1995.

[40] W. H. Coulter, "Means for counting particles suspended in a fluid." US patent 2656508, 1953.

[41] U. D. Larsen, G. Blankenstein, and J. Branebjerg, "Solid state sensors and actuators," *Transducers '97 International Conference*, June 16–19, Chicago, IL, 1997.

[42] K. Cheung, S. Gawad, and P. Renaud, "Impedance spectroscopy flow cytometry: On-chip label-free cell differentiation," *Cytometry A*, 65A, pp. 124–132, 2005.

[43] M. Koch, A. G. R. Evans, and A. Brunnschweiler, "Design and fabrication of a micromachined coulter counter," *J. Micromech. Microeng.*, 9, pp. 159–161, 1999.

[44] A. Jahn, W. N. Vreeland, M. Gaitan, and L. E. Locascio, "Controlled vesicle self-assembly in microfluidic channels with hydrodynamic focusing," *J. Am. Chem. Soc.*, 126(9), pp. 2674–2675, 2004.

[45] R. Scott, P. Sethu, and C. K. Harnett, "Three-dimensional hydrodynamic focusing in a microfluidic Coulter counter," *Rev. Sci. Instrum.*, 79, p. 046104, 2008.

[46] S. Zheng, M. Liu, and Y. C. Tai, "Micro Coulter counters with platinum black electroplated electrodes for human blood cell sensing," *Biomed. Microdevices*, 10, pp. 221–231, 2008.

[47] R. Rodriguez-Trujillo, C. A. Mills, J. Samitier, and G. Gomila, "Low cost micro-Coulter counter with hydrodynamic focusing," *Microfluid Nanofluid.*, 3, pp. 171–176, 2007.

[48] X. Cheng, Y. S. Liu, D. Irimia, U. M. Demirci, L. Yang, L. Zamir, W. R. Rodriguez, M. Toner, and R. Bashir, "Cell detection and counting through cell lysate impedance spectroscopy in microfluidic devices," *Lab Chip*, 7, pp. 746–755, 2007.

[49] D. Huh, W. Gu, Y. Komotani, J. B. Grotberg, and S. Takayama, "Microfluidics for flow cytometric analysis of cells and particles," *Physiol. Meas.*, 26, pp. R73–R98, 2005.

[50] S. M. Kim, S. H. Lee, and K. Y. Suh, "Cell research with physically modified microfluidic channels: A review," *Lab Chip*, 8, pp. 1015–1023, 2008.

[51] C. E. Sims and N. L. Allbritton, "Analysis of single mammalian cells on-chip," *Lab Chip*, 7, pp. 423–440, 2007.

[52] A. Ozcan and U. Demirci, "Ultra wide-field lens-free monitoring of cells on-chip," *Lab Chip*, 8, pp. 98–106, 2008.

[53] T.-W. Su, S.-O. Isikman, W. Bishara, D. Tseng, A. Erlinger, and A. Ozcan, "Multi-angle lensless digital holography for depth resolved imaging on a chip," *Optics Express*, 18(9), pp. 9690–9711, 2010.

[54] T.-W. Su, S. Seo, A. Erlinger, and A. Ozcan, "High-throughput lensfree imaging and characterization of a heterogeneous cell solution on a chip," *Biotechnol. Bioeng.*, 102(3), pp. 856–868, 2009.

[55] S. Seo, T.-W. Su, A. Erlinger, and A. Ozcan, "Multi-color LUCAS: Lensfree on-chip cytometry using tunable monochromatic illumination and digital noise reduction," *Cell. Mol. Bioeng.*, 1(2–3), pp. 146–156, 2008.

[56] X. Cheng, D. Irimia, M. Dixon, K. Sekine, U. Demirci, L. Zamir, R. G. Tompkins, W. Rodriguez, and M. Toner, "A microfluidic device for practical label-free CD4+ T cell counting of HIV-infected subjects," *Lab Chip*, 7, pp. 170–178, 2007.

[57] UNAIDS, "AIDS epidemic update: December 2006." Available: http://www.unaids.org/en/HIV_data/epi2006/default.asp (accessed August 21, 2008).

[58] S. Seo, T.-W. Su, D. K. Tseng, A. Erlinger, and A. Ozcan, "Lensfree holographic imaging for on-chip cytometry and diagnostics," *Lab Chip*, 9, pp. 777–787, 2009.

10

ADVANCED TECHNOLOGIES FOR REAL-TIME MONITORING AND CONTROL IN BIOMICROFLUIDICS

Francesca Sapuppo, Marcos Intaglietta, and Maide Bucolo

10.1 INTRODUCTION

The characterization and control of fluid and particle motion in microfluidic systems have, for some time, been active areas in the industrial field for integrated circuits (ICs), such as cooling and inkjet printing. In recent years, however, applications of microfluidics in the biomedical research field have become increasingly important. Such applications regard both *in vivo* biological systems, oriented to research in the microcirculation environment, and *in vitro* systems for the development of integrated devices for the analysis of biological fluids and particles (lab-on-chip [LoC]), and they are referred to here as biomicrofluidic applications.

On one hand, the study of biological microfluidic systems, such as the microcirculatory one, finds application in the diagnostics of pathologies, such as retinal abnormalities, hypertension, and cancer, which can be characterized through the analysis of angiogenesis phenomena and modifications in microcirculation conditions. In this environment, microcirculation is the *in vivo* key application and involves experimental preparation on laboratory animals and observation of microfluidic phenomena. In particular, animals, such as hamsters and rats, are at the basis of the experimental results presented throughout this chapter. As an example, current experimental studies associated with the development of

CMOS Biomicrosystems: Where Electronics Meet Biology, First Edition. Edited by Krzysztof Iniewski.

artificial blood [1,2] require detailed information on blood particles, their motion, and their interaction with the surrounding environment. In this scenario, the knowledge of processes such as oxygen consumption and nutrient delivery (blood exchange), as well as changes in microvascular parameters, is considered fundamental for the interpretation of the cause–effect relationship in the complex vascular regulatory system. Moreover, the development of real-time methods to model such processes and to determine structural characteristics, such as microvessel network maps and microvessel diameter, as well as the assessment of functional parameters such as capillary density, flow velocity, and red blood cell (RBC) density, have been active areas of research in microhemodynamics. The characterization of quantitative and qualitative parameters of the microvascular network is important in microhemodynamic research fields, since the information on blood flow behavior can be used for the development of an analytical characterization of the microcirculatory regulatory system.

On the other hand, the study of *in vitro* microfluidic devices to carry out biomedical research and clinical diagnostics is motivated by their significant advantages. First, the volume of fluids within microdevice channels is very small, usually in the order of nanoliters, and therefore, the amount of reagents and analytes can also be maintained at a minimum. These specifics, along with a fast time response and IC technology, are the basis for the creation of integrated, portable, real-time clinical and diagnostic devices (LoC) and also of micro total analysis systems (μTAS) [3–5], which avoid time-consuming laboratory analysis procedures. They consist of polymer or silicon micrometric devices and exploit the interaction between properties of fluids, particles, and devices, such as geometry, electromagnetic fields, and mechanic forces, for direct, active manipulation of microfluidic phenomena. In this case, the characterization of artificially induced particle transport is fundamental, since it is the basis for the extraction of significant information from the analysis of physiological components (e.g., DNA chip). The research environment related to microfluidic systems thus involves a variety of processes occurring on the micrometric scale; these can be due to continuous or discrete flows of fluids, gases, particles, droplets, emulsions, suspensions, and so on. In order for them to be functional, the physics of such phenomena must be known; they also need to be dynamically monitored and controlled throughout the entire spatial domain, often consisting of complex networks of microchannels, reservoirs, or active areas. The choice and characterization of materials, the geometry of the devices, the physical and chemical properties of the fluids and particles (i.e., dielectric, ferrofluids), and the manipulation and control methods and technologies are directly related to the type of processes to be generated and to the applications addressed.

After a brief overview of the existing *in vivo* and *in vitro* microfluidic systems and the issues related to the characterization of their spatial and temporal dynamics, such as movement of fluids, gases, and particles, applications of interest and case studies in biomicrofluidics will be discussed.

We focus the attention on advanced methodologies and technologies related to monitoring and controlling issues in biomicrofluidics, in particular, giving detailed descriptions of two case applications. The first is a real-time implementation of a pointwise flow measurement system based on standard optics and PC-based hybrid analog–digital (AD) processing technology; it represents an example of usage of standard technologies for approaching advanced methodological issues. The second is a review of applications of cellular nonlinear/neural networks (CNNs), a parallel analog architecture for real-time image processing and parameters extraction in biomicrofluidics. Finally, some perspectives on polymeric technology applications addressing biomicrofluidic monitoring and control issues are given, showing also a feasibility study on micro-optical interfaces for flow monitoring.

10.2 BIOMICROFLUIDICS SYSTEMS AND RELATED ISSUES

A brief description of existing microfluidic systems related to both *in vivo* microcirculation and *in vitro* devices for biomedical applications may give an idea on the parameters and on the microfluidic phenomena of interest, and also allows to focus the attention on the stimuli that such applications give to the research field and, in particular, to the ones related to the monitoring and control systems.

10.2.1 *In Vivo* Microcirculation

The microfluidic study in *in vivo* systems is mainly related to the microcirculation environment [2]. The fluids and particles we are dealing with are, therefore, blood and its components, and the channels are represented by microvessels.

The parameters and issues we face in the microcirculation research field involve biological phenomena that occur from the micrometric to the nanometric scales. Red blood cells (RBCs) range in size from 5 to 8 μm and have a very thin membrane (~10 nm) that encapsulates the cytoplasm. The ability of the cells to deform largely determines the ease with which blood flows through the microvessels. In general, cell deformation and, therefore, their flow pattern is a complex phenomenon caused by external forces acting over the surface of the membrane. The rate and extent of deformation depend on the intrinsic and extrinsic properties of the cell. Other critical factors in this study are represented by the thinness of the vessels and by the overall surface that their wall offers to blood exchanges. The former is the ratio of the capillary section (in the order of 10^{-12} m) to its length (~10^{-3} m); the latter is represented by the product of the circumference (~10^{-6} m) and the length (~10^{-3} m). Another critical point in the study of microcirculation phenomena is represented by the adhesion surface between the RBC and vessel walls, where all the functions of the blood are accomplished.

The information extracted regarding the velocity and the passage of RBCs in the vessels represents a fundamental step for the evaluation of shear stress and, therefore, of the mechanochemical interactions between blood and the luminal endothelial cell surface. These biological phenomena take place at scales of a few hundred nanometers, corresponding to endothelial extracellular particle thickness sensitive to the plasma shear stress [6].

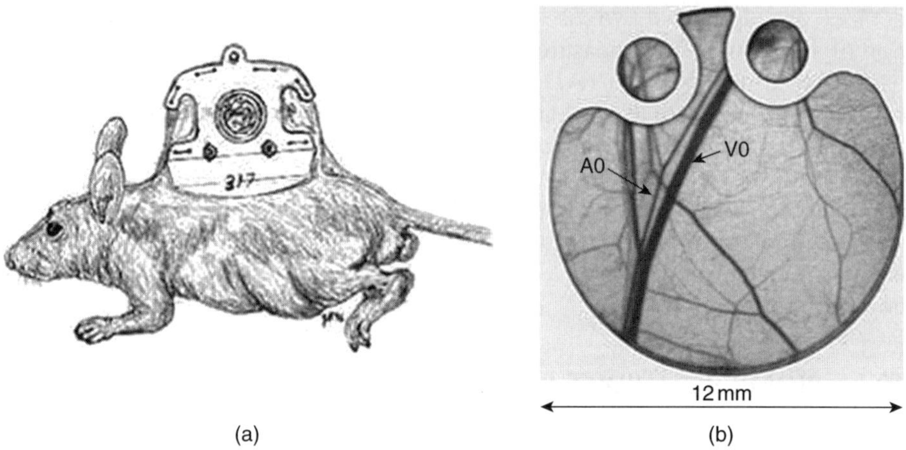

Figure 10.1. Hamster skinfold chamber. (a) Drawing of a preparation on hamster dorsum skin. (b) Visualization of microvessels in a chamber.

In this scenario, the knowledge of processes such as oxygen consumption and nutrients delivery (blood exchange), as well as microvascular parameter changes, is considered fundamental for the interpretation of the cause–effect chain in the complex vascular regulatory system; the development of real-time methods to model such processes and to determine structural characteristics, such as microvessel network maps and microvessel diameter, as well as the assessment of functional parameters such as capillary density, flow velocity, and RBC density, is an active area of research in microhemodynamics [7, 8].

Microcirculation can be observed in surgical animal preparations, which are minimally invasive experimental solutions for nondestructive and noninterfering observation of microfluidic *in vivo* phenomena [9]. Images of microvessels can be obtained through a transparent window chamber surgically implanted in the dorsum of hamsters (Fig. 10.1a). The tissue can be transilluminated and observed by means of intravital microscopy.

Figure 10.1b shows a sample and conventional *in vivo* preparation of the hamster window chamber model with visible arteriole (A0) and vein (V0). Other vessels (A1 branches off A0; A2 branches off A1; A3 branches off A2; A4 branches off A3), capillaries (defined as vessels with single red cell transit), and venules can only be classified under the intravital microscope.

10.2.2 *In Vitro* Devices

A variety of devices currently exist on the market [10, 11], such as the μTAS, also called LoC [3–5]. These systems are being designed to analyze great numbers of data simultaneously, such as the data obtained from genomic, proteomic, and metabolic studies, mainly *in vitro*. At the same time, the systems are also suitable for the analysis of environmental gases and fluids, as well as food and water. They perform different analyses on a one-device solution, whose technology mainly includes the integration of polymers for the microfluidics part and standard metal-based and silicon-based technology for sample treatment, detection, and the control units.

In microfluidics-related literature, the physics of small volumes (nanoliters) of fluids is parameterized by a series of dimensionless numbers expressing the relative importance of various physical phenomena. Specifically, parameters are considered as the Reynolds number, Re, addressing the relative importance of inertial effects to viscous forces. Other physical and chemical properties, such as flow rates, velocity fields, particle concentration, flow patterns, and chemical components profiles, are also central in the study of microfluidics applications [12].

In this chapter, we mainly focus on spatiotemporal dynamics that take place in microflow and that involve phenomena such as droplet formation and mixing processes considering multiple-phase fluids.

The droplet formation is related to emulsion generation as dispersions of micrometric droplets in liquids. Such emulsions are widely used in a number of industrial domains, such as pharmaceutical [13], cosmetic [14], and food industries [15]. In particular, droplets generated by immiscible fluids, such as oil and water, or bubbles generated as gases in liquids in microfluidic devices are considered of particular interest because they become popular for generating nanovolume vessels for biochemical assays.

The mixing process is also a focal point in the implementation of most microfluidic applications [16] since the need arises for mixing small volumes of fluids in low Reynolds number flows.

In the theoretical case of a single fluid and small Reynolds numbers, given the length scales of typical microfluidic applications (ca. 100 μm), flows are inertialess, and what happens for finite Reynolds numbers is, in general, of limited applicability. Suffice it to say that there are many methods to enhance mixing in microflow [16]: electroosmotic effects, use of patterned walls, and systems based on the use of geometric reorientations (as in passive mixers), and so on.

In cases of multiple-phase flow, microfluidic systems can rely on the use of mixing by chaotic advection inside droplets in microfluidic channels to perform kinetic measurements with high temporal resolution and low consumption of samples. The system is based on the use of immiscible fluids to form and transport droplets containing multiple reagents through a winding channel microfluidic network [17], and finds many applications.

In microfluidic technology, droplets of micrometer sizes are mainly produced and individually manipulated [18, 19], but it is also possible to control the droplet flow using input-flow variability.

A sample setup for a basic *in vitro* experimentation and droplet or bubble flow can be represented by a simple system consisting of a polymer-made microfluidic chip with microchannels, and external mechanic microfluidic pumps (twin pump slide) that allow control of the input flow rate in the chip through an electronic pump control acting on the frequency (Fig. 10.2).

The commercial solution provided by ThinXXS (Zweibrücken, Germany) [20], in particular, is made of a microfluidic chip with different sizes of serpentine

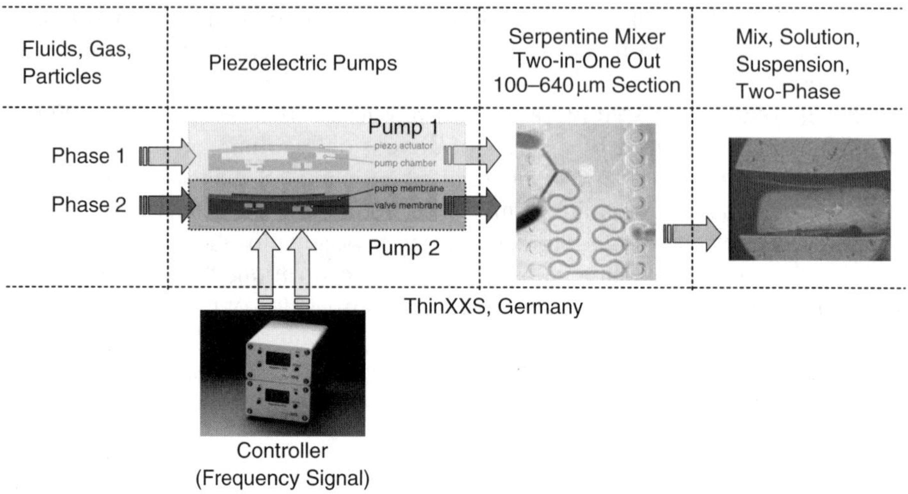

Figure 10.2. Sample *in vitro* system setup: two inputs, piezoelectric pumps, electronic pump controller, serpentine mixer with Y-junction, image of the microchannel (Copyright © 2008 IEEE).

mixers (snake mixer slide). Each of them consists of a Y-junction with two inputs and one output in order to allow two-phase flow generation and mixing. Their section varies from 100 to 640 μm, and the volumes they can handle vary from 0.2 to 50 μL. The maximum size of the particles that can be handled is 5 μm. The polymer used in this device is cyclic olefin copolymer (COC); it provides good optic, thermal, and chemical properties.

10.2.3 Methods and Technologies for Monitoring and Control

In microfluidics applications related to the biomedical field both in the *in vivo* and *in vitro* environments, the movement of fluids, gases, droplets, and particles can be studied, considering the optical information provided. Monitoring issues concerning particles' or droplets' temporal dynamics, velocity, and spatial flow patterns in microchannels can thus be addressed using the standard optoelectronic technology, allowing the acquisition and analysis of optical signals related to pointwise information or two-dimensional (2D) optical maps (images) as full-field information.

Monitoring is a preliminary and preparatory step for noninvasive control and actuation of spatiotemporal dynamics in microfluidic devices. Different approaches are possible: on one hand, the possibility of tuning the microfluidic flow behavior, exploiting a multidynamic external input flow, and changing parameters such as frequency, magnitude, shape, and volume fraction of different fluid components; on the other hand, we may exploit, in a noninvasive fashion, the direct action of optical or thermal effects on the internal physical or chemical parameters of the microfluidic flow.

Generation and control of droplet formation are approached in standard processes using high shear forces, for example, by rotor–stator, ultrasonic, or high-pressure homogenizing systems. As the fundamental droplet formation mechanism is of a statistical nature, however, these methods lead to broad droplet-size distributions. Droplets can also be generated by exploiting microchannel geometries, such as the surface tension controlled break-off issued from tiny orifices. In general, several microfluidic methods have been experimented, exploiting different geometries and temporal dynamics of input flows for the production of droplets [18, 19]. The controlled manipulation of droplets in the confining rigid boundaries of microchannels can also be applied for additional unit operations in droplet processing, like splitting, merging, or separation.

As an alternative to control droplet or particle motion, the effects of temperature and flow rate of aqueous fluid on droplet formation and size manipulation have been studied in literature [21]. At constant flow rates of the two liquids, different droplet breakup regimes can be observed and their transition capillary numbers can be identified, as well as their temperatures. The heat generated by integrated microheaters changes the droplet formation process. Increasing the temperature enlarges the size of the droplets significantly.

The optical control is based on the same principle, since it exploits the thermal effect due to optical signals to manipulate and react small droplets [22].

10.2.3.1 Pointwise Monitoring. Automated measurement methods are desirable for obtaining flow punctual information in microfluidics, and several methods exist for determining the two-phase or suspension flow in microchannels. Optical Doppler intravital velocimetry [23] is used in microfluidic channels of all sizes, although difficulties arise in measuring flow in channels placed in different layers.

Another method based on optical properties of the fluids is the dual-slit methodology [24], which, in particular, is based on the optical contrast between different phases of the fluids or between fluids and particles. It is a practical real-time method for measuring the transit time of RBCs between two optical windows, thus obtaining blood flow and, in general, particle velocity with a spatial resolution that is flexible and dependent on the optical characteristics of the setup. Compared with other methodologies, the implementation of the dual slit is based on a relatively simple experimental setup, involving optical magnification and standard optical detection (photodetectors) and on processing units performing the real-time cross-correlation between signals and, therefore, the velocity evaluation.

Conventional approaches for the implementation of the velocity measurement through the dual-slit and the cross-correlation processing use hardware solutions to yield real-time signal processing. Examples of this implementation are the Hewlett Packard (HP) Digital Correlator (Model 3721A, Delta T Product, Tucson, AZ) and the Velocity Tracker (Mod-102 B, Vista, Inc., San Diego, CA).

The HP Digital Cross-Correlator (no longer manufactured) provides, as its output, a cross-correlation curve that is represented by a vector of 100 points displayed via a shift register. The maximum of the correlation curve is obtained using a peak detector that finds the time delay corresponding to the maximum correlation and gives an analog output voltage proportional to the velocity. One of the drawbacks of this system is that the amplification applied to upstream and downstream signals affects the correlogram shape and amplitude, and, therefore, the peak detection process.

The Vista Velocity Tracker is based on the multiplication of the upstream signal, delayed in a tapped digital delay line, with the downstream signal, by adjusting the system sampling frequency, which becomes directly proportional to the flow velocity. This process requires an initial manual tuning of the frequency, which controls the delay line, in order to place the point of maximum correlation in the range of the feedback control that governs the system. The measurements obtained with this system are strongly dependent on the skill of the operator, since detection of the correlation peak and, therefore, of the velocity is based on a visual understanding of the cross-correlogram display.

In this chapter, we will present hybrid analog-digital (AD) real-time implementations of the dual slit through cross-correlation between optical signals, which uses standard optics and PC technology, as described in [25].

10.2.3.2. Full-Field Monitoring. Different methods to generate 2D profiles of microfluidic flow value are considered convenient to find spatiotemporal dynamics in microfluidic environments.

An example of technique based on the Doppler principle is enhanced high-resolution laser Doppler imaging (EHR-LDI); it performs flow measurement using the principle of light scattering and the spectral analysis of the scattered signal [26, 27]. The velocity measurement is not absolute, but relative particle velocities over the period of measurement can be obtained.

As an alternative, a really common method is the particle imaging velocimetry (PIV) algorithms, which is central processing unit (CPU)-based implementation applied on digital images. It therefore requires acquisition through high-speed digital camera setup. PIV assesses particle flow velocity profiles and has been used extensively in experimental fluid mechanics. A number of variations of the PIV methods have been reported, such as the 2D cross-correlation method, the particle-tracking method, and the iterative correlation method [28, 29]; they are based on the analysis of images outputting the velocity information for each point in the 2D space.

As an alternative to digital standard processing technology, such image-based analysis can be performed through real-time analog technology such as CNNs [30, 31]; it represents the technological solution that will be described in some applications in this chapter.

They perform real-time extraction of the microfluidic parameters using the optical information transduced with an integrated vision system based on the CNN and a technology called Focal Plane Processor (FPP; Eye-RIS Vision

System v1.1.7, Anafocus, Seville, Spain) [32, 33], which provides a support for CNN-based real-time image processing.

The spatial distribution and the temporal dynamics of microfluidic phenomena can be mapped through the distributed CNN structure and their analog processing features with a spatiotemporal scale, which is comparable with the scale of biological phenomena thanks to the highly integrated technology, the high-speed parallel processing, and the low-circuit time constant. Real-time algorithms extracting parameters, such as velocity, particle concentration, microchannel functional map, and others, are present in literature [6, 34–36].

Furthermore, the application of the CNNs in image processing analysis allows the automatization of the real-time monitoring system. It overcomes manual detection and operator-dependent characterization of microvascular parameters, which are both subjective and can, therefore, lead to interobserver variability and conflicting results.

10.3 POINTWISE FLOW MONITORING

A real-time, noninvasive measurement system is discussed here for the automatic, continuous, and real-time measurement of two-phase fluid flow velocity in microchannels. The system presents flexibility in relation to the various experimental working conditions: it can give average velocity values and continuous readings, and it provides a method for recording results and tuning parameters, and a user-friendly interface. Optimization of the implementation in terms of memory allocation and execution time makes the system portable and scalable to different technologies, yielding to more compact solutions.

Such system is based on the dual-slit methodology and is implemented using standard optics and hybrid AD standard technology, meeting the requirement of noninvasiveness and real-time performance. Such implementation overcomes some disadvantages of conventional analog measurement systems. Moreover, system automation makes measurements independent of user experience and void of operator-related biases.

Applications to the microcirculation environment for the dynamic measurement of blood flow show the effectiveness of such monitoring systems.

10.3.1 Methodology and Setup

The dual-slit methodology is based on the use of two optical windows (slits) that are positioned on the microscopic image of the vessel on its centerline, to record the light fluctuations produced on each window by the passage of RBCs flowing through the vessels [24]. The separation between windows causes the downstream signal to be delayed with respect to the upstream signal. RBC velocity is given by the ratio of the window separation, a property of the optical system, and the delay between signals, which has to be measured. The cross-correlation technique is a practical method for measuring delay between noisy signals based on the computation of the cross-correlation function, which has a maximum at the most probable delay between the two signals.

In the dual-slit methodology, the optical windows can be implemented using analog optic sensors (photodiodes, photomultipliers) applied directly to the microscope-magnified image. These techniques yield two analog voltage signals proportional to the light intensity changes at the slits. The number, size (width [w] and height [h]), and distance (slit separation [SS]) of the slits can be varied to modify the spatial resolution of the system and the accuracy and sensitivity of the delay measurement. Figure 10.3 shows the functioning scheme of the dual-slit methodology for *in vivo* experimentation on animal preparations.

10.3.2 Real-Time Velocity Measurement

The real-time solution is implemented using a system based on conventional dual-slit velocimetry, where the data are processed using a hybrid AD approach, implemented by an AD cross-correlator (AD-XCORR) [25]. The system exploits hardware–software resources implementing an optimized recursive cross-correlation function through a software algorithm in a real-time environment (Fig. 10.4).

The calibration of the system, parametric studies, and theoretical calculation of the measurement uncertainty are performed using a known velocity target. These characterizations made it possible to determine the detectable velocity range (0.3–120 mm/s), to define the relative uncertainty in the measurement, and to assess the sensitivity of the system to critical parameters, such as the gain applied to the analog input signal and the integration time in the cross-correlation algorithm.

A statistically meaningful number of *in vivo* measurements shows that the results are consistent with those obtained by the conventional analog cross-correlator (Vista system) [25].

Moreover, the dynamic response of the AD-XCORR is tested during a real case of study represented by a hypoxia experiment. The velocity changes detected are consistent with the physiological response of the animal (Fig. 10.5).

10.4 CELLULAR NONLINEAR NETWORKS-BASED FULL-FIELD MONITORING

The use of CNN represents a breakthrough in microhemodynamics.

The system architecture, including the circuital implementation, as well as the mathematical models, is basic for the understanding of the processing unit, making clear how the very large-scale integration (VLSI) analog technology (Eye-RIS Vision System) represents an extremely suitable solution for the computation and analysis of images. Moreover, a significant sample application on DNA microarray is presented to introduce the versatility of CNN-based solutions in the biomedical field.

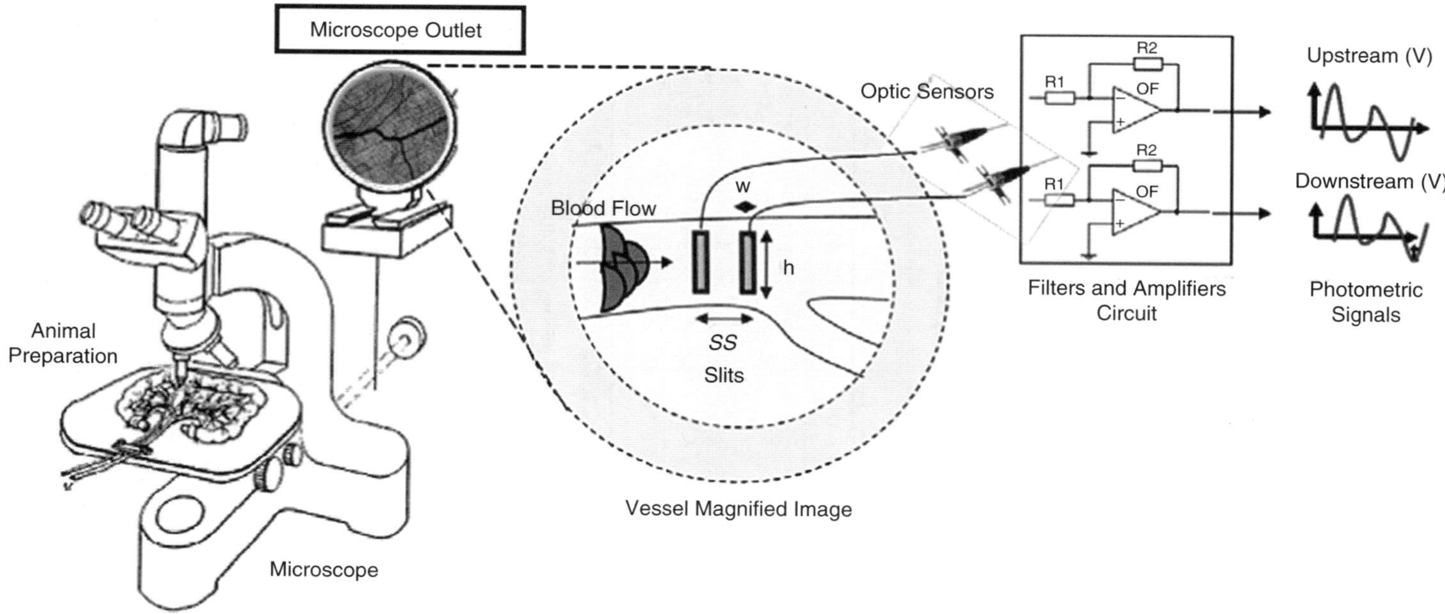

Figure 10.3. Scheme of the dual-slit methodology implementation for *in vivo* experimentation on animal preparations (Copyright © 2007 IEEE).

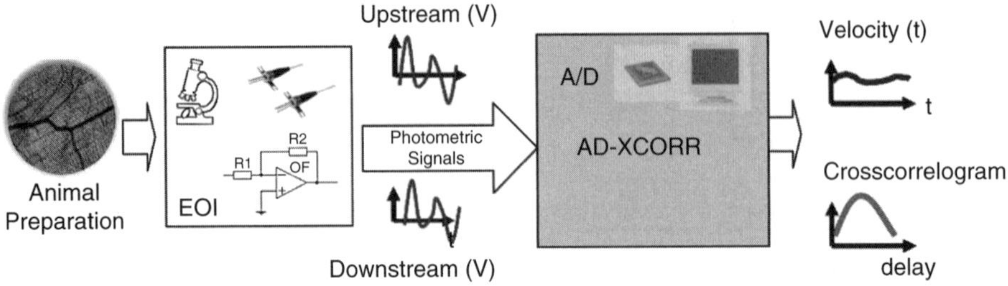

Figure 10.4. Dual-slit-based velocity system. Electro-optic instrumentation (EOI) that converts the changes in optical density due to the passage of red blood cells into voltage signals (see Fig. 10.3) (Copyright © 2007 IEEE).

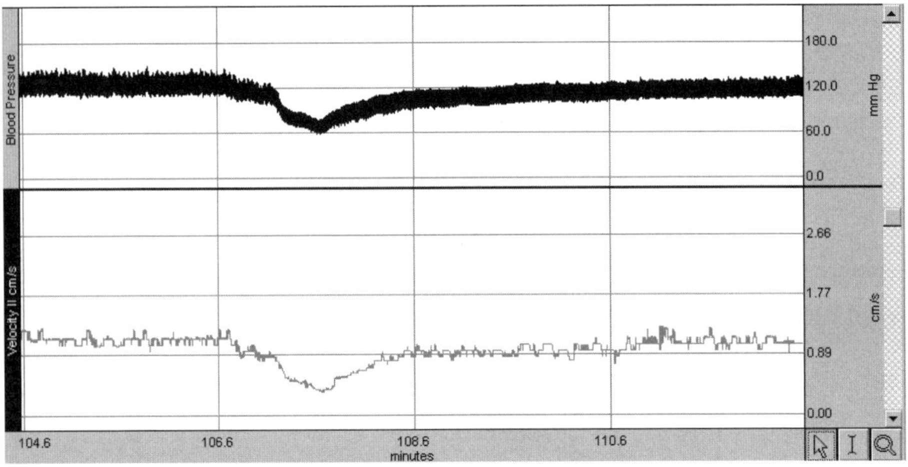

Figure 10.5. Hypoxia experiment and monitoring of a 40-μm arteriole: blood pressure reading (mm Hg) at the top; AD-XCORR velocity reading (cm/s) at the bottom (Copyright © 2007 IEEE).

10.4.1 System Architecture and Model

CNNs were introduced in 1988 by Chua and Yang [30, 31]. These networks are arrays of identical dynamical and continuous systems, named cells, with only local interactions that can be programmed by the so-called template matrices. The intrinsic structure of the CNN architecture makes them suitable for spatially distributed input processing, such as image processing. Each cell can be seen as the processing unit for an element of the array of inputs, such as the pixel intensity value. The connection between cells determines the processing of each pixel as a function of the neighboring pixel values and the input image. This allows the implementation of the filters and kernels on which traditional image processing is based and, in general, the processing of information through a spatially distributed partial differential equation (PDE) solution.

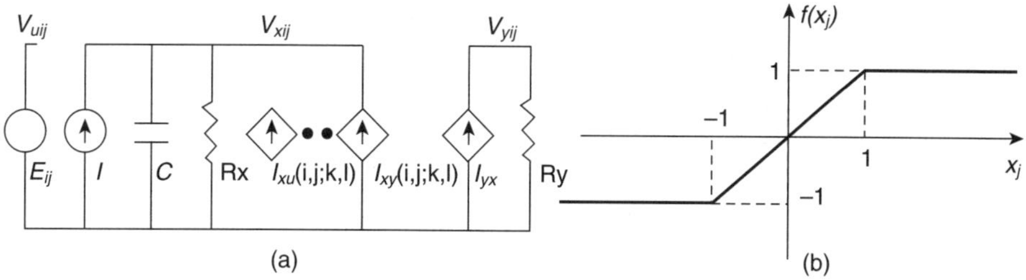

Figure 10.6. (a) Electrical scheme of a single cell. (b) Characteristic of the output nonlinearity.

Consider an $M \times N$ CNN, with $M \times N$ cells arranged in M rows and N columns; the cell on the ith row and jth column is denoted by $C(i,j)$.

A typical example of a cell $C(i,j)$ of a CNN is shown in Figure 10.6a, where u, x, and y denote, respectively, the input, the state, and the output; the node voltage v_{xij} of $C(i,j)$ is the state of the cell. The node voltage v_{uij} is called the input of the cell $C(i,j)$ and the node voltage v_{yij} is its output. The basic electrical circuit of each cell $C(i,j)$ contains one independent current source I; one linear capacitor C; two linear resistors Rx and Ry; $2 \times r$ linear voltage-controlled current sources coupled to the neighboring cells via the controlling input voltage v_{ukl}, and the feedback from the output voltage v_{ykl} of each neighboring cell $C(k,l)$, where r is the number of neighboring cells.

For all cells, $C(k,l) \in N_r(i,j)$, $I_{xy}(i,j;k,l)$, and $I_{xu}(i,j;k,l)$ are linear voltage-controlled current sources with the characteristics in (Eq. 10.1):

$$\begin{aligned} I_{xy}(i,j:k,l) &= A(i,j;j,l)v_{ykl} \\ I_{xu}(i,j:k,l) &= B(i,j;j,l)v_{ykl} \end{aligned}. \qquad (10.1)$$

The only nonlinear element in each cell is a piecewise linear voltage-controlledcurrent source $I_{yx} = (1/R_y) f(v_{xij})$ with characteristic $f(x_{ij})$ as shown in Figure 10.6b.

The dynamics of a CNN circuit is governed by a set of equations, called state equations, output equation, input equations, constraint condition, and parameter assumption. They represent the fundamental description of the state and output of each cell and fix some conditions and constraints on the input and on the parameter values in the analog circuit.

The dynamics of CNNs have both output feedback and input control mechanisms. The output feedback effect depends on the interactive parameter $A(i,j;k,l)$, referred to as feedback operator; the input control effect depends on $B(i,j;k,l)$ termed as the control operator. The time constant of circuit dynamics $\tau = CRx$ is in the order of 10^{-7} seconds. These equations and their hardware implementation constitute the basis for a new paradigm of analogic cellular computing whose most advanced implementation is the so-called CNN Universal Machine (CNN-UM).

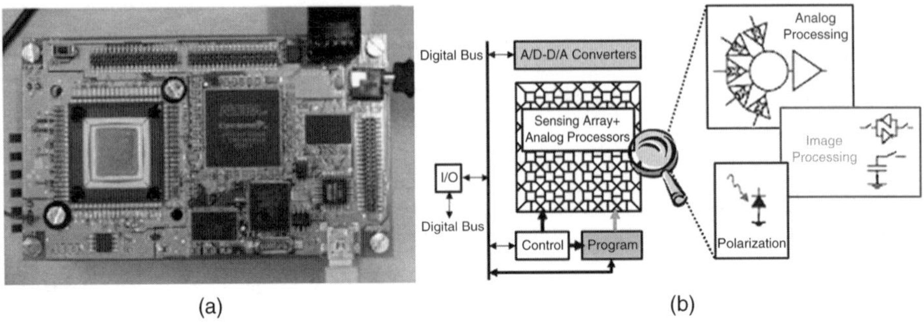

Figure 10.7. The Eye-RIS Vision System (a). PCB Processing Module. (b) CNN Device Schematic.

As shown in the state equation (Eq. 10.2), each cell of the CNN interacts directly with the neighboring cells by means of the programmable template parameters (A and B), which correspond to the synaptic weights of a network structure:

$$C\frac{dv_{xij}(t)}{dt} = -\frac{1}{R_x}v_{xij}(t) + \sum_{C(k,l)\in N_r(i,j)} A(i,j;k,l)v_{ykl}(t) + \sum_{C(k,l)\in N_r(i,j)} B(i,j;k,l)v_{ukl}(t) + I_{bias}.$$

$$1 \le i \le M; 1 \le j \le N \tag{10.2}$$

Thus, several processing tasks and full algorithms can operate in the CNN by setting the initial conditions for the CNN state variables and for the input values. Both of these stand for the operands of CNN processing and can be represented as images.

Programming by template, together with the analog operating mode, provides the opportunity to perform complex algorithms in a short time and at a very high computing speed, compared with digital microprocessing technologies [37]. The complexity in performing image processing with such technology is, therefore, to find a suitable and optimized sequence of templates carrying out the desired image transformations.

10.4.2 The Eye-RIS Vision System Solution

The possibility to integrate the CNN-based FPP device in the optical path without using an AD interface is considered by using the Eye-RIS Vision System [32, 33] (Fig. 10.7a).

The Eye-RIS is a multiprocessor system made up of two processors, namely the ACE16kv2 (Anafocus) FPP and the Nios II Digital Microprocessor (Altera, San Jose, CA). ACE16kv2 FPP acts as an image coprocessor. FPPs are meant for parallel processing of 2D data sets. They employ one processor per data channel.

Processing tasks are carried out by making the processors, and hence the data, interact. Besides, each processor in the ACE16kv2 chip is merged with an optical sensor. This means that each pixel can both sense the corresponding spatial sample of the image and process these data in close interaction and cooperation with other pixels (Fig. 10.7b).

ACE16kv2 is massively parallel, performing operations simultaneously in all of its cells. It mainly processes images in the analog domain; consequently, no AD conversion is required. Therefore, the FPP generally works with no intervention of the Nios II processor, which is involved at the programming step and for logic functionality implementation. All of these features allow the Eye-RIS Vision System to process images at ultrahigh speed and with very low power consumption.

10.4.3 An Application in Biology: The DNA Chip

Since even the most powerful microscope is unable to distinguish among genes, new methodologies are required to gain the global gene expression profile. A fundamental issue involves finding the right technology that can monitor in parallel all the DNA sequences and has the right sensibility to detect the different levels of gene expression.

Working on the microarrayed DNA chip technology, a fundamental issue concerns the methodologies to process these images. The most important characteristics to draw from fluorescence images are the assessment of the hybridization degree, which is proportional to the intensity of each color spot.

Nothing more than matrix-arranged analog image processing devices, such as the CNN system, can efficiently handle this real-time image processing task. In fact, when implemented as a mixed-signal VLSI chip, the CNN-UM is capable of image processing at rates of trillions of operations per second with very small size and low power consumption. Moreover, the array system obtained from the integration of an adaptive multisensor array in the CNN-UM (sensor plus computer) offers unprecedented capabilities [38].

To conduct a good experiment, a prefiltering procedure of the fluorescence images is the first fundamental phase necessary to extract the necessary information. First, the background noise has to be efficiently cleaned off. Since the spots are ordered into a grid, if there are spots overlapping the grid, they then have to be deleted in order not to obtain wrong results; it is better to delete some spots than to run the risk reading the information wrongly. From this point of view, the image also has to be cleaned from patches as well as from irregular spots that are bigger or smaller than a given threshold. After these filtering operations, the following phase regards the intensity analysis of each remaining fluorescent channel.

Since the image resulting from a confocal microscopic reading is a color image, it is necessary to split it into the three basic color images: red, green, and blue. Since two fluorescent materials are used in the example reported here, namely red and green, it is necessary to split the original image into only these two basic color components, obtaining two grayscale images representing the intensity level for each basic color.

In order to perform the microarray analysis via the CNN architecture, a procedure is fully developed, including both the prefiltering and the intensity analysis phases, describing each step of the algorithm in terms of single operations carried out by the CNN library templates [37]. The fundamental steps of the analysis procedure are sketched in the flowchart depicted in Figure 10.8.

10.5 CELLULAR NONLINEAR NETWORKS APPLICATIONS IN BIOMICROFLUIDICS

The characterization of quantitative and qualitative parameters of the biomicrofluidic properties and network is important in this research field, since the information on blood and, in general, on particle flow behavior can be used for the development of an analytical characterization of microfluidic monitoring and control systems.

CNN algorithms are described here for microcirculation parameter observation and analysis [6, 34, 35], and for two-phase flow in microchannel analysis [36].

10.5.1 Microcirculation Monitoring

Images of microvessels are obtained during *in vivo* experiments from a transparent window chamber surgically implanted in the dorsum of hamsters [9].

In order to perform experimental application of CNN in microcirculation environment, the tissue of the hamster skinfold chamber (Fig. 10.2) is transilluminated and observed by means of intravital microscopy. Images are recorded using a black and white CCD analog camera (Cohu 4815–2000, San Diego, CA) and digitalized at a framing rate of 30 fps with a resolution of 320 × 240 pixels. *In vivo* experiments are here represented by tape-recorded images using the analog CCD camera placed as output of the microscope and an analog videocassette recorder (VCR). The analog video is successively digitalized and fed into the CNN-based algorithm to be tested.

Frame sequence analysis is carried out via a fully developed CNN-based procedure, where each step of the algorithm is described and executed in terms of single operations carried out by the templates contained in the CNN library template. The algorithm is written using a dedicated programming language capable of managing the internal memories and the 4096 internal instructions of the chip directly; it also operated with a frame grabber to acquire the image directly from the camera.

New optimized CNN-based algorithms are designed in order to extract microhemodynamic parameters from the images produced by microscopic equipment. The algorithms can be first implemented via software simulation, using the standard templates that can be found in Roska et al. [37], and then via hardware on an ACE16kv2 CNN chip.

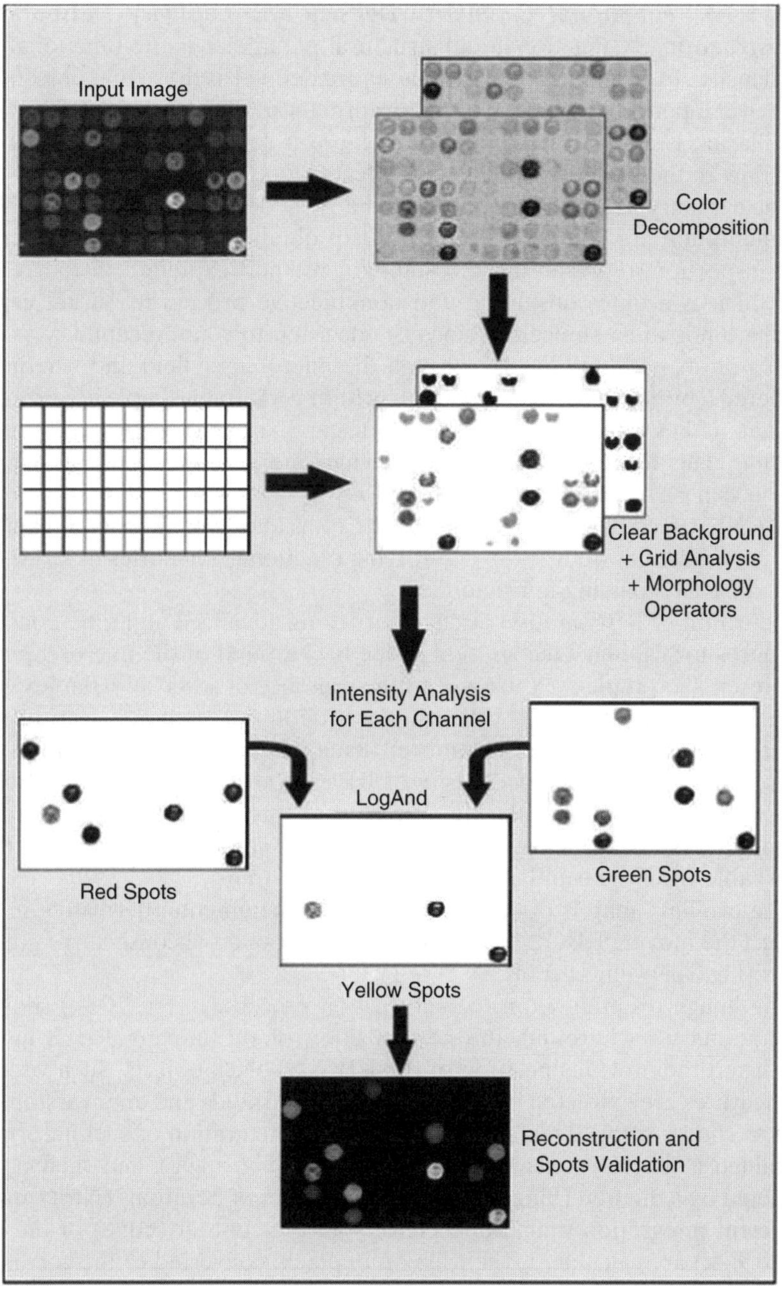

Figure 10.8. CNN-based image flow for DNA microarray real-time analysis (Copyright © 2002 IEEE).

10.5.1.1 Functional Capillarity Density and Capillary Network Map Reconstruction. A functional and structural parameter is the functional capillarity density (FCD). Functional capillaries are defined as those that present RBC transit over a period of 30 seconds. This information, along with the RBC velocity in capillaries, is considered important because it represents a measurement of blood flow in the tissue and, therefore, affects blood exchanges at that level.

The capillary network mapping and the FCD determination by CNN-based instrumentation are compared with the map obtained by an expert operator's visual inspection, representing the standard method exploited in the research field, which is a time-consuming and nonobjective procedure. Visual capillary counting tends to be subjective, since the decision to count a capillary as a unit depends on its path and length through the microscopic field and whether it is considered a branch. The CNN-based algorithm performs a simple length calculation that makes counting independent of capillary position, orientation, and branching. This technology therefore provides more objective results, which, in addition, can be obtained in real time.

A CNN-based algorithm measures the total length of functional capillaries in the microvascular network by identifying functional capillaries in terms of the motion of RBCs passing in single file.

Detection of RBC motion circumvents using advanced and time-consuming procedures for capillary recognition in the background of the microscopic field. This presents a complex texture due to the presence of adiposities, larger vessels, and different tissue layers at different focal depths. A common feature is that they are all stationary, since the hamster tissue window is fixed in relation to the microscope. Furthermore, the density of RBCs in the larger vessels is such that their image is a dense mass, lacking the features presented by individual RBCs, thus, larger vessels also have the appearance of a stationary object, readily distinguishable from the capillaries with single RBC transit.

The capillary map is obtained using a time incremental derivative in order to detect the moving RBCs, implemented by subtraction of consecutive grayscale images (Fig. 10.9—input frame N, N + 1).

The image resulting from the subtraction procedure (Fig. 10.9c) shows the gray stationary background, the new position of the moving RBCs as white objects, and the previous position of the RBCs as black objects. Such white and black features are extracted by using different thresholds and an inversion of the image as shown respectively in Fig. 10.9d and e. Integration of such information is obtained (Fig. 10.9f), cleaned from the noise (Fig. 10.9g), and then partially completed by a dilation (Fig. 10.9h). After a number of iterations (Ntot) ensuring a sufficient integration time, the skeleton and the contour (edge) of the image (Fig. 10.9i–k) are calculated. Their length in pixels, converted to the corresponding measurement in meters, represents a quantitative indicator for the functional capillary map.

The two procedures give different but consistent measurements of the total length of capillaries, and the results are compared by halving the length of the capillary edges. The total length resulting from the algorithm is expressed in

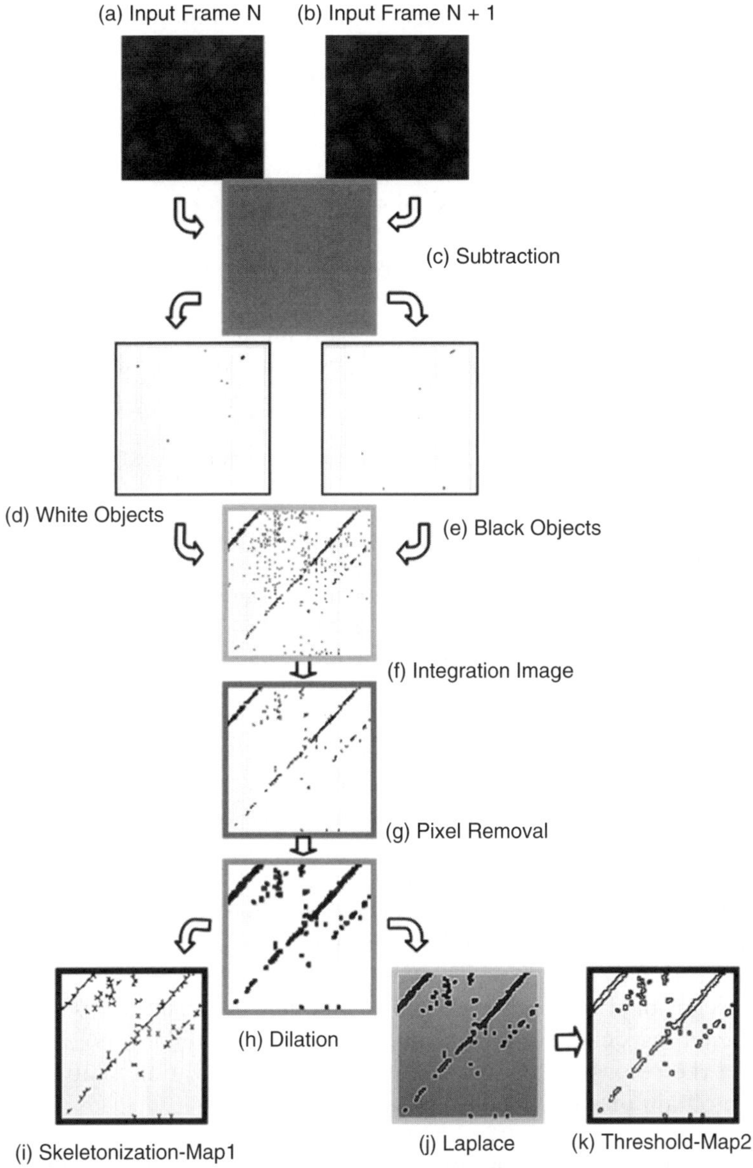

Figure 10.9. Capillaries network mapping and functional capillary density calculation through Eye-RIS FPP: (a,b) consecutive frames of the microvascular field; (c) grayscale image resulting from two frames subtraction; (d) threshold to extract the white objects; (e) threshold to extract black objects; (f) integration in time of the moving objects; (g) pixel removal; (h) dilation; (i) skeletonization; (j) Laplace filter; and (k) threshold for edge detection (Copyright © 2008 IEEE).

numbers of pixels and can be converted into an actual length by using the size of the field and the pixel resolution of the pictures.

In this case, the length measured after skeletonization is 1.751 mm, and after edge detection is 1.375 mm. As expected, the difference between the two results is higher than the one between the results in simulation because of noise effect.

Through the described algorithm, it is possible to trace most of the capillaries. However, because of the highly noisy images and of the nonideality of the experimental conditions, these maps are often missing some parts. It is therefore necessary to reconstruct the capillary map from images representing an incomplete capillary network.

To reconstruct the capillary map, the algorithm takes as input an image representing an incomplete capillary network map and the following sequence of operations is carried out:

1. Direction detection templates are applied in order to extract the network directional contribution at 0°, 45°, 90°, and 135°.
2. Digital pixel counting of each component is used to discriminate the main contributions.
3. The two main direction components are evaluated.

Figure 10.10 relates the results obtained on the ACE16kv2 chip. Figure 10.10a in particular shows how a direction detection, which gives the network components at 0°, 45°, 90°, and 135°, can be used to evaluate the main direction components by counting the pixels relative to the images representing sections of the image at the given direction.

It shows how 45° and 90° can be considered the two main components (MAX1, MAX2), which are at the basis of the actual reconstruction of the network. The next step is the actual reconstruction of the network (Fig. 10.10b) starting from the incomplete capillary map and the obtained main direction component through the application of a growth template on the starting network map in the two main directions. This step is repeated for a number of iterations that is proportional to the directional component contribution to obtain the reconstructed map; concave filler template is applied to fill out possible discontinuities in the reconstructed network and the skeletonization template may be applied to obtain the actual length of the capillary map.

10.5.1.2 Red Blood Cell Density and Velocity. The evaluation of RBC velocity is based on the idea that, assuming a track of moving RBCs can be traced for a certain period of time, the faster the particles move, the longer the track they will leave, and that the more often they pass in a microscopic window, the darker the track left will be. This concept is also clearly related to the number and therefore density of RBCs in the capillary network, as we will describe in the calibration curve in Figure 10.11b.

Starting from either the black object (Fig. 10.9e) or the white object (Fig. 10.9d) images (in the capillary map length algorithm), the following sequence of operations and templates is applied:

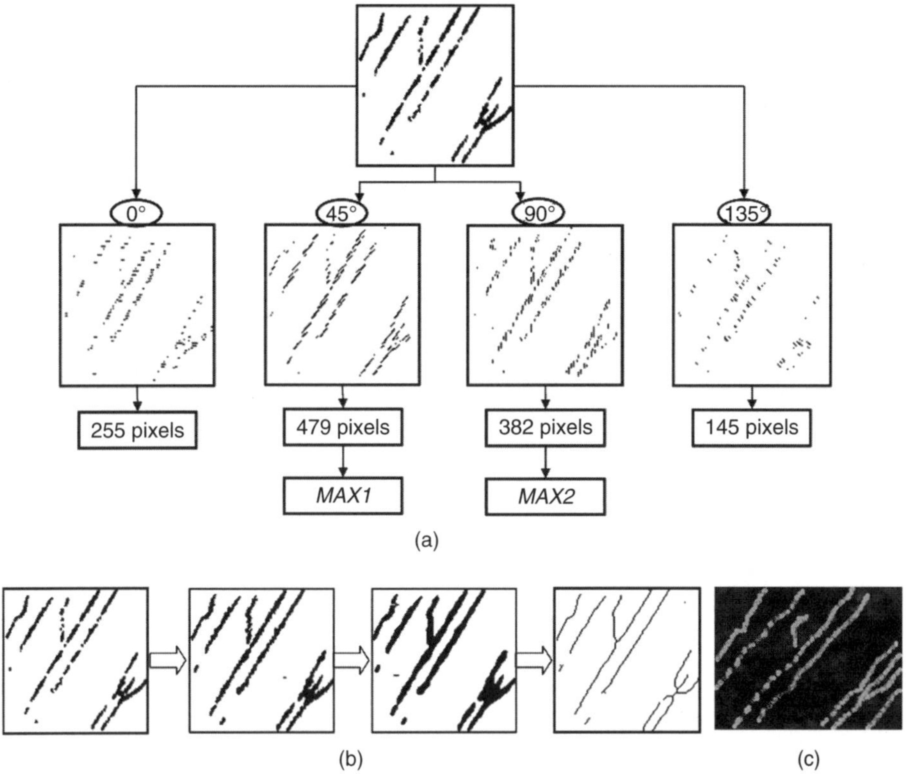

Figure 10.10. (a) CNN-based algorithm for the detection of the capillary direction. (b) CNN-based capillary network reconstruction. (c) Capillary network obtained by visual inspection. (Copyright © 2006 IOP).

1. Conversion and rescaling from binary to grayscale is performed in order to obtain an image (grayscale moving object) in which the moving objects are shown with the lowest grayscale level and allowing the maximum resolution in the RBCs tracks representation.
2. Integration through sum of the moving objects scaled image in order to trace the RBC path over a certain period of time (grayscale RBC tracks image).
3. The average intensity value of the integrated grayscale image is computed to give a measure of the RBC velocity (RBCV); this information had to be combined with the RBC density in order to obtain an objective measurement of velocity.
4. RBC density could be obtained as the ratio between the number of black pixels in the black object image and the number of black pixels in the complete map image.

The black object image is chosen and rescaled as the minimum level of grayscale image. Integration is performed in order to trace the path of the RBCs over

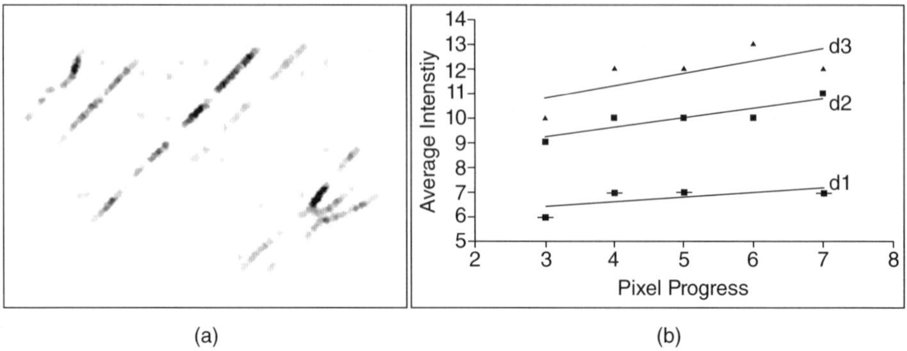

Figure 10.11. (a) CNN-based RBC grayscale tracks. (b) Linear trend in the calibration curves. (Copyright © 2006 IOP).

a certain time period that is given by the ratio between the total number of frames considered and the frame rate (30 fps).

In order to find a relationship between the average intensity value of the RBC tracks image and the actual velocity of the RBCs in capillary, a calibration is performed, taking into account the RBC density, which can easily be determined by the ratio between the number of black pixels in the image representing the moving particles and the number of black pixels in the entire capillary map image.

Image sequences are created artificially to emulate the blood flow in capillaries with known density and known velocity expressed pixel progression within the frame interval.

The particle density and the pixel progression in the frame interval are varied, and the average intensity of the particle tracks grayscale image resulting from the newly developed algorithm is measured.

The algorithm is implemented on the ACE16kv2, and the resulting image representing the RBC tracks is shown in Figure 10.11a.

Calibration curves (Fig. 10.11b) are calculated on the ACE16kv2 for three different particle densities ($d1 < d2 < d3$). The increasing offset of the curves is due to the fact that particle density strongly affects the particle tracks. This is clear if we think that the more particles pass through a certain area over a period of time, the higher their track intensity will be. The calibration curves show different slopes according to the different pixel progressions. The pixel progression represents the path of a particle over time between consecutive frames. It can be converted into velocity (m/s) by multiplying it by the size in meters of the pixel (in this case 1.1 μm) and by the frame rate (30 fps). An average velocity value is thus associated with each image. A general law could be obtained to extrapolate the velocity of the particles that do not fall into the characterized space in terms of velocity and density.

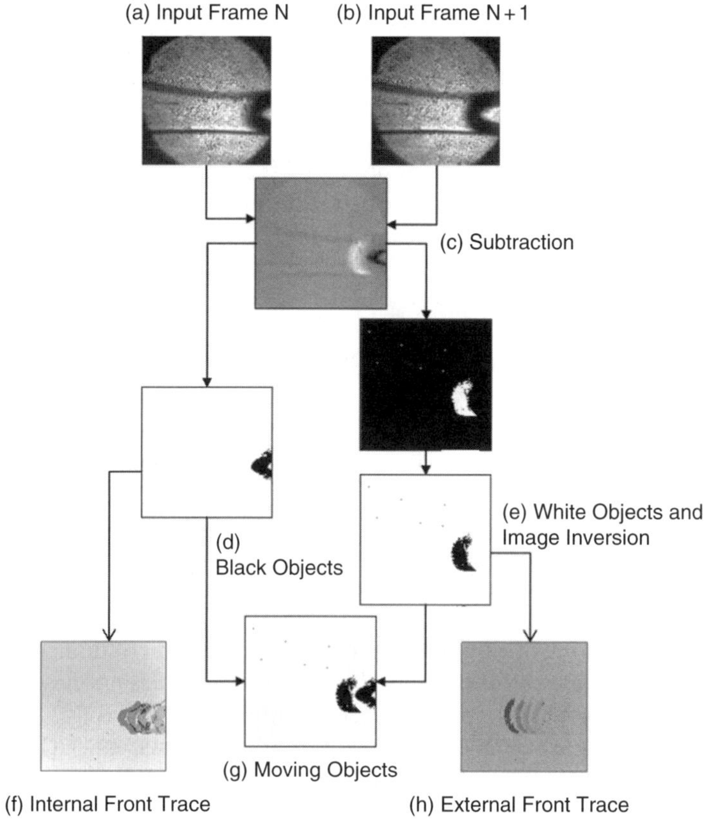

Figure 10.12. (a,b) Consecutive frames of the air bubble movement. (c) Subtraction image. (d) Threshold to extract the old position (black objects). (e) Threshold to extract the new position (white objects) and binary image inversion. (g) Logic AND of the binary images extracting the moving object and the channel map. (f,h) Grayscale bubble front traces. (Copyright © 2008 IEEE).

10.5.2 Two-Phase Flow in *In Vitro* Microchannels

The microfluidic phenomena analyzed here are related to two-phase flow in microchannels. Water and air are pumped through piezoelectric micropumps into the two inlets of a serpentine mixer (ThinXXS [20]) and flow patterns are extracted using the CNN-based algorithms. The chosen channel has a section of 640 μm. The two-phase flow images are magnified through a customized optic system and sensed, exploiting the Eye-RIS Vision System positioned in the optic path.

The air bubbles are traced, monitored, and analyzed; the channel profile is obtained using the CNN-based algorithm. The exposure time for optic acquisition via FPP is 20 ms. The image results obtained for a single cycle of the algorithm on the Eye-RIS FPP are shown in Figure 10.12. These operations are then repeated for all the consecutive acquired images, and the channel map is traced.

It is also possible to analyze the shape and flow patterns of the bubbles using different thresholding procedures and scaling the black and white images. Figure 10.12f,h shows the flow pattern of the bubbles in a dynamic way, presenting differences between the internal front of the bubble and its external front, respectively. Image processing analysis on the bubble flow can yield to the monitoring of the velocity and to the modeling of the phenomena. Such algorithms, characterizing the bubble shapes through differentiation, could be improved by using CNN-based feature extraction and bubble model generation, which would allow the process of classification of bubble patterns.

10.6 DISCUSSION ON FUTURE TRENDS

Huge advances have been made in silicon-based technology over the last few decades and different strategies of integration of microfluidics with electronics have been widely explored and exploited [3–5]. Such technology, therefore, no longer represents a mysterious tool for the realization of chemical [39], temperature [40], and optical sensors [41]. Silicon-based advances have also been made in the realization of a wide range of electromechanical sensors and actuators using micro-electromechanical systems (MEMS) [42, 43], in the integration of analog and digital control circuitry oriented not only to microfluidics-based analysis, but also to widely diffused general-purpose electronics applications.

The use of polymers in the implementation of microfluidics, such as poly(dimethylsiloxane) (PDMS) and poly(methylmethacrylate) (PMMA) [44, 45], and of alternative materials, such as semiconductive [46] or conductive polymers [47], or electroactive materials [48–51], is also presently becoming well known, since this continues to address the same tasks as the elements used in silicon technology, such as semiconductors and conductors, or materials having mechanical properties sensitive to electrical fields. Several studies in literature show how polymers can be sensitive to pressure [52], light [53], temperature [54], or chemicals [55], leading to the idea of polymeric sensors, as well as polymeric smart materials [47–49] whose electromechanical properties can be modified and tuned by applying voltage yield to polymeric actuators.

Finally, recent literature has shown the prospects for the realization of polymeric electronic components [56] and transistors [57, 58]. In detail, it has been illustrated how polymers, whose charge mobility can be varied and tuned by their electrical conditions, are at the basis of polymeric electronic components fabrication.

The polymeric technology, among its potentialities, also presents the possibility for optical information characterization in microfluidic application [59]. Polymeric devices can, in fact, be designed, exploiting the optical properties of polymers and the integration with fiber optics and standard optical detectors. Such approach would represent a step forward in solving world-to-microfluidic chip interfaces issues, since ad hoc designed polymeric micro-optic devices would allow the creation of disposable interfaces to be directly positioned on the microfluidic devices and, in particular, on *in vivo* microcirculation areas, respecting the spatial and temporal resolution of the phenomena under investigation and fulfilling the requirement of noninvasiveness and biocompatibility. The development of a simple device to be directly superimposed on the backlit skin of animal experimental preparations or on microfluidic transparent devices, for the characterization of optical information, would allow laboratories without specific and advanced optical equipments to start research activity on fluid and particle flow in microfluidic processes. Furthermore, the design of such polymeric optical interfaces, representing a miniaturization and optimization of standard optical setup designs, would make them portable and easily attachable to animal skin or, in general, to microfluidic devices, thus providing the opportunity for a continuous monitoring of the flow processes independently of the experimental setup and disturbances.

If we consider the medium- to long-term path of this new trend in technology, we can foresee that they will also lead to the development of devices providing integration of electrodes and optical-controlled signal generators, all in polymeric technology. This vision, along with the possibility of integrating pointwise or 2D full-field design, would also lead to the possibility of actuating forces, due to optical, thermal, or other noninvasive effects, on a punctual area or on a 2D one, exploiting ad hoc polymeric microinterface devices.

Here, we present a feasibility study showing the effectiveness of a sample micro-optical interface addressing monitoring issues and exploiting optical properties of fluids. In particular, we want to show the possibility of miniaturization of existing or new optical setups and technologies for monitoring and control in microfluidics using integrated polymeric micro-optical interfaces.

10.6.1 A Polymeric Solution for Biomicrofluidic Monitoring

An example of a miniaturized device for the detection of optical information for microfluidic phenomena in *in vivo* and *in vitro* experimentations is reported here. The device presented here gives the possibility of replacement for standard intravital microscopy, working as a disposable interface between live tissue in animal preparations and the processing technology oriented to optical signal characterization; it is therefore applicable in microcirculation experimentation and in modular micro total analysis devices. The microdevice is based on multilayer PDMS technology, having good optical properties, transparency, and biological compatibility, and is a miniaturized evolution of the optical setup for flow monitoring, offering a valid alternative to microscopy and other integrated velocimetry optical methods (Fig. 10.13a). It thus provides a practical solution for integration, requiring, in fact, only standard transillumination to backlight the microfluidic process and the acquisition of local information through fiber optics and photosensors.

In detail, the device exploits the PDMS–air interface and their difference in refraction index ($n_{air} = 1$, $n_{PDMS} = 1.41$) to generate total internal reflection (TIR) [60]. As described in Figure 10.13b, the basic idea of the optical interface design

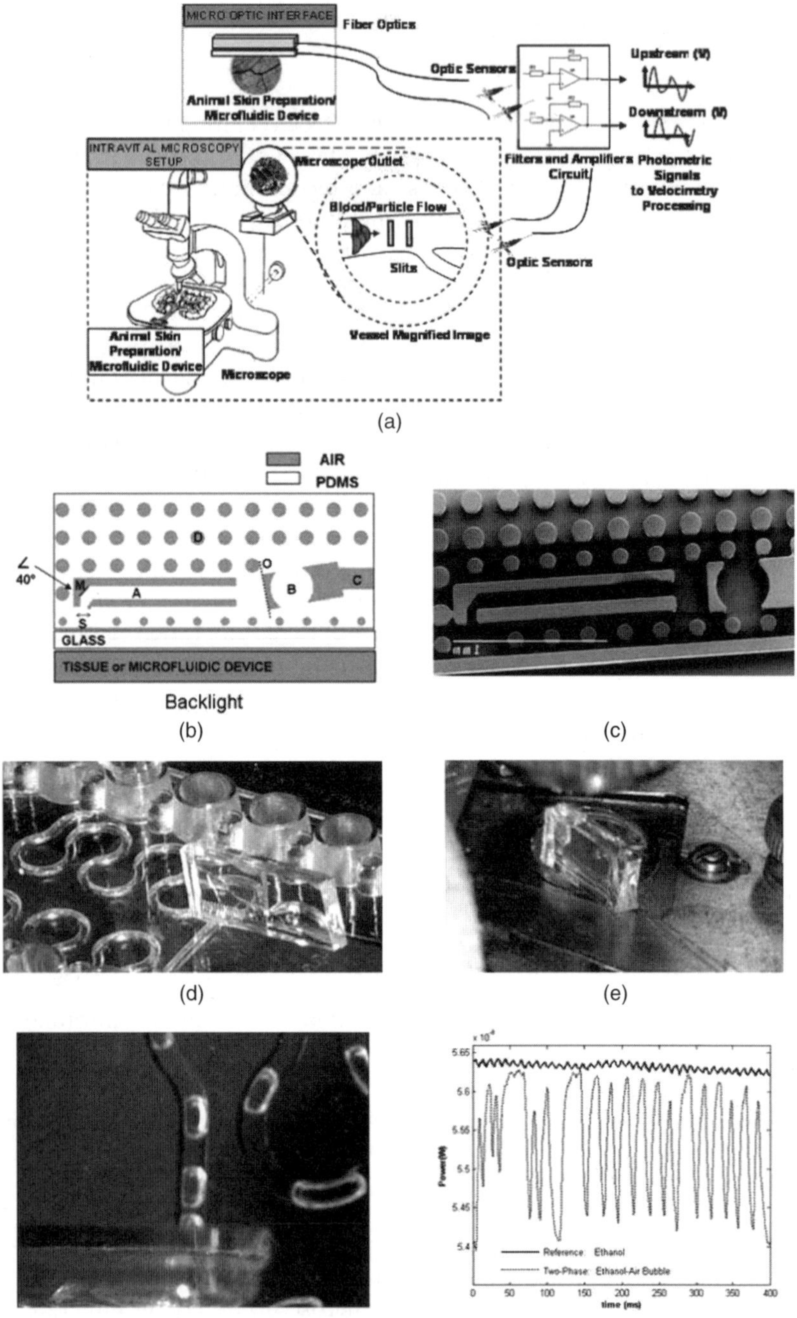

Figure 10.13. Dual-slit implementation with the micro-optical interface and with intravital microscopy. (a) Global system. (b) Section of the device (2.5 × 1 mm). S: slit; M: 40° mirror; A: optical waveguides (100-μm width); B: biconvex focusing lens; C: fiber-optic insertion and alignment; D: pillar mirror; O: output surface. (c) SEM image of the device SU-8 master. (d) PDMS device on a microfluidic mixer. (e) PDMS device on hamster skinfold preparation. (f) Microscopic upper view of the micro-optical interface superpositioned on a microfluidic chip with bubble flow. (g) Light power dynamic response to bubble dynamics.

is to acquire the optical information inside the slits (S), to bend the rays using the micro-air mirror (M), to confine the light using a waveguide (A), and to direct it to the output fiber-optic insertion (C), passing through a focusing biconvex lens (B) that corrects the numerical aperture of the rays.

In the design process, ray-tracing simulations can provide a valid support for optimization of the optical paths and for refraction index tolerance testing.

Fabrication through micromolding techniques [61] using SU-8 masters (MicroChem, San Jose, CA) can be considered a convenient technological solution as the smoothness of the walls allows for efficient implementation of optical interfaces as shown in the scanning electron microscope (SEM) images in Figure 10.13c.

The pictures in Figure 10.13d,e show the possibility of positioning the polymeric device on microfluidic chips or on *in vivo* animal preparations for flow characterization. Such setups can be completed by including fiber optics as light source for transillumination and also for collecting light and interfacing the device with an optical detector, in this case, a power meter (Model 1930 F-SL, Newport, Irvine, CA). In particular, Figure 10.13f shows the working condition of the microfluidic chip, where air bubble passage in ethanol can be visualized along with the superposed micro-optical interface. The dynamic response to the optical information related to the bubbles passage in the microfluidic chip can be visualized as light power variation in the power meter (Fig. 10.13g) giving a measure of the efficiency of the device in optical characterization of flow.

ACKNOWLEDGMENTS

The results related to the section entitled "Polymeric Solution for Biomicrofluidic Monitoring" were obtained through the access to the CNM-IMB "integrated nano and microelectronics clean room (ICTS), which was partially supported by the GICSERV program, funded by the ICTS Access Program of the Spanish Minister of Science and Innovation.

REFERENCES

[1] H. G. Klein, "Oxygen carriers and transfusion medicine," *Artif. Cells Blood Substit. Immobil. Biotechnol.*, 22, pp. 123–135, 1994.

[2] M. Intaglietta, "Microcirculatory basis for the design of artificial blood," *Microcirculation*, 6(4), pp. 247–258, 1999.

[3] S. J. Lee and S. Y. Lee, "Micro total analysis system (μ-TAS) in biotechnology," *Appl. Microbiol. Biotechnol.*, 64, pp. 289–299, 2004.

[4] A. van den Berg and T. S. J. Lammerink, *Micro Total Analysis Systems: Microfluidic Aspects, Integration Concept and Applications, Topics in Current Chemistry*, Vol. 194. Berlin; Heidelberg: Springer Verlag, 1998.

[5] R. Daw and J. Finkelstein, "Insight: Lab on a chip," *Nature*, 442(7101), pp. 367–418, 2006.

[6] F. Sapuppo, M. Bucolo, M. Intaglietta, L. Fortuna, and P. Arena, "Cellular nonlinear network: Real-time technology for the analysis of microfluidic phenomena in blood vessels," *Nanotechnology*, 17, pp. S54–S63, 2006.

[7] A. Sckell and M. Leunig, "The dorsal skinfold chamber: Studying angiogenesis by intravital microscopy," *Methods Mol. Biol.*, 467, pp. 305–317, 2009.

[8] M. D. Menger, M. W. Laschke, and B. Vollmar, "Viewing the microcirculation through the window: Some twenty years experience with the hamster dorsal skinfold chamber," *Eur. Surg. Res.*, 34(1–2), pp. 83–91, 2002.

[9] B. Endrich, K. Asaishi, A. Gotz, and K. Messmer, "Technical report: A new chamber technique for microvascular studies in unanaesthetized," *Res. Exp. Med. (Berl.)*, pp. 125–134, 1980.

[10] S. Aldridge, "'Agilent Advances Lab-on-a-Chip Technique' technology offers automated separation and analysis," *Genetic Engineering and Biotechnology News*, Vol. 25(2), 2005.

[11] Veredus Laboratories, "Veredus Laboratories' VereFlu™ chip tested successfully on clinical samples of pandemic H1N1-2009 influenza strain," Press Release, Singapore, July 6, 2009.

[12] T. M. Squires and S. R. Quake, "Microfluidics: Fluid physics at the nanoliter scale," *Rev. Mod. Phys.*, 77, pp. 977–1025, 2005.

[13] S. Tamilvanan and S. Benita, "The potential of lipid emulsion for ocular delivery of lipophilic drugs," *Eur. J. Pharm. Biopharm.*, 58, pp. 357–368, 2004.

[14] O. Sonneville-Aubrun, J. T. Simonnet, and F. L'Alloret, "Nanoemulsions: A new vehicle for skincare products," *Adv. Colloid Interface Sci.*, 108–109, pp. 145–149, 2004.

[15] Y. Cho and J. Park, "Evaluation of process parameters in the O/W/O multiple emulsion method for flavor encapsulation," *J. Food. Sci.*, 68(2), pp. 534–538, 2006.

[16] V. Hessel, H. Löwe, and F. Schönfeld, "Micromixers—A review on passive and active mixing principles," *Chem. Eng. Sci.*, 60(8–9), pp. 2479–2501, 2005.

[17] M. R. Bringer, C. J. Gerdts, H. Song, J. D. Tice, and R. F. Ismagilov, "Microfluidic systems for chemical kinetics that rely on chaotic mixing in droplets," *Phil. Trans. R. Soc. Lond.*, 362, pp. 1087–1104, 2004.

[18] D. R. Link, S. L. Anna, D. A. Weitz, and H. A. Stone, "Geometrically mediated breakup of drops in microfluidic devices," *Phys. Rev. Lett.*, 92(5), pp. 054503.1–054503.4, 2004.

[19] P. Garstecki, I. Giltin, W. DiLuzio, G. M. Whitesides, E. Kumacheva, and H. A. Stone, "Formation of monodisperse bubbles in a microfluidic flow-focusing device," *Appl. Phys. Lett.*, 85, pp. 2649–2651, 2004.

[20] H. Brüning and T. Stange, "Towards Lab-on-a-Chip devices for personalised medication and diagnostics," *Medical Device Technology*, 15(10), pp. 40–41, 2004.

[21] S. H. Tan, S. M. S. Murshed, N. T. Nguyen, T. Neng Wong, and L. Yobas, "Thermally controlled droplet formation in flow focusing geometry: Formation regimes and effect of nanoparticle suspension," *J. Phys. D Appl. Phys.*, 41, pp. 165501–165508, 2008.

[22] D. G. Grier, "A revolution in optical manipulation," *Nature*, 424, pp. 810–816, 2003.

[23] J. L. Borders and H. J. Granger, "An optical Doppler intravital velocimeter," *Microvasc. Res.*, 27, pp. 117–127, 1984.

[24] H. Wayland and P. C. Johnson, "Erythrocyte velocity measurement in microvessels by a two-slit photometric method," *J. Appl. Physiol.*, 22(2), pp. 333–337, 1967.

[25] F. Sapuppo, M. Bucolo, M. Intaglietta, P. C. Johnson, L. Fortuna, and P. Arena, "An improved instrument for real-time measurement of blood velocity in microvessels," *IEEE Trans. Instrum. Meas.*, 56(6), pp. 2663–2671, 2007.

[26] M. Linden, H. Golster, S. Bertuglia, A. Colantuoni, F. Sjoberg, and G. Nilsson, "Evaluation of enhanced high-resolution laser Doppler imaging in an in vitro tube model with the aim of assessing blood flow in separate microvessels," *Microvasc. Res.*, 56, pp. 261–270, 1998.

[27] H. Golster, M. Linden, S. Bertuglia, A. Colantuoni, G. Nilsson, and F. Sjoberg, "Red blood cell velocity and volumetric flow assessment by enhanced high-resolution laser Doppler imaging in separate vessels of the hamster cheek pouch microcirculation," *Microvasc. Res.*, 58, pp. 62–73, 1999.

[28] K. Tsukada, H. Minamitani, E. Sekizuka, and C. Oshio, "Image correlation method for measuring blood flow velocity in microcirculation: Correlation 'window' simulation and in vivo image analysis," *Physiol. Meas.*, 21, pp. 459–471, 2000.

[29] Y. Sugii, S. Nishio, and K. Okamoto, "*In vivo* PIV measurement of red blood cell velocity field in microvessels considering mesentery motion," *Physiol. Meas.*, 23, pp. 403–416, 2002.

[30] L. O. Chua and L. Yang, "Cellular neural networks: Theory," *IEEE Trans. Circuits Syst. I*, 35, pp. 1257–1272, 1988.

[31] L. O. Chua and L. Yang, "Cellular neural networks: Applications," *IEEE Trans. Circuits Syst. I*, 35, pp. 1273–1290, 1988.

[32] G. Liñán, R. Domínguez-Castro, S. Espejo, and A. Rodríguez-Vázquez, "ACE16K: An advanced focal-plane analog programmable array processor," *European Conference on Circuit Theory and Design*, 2001, pp. 345–348.

[33] Eye-RIS Vision System v1.1.7, "Evaluation kit: Hardware description," AnaFocus (all rights reserved), 2006.

[34] F. Sapuppo, M. Bucolo, M. Intaglietta, L. Fortuna, and P. Arena, "Cellular nonlinear networks for microcirculation applications," *Int. J. Circuit Theory Appl.*, 34(4), pp. 471–488, 2006.

[35] F. Sapuppo, M. Bucolo, M. Intaglietta, L. Fortuna, and P. Arena, "Cellular nonlinear network in microcirculation characterization," *J. FDE—Special Issue on CNN: Theory and Applications*, 13, pp. 23–42, 2006.

[36] F. Sapuppo, M. Intaglietta, and M. Bucolo, "Bio-microfluidics real-time monitoring using CNN technology," *IEEE Trans. Biomed. Circuits Sys.*, 2(2), pp. 78–87, 2008.

[37] T. Roska, L. Kek, L. Nemes, A. Zarandy, and P. Szolgay, "CSL CNN Software Library (Templates and Algorithms)", vers. 1.1, Budapest, Hungary, 2000. Available: http://lab.analogic.sztaki.hu/Candy/csl.html.

[38] P. Arena, M. Bucolo, L. Fortuna, and L. Occhipinti, "Cellular neural networks for real-time DNA microarray analysis," *Engineering in Medicine and Biology Magazine*, Vol. 21(2), pp. 17–25, 2002.

[39] A. Hierlemann, D. Lange, C. Hagleitner, N. Kerness, A. Koll, O. Brand, and H. Baltes, "Application-specific sensor systems based on CMOS chemical microsensors," *Sens. Actuators B Chem.*, 70(1–3), pp. 2–11, 2000.

[40] M. C. Weng and J.-C. Wu, "A temperature sensor in 0.6 μm CMOS technology," *ASICs: The First IEEE Asia Pacific Conference*, 1999 AP-ASIC apos; Vol. 99, pp. 116–119, 1999.

[41] F. Xiaoyue, K. S. Hsiao, V. P. Chodavarapu, A. H. Titus, and A. N. Cartwright, "Colorimetric porous photonic bandgap sensors with integrated CMOS color detectors," *IEEE Sens. J.*, 6(3), pp. 661–667, 2006.

[42] R. Bogue, "MEMS sensors: Past present and future," *Sens. Rev.*, Vol. 27(1), pp. 7–3, 2007.

[43] T. Nan-Chyuan and S. Chung-Yang, "Review of MEMS-based drug delivery and dosing systems," *Sens. Actuators A Phys.*, 134(2), pp. 555–564, 2007.

[44] J. C. Mcdonald and G. M. Whitesides, "Poly(dimethylsiloxane) as a material for fabricating microfluidic devices," *Acc. Chem. Res.*, 35(7), 2002.

[45] L. Brown, T. Koerner, J. H. Horton, and R. D. Oleschuk, "Fabrication and characterization of poly(methylmethacrylate) microfluidic devices bonded using surface modifications and solvents," *Lab Chip*, 6, pp. 66–73, 2006.

[46] G. Hadziioannou and P. F. V. Hutten, *Semiconducting Polymers*. Weinheim, Germany: Wiley-VCH, 2007.

[47] T. A. Skotheim, R. L. Elsenbaumer, and J. R. Reynolds, *Handbook of Conducting Polymers*, Vols. 1, 2. New York: Marcel Dekker, Inc., 1998.

[48] C. Bonomo, L. Fortuna, P. Giannone, S. Graziani, and S. Strazzeri, "Motion sensors and actuators based on ionic polymer-metal composites," in *Device Applications of Nonlinear Dynamics,* S. Baglio and A. Bulsara, Eds. Berlin/Heidelberg, Germany: Springer, 2006, pp. 83–99.

[49] S. J. Kim, H. I. Kim, S. J. Park, I. Y. Kim, S. H. Lee, T. S. Lee, and S. I. Kim, "Behavior in electric fields of smart hydrogels with potential application as bio-inspired actuators," *Smart Mater. Struct.*, 14, pp. 511–514, 2005.

[50] G.-H. Kwon, J.-Y. Park, E.-J. Lee, M. L. Frisk, D. J. Beebe, and S.-H. Lee, "Microchip-based polymeric multifunctional microbots," *Proceedings of µTAS*, 2008, Vol. 2, pp. 1977–1979.

[51] J. H. Burroughes, D. D. C. Bradley, A. R. Brown, R. N. Marks, K. Mackay, R. H. Friend, P. L. Burns, and A. B. Holmes, "Light-emitting diodes based on conjugated polymers," *Nature*, 347, pp. 539–541, 1990.

[52] I. Manunza and A. Bonfiglio, "Pressure sensing using a completely flexible organic transistor," *Biosens. Bioelectron.*, 22(12), pp. 2775–2779, 2007.

[53] S. Sugiura, A. Szilágyi, K. Sumaru, T. Takagi, M. Zrínyi, and T. Kanamori, "On-demand fluid control on microchip by micro-patterned light irradiation using photo-responsive hydrogels," *Proceedings of µTAS*, 2008, Vol. 2, pp. 838–840.

[54] N. E. Reticker-Flynn, H. W. Lee, and S.-G. Kim, "Printing of temperature-sensitive hydrogels for compact microfluidic valves," *Proceedings of µTAS*, 2008, Vol. 2, pp 1027–1029.

[55] C.-H. Weng, W.-M. Yeh, K.-C. Ho, and G.-B. Lee, "A microfluidic system utilizing molecularly imprinted polymer films for amperometric detection of morphine," *Sens. Actuators B Chem.*, 121(2), pp. 576–582, 2007.

[56] H. A. Stone, "Tuned-in flow control," *Nature Phys.*, 5, pp. 178–179, 2009.

[57] L. Fortuna, M. Frasca, M. Gioffre, M. Rosa, N. Malagnino, A. Marcellino, D. Nicolosi, L. Occhipinti, F. Porro, G. Sicurella, E. Umana, and R. Vecchione, "On the way to plastic computation," *Circuits and Systems Magazine, IEEE*, Vol. 8(3), pp. 6–18, 2008.

[58] H. Sirringhaus, "Device physics of solution-processed organic field-effect transistors," *Adv. Mater.*, 17(20), pp. 2411–2425, 2005.

[59] M. Bucolo, V. J. Cadarso, J. Esteve, L. Fortuna, A. Llobera, F. Sapuppo, and F. Schembri, "A disposable micro-electro-optical interface for flow monitoring in bio-

microfluidics," *Proceeding of the 12th Conference on Miniaturized Systems of Chemistry and Life Science*, 2008, pp. 1579–1581.

[60] A. Llobera, S. Demming, R. Wilke, and S. Büttgenbach, "Multiple internal reflection poly(dimethylsiloxane) systems for optical sensing," *Lab Chip*, 7(11), pp. 1560–1566, 2007.

[61] D. C. Duffy, J. Cooper McDonald, O. J. A. Schueller, and G. M. Whitesides, "Rapid prototyping of microfluidic systems in poly (dimethylsiloxane)," *Anal. Chem.*, 70, pp. 4974–4984, 1998.

11

MONITORING OF STEM CELL CULTURE PROCESS USING ELECTROCHEMICAL BIOSENSORS

Xicai Yue and Emmanuel M. Drakakis

The fundamental bottleneck of any bioprocess is the lack of real-time, online, *in situ*, quantitative information with respect to cellular behaviors in cultures. As a result, control, optimization, and scale-up of bioprocesses are essentially manual (empirical)—which results in suboptimal productivity (i.e., inadequate cell expansion) and product quality. To harness the immense potential of stem cells in terms of their plasticity and expansion capabilities, the physiological activity in relation to local culture parameters, such as pH, dissolved oxygen, nutrient and metabolite concentrations, and growth factor concentrations, needs to be recorded quantitatively with the required level of accuracy. Subsequently, it must be evaluated in a biologically meaningful manner, which, in the long run, could allow the systematic development of clinically relevant culture systems and methodologies, leading to the engineering of reproducible, well-characterized, regenerated "designer" tissues and organs that meet the strict regulatory criteria for clinical applications.

This chapter will discuss the acquirement of quantitative information of the cell culture bioprocess using biosensors. The interfacing for a single potentiometric biosensor and for a single amperometric biosensor will be discussed first, and then followed by the discussion on building a multichannel, PC-based, real-time measurement system. An up to 128-channel, multiparametric physiological measurement system for monitoring stem cell culture process will be described in detail as an example.

CMOS Biomicrosystems: Where Electronics Meet Biology, First Edition. Edited by Krzysztof Iniewski.

11.1 INTRODUCTION

Stem cells, with the properties of self-renewal and multilineage differentiation, are capable of developing a diverse range of specialized cell types, such as bone, cartilage, neural cells, pneumocytes, muscle, skin, endothelial cells, epithelial cells, and hepatocytes, used in tissue engineering, cellular therapies, and drug screening. Bone marrow has been used to treat leukemia. A wider variety of diseases can be potentially treated including cancer, Parkinson's disease, juvenile diabetes, spinal cord injuries, and autoimmune diseases. Additionally, it can be used for the repair of the retina and damaged muscles [1–4].

In general, stem cell culture process is far from being automated, which means that, on one hand, the culture process can hardly be optimized, and, on the other hand, there is inadequate quantitative information to assess the quality of cultured cells, although some attempts [5–7] have been made. The absence of quantitative information on culture process results in low quality, yield, and purity of the final product.

The most common setup for online, *in situ* measurement for cell culture is the incorporation of flow injection analysis (FIA), where a sample is taken from the bioreactor while the sensor is not in direct contact with the culture media. This method reduces the interaction between the biosensor and the culture media but produces readings of a delayed response since the physiological data are measured at the outlet of the sample port, therefore, not providing real-time information that would enable prompt adjustments to the ongoing process [8, 9]. To acquire real-time information, sensors must be directly exposed in the culture media so that culture parameters can be measured at any time [10]. The real-time monitoring of the stem cell culture process, with different types of biosensors being directly inserted to the culture media to acquire as much temporal physiological information as possible across the area of the culture, offers a solution as far as the determination of the dynamic behavior of the culture is concerned.

The most important physicochemical parameters in the case of the stem cell culture are pH, oxygen and carbon dioxide tension, and temperature. These parameters affect the cell expansion rate and cell population. Glucose, glutamine, lactate, and ammonia are nutrient and metabolite parameters that determine cell growth, differentiation, and cell death. The parameters for cytokines are stem cell factor (SCF), Flt3 ligand (FL), interleukin (IL)-3, -6, -11, and thrombopoietin (TOP), which help in the proliferation and differentiation of cells toward specific lineage. To harness a set of parameters for stem cell culture, multiparametric monitoring systems are needed to record those parameters simultaneously.

Electrochemical biosensors are normally based on enzymatic catalysis of a reaction that produces or consumes electrons. Hence, they can be used for the detection of nutrients and metabolites of cell cultures [11]. By means of different types of biosensors, a set of stem cell culture parameters can be monitored. In the following sections, we focus on how to interface biosensors to measure physiological parameters of stem cell cultures (and cell cultures in general) and how to build a measurement system for such a monitoring.

11.2 BUILDING A MEASUREMENT SYSTEM

11.2.1 Generic Physiological Measurement System

The block diagram of a PC-based, multichannel stem cell culture monitoring system is shown in Figure 11.1, which is based on the established data acquisition technologies [12, 13]. Analog signals from different types of sensors are input into their corresponding data acquisition boards where they are conditioned and then converted to digital form. The digitized signals are then transferred to the PC. The acquired data are displayed on the PC screen in real time for long-term monitoring. The PC is also used for further data processing and data management.

The whole data acquisition process is fully controlled by the PC via high-speed computer interfaces, such as a peripheral component interconnect (PCI) bus [14] or a universal serial bus (USB) [15]. To minimize measurement deviation, calibration processes controlled by the PC are run before each measurement trial to preserve the measurement accuracy.

The function blocks of the multichannel data acquisition module in Figure 11.1 are shown in Figure 11.2. High-impedance (impedance of up to $10^9\,\Omega$) voltage signals from potentiometric sensors are amplified (or buffered) with high-input impedance amplifiers to convert these signals to low-impedance voltage signals. High-impedance, low-current signals (current signal drop to 10^{-10} A) from amperometric sensors are converted to low-impedance voltage signals with high-input impedance, low-leakage current amplifiers. Direct current (DC) imperfections [16], such as bias current, offset current, and offset voltage, can be restricted within the measurement tolerance. Otherwise, a calibration process is applied to

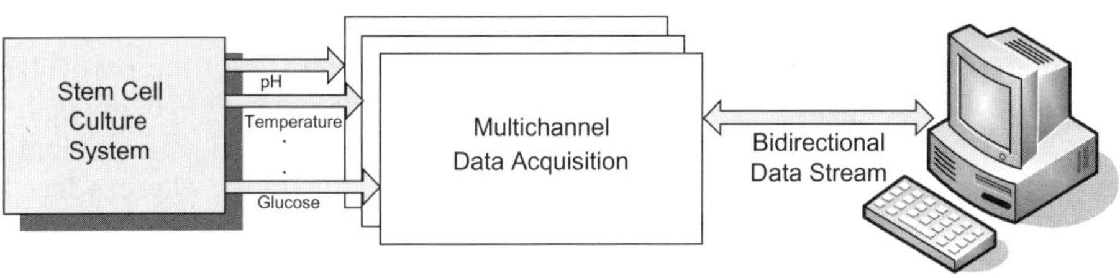

Figure 11.1. Generic blocks of a stem cell culture monitoring system.

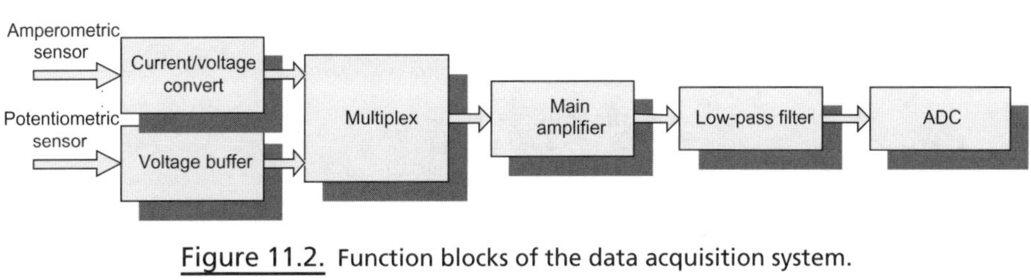

Figure 11.2. Function blocks of the data acquisition system.

eliminate these DC imperfections. The low-impedance voltage signals are multiplexed, amplified, and then level shifted to the input range of the analog-to-digital converter (ADC). A low-pass filter is placed between the amplifier and the ADC to eliminate aliasing errors and to improve signal-to-noise ratio (SNR) in the noisy measurement environment.

11.2.2 Measurement Requirements

11.2.2.1 Measurement Accuracy versus Electrical Specifications. The original requirements for the monitoring of a stem cell culture are given as physiological parameter values determined from a biological perspective. These physiological measurement requirements can be translated to electrical specifications of the measurement system, such as signal dynamic range and measurement accuracy, according to the specifications of the sensor type and sensor manufacturing technology.

Electrochemical biosensors can be classified into two categories: potentiometric and amperometric sensors. The physiological measurement requirements for those two types of sensor are translated in different ways.

Potentiometric Sensors. This type of sensors is screenprinted and conducts polymer immunoassays. A potentiometric sensor has two electrodes: the sensing and the reference electrodes. The electrical signal (E) is produced by electrochemical and physical changes in the conducting polymer layers due to changes occurring at the surface of the sensor. Theoretically, for commonly used ion-selective electrodes (ISEs), E is expressed by Nernst's equation [17] as

$$E = E^0 + (RT/nF)\cdot \ln(C), \tag{11.1}$$

where E denotes the total potential (in millivolts) developed between the sensing and reference electrodes, E^0 is a constant that is characteristic of the particular ISE and reference pair, R is the gas constant (8.314 J/degree/mol), T denotes the absolute temperature, n is the signed charge of the ion (for pH and ammonia measurement, $n = 1$), F is the Faraday constant (96,500 C/mol), whereas $\ln(C)$ denotes the logarithm of the concentration of the measured substance in moles.

A typical example of a potentiometric sensor is that of pH, which measures the acidity or alkalinity of a solution. A pH sensor can be considered as a millivolt-level voltage source with a series source resistance dependent on the electrode's composition and configuration. The pH value, which equals to $\log(C)$, is calculated from Equation 11.1 as

$$\text{pH} = \frac{E - E^0}{2.303RT/F}. \tag{11.2}$$

The constant 2.303 is the conversion factor from natural to base-10 logarithm. At room temperature (25°C), $2.303RT/F$ is a constant of 59 mV, which means that the pH sensor signal changes by 59 mV/pH unit, and therefore, a measurement requirement of ±0.1 pH units translates linearly to a voltage resolution of ±5.9 mV.

Another typical potentiometric sensor is the ammonia sensor, which has a logarithmic response. From Equation 11.1, the measured voltage E is linear to $\ln(C)$, where C is the ammonia concentration to be measured. The ammonia concentration C can then be derived from Equation 11.1 as

$$C = e^{\frac{E-E^0}{RT/nF}}, \tag{11.3}$$

where $n = 1$. For a required measurement resolution of ammonia concentration (dC), the electronic measurement specification for a maximum measurement error (dE) is derived from Equation 11.3 as

$$dC = e^{\frac{E-E^0}{RT/F}} \cdot \frac{1}{RT/F} \cdot dE = C \cdot \frac{1}{RT/F} \cdot dE. \tag{11.4}$$

The relation (Eq. 11.4) shows that the electrical measurement specification (dE) is not only dependent on the needed or targeted physiological measurement resolution dC, but that it is also dependent on the concentration C itself:

$$dE = \frac{\frac{RT}{F} dC}{C}. \tag{11.5}$$

In room temperature, RT/F is a constant of 26 mV. For a required measurement resolution within the concentration variation range, a voltage resolution can be obtained from Eq. 11.5 using the maximum anticipated concentration value within the concentration variation range. For example, to measure ammonia concentration in the range from 0.5 to 5 mM with a resolution of ±0.1 mM, 5 mM is used to calculate the voltage measurement resolution and results in ±0.52 mV.

Amperometric Sensors. Amperometric sensors can be modeled as high-impedance, nanoampere-level current sources. An amperometric sensor can measure either the current, which is the rate of flow of electrons proportional to the analyte concentration at a fixed potential (often termed "excitation"), or the potential at zero current, which has a logarithmic response.

The typical amperometric sensor is the glucose sensor, which has three electrodes: the reference electrode, the counter electrode, and the working electrode. To measure the current signal sourced from the working electrode, an "excitation" voltage signal is applied between the reference electrode and the working electrode. For a sensor with an output current range from 100 pA to 10 nA, the ±0.5 mM measurement resolution requirement when the glucose concentration varies in the range of 0–35 mM can be linearly translated to a current resolution of 140 pA.

11.2.2.2 Requirements Relevant to System Control. The required measurement interval is the most important parameter that directly affects the control technology being used in the measurement system especially when real-time control is needed.

It is a basic requirement in a continuous monitoring system that the control process for acquiring data and the basic processing of the acquired data (such as data saving) should be completed before the subsequent measurement.

For multichannel applications, the measurement time slot for a specific channel is significantly reduced. For 32 channels with 1-minute measurement intervals for each channel, a 1.875-second time slot is available to measure a given channel. The measurement time slot is reduced to 0.47 seconds for 128 channels if the same measurement interval for each channel is applied. Relevant discussion on system control can be found in Section 11.2.4.

11.2.3 Signal Conditioning

11.2.3.1 Potentiometric Biosensor and Interfacing. The potentiometric biosensor can be considered as a millivolt-level voltage source with a series source resistance varying between 10^6 and $10^9\,\Omega$. This high-impedance voltage signal is sensitive to interferences. Therefore, it should be transferred to a low-impedance signal.

As shown in Figure11.3a, high-impedance voltage signals from potentiometric biosensors are directly applied to a high-input impedance, noninverting amplifier to output a low-impedance voltage signal. A field-effect transistor (FET) input stage amplifier whose input impedance is higher than $10^{13}\,\Omega$ is used (i.e., more than 10^4 times of the biosensor impedance). An accurate voltage gain $A_v = 1 + R_2/R_1$ is obtained with a low offset voltage, low drift, low-input bias current, and low-input offset current (in the order of femtoamperes) amplifier.

The biosensor is connected to the amplifier in the single-ended form, shown in Figure 11.3a, with the reference electrode directly connected to the ground. However, in more common cases, both electrodes of the sensor are connected to the amplifier as a differential input signal to reduce the common mode interference while the culture media is connected to the ground. In such a case, a FET-input instrumentation amplifier (IA) with extremely low-input bias current is adopted, as shown in Figure 11.3b. A gain of $A_v = 1 + 2R/R_g$ is obtained.

The relationship between the source impedance and the input impedance of the amplifier in Figure 11.3b is shown in Figure 11.3c. As the input impedance R_{in} of the IA is in series with the source impedance R_s, the differential voltage applied to the amplifier is calculated by

$$V_{in} = \left(\frac{R_{in}}{(R_{in} + R_s)} \right) V_s, \tag{11.6}$$

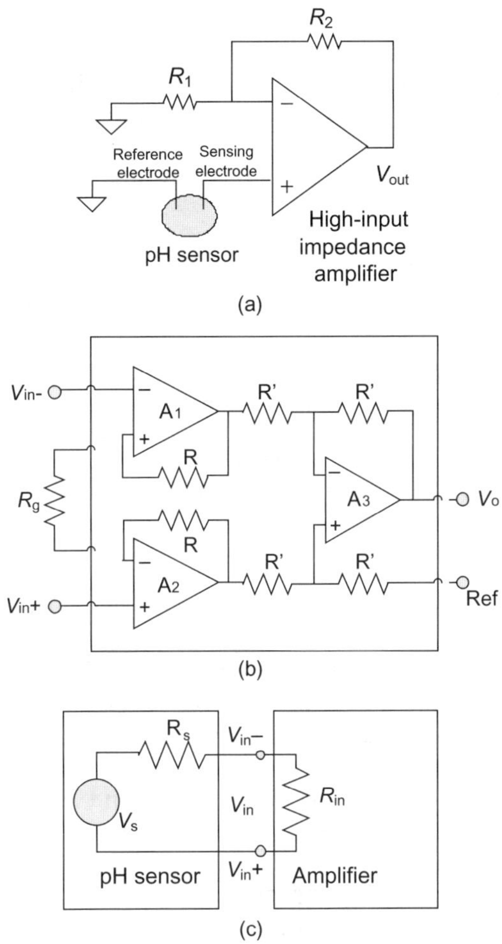

Figure 11.3 (a) pH sensor interfaced by a noninverting amplifier. (b) Instrumentation amplifier for interfacing differential input signal. (c) Relationship between the source impedance and the input impedance of the amplifier.

where V_s denotes the voltage signal produced by the sensor and V_{in} denotes the input signal to the amplifier, which equals $(V_{in+} - V_{in-})$. If $R_{in} = 99R_s$, there is a 1% difference between V_{in} and V_s. Therefore, to obtain an accurate measurement result, R_{in} should be bigger than R_s to make $R_{in}/(R_{in} + R_s) \approx 1$. Equation 11.6 can be viewed from another way. If R_{in} is small, then the current flowing through the sensor $I = V_s/(R_{in} + R_s)$ will be relatively big. However, it is expected that there is no current flowing through the reference electrode of a potentiometric sensor.

As mentioned previously, the ammonia sensor is a potentiometric sensor that has a logarithmic response; therefore, it would be reasonable to consider the direct acquisition of ammonia concentration using an exponential amplifier, which exploits an electronic component that has a logarithmic relationship between the voltage applied to it and the current flowing through it (e.g., the semiconductor diode). However, in practice, it is difficult to implement.

An exponential amplifier can be implemented by replacing R_1 in Figure 11.3a with a diode and connecting the sensor signal to this diode. The relationship between the input voltage V_{in} and the output voltage V_{out} of the exponential amplifier is

$$V_{out} = -RI_s\left(e^{\frac{V_{in}}{KT/q}} - 1\right), \tag{11.7}$$

where I_s is the saturation current of the diode. Similar to RT/F in Equation 11.1, KT/q in Equation 11.7 equals 26 mV at room temperature.

When $e^{\frac{V_{in}}{KT/q}} >> 1$, V_{out} relates V_{in} exponentially: $V_{out} = -RI_s e^{\frac{V_{in}}{KT/q}}$. However, in order for this exponential relationship to hold, V_{in} must be larger than KT/q (26 mV at room temperature). Unfortunately, as the output of the potentiometric sensor may stay in tens of minivolts, the assumption $e^{\frac{V_{in}}{KT/q}} >> 1$ is not always satisfied, and consequently, V_{out} and V_{in} do not share an exact exponential relationship.

One further disadvantage of using exponential amplifiers is that the measurement accuracy of the amplifier is sensitive to temperature through the KT/q exponent. To achieve a highly accurate measurement, an extra accurate temperature measurement is needed. Therefore, in practice, ammonia sensors are interfaced as shown in Figure 11.3a or Figure 11.3b and the measured data are nonlinearly transformed into ammonia concentration via post-data processing.

11.2.3.2 Amperometric Biosensor and Interfacing. Amperometric sensors are current sensors with three electrodes. As shown in Figure 11.4, two electrodes (counter [C] and reference [R]) of the sensor are connected to the excitation amplifier, which is connected as a voltage follower, to set the excitation voltage between the working electrode (W) and the reference electrode and, in the mean time, to guarantee that there is no current flow through the reference electrode. The working electrode is connected to the main amplifier, which is

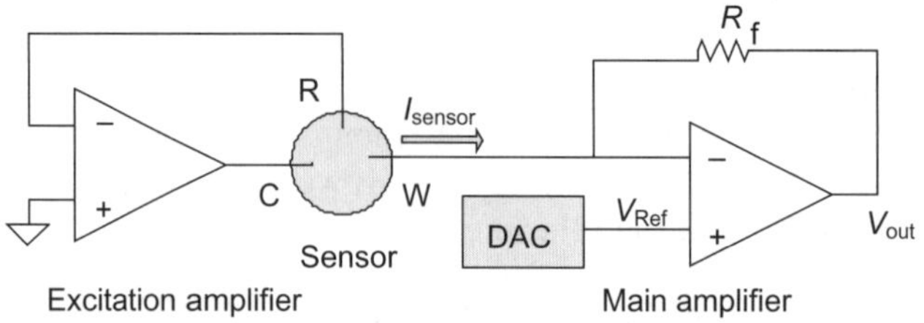

Figure 11.4. Amperometric sensor interfacing.

usually an electrometer-operating amplifier with very low-input bias current (in the order of femtoamperes) to convert the current signal, which flows between the counter and the working electrode to an output voltage signal of V_{out}.

As the current signal from the sensor is very weak, signal flowing through the working electrode (W) is measured by means of a low-input leakage current, low-offset voltage, and high-input impedance main amplifier. A precise, large resistance R_f (in the order of megaohms or even larger for measuring the tiny current) is used in the feedback branch of the main amplifier. Special effort is undertaken to avoid interference. Sensors are connected to electronic circuits via shielded high-insulation coaxial cables. For very weak current measurement, triaxial cable (which has an extra guard layer being driven at the same potential as the inner signal conductor) is adopted to prevent the leakage current from the outer shield layer (which is usually connected to the ground) to the inner signal conductor. Although characteristic impedance is used as an important transmission parameter, a shielded high-insulation cable without consideration of its characteristic impedance is adequate to provide a low-leakage and low electrical interference transmission as the frequency of biosensor signals is lower than 200 Hz, while characteristic impedance is a high-frequency parameter of the transmission line. At the printed circuit board (PCB) level, a guard ring is placed near the input of the main amplifiers to minimize interference caused by spurious, undesired signals. If multilayer PCB is used, it is recommended to place a ground plane beneath the main amplifier to reduce the leakage current from other layers of the PCB. A digital-to-analog converter (DAC) and a low offset voltage amplifier are used to set the sensor excitation voltage levels precisely. The sensor current is calculated from the measured voltage signal V_{out} by

$$I_{sensor} = \frac{(V_{Ref} - V_{out})}{R_f}, \tag{11.8}$$

where V_{Ref} denotes the excitation voltage setting by the DAC.

By choosing different R_f values, different amperometric sensors of different output current ranges can be interfaced. Another way to interface different amperometric sensors is to use a fixed R_f in the current-to-voltage conversion stage followed by a programmable gain amplifier (PGA) stage for further amplification, which will be discussed in Section 11.2.3.3.

An alternative way to apply the excitation signal is to swap the polarity of the working electrode and the reference electrode: DAC output is connected to the noninverting input of the excitation amplifier of Figure 11.4 and the noninverting input of the main amplifier is connected to the ground. The advantage of this configuration is that it is simple to calculate the current from the measured voltage without using V_{Ref}:

$$I_{sensor} = -\frac{V_{out}}{R_f}. \tag{11.9}$$

However, it is common sense to set the reference point to the ground in practice, especially in applications with different types of sensors where more than one reference electrode from different types of sensors are employed.

It seems that the current is independent to V_{Ref} in Equation 11.9, while it is dependent to V_{Ref} in Equation 11.8. In practice, V_{Ref} is not arbitrary for a given sensor. It has an optimal value that is determined by the sensor type, and therefore, this specific value should be used as V_{Ref} for the specific sensor. To ensure the system's high measurement accuracy, this excitation voltage set by the DAC shown in Figure 11.4 can be remeasured during a calibration phase.

Although there are no regulations on discharging the culture media, it is recommended to change the polarity of the excitation to avoid the long-term charging of the culture media when amperometric sensors are employed for bioprocess monitoring. Without this discharging process, the culture media might be polluted by the continuous charging from the excitation voltage of the amperometric sensors within the several-day-long culture process. The discharging of the culture media can be implemented between two adjoining measurements as each amperometric sensor has its optimal excitation voltage, which means that after reversing the polarity of the excitation voltage for discharging, the excitation voltage should be set back again to prepare for the next measurement.

11.2.3.3 Measurement Flexibility. Different types of sensors have different output signal ranges and different measurement resolutions. Even for the same type of sensors, Equation 11.1 shows that the output signal E changes with the variable E^0, which is specific to different sensors. Sensor output signal range affects the gain setting of the amplifier, which is supposed to be adjusted to its maximum to improve the SNR. For example, in pH measurement, 1-pH unit change in culture media corresponds to a 59-mV voltage span of the sensor output. If the output signal is in the range of 0–59 mV, for a ±5 V supplied rail-to-rail output amplifier, the maximum gain that can be set is about 80. If the output of the sensor changes from −29 to 30 mV, then the maximum gain can be set to 160. A measurement system, which can interface to a wide variety of sensors, should be characterized by adaptable amplification. A measurement system capable of interfacing with different biosensors can be achieved using PGAs, which enable the dynamic range of the input signal to be adaptable through the setting of the gain and the offset of the PGAs. As PGAs usually do not have high-input impedance, an impedance conversion stage is placed in front of the PGAs. Figure 11.5 shows such a solution implemented using a voltage buffer and two stages of PGAs.

A potentiometric sensor signal is connected to the voltage buffer to convert the high-impedance signal to a low-impedance one. The gain of this buffer can be set to 1 to keep the maximum dynamic range of the input signal (reaching as high as the power supply) or can be set to a larger value for a high SNR when the input signal is very weak. By combining the gain and offset settings in both PGA stages, an appropriate gain can be achieved to meet the measurement requirements for the specific sensor. The pair of PGAs in Figure 11.5 plays

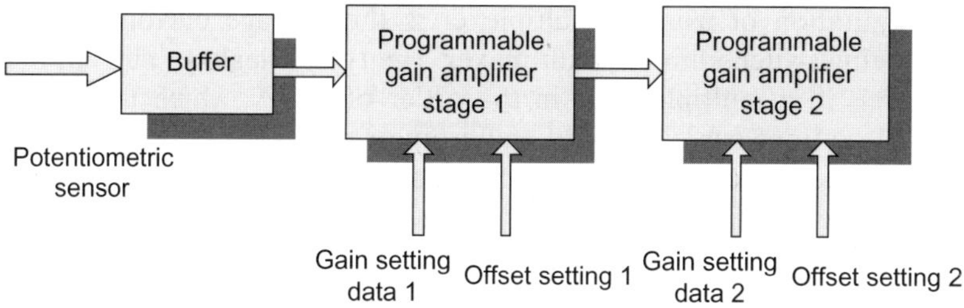

Figure 11.5. Flexible gain stages for potentiometric sensors.

different roles. The PGA in the first stage is mainly used to adjust the output of PGA_1 to zero offset so that a maximum gain can be set by the second PGA. The second stage is the main amplification stage, and the output of the PGA should be level shifted to the input range of the ADC, which is usually of a single polarity (positive) while the sensor signal is usually bipolar. An advantage of using two stages of PGAs is that the whole circuit has more gain steps as the gain in each PGA stage can be selected from 1, 2, 4, to 2^N. Two PGA stages increase the numbers of selective gains from N to 2N and expand the maximum gain from 2^N to 2^{2N}.

The gain of each PGA can be set digitally by logic switches or by data registers, and the offset of the PGA can be set by means of a voltage reference or using a DAC for a more flexible and a more precise setting. At system level, a flexible way of PGA setting is to send the gain and offset setting data via the host PC before a measurement or even within the measurement process for adjusting the PGAs promptly, which makes the whole measurement system programmable and able to meet different measurement requirements.

11.2.3.4 Sensor Array and Multichannel Measurement. The interfacing with a single sensor has been discussed. Usually, in practical applications, several types of sensors are concurrently used in a multisite or a sensor array configuration.

Multichannel measurements can be achieved by replicating each individual circuit for a single channel or using a multiplexer to share the common circuits. A multiplexer can be added between the sensor output and the ADC input (refer to Fig. 11.2). The exact position depends on measurement requirements.

The choice of multiplexer positioning for both the potentiometric and amperometric sensors is a trade-off between measurement accuracy and circuit complexity. The safest position to place the multiplexer is before the ADC where sensor signals from each measurement channel have already been conditioned individually. Hence, a multiplexer can be used to share the ADC. However, in most cases, the position of the multiplexer can be moved as close as possible to the sensor to avoid duplicating the identical signal-conditioning circuits of each measurement channel. The closer the multiplexer to the sensors is, the simpler the signal-conditioning circuit becomes.

The limitation of using a multiplexer is the leakage currents among the channels being switched in the multiplexer. The typical leakage current between two channels in a multiplexer is in the order of 10^{-9} A, which is equal to the order of the output current of the amperometric sensor or the potentiometric sensor when the source impedance of the sensor reaches $10^9\ \Omega$. For potentiometric sensors, the position of the multiplexer is after the voltage buffer, which converts the high-impedance signal to a low-impedance one, as for low-impedance signals, the multiplexer current leakage can be neglected. For amperometric sensors, the multiplexer should be placed after the current-to-voltage converter. However, there is a motivation to share the current-to-voltage converter in amperometric sensor measurement. As the feedback resistor R_f in Figure 11.4 is usually very large, it is not cost-effective to use many precise, large-valued resistors. Furthermore, as shown in Figure 11.6, to achieve high measurement accuracy, a resistor network composed of R_1 to R_3 is adopted in the main amplifier to select among different resistors so that the current-to-voltage converter can output a maximum voltage signal for a specific current range to improve the SNR. If each measurement channel has its own current-to-voltage converter with a resistor network, then many resistors and switches (S_1 to S_3 in Fig. 11.6) for switching these resistors are needed.

One of the solutions to share the current-to-voltage conversion main amplifier with a resistor network is to build up a multiplexer with high specification relays that have more than $10^{12}\ \Omega$ insulation resistance (leakage current in the order of femtoamperes). This solution is shown in Figure 11.6.

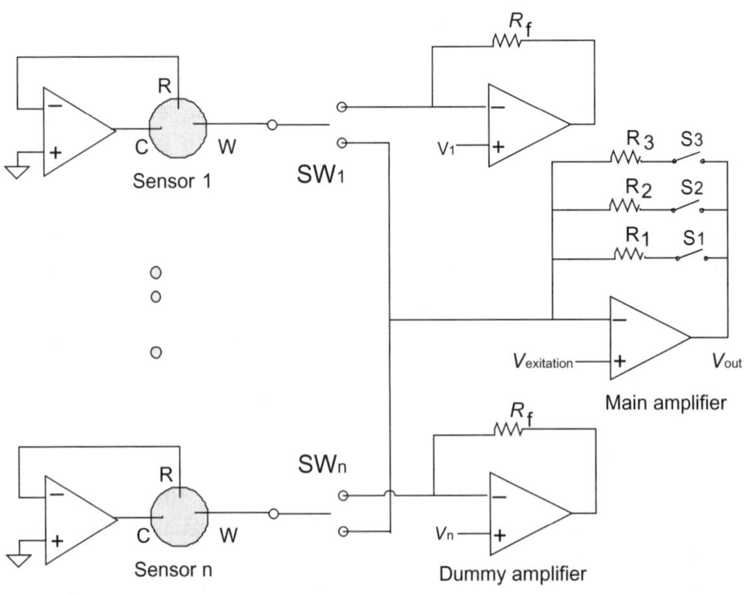

Figure 11.6. Switching of amperometric sensors via relays.

The reason of using crossover switches (SW_1 to SW_n in Fig. 11.6) rather than on–off switches is because if sensors are switching to the voltage-to-current converter (main amplifier shown in Fig. 11.6) using on–off relays, they are not working continuously. In this case, a sensor is only active when it is taking a measurement. As it can take minutes to set up a stable current for an amperometric sensor, directly switching a sensor between on and off states will affect the measurement accuracy or prolong the measurement process if the same measurement accuracy is maintained. Crossover switches and lower specification dummy current-to-voltage conversion circuits for each measurement channels are applied to keep those sensors that do not measure operating continuously.

Another type of sensor that calls for multichannel measurement is the multianalyte sensor [18], which integrates several different types of sensors as one physical sensor with multioutput electrodes. As it comes out with both amperometric and potentiometric output signals with the same reference electrode, special attention should be paid to avoid short circuits via the reference electrode when circuits shown in Figure 11.3a and circuits with the excitation voltage applied in the noninverting input of the excitation amplifier (refer to Fig. 11.4) are working together.

11.2.4 Analog-to-Digital Converters

The ADC that converts a continuous analog signal to discrete digital code is the final stage in Figure 11.2 of the data acquisition block diagram, and therefore, it is the last factor to determine the system accuracy. The voltage resolution of an N-bit ADC equals $V_{\mathrm{Ref}}/2^N$, where V_{Ref} is a voltage reference or the full-scale measurement range (FSR). According to the structure, there are several types of ADC, such as flash ADC, pipeline ADC, and sigma–delta (Σ–Δ) ADC, providing up to a 24-bit resolution suitable for different applications, such as high-speed or low power consumption.

Selection of an ADC block for stem cell culture monitoring is straightforward. DC accuracy parameters of offset error, full-scale error, differential nonlinearity (DNL), and integral nonlinearity (INL) are prevalent, while dynamic parameters, such as total harmonic distortion (THD), and timing parameters, such as aperture jitter, are less important, since in the case of cell cultures, sensor signals are relatively static DC-like signals. In practice, 3–4 extra bits should be added to the numbers of the required bits. For example, to measure a 0- to 5-V input range signal with a resolution of 2 mV, a 12-bit ($N \geq \log 2^{5000/2}$, where N is an integer) ADC is required in theory, and in practice, a 16-bit ADC is applied to ensure that 12 effective bits (in the worst case, a 16-bit ADC can have a total DC error of about 10 least significant bits [LSBs]).

Another consideration when choosing an ADC is its convenience to use. Some ADC chips include a voltage reference, a multiplexer, or even some signal-conditioning circuits, and on-chip PGAs. An ADC with a parallel interface is easier to use than a serial interface, where single-bit data come with an extra clock.

11.2.5 Temperature Compensation and System Calibration

In an electronic measurement system, the measurement performance drifts with temperature. To compensate for the performance variation with temperature, the temperature should be measured in real time with high accuracy. Furthermore, Equation 11.1 shows that the output of the potentiometric sensor is temperature dependent, and therefore, the measurement results should be temperature compensated.

There are three basic types of temperature sensors: the thermocouple, the thermistor, and the resistance temperature detector ([RTD] also called resistance thermometer). The RTD is the most stable and accurate device. To measure temperature using RTD, a small constant current (such as 1 mA) is applied, and the voltage drop across the RTD is measured. As shown in Figure 11.7, a four-wire RTD (two wires for carrying the "sense" current and two for measuring the voltage across the element) is adopted to measure the value of the RTD resistance in the most accurate and reliable way. The constant current is produced by a precision voltage reference and an ultraprecise resistor R_1. A low-leakage current (in the orders of picoamperes) amplifier is used for the current-to-voltage conversion. The potential difference across pins 2 and 3 of the four-wire RTD is amplified by means of an IA, such as a PGA, and the output of the PGA can be further amplified by the subsequent PGA, as shown in Figure 11.5. Note that unlike the current-to-voltage conversion used for biosensors (see main amplifier in Fig. 11.4), high-input impedance amplifiers are not necessary in this case.

The resistor value can be calculated by the measured voltage signal and the known current source value. The temperature is calculated by the Callendar–Van Dusen equation [19],

$$R_T = R_0[1 + \alpha T + \beta T^2]\ (T > 0°C), \tag{11.10}$$

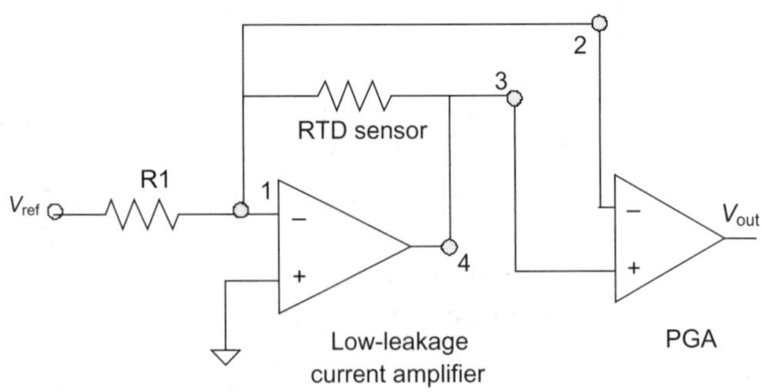

Figure 11.7. RTD sensor (ohmic) signal conditioning/interfacing diagram.

where R_T denotes the measured resistance value (in ohms) at a temperature (T) in °C, R_0 is a known resistance value, whereas α and β are known constants specific to the RTD sensors. For the used PT100, $R_0 = 100\,\Omega$, $\alpha = 3.9083E\text{-}3$, and $\beta = -5.775E\text{-}7$.

Since the leakage currents of both amplifiers can be in the order of 10^{-12}, the wire resistors of the four-pin RTD have no significant effect on the accuracy of the test result. Although the values of the wire resistors are higher compared with the required measurement resolution of $0.038\,\Omega$ when ±0.1°C accuracy is required [9], wire resistors between pin 1 and the RTD sensor, and between pin 4 and the RTD sensor, do not contribute to the measured voltage. Meanwhile, the resistors between pin 2 and the RTD sensor, and between pin 3 and the RTD sensor, contribute a limited error voltage as there is almost no current flow through them.

Apart from the temperature compensation, the measurement system should be calibrated to remove system errors, such as the voltage offset caused by the DC imperfection of the amplifier and the nonlinearity of the ADC. The simplest way to calibrate the system error is to measure the output signal with a zero input of the potentiometric sensor interface circuit shown in Figure 11.3a,b or the amperometric sensor interface circuit shown in Figure 11.4. The measured voltage is treated as the system voltage offset, which would be deducted from the ordinary measurement results for calibration. For more accurate calibrations, a set of precise, known signals can be applied to the system, forming a set of calibration curves.

11.2.6 Personal Computer-Based Measurement Control

11.2.6.1 Measurement Control. The measurement system is usually controlled in real time, which means that all control procedures for a measurement should be accomplished before the next measurement. PC-based real-time measurement control can be implemented either by a PC running a real-time operating system (RTOS) or by a PC that is linked to a real-time embedded microcontroller/field programmable gate array (FPGA)/digital signal processor (DSP).

An RTOS is a multitasking operating system intended for real-time applications. Typical PC running RTOS are Microsoft Windows CE and QNX (QNX Software Systems, Ottawa, Canada). With RTOS, the measurement system can be directly controlled in real time by a PC when control tasks are not too complicated. The commonly used real-time control configuration is the use of a PC with an embedded microcontroller/FPGA/DSP. In this case, a PC running under an ordinary operating system, such as Windows, is working as a measurement control console to input control commands to the embedded microcontroller/FPGA/DSP where real-time control performs, and to display measurement results sent back from the embedded microcontroller/FPGA/DSP. From the point of real-time control, there is no significant technical difference among the microcontroller, FPGA, and DSP, although FPGA and DSP are more powerful in real-time signal processing than the microcontroller. FPGA and DSP can work in a higher frequency (hundreds of megahertz) than that of the microcontroller, enabling a more precise real-time control. Furthermore, FPGA and DSP have the ability to provide programmable logic and other digital circuits useful for a measurement system.

The main benefit of using a PC with an embedded microcontroller/FPGA/DSP for measurement control is to keep the whole measurement system fully programmable [20]. Measurement parameters can be downloaded from the PC to the reprogrammable microcontroller/FPGA/DSP before the measurements, which makes the full measurement system flexible in setting types of applied sensors and their input signal ranges, changing measurement interval and selecting proper calibration methods. As real-time measurement control is performed in the microcontroller/FPGA/DSP, the host PC has more power to deal with offline post-data processing to transform the measured raw electrical data to a biologically meaningful format or to appropriately visualize the raw data. The wide variety of readily available PC application software packages (such as LabVIEW [21]; National Instrumentation, Austin, TX) simplify the instrumentation graphic user interface (GUI) design, and the technical developments in USB makes the link between PC and microcontroller/FPGA/DSP more compact.

11.2.6.2 Wireless Sensor Network. A wireless sensor network (WSN) consists of a set of wireless sensor nodes. Each node is typically equipped with a microcontroller and a radio transceiver or other wireless communications devices. WSNs are usually powered by battery and operate in the industrial, science, and medicine (ISM) band acting as short-range radio systems.

When using a number of three-dimensional (3D) bioreactors operating in parallel, it is impractical to connect a multitude of wired sensors to the measurement system. Wireless connections facilitate the parallel monitoring of many bioreactors. As the communication speed of the wireless node is not too high, a small number of channels can be controlled by a wireless node, and a WSN is formed when a large number of measurement channels are needed. An alternative way to configure the WSN for stem cell culture is to combine each sensor (or a sensor array) with a wireless node forming an intelligent wireless sensor system. Each wireless node can have one potentiometric measurement channel and one amperometric measurement channel so that a biosensor, regardless of type, can be interfaced with it. Figure 11.8 shows a WSN scheme for cell cultures based on Texas Instruments' CC2511 chip (low-power system-on-chip [SoC] with microcontroller, memory, 2.4-GHz RF transceiver, and USB controller).

11.3 STEM CELL CULTURE PROCESS MONITORING

An up to 128-channel physiological measurement system has been employed for the monitoring of the stem cell culture process [9]. The monitoring system can interface with potentiometric and amperometric sensors monitoring pH, ammonia, CO_2 tension, oxygen tension, and glucose concentration. It is a fully

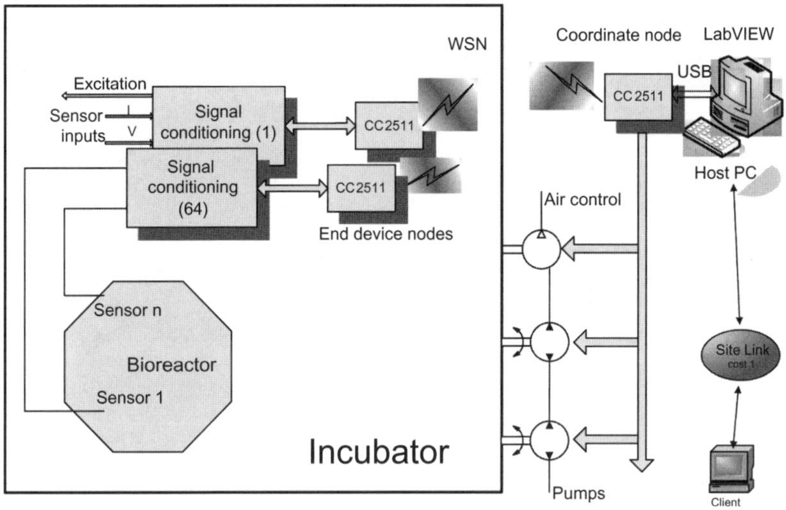

Figure 11.8. Stem cell culture monitoring using WSN. LAN: local area network.

programmable system. The types of sensor and the numbers of measurement channels for each type of sensors can be configured to meet specific measurement requirements. Measurement parameters, such as the "measurement time interval," can be set during the measurement process. The dynamic range of the input signal is adjustable; therefore, the system can be flexible enough to interface with new types of sensors for new applications in the field of stem cell culture or other bioprocess monitoring.

11.3.1 Sensors and Bioreactors

The newly designed bioreactor has the form of a standard six-well plate culture system (diameter of 3.3 cm and depth of 1.0 cm). It has two inlet and two outlet ports for perfusion, allowing cells to obtain fresh media and prevents them from escaping the bioreactor. The cover of the bioreactor is made from poly(dimethylsiloxane) (PDMS), which allows sensors to be conveniently inserted into the culture media at any location required.

Sensors that are continuously exposed in culture media for several days must have good biocompatibility properties [22–24], as the interaction between the sensor and the culture media can raise undesirable issues, such as protein adsorption and cell adhesion, which reduce both the sensitivity and the longevity of the sensor. The sensors used for monitoring have been optimized in terms of biocompatibility while their miniaturization (dimensions of the sensing part are 0.2 mm; see Fig. 11.9) ensures that a number of them can be easily placed within the newly developed small-quantity bioreactor. A bioreactor with sensors installed for stem cell culture [9] is shown in Figure 11.9.

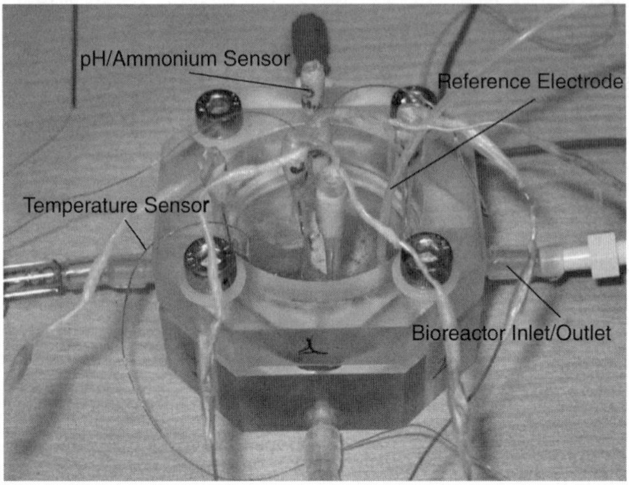

Figure 11.9. Bioreactor with placed sensors.

It is worth mentioning that in practice, even with good sensor biocompatibility, the culture media itself is often complicated. The sensor's property varies when exposed to physiological solutions for a long period of time. A post-culture calibration process is applied to deal with the variations, and therefore, in practice, the measured electrical data are translated into physiological ones via relations such as Section 11.2.2.1 with their parameters modified by the calibration process. This suggests that the relevant electrical specifications of the measurement system should be adjusted for cell culture applications. An example is that the ISE pH sensor exposed in culture media displayed a linear but sub-Nernstrian response [24, 25]. The calibration process reveals a 30-mV/pH-unit slope in cell culture for pH measurement rather than a 59-mV/pH-unit slope, which is calculated via Equation 11.2. This means that if the measurement requirement of the pH level is 0.1 units, then the electrical accuracy of the measurement system should be adjusted from 5.9 mV for normal pH measurement to 3 mV for monitoring the cell culture process.

11.3.2 Measurement System

The measurement system is composed of up to eight measurement modules, each with 16 measurement channels, as shown in Figure 11.10. There are three types of measurement modules: potentiometric, amperometric, and temperature modules. Each module has its own preamplifier (voltage buffer for the potentiometric module and current to voltage converter for the amperometric module), a multiplexer, a low-pass filter, a PGA, and a 16-bit ADC providing a theoretical 76-μV voltage resolution at 5-V power supply. In practice, this degrades close to 1 mV. However, the measurement resolution of a module can be set to much less

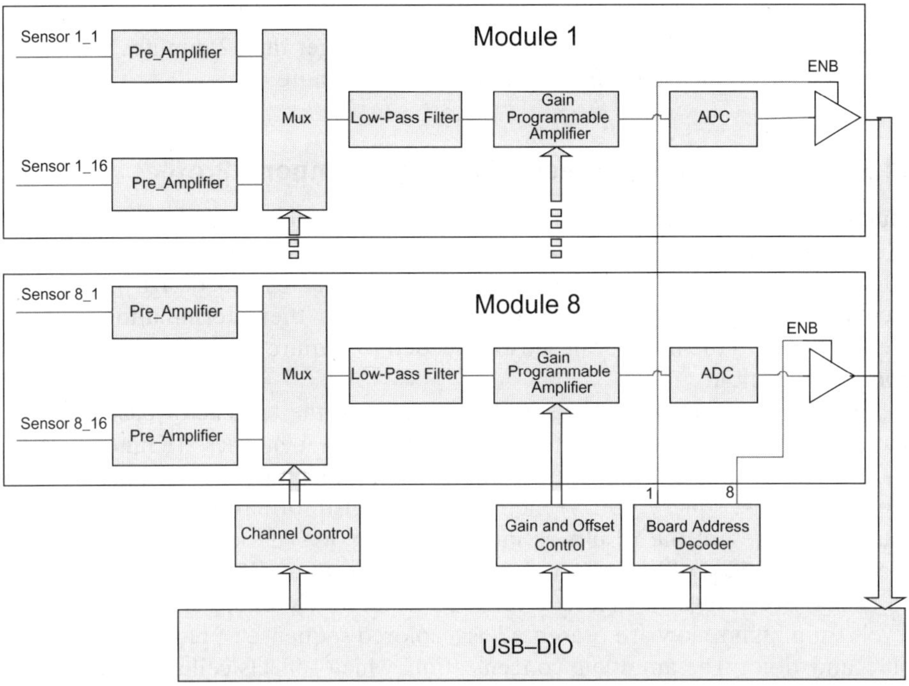

Figure 11.10. Block structure of the measurement system. ENB: enable.

than 1 mV, depending on the gain settings of the preamplifier and the PGA (maximum gain = 4096). The measurement modules are connected to the host PC via a USB–digital input/output (DIO), which converts command signals from USB format to input–output (I/O) signals for measurement control and also converts measured raw data to USB signal format for sending measured data back to the PC.

To measure a target sensor, the module address is sent to all data acquisition modules. The ADC in the data acquisition module whose address setting matches the sent module address signal is enabled. The channel selection data and PGA setting data are sent and the measured data is transferred from the enabled ADC to the host PC. National Instrumentation LabVIEW® 7.1 is adopted as the programming language. All the system details can be configured: the number of modules can be set; each module can be set to its corresponding type or be turned off. Measurement channels in the modules can be turned on or off. The measurement interval can be set to control the recording data size during a several-day-long culture process. For amperometric sensors, the excitation voltage can be set for each individual channel. The maximum measurement control time for a measurement is about 1 second; therefore, the minimum measurement interval for an individual sensor is about 2 minutes when all 128 measurement channels are active. Real-time control has been achieved for a total of 128 channels when the measurement interval for each channel is set to longer than 2 minutes. For a 64-channel measurement, this figure can be set to 1 minute.

11.3.3 Design of Experiment and Spatiotemporal Profiles

Multichannel measurements can be exploited for multiparametric/multisite parallel measurements or for the acquisition of spatiotemporal information at the same site. To acquire spatiotemporal information, statistical design of experiments (DOEs) [26–28] is used as a methodology for the determination of both the location and the number of sensors needed to acquire the desired spatiotemporal information.

Nine sensors are placed at specific positions of the bioreactor based on the statistical DOE as shown in Figure 11.11a. Using Umetrics' (Kinnelon, NJ) MODDE7—a software package for analysis and modeling of spatial profiles via the partial least squares (PLS) method to extract information from large data sets and to present the results as interpretable plots—the contour plots of the spatial ammonia profiles collected during the culture at different times (24 and 48 hours) are shown in Figure 11.11b and c, in which ammonia concentration levels from high to low are presented as a colored sequence of red, yellow, green, blue, and violet. The ammonia concentration, which reflects cellular activities in culture variations with time, can be observed by comparing different subfigures, which present the spatial ammonia concentration information at any specific given moment. These results mean that the temporally evolving spatial ammonia concentration profiles within the bioreactor can be obtained via the DOE-based, real-time, multichannel measurements. Hence, the cell culture monitoring system is capable of providing a more detailed culture information.

11.4 SUMMARY

- The output of the stem cell culture process is significantly influenced by culture parameters. Therefore, using potentiometric and amperometric biosensors to monitor the state of the culture provides quantitative real-time multiparametric information, which can be useful for the optimization of the stem cell culture process.
- Biological measurement requirements for stem cell culture process can be translated into electronic specifications in linear or nonlinear ways according to the underlying sensor physics. However, in practice, the measured electrical data are translated into physiological one via relations modified by calibration process.
- Using a PC as the console and a microcontroller/FPGA/DSP to perform real-time control, programmable components, such as PGAs, in data acqui-

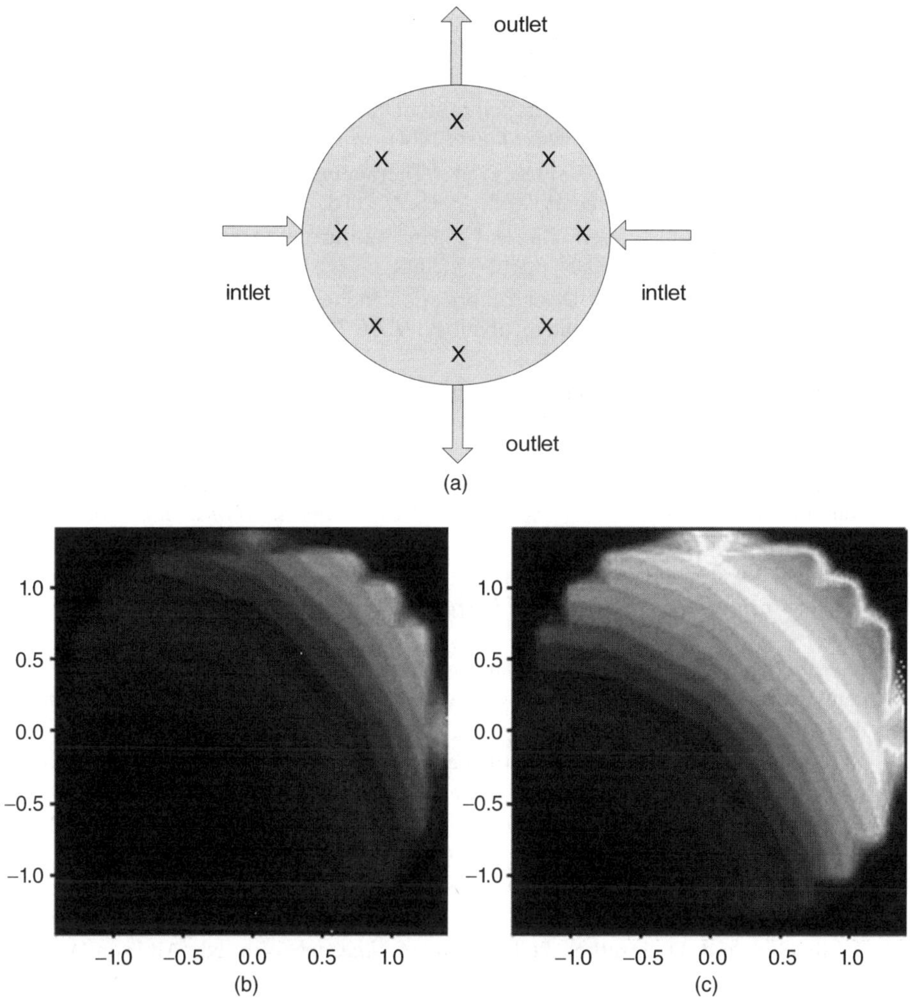

Figure 11.11. Temporally evolving spatial ammonia concentration profile at different times from cell's seeding. (a) Positions of the nine sensors; (b) after 24 hours; (c) after 48 hours.

sition circuits can be used to keep the whole measurement system flexible and capable of accepting new types of sensors.

- Real-time multichannel measurements can provide spatiotemporal, physiological information of the cell culture process.

REFERENCES

[1] F. M. Watt and B. L. M. Hogan, "Out of Eden: Stem cell and their niches," *Science*, 287, pp. 1427–1430, 2000.

[2] L. E. Fenno, L. M. Ptaszek, and C. A. Cowan, "Human embryonic stem cells: Energing technologies and practical applications," *Curr. Opin. Genet. Dev.*, 18, pp. 324–329, 2008.

[3] M. F. Corsten and K. Shah, "Therapeutic stem-cell for cancer treatment: Hopes and hurdles in tactical warfare," *Lancet Oncol.*, 9(4), pp. 376–384, 2008.

[4] A. E. Ting, R. W. Mays, M. R. Frey, et al., "Therapeutic pathways of adult stem cell repair," *Crit. Rev. Oncol. Hematol.*, 65, pp. 81–93, 2008.

[5] F. Ulloa-Montoya, C. M. Verfaillie, and W. Hu, "Culture systems for pluripotent stem cells," *J. Biosci. Bioeng.*, 100(1), pp. 12–27, 2005.

[6] J. Chrisen and A. Andreou, "Design, fabrication and tesing of a hybrid CMOS/PDMS microsystem for cell culture and incubation," *IEEE Trans. Biomed. Circuits Syst.*, 1(1), pp. 3–18, 2007.

[7] E. Cimetta, E. Figallo, C. Cannizzaro et al., "Micro-bioreactor arrays for controlling cellular environments: Design principles for human embryonic stem cell applications," *Methods*, 47, pp. 81–89, 2009.

[8] M. Lim, H. Ye, N. Panoskaltsis, et al., "Intelligent bioprocessing for haemotopoietic cell culture using monitoring and design of experiments," *Biotechnol. Adv.*, 25(4), pp. 353–368, 2007.

[9] X. Yue, E. M. Drakakis, M. Lim, et al., "A real-time, multi-channel monitoring system for stem cell culture process," *IEEE Trans. Biomed. Circuits Syst.*, 2(2), pp. 66–77, 2008.

[10] X. Yue, E. M. Drakakis, A. Mantalaris, et al., "Generation of spatio-temporal concentration profiles for cell culture systems: A case study in ammonia," *Measurement*, 43(2010), pp. 1207–1216, 2010.

[11] M. Pohanka and P. Skladal, "Electrochemical biosensors—Principles and applications," *J. Appl. Biomed.*, 6, pp. 57–64, 2008.

[12] R. B. Northrop, *Introduction to Instrumentation and Measurements*, 2nd ed. Boca Raton, FL: CRC Press, 2005.

[13] J. Park and S. Mackay, *Practical Data Acquisition for Instrumentation and Control Systems*. Oxford: Elsevier, 2003.

[14] T. Shanley and D. Dzatko, *PCI System Architecture*, 4th ed. Addison-Wesley, 1999.

[15] D. Anderson, *Universal Serial Bus System Architecture*, 2nd ed. Addison-Wesley, 2001.

[16] A. R. Hambley, *Electronics*, 2nd ed. Prentice Hall, 2000.

[17] J. Wang, *Analytical Electrochemistry*, 2nd ed. John Wiley & Sons, 2000.

[18] E. Hwang, D. Pappas, A. Jeevarajan, et al., "Evaluation of the paratrend multi-analyte sensor for potential utilization in long-duration automated cell culture monitoring," *Biomed. Devices*, 3(6), pp. 241–249, 2004.

[19] D. Garvey, "So, what is an RTD?" *Sensors*, 16(8), pp. 39–42, 1999.

[20] X. Yue and C. McLeod, "FPGA design and implementation for EIT data acquisition," *Physiol. Meas.*, 29, pp. 1233–1246, 2008.

[21] G. W. Johnson and R. Jennings, *LabVIEW Graphical Programming: Practical Applications in Instrumentation and Control*, 3rd ed. New York: Mc Graw-Hill, 2001.

[22] M. J. Madou, *Fundamentals of Microfabrication: The Science of Miniaturization*. Boca Raton, FL: CRC Press, 2002.

[23] G. Voskerician and J. Anderson, "Sensor biocompatibility and biofouling in real-time monitoring," in *Wiley Encyclopedia of Biomedical Engineering*. John Wiley & Sons, 2008.

[24] A. Radomska, S. Suckect, H. Ye, et al., "Biocompatible ion selective electrode for monitoring metabolic activity during the growth and cultivation of stem cells," *Biosens. Biochem.*, 24(2008), pp. 435–441, 2008.

[25] O. T. Guenat, S. Generelli, N. F. Rooij, et al., "Development of an array of ion-selective microelectrodes aimed for monitoring of extracellular ionic activities," *Anal. Chem.*, 78(1), pp. 7453–7460, 2006.

[26] N. H. Kim, M. H. Choi, S. Y. Kim, et al., "Design of experiment (DOE) method considering interaction effect of process parameters for optimization of copper chemical mechanical polishing (CMP) process," *Microelectron. Eng.*, 83(3), pp. 506–512, 2006.

[27] R. H. Myers and D. C. Montgomery, *Response Surface Methodology: Process and Product Optimization Using Designed Experiments*. John Wiley & Sons, Inc., 2002.

[28] M. Lim, H. Ye, X. Yue, et al., "Towards information-rich bioprocessing: Generation of spatio-temporal profiles through the use of design of experiments to determine optimal number and location of sensors—An example in thermal profiles," *Biochem. Eng. J.*, 40, pp. 1–7, 2008.

PART III

EMERGING TECHNOLOGIES

12

BUILDING INTERFACES TO DEVELOPING CELLS AND ORGANISMS: FROM CYBORG BEETLES TO SYNTHETIC BIOLOGY

Hirotaka Sato, Daniel Cohen, and Michel M. Maharbiz

12.1 INTRODUCTION

The world today stands on the brink of a technological revolution, much as we did in the mid-20th century when we began to build complex electronic devices. Across the globe, engineers are beginning to modify, redesign, and synthesize completely new functions into living cells, a nascent discipline currently called *synthetic biology* [1, 2]. While this will undoubtedly have a broad impact on medicine and human health, it will have, we believe, an even larger impact on basic technology: Humans will soon begin to make machines with the same processes that nature uses to make organisms. This may be a scary proposition, but it is almost unavoidable. More machines—beetles, trees, bacteria, people—are made using nature's fabrication technologies than using our factories. Nature's machines heal themselves when hurt. They rely on components developed over millions of years of evolution. They are often robust in the face of communication noise. They rely on a versatile palette of materials we do not fully understand. As we begin to exploit these and other facets of nature's fabrication

CMOS Biomicrosystems: Where Electronics Meet Biology, First Edition. Edited by Krzysztof Iniewski.

paradigms—with all the ethical implications that this entails—the concurrent development of sophisticated, high-resolution interface technologies to cells, tissues, and organisms will be paramount [3]. It is also likely that many of these interface technologies will be increasingly within the reach of sophisticated non-professionals; indeed, the two examples in this chapter were chosen specifically because they were built primarily with off-the-shelf components.

12.1.1 Synthetic Biology and Interfaces

Synthetic biologists aim to build biological components and assemble them into integrated systems to accomplish many particular tasks [1,2]. Usually, this focuses on designing biological parts (using genes, ribosome binding sites, terminators, etc.), assembling these parts using automated procedures, and integrating them into cells to generate new functions. For example, at the University of California, Berkeley, and the University of California, San Francisco, researchers are working on tumor-killing bacteria [4] and on cells that may one day efficiently produce biofuels [5]. A common feature in all of these efforts is the generation of cells that carry out a single given task at optimum efficiency. For example, *Escherichia coli* bacteria may be designed to produce a biofuel at maximum output when grown in liquid culture to high density. Similarly, a population of 10^{10} identical bacteria might be required to invade a tumor and ultimately destroy it. The underlying assumption of this approach is that entire, genetically identical populations of cells will all behave in the exact same way.

In multicellular organisms, however, cells in a population each behave differently and coordinate this different behavior to generate complex structures (i.e., a liver). A single cell is constantly transducing information from the micro-environment around it. Various genetic pathways allow it to sense and respond to stimuli with webs of chemical feedback. These pathways enable cells to coordinate with each other: By exchanging chemical or mechanical information, cells can influence the states of the cells near them and change what they do (Fig. 12.1). A central paradigm in developmental biology, which we have explored in our lab in the context of microchemical interfaces [6–10], is that a small set of chemical signals, called *morphogens*, is exchanged between cells as a way of coordinating complex behavior (such as building the distinct regions of a liver). These morphogens induce the coordinated differentiation of several different cell types of a functional organ along a gradient, generating the normal neighbor relationships of a functional organ. In the developing neural tube of the early embryo, for example, gradients of certain secreted morphogens define the fate of cells along the dorsoventral axis (Fig. 12.2).

Two key points are fundamental:

1. *Synthetic* chemical communication pathways can be introduced into bacterial cells, enabling them to communicate with each other in new ways. This communication can lead to the generation of different populations of cells, each doing different things.

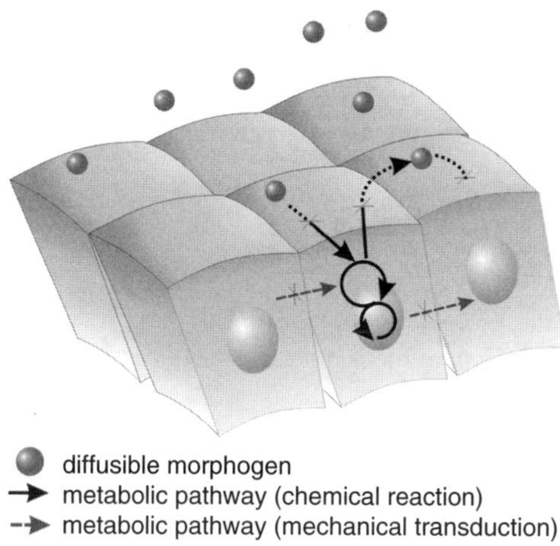

Figure 12.1. Cells in multicellular organisms communicate via the exchange of chemical and mechanical signals.

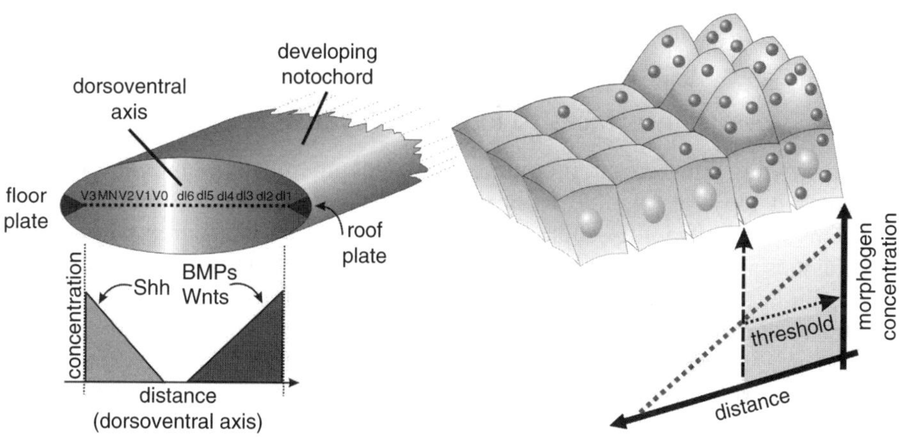

Figure 12.2. Lessons from developmental biology. (Left) Mammalian neuronal cell fate along the dorsoventral axis is specified in the early embryonic neural tube by three sets of morphogens (the Shh, BMP, and Wnt families of molecules). Shh is secreted by the notochord and the floor plate; BMPs and Wnts are expressed by the roof plate [11, 12]. (Right) Certain threshold concentrations of diffusible signals will trigger "all-or-nothing" developmental changes (here conceptualized by a change in cell shape at high concentrations of signal molecules).

2. We can build microsystems (with appropriate sensors, actuators, and feedback) to communicate with these synthetic pathways. This will enable the patterning and control of gene expression in the synthetic biological systems and be the cornerstone of a completely new way of making synthetic, multicellular machines (Fig. 12.3).

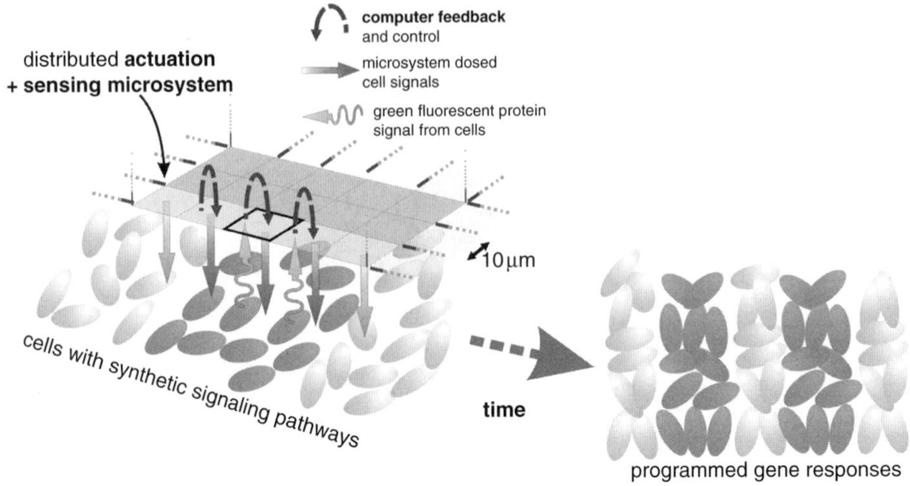

Figure 12.3. A microsystem capable of chemically dosing signal molecules and sensing cell responses across an entire cell culture with high resolution, coupled with cells modified with synthetic pathways, would enable the top-down design of synthetic multicellular systems.

12.1.2 Cyborgs: Interfaces to Existing Organisms

Beyond creating interfaces to modulate the developmental dynamics in synthetic constructs, it may be possible to obtain novel form or function from existing organisms with sufficiently sophisticated interfaces. This is an area with major ethical issues, many of which have been discussed for a long time, but which are once again in the fore given the power of today's technologies.

One such example is the remote control of insect flight (and more generally, locomotion and behavior) via implanted neural stimulators. If it were possible to remotely control the flight of insects, and receive information from onboard sensors, there would be many applications. In biology, the ability to control insect flight would be useful for studies of insect communication, mating behavior, and flight energetics, and for studying the foraging behavior of insect predators such as birds, as has been done with terrestrial robots [10]. In engineering, electronically controllable insects could be useful models for insect-mimicking micro- and nano-air vehicles (M/NAVs) [13–15]. M/NAVs are the subject of intense research and development [16–19]. Despite major advances, M/NAVs still present significant trade-offs between payload mass, flight range, and speed. Currently, the principal limiting factors are the energy and power density of existing fuel sources and the flight dynamics of very small flyers. Insects have flight performance (as measured by distance and speed vs. payload and maneuverability) as yet unmatched by man-made craft of similar size. Moreover, both the flight dynamics and the neurophysiology of these organisms are increasingly well understood [20–30]. Furthermore, tetherless, electrically controllable insects themselves could be used as M/NAVs and serve as couriers to locations not easily accessible to humans or terrestrial robots.

12.2 EXAMPLE INTERFACES

12.2.1 Technologies for Interfacing with Developing Cells

This section presents results covered in more detail in Cohen et al. [31]. The development of methods that introduce spatiotemporal perturbations into developing, multicellular systems via soluble molecules has a long history [32–35] and a rich, recent body of literature. Specifically, advances in microfluidics [6, 36–39] and biochemistry [40–43] are beginning to open the door to direct modulation of developmental pattern formation at the spatial and temporal scale of the cell's control circuitry. Such devices can provide spatially rich, real-time input–output (I/O) signals to bias developing cells into specific phenotypes. In the context of synthetic biology, such interfaces would add a degree of control over the pattern formation dynamics in multicellular structures that are expressing genetic circuits intended to coordinate activity through soluble molecules [44]. Initial efforts in building synthetic multicellular constructs have already begun [6, 26–45], and as these mature, a robust chemical interface will be invaluable in addressing and biasing the development of patterns. In the context of control theory, such devices would allow an exploration of equilibria, stability criteria, and the nonlinear dynamics of regulatory circuits. In the context of regenerative medicine and tissue engineering, these devices could potentially provide active, spatiotemporal control of morphogenesis [19, 46, 47].

While it is clear that these applications call for systems capable of high-resolution dosing of multiple chemicals onto ensembles of cells, it is less clear how best to achieve this in a way that is low cost, versatile, and open source. Although a number of microfluidics-based attempts have been published [16, 20, 48, 49], all have limitations in resolution, complexity of fabrication, or ease of use. As an alternative to microfluidics, we considered inkjet technology. To date, inkjets have been incorporated into a variety of biological techniques including direct cell printing for patterning [50, 51] and tissue engineering [52–55]; assorted cell factor printing to regulate cell positioning and behavior [36, 56–58]; and DNA microarray fabrication [59]. This list demonstrates the versatility of the platform, although inkjets have yet to be used for active regulation of cellular behavior. Commercialized research-grade inkjet systems such as the Fujifilm Dimatix system exist but cost orders of magnitude more than consumer-grade printers and usually only print one ink at a time.

This section presents the adaptation of a commercial, low-cost, piezoelectric inkjet printer and commercial compact discs (CDs) for use as a chemical interface system designed to actively regulate cellular development (Fig. 12.4). The printer is capable of addressing up to six different soluble chemicals, and subsequently delivering precise doses of these chemicals to cell cultures at 226 dots/mm. The platform can be integrated with in-line microscopy to acquire data at specific

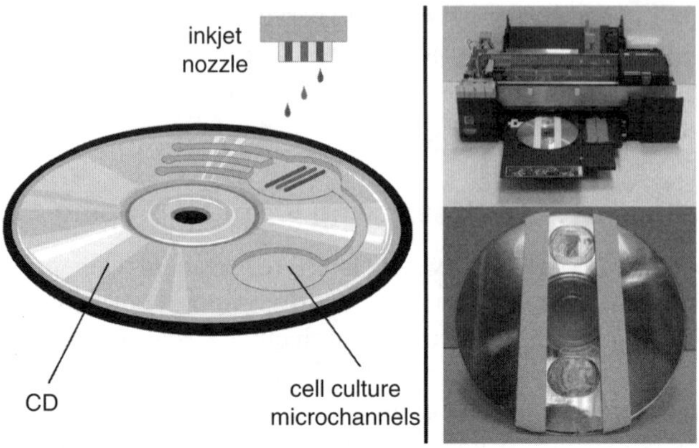

Figure 12.4. Overview of the printing system. (Left) The dosing method is based on the delivery of multiple chemical compounds from piezoelectric printer heads onto specially prepared compact disc (CD) templates modified so as to support thin layers of microbial agar cultures. (Upper right) The printer used was the Epson R280, here shown being loaded with a modified CD. (Lower right) Close-up of a modified CD with stand-offs for the print rollers and two LB/X-Gal agar cultures of *E. coli* dosed with lactose patterns (which induce the characteristic X-Gal blue color; see text).

time points post dosing. Additionally, no custom software is required for our approach, making the whole system simple and user-friendly. While the CD platform is compatible with the rich toolset of polymer microfluidics [60, 61] and could be adapted into a more sophisticated device in future work, its use here was solely as a convenient, readily modifiable substrate that was compatible with the printer. In this study, we used the inkjet to control the spatiotemporal reaction–diffusion dynamics of gene expression in the *lac* regulatory system by printing specific patterns of lactose and glucose onto a field of *E. coli*.

The choice of *lac* was deliberate, in that it is commonly used in synthetic biology and has a number of interesting control features, including well-characterized feedback and multiple stable states [62–64]. We were surprised by the observation that the bistable nature of the *lac* operon's feedback system, when perturbed with patterns of lactose (inducer) and glucose (inhibitor), can lead to coordination of cell expression patterns across a field in ways that mimic motifs seen in developmental biology. Examples of this behavior include sharp gene expression boundaries and the generation of traveling waves of mRNA expression from single "trigger" patterns. In this context, lactose and glucose are analogous to morphogens in a developing system, with the *lac* operon acting as a general template for exploring how spatially graded perturbations generate rich behavior in bistable circuits. This is especially interesting given that *lac*, while employing positive and negative feedback and diffusible molecules (i.e., glucose and lactose) is not usually considered a reaction–diffusion system capable of generating pattern (there is a vast literature on Turing-type and other reaction–diffusion systems and their applications to biological pattern formation [65]).

12.2.1.1 The Lac Operon. The *lac* operon is one of the best studied regulatory pathways in microbes [43–45, 66]. In *E. coli*, kinetic data for the entire *lac* operon is available and robust models have been developed [43, 44]. Moreover, the dynamics of the system are well understood, and are known to exhibit multiple stable points [43–45]. In the canonical *lac* system (Fig. 12.2), extracellular lactose is taken up by lactose permease, where it is converted to allolactose by β-galactosidase. Allolactose upregulates the production of both β-galactosidase and lactose permease by inhibiting the repressor of the *lac* promoter. This lactose-based, positive feedback loop has been shown to be bistable [43]; below a certain lactose threshold, little lactose is converted, while above this threshold the system jumps to a much higher consumption rate. Glucose acts to inhibit the conversion of lactose by lowering the transcription rate of the *lac* operon via cAMP and the catabolite repressor protein (CRP). This acts as negative feedback for the conversion of lactose. Both lactose and glucose are soluble in the extracellular space. In Cohen et al. [31], we chose to model the *lac* system as a set of coupled differential equations, following models developed by Ellis-Davies and Basu et al. [43, 44]. The well-known X-Gal assay [67] introduces the soluble X-Gal compound, which is cleaved by β-galactosidase into 5-bromo-4-chloro-3-hydroxyindole, in turn oxidizing to 5,5′-dibromo-4,4′-dichloro-indigo, finally resulting in an insoluble blue product. This allows for β-galactosidase activity to be assayed.

12.2.1.2 Modification of an Inkjet Printer into a Chemical Interface for Cells. For a detailed description of the modifications made to the Epson R280, see Cohen et al. [31]; an abbreviated summary is presented here. The necessary modifications require commonly available tools and several hours to complete, and the procedure should be adaptable to a number of different printer types.

We used an Epson R280 inkjet for a number of reasons. Epson printers use piezoelectric printheads, as opposed to thermal jet heads. While both types of printhead would probably suffice for our experiments, the mechanical nature of piezoelectric heads means that they can safely print a greater array of chemicals, and they do not impose temperature fluctuations on the printed fluid. Additionally, the R280 has the ability to print on rigid substrates (CDs), which is not a common feature. Finally, the whole system is low cost (~$100) and widely available.

There are three fundamental challenges related to manipulating a printer: loading customized inks, uniquely specifying which inks are used during a print job, and interfacing with the biological substrate. Using an Epson R280 printer, we were able to load our lactose and glucose inks by interrupting the ink-charging process and manually injecting, by syringe pump, our solutions into specific color reservoirs (300 mM of lactose, 500 mM of glucose). This technique requires no manipulation of the ink cartridges themselves; we inject ink downstream of the cartridge, meaning that the printer functions completely normally but prints the injected solutions rather than ink. This is simpler, less damaging to the printer, and does not require the use of third-party hardware. Having primed the printer, the final step was to prepare it to accept a cell-bearing substrate.

We took advantage of the R280's ability to print directly onto the surface of CDs and milled 800-μm-deep wells directly into the surface of the CDs. The size, geometry, and position of these wells were selected so as not to interfere with any of the printer's mechanisms (feed rollers or carriage drive system). By using a CD template in Adobe Photoshop, it was possible to create any planar pattern, uniquely specify the inks to be used, and print directly into the wells.

We used sterile shim stock to cut out individual pieces of cell-bearing agar and transfer them to the appropriate wells on the surface of the CD. The CD was then loaded into the printer, and the print job sent. No run lasted longer than 2 minutes, and at no point did the cells come into contact with any components of the printer, which had previously been sterilized with 70% ethanol. Post printing, the agar slices were transferred to hydrated petri dishes, placed in the incubator, and observed over a period of 15 hours.

12.2.1.3 Chemically Addressing Cells with the Inkjet Printer. There were three key goals for this work. First, we aimed to demonstrate the feasibility of using a commercial inkjet printer as a microdosing chemical interface for cellular systems. Second, we wished to determine whether inclusion of diffusion terms into a partial differential equation (PDE) model of the *lac* operon would predict the gene expression patterns generated by the printer. Lastly, we hoped to explore the types of morphogenetic-like behaviors that could be induced solely through direct, chemical manipulation of the *lac* operon.

We first characterized the resolution and pattern-formation capability of the printer system. As we were printing into hydrated agar (which would allow for diffusion of any dosed molecule), we could not rely on the resolution specifications of the printhead. Concentrated lactose (300 mM) was printed in parallel bars of varying widths onto samples and the resulting X-Gal pattern was recorded (Fig. 12.5). By fitting this data (in addition to the transient data presented in Fig. 12.9) to our finite element reaction–diffusion model, the effective diffusion rates were calculated. Below a certain width of printed inducer, diffusion reduces the peak concentration, and the *lac* operon never switches to its ON state (note how the fourth bar shows a marked, nonlinear decrease in induction). For 300 mM of lactose and our agar formulation, features smaller than 700 μm tended not to visibly induce. Thus, by varying the diffusion constant of the medium and the concentration of dosed inducer, the exact minimum width of an induced feature can be precisely controlled. A demonstration of this involves using half-toning to produce size-graded, two-dimensional features across a field of cells (Fig. 12.6). Here, we see blurring between closely spaced, large features, but more well-defined, smaller features. The implication is that, by taking the transport characteristics of a system into account, we can modulate how features interact with each other.

Bearing this in mind, Figure 12.7 shows a cross-hatched pattern used to test the uniformity of response (and resolution) across a large field of cells. This also demonstrates how the inkjet, in conjunction with a bistable circuit, can be readily used to produce sharp boundaries enclosing noninduced material even in the

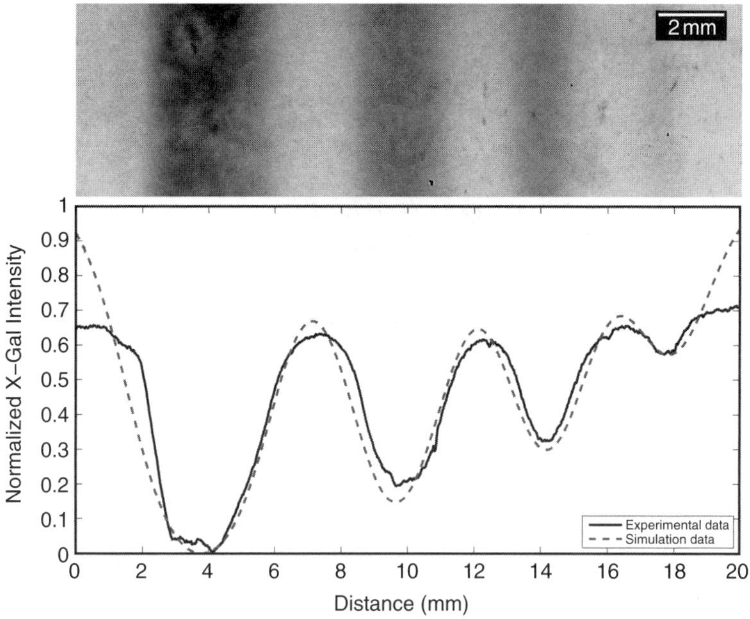

Figure 12.5. Resolution test varying only the width of the printed region. Four bars of lactose were printed with the following widths (left to right): 3.5 mm; 2.0 mm; 1.5 mm; and 0.75 mm. Note the abrupt transition to a low level of induction at the 0.75-mm bar. Further note the close agreement between the empirical data and the simulation. The discrepancies at the boundaries are a result of the optical properties of the agar at the boundaries of the sample that were not taken into account in the simulation.

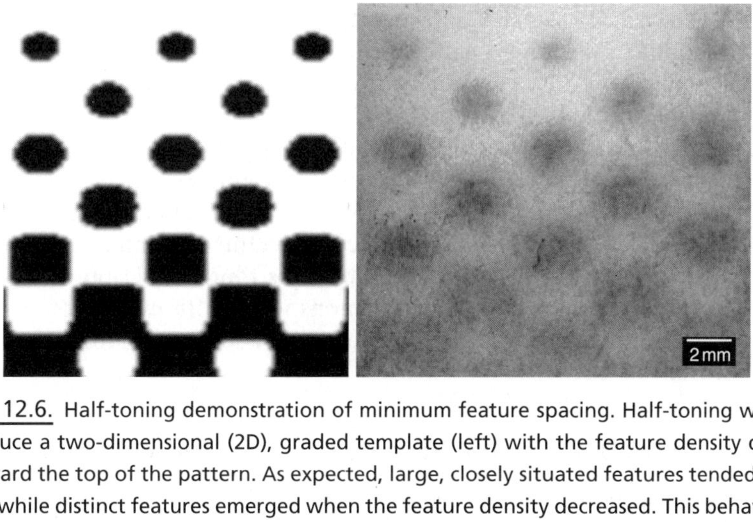

Figure 12.6. Half-toning demonstration of minimum feature spacing. Half-toning was used to produce a two-dimensional (2D), graded template (left) with the feature density decreasing toward the top of the pattern. As expected, large, closely situated features tended to blur (right), while distinct features emerged when the feature density decreased. This behavior can be used to modulate feature interaction as a function of geometry and the transport properties of the medium.

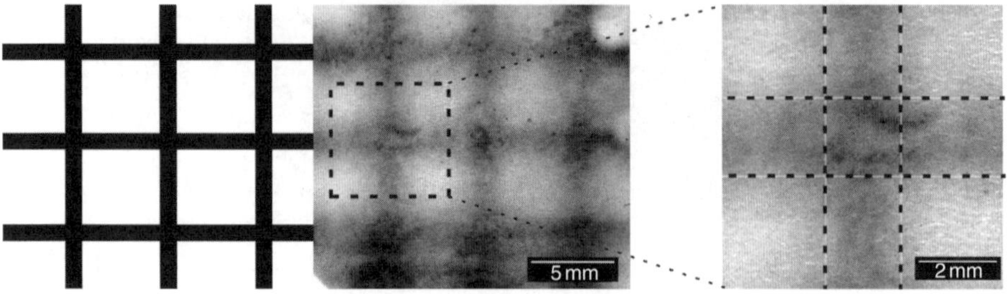

Figure 12.7. Cross-hatched lactose resolution testing. The printer template and corresponding induction profile are shown (left), alongside a close-up of a junction that demonstrates the sharp drop-off in induction that occurs despite diffusion.

presence of an interdiffusion zone. By taking advantage of bistability in the presence of weak gradients, we can achieve fairly sharp boundaries (see Fig. 12.7), a motif observed in embryonic developmental programs [68]. Given a well-tuned simulation tool, it is possible to design and print almost any induction pattern within the resolution constraint given above.

Figure 12.8 shows the development of X-Gal pattern over time subsequent to lactose induction. Time-lapse microscopy was performed within an incubator, with images being taken every 20 minutes for a period of 3 hours. This data was used to fit the finite element model diffusion rates. Typically, induction becomes visible by eye after 45 minutes, and will then plateau at around 1.5 hours.

Figure 12.9 shows results for a piecewise continuous lactose gradient across a field of cells. Working from a grayscale image generated on a commercial drawing program (CorelDraw 11.0), the first bar contains 0.24 M lactose, and each subsequent bar is 20% less concentrated than the previous bar. Such a pattern is not easily attainable without a patterning device, such as the inkjet, and demonstrates the ability to produce customized, finely controlled patterns, in turn allowing fine control of cellular behavior.

We took advantage of the R280's ability to print multiple types of ink by creating patterns composed of both lactose and glucose. Specifically, we first printed a large, uniform field of lactose (300 mM) over an entire sample, immediately followed by a narrow bar of glucose (550 mM) printed on top of the lactose (Fig. 12.10). Glucose is an exceptionally strong transcriptional inhibitor for the *lac* promoter, and this effect is demonstrated both by the complete lack of induction underneath the glucose bar, and the graded level of induction propagating out from that spot.

12.2.2 Cyborg Beetles: The Remote Control of Insect Flight

The following section is an abridged version of the results presented in Sato et al. [69]; that publication also contains a number of videos demonstrating the results discussed below.

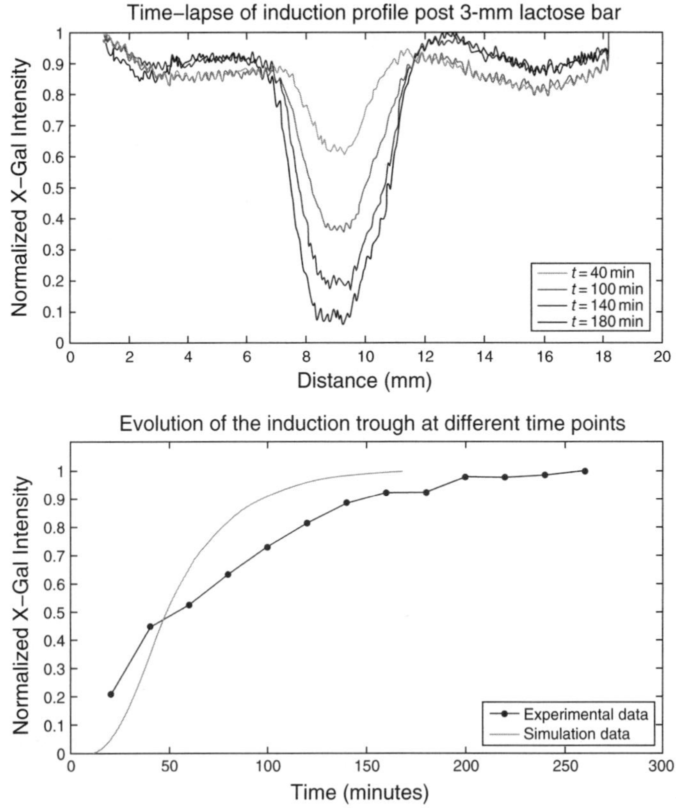

Figure 12.8. Transient lactose induction profiles. A single bar of lactose was printed and observed for 3 hours. Induction profiles taken at various time points are presented (left) alongside the corresponding rate curves. Each data point in the empirical rate curve comes from averaging the intensity across the trough of the corresponding induction profile.

Flight control of insects ideally requires the triggering of flight initiation and cessation as well as the free-flight adjustment of orientation with three degrees of freedom [23]. These flight parameters are controlled by insects via modulation of the wing movements using flight muscles. Insects exhibit two major categories of flight muscular control [24]. Some insects, such as dragonflies and locusts, possess synchronous flight muscles, which oscillate under direct flight control with one-to-one matches between neuronal stimulus episodes and wing muscle contractions. Other species (e.g., fruit flies, beetles) possess asynchronous flight muscles, which oscillate under indirect control. Of these, beetles are among the largest; *Cotinis texana* (ca. 2 cm, 1 g) and *Mecynorrhina torquata* (ca. 6 cm, 8 g) were large enough to carry implanted or dorsally mounted microsystems presented here. In these species, motor neurons to the flight muscles fire at much lower frequencies than the wing oscillation frequencies, and neuronal output

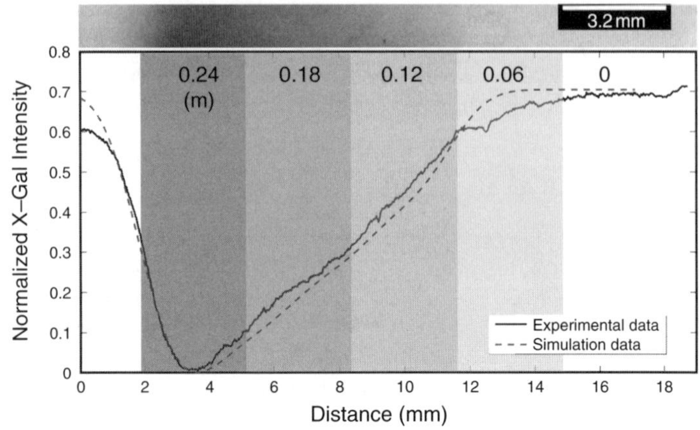

Figure 12.9. Piecewise continuous lactose gradient profile. Here, a 5-bar (3.2 mm/bar), piecewise continuous lactose gradient was printed, where the numbers across the bars represent the concentration of lactose printed in that bar. Again, we see very close agreement between the empirical data and the simulation.

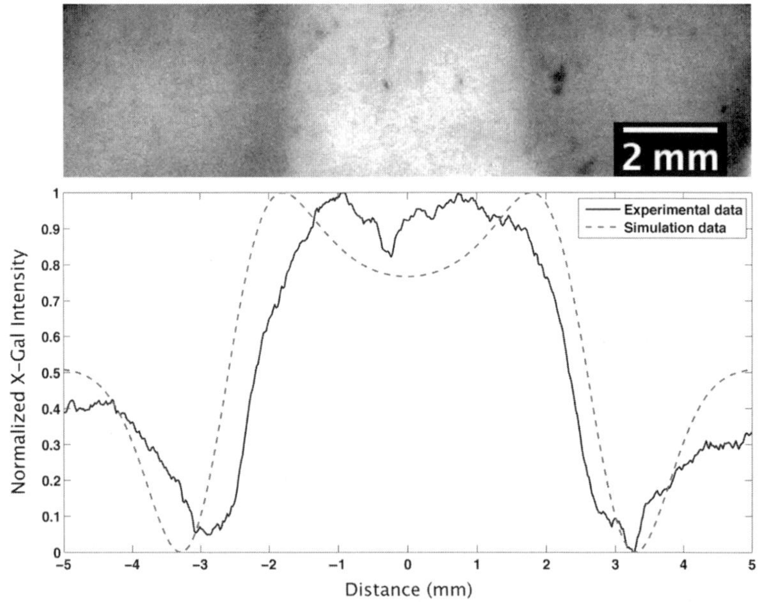

Figure 12.10. Activator–inhibitor printing with lactose and glucose. Lactose was first printed over an entire field of cells. Following this, a glucose bar, measuring 2.2-mm wide, was printed down the center of the field, on top of the lactose. Glucose is an inhibitor, while lactose is an activator. The result, with which the simulation agrees, is a region of repressed *lac* operon activity framed by dark boundaries. The increased induction arises because the lactose under the glucose is not consumed and therefore diffuses laterally, increasing the amount of lactose available for consumption along the boundaries.

serves to turn flight on and off, and to modulate power, but not to directly control each flight muscle contraction [25–27].

The neural control of flight initiation and cessation has not been studied in beetles and is not perfectly understood in any insect. However, in both locusts and fruit flies, there is evidence that visual, auditory, or wind stimulus of receptors can lead to output from the brain that can initiate and modulate flight via giant fiber interneurons [28, 29]. Therefore, we chose to attempt to start, stop and modulate wing oscillations using direct electrical stimulus of the brain. Turns require asymmetric kinematics of the wings, which are accomplished by flight muscles such as the basalar muscles [30]. We attempted control of turns by direct electrical stimulus of the basalar muscle, one of the major indirect flight muscles of these beetles [26, 27, 70].

For our initial experiments, we developed a system capable of tetherless control of beetle but without wireless communication. We preprogrammed flight instructions using a microcontroller (Fig. 12.11A; Texas Instruments, MSP430F2012IPWR, 63 mg, 5.0 mm × 4.5 mm × 1.0 mm) powered by a rechargeable lithium ion coin battery (Panasonic, ML614, 3.0 V, 160 mg, Ø6.8 mm × 1.4 mm, 3.4 mAh), which were mounted on the pronotum. For fully wireless communication and on-demand command of flight instructions to beetle, we developed a miniaturized radio frequency (RF) system that used two CC2431 microcontrollers (6 × 6 mm, 130 mg, 2.4 GHz); one acting as the beetle-mounted RF receiver (Fig. 12.11B) and one as computer-driven RF transmitter base station. The RF receiver was powered by another rechargeable lithium ion battery (Micro Avionics, 3.9 V, 350 mg, 8.5 mAh). Electrical signals generated by either microcontroller drove steel wire electrodes (Ø125 μm) implanted into the brain, optic lobes, and basalar muscles (implant sites 1, 2, and 4 in Fig. 12.11, respectively). A common counterelectrode for the brain and basalar muscle stimuli was implanted into the posterior pronotum (implant site 2 in Fig. 12.11). In the following sections, polarity type (positive or negative) in stimulus will be expressed as potential versus the counterelectrode unless extra explanation is added.

12.2.2.1 Modulation of Wing Oscillations. For *C. texana*, alternating positive and negative potential pulses between an electrode implanted into the brain and a counterelectrode implanted into the posterior pronotum of the adult insect reproducibly generated flight initiation and cessation with success rate of 56% in fully tethered and weakly tethered *Cotinis* beetles as shown in Figure 12.12. To attempt to distinguish whether the positive or negative pulses were more important in flight initiation, we compared three different types of electrical stimuli: trains of alternating negative and positive potential pulses, trains of positive potential pulse, and trains of negative potential pulses as shown in Figure 12.13. Positive potentials, whether alone or alternating with negative pulses, initiated flight but negative potential pulses alone did not. Positive pulses and alternating positive and negative pulses were equally effective in eliciting flight: five of nine and four of nine insects initiated the flight in response to the stimulation, respectively. Data on stimulated flight bouts in individual *C. texana*

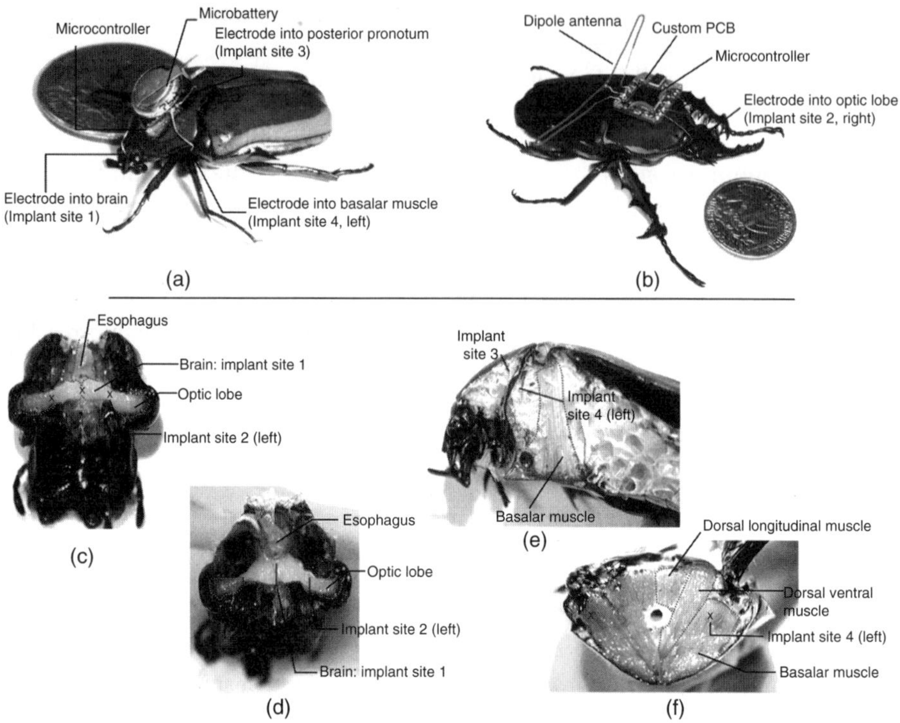

Figure 12.11. (A) Tetherless flight control system (~230 mg) mounted on *Cotinis texana* (green June beetle) via beeswax next to a U.S. $0.25 coin. A microcontroller provided potential pulses to four stimulating wire electrodes (Ø125 μm) implanted into the brain, left and right basalar muscles, and posterior pronotum (counterelectrode). (B) Radio flight control system (~1.3 g total) mounted on *Mecynorrhina torquata* via beeswax next to a U.S. $0.25 coin. The system consisted of a microcontroller, a custom printed circuit board (PCB), a dipole antenna, microbattery and stimulating wire electrodes (Ø125 μm) implanted as in *Cotinis*. Front (C) and tilted(D) views of dissected *Cotinis* beetle head showing the brain stimulator at implant site 1 and optic lobe stimulator at implant site 2. The brain stimulator was implanted along the rostral-caudal midline of the head, at the center between the left and right compound eyes. Implant site 2 was at the interior edge of each compound eye. (E) Sagittal section of thorax showing the counterelectrode at implant site 3 and the basalar muscle stimulator at implant site 4. (F) Cross section of the mesothorax showing the basalar muscle stimulator sites (implant site 4 on left and right sides). The basalar muscle stimulator was implanted from midway between the sternum and notum of the mesothorax to a depth of approximately 1 cm in the rostral-caudal direction on either left or right side of the insect. The blue letters X and bars indicate implant sites and approximate implant length, respectively. *Mecynorrhina torquata* has nearly identical, scaled anatomy to *Cotinis texana*.

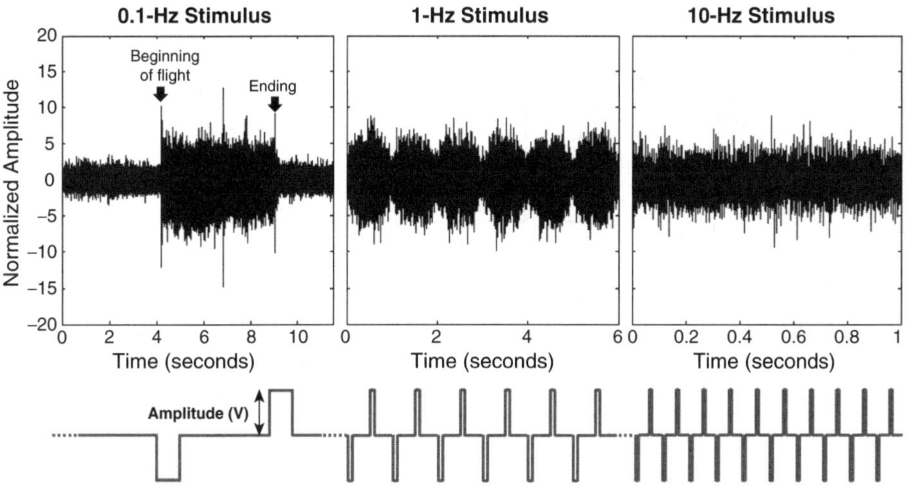

Figure 12.12. Initiation and cessation control of *Cotinis texana* beetle tethered flight. (Top) Audio recordings of tethered beetle. (Bottom) Applied potential to the brain (with counterelectrode in posterior pronotum). The applied potential waveform is identical to that in Fig. 12.13A, but the frequency varied. As the period between pulses decreased, the beetle was incapable of fully starting or stopping wing oscillation, and audio amplitudes were modulated by the stimulus frequency. Audio amplitudes were normalized to the average audio amplitude during normal, sustained flight.

are summarized in Table 12.1. In the case of alternating polarity pulses, the majority of flight initiations happened either during or immediately after the negative potential pulse (following a positive pulse) was applied to the beetle brain (see the columns named DN, AN, DP, and AP in Table 12.1). For each insect, there was a median amplitude threshold for flight initiation being 3.2V; below this voltage, legs stretched or contracted but flight did not start. Legs folded inward during negative pulses and extended (flight position) during positive pulses, which suggests that positive pulses activate at least some of the complex motor patterns of flight initiation, while negative pulses activate an opposite set of muscle activations. In the weakly tethered and fully untethered conditions, some *C. texana* collapsed briefly when stimulated, which indicates that the stimulus caused not only muscle movement coordinated with wing oscillation but also uncoordinated muscle movement associated with generalized neural depolarization.

Given the initial data from *Cotinis*, we chose to extend this study to control of beetles in free flight; this required a slightly larger beetle to carry our radio-equipped system (RF receiver + battery = 1331 mg). We chose the *M. torquata* beetle, which has enough payload capacity of ca. 2.4 g. Prior to the free flight experiment, we tested *M. torquata* in tethered situation to determine optimal stimulus conditions. For *M. torquata*, the same waveforms as that for *C. texana*, but at higher frequency, were applied between two electrodes implanted into the

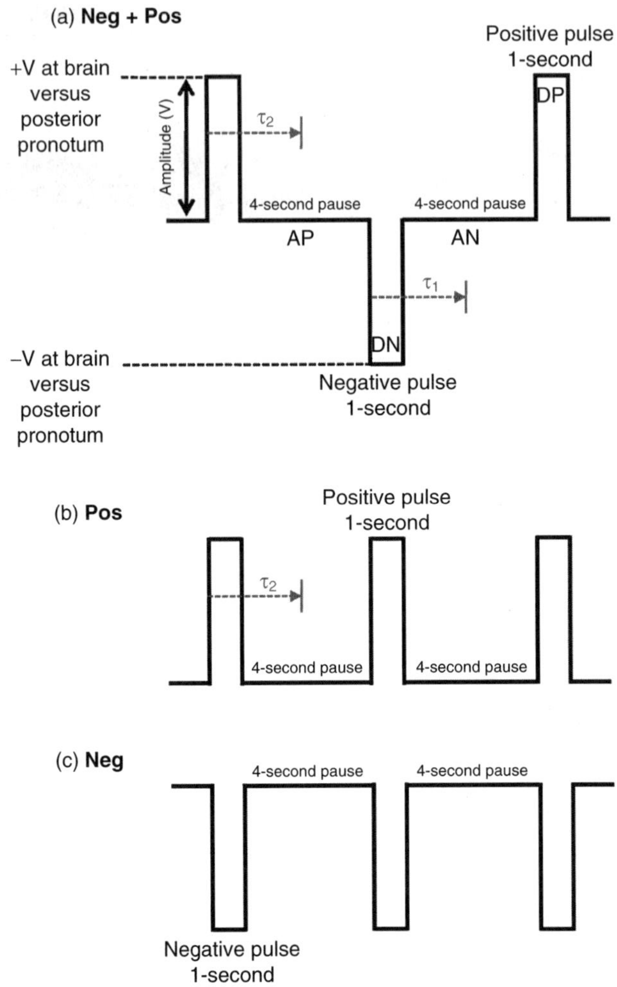

Figure 12.13. Three types of pulse trains (stimulus protocols) were investigated to elicit flight. (A) Neg + Pos: alternating 1-second duration positive and negative pulses. (B) Pos: 1-second duration positive pulses (C) Neg: 1-second duration negative pulses. Pulse amplitude was swept from 0.1 to 5.0 V in 100-mV increments when testing for the amplitude threshold. Delay, τ_1 or τ_2, is response time from beginning of positive or negative potential pulse to beginning of wing oscillation, respectively. See Sato et al. [69] for data on stimulated flight bouts in all tested *Cotinis texana*.

left and right optic lobes. Implantation into the optic lobe yielded a much higher success rate and, unexpectedly, did not affect the beetle's ability to steer in free flight (see below; Fig. 12.14). All 10 insects tested initiated flight in response to stimulation, with the median number of stimulations required to initiate flight being 19 (range 1–59; one stimulation was 10 msec as shown in Fig. 12.14B), and

TABLE 12.1. Data on Stimulated Flight Bouts in Individual *Cotinis texana*

	Insect	Weight (g)	Amplitude Threshold (V)	Number of Flight Bouts	Total Flight Duration (seconds)	Duration of a Single Flight (seconds)	τ_1 (seconds)	τ_2 (seconds)	DN	AN	DP	AP
Neg + Pos	1	1.27	3.2	13	7.9	0.5	0.5	–	11	2	0	0
						0.2–1.4	0.2–3.7					
	2	1.12	2.1	3	273.7	1.7	1.2	0.8	1	0	2	0
						1.3–270.6		0.8–0.9				
	3	0.98	1.6	1	6.2		1.0	–	0	1	0	0
	4	0.83	3.6	1	1793.1		2.0	–	0	1	0	0
	5	0.90	4.7	1	177.7		0.0	–	1	0	0	0
Pos	6	0.90	2.9	92	774.0	2.5	–	1.4	–	–	0	92
						0.5–233.3		0.5–1.8				
	7	1.12	2.1	22	46.1	2.0	–	1.0	–	–	8	14
						1.0–3.7		0.5–4.1				
	8	0.73	3.7	6	5.4	0.6	–	1.6	–	–	0	6
						0.5–2.1		1.4–4.8				
	9	0.88	4.0	5	348.1	58.8	–	1.6	–	–	0	5
						22.0–148.6		1.5–2.0				

Alternating positive and negative potential pulse trains (0.1 Hz, called "Neg + Pos" in the first column) and positive pulse trains (0.2 Hz, called "Pos") were applied to the insects as described in the text and in Figure 12.13. For each of the two types of stimuli, nine beetles were implanted and tested as described. Four of nine and five of nine beetles did not fly in Neg + Pos and Pos protocols, respectively. For single flight duration, median (upper numbers) and range (lower numbers) are shown if applicable. Medians and ranges of response times, τ_1 and τ_2, are also shown in the same manner. τ_1 and τ_2 are defined in Figure 12.13. DN: number of flight bouts that began during negative pulse, AN: number of flight bouts that began after negative pulse, DP: number of flight bouts that began during positive pulse, AP: number of flight bouts that began after positive pulse (see Fig. 12.13).

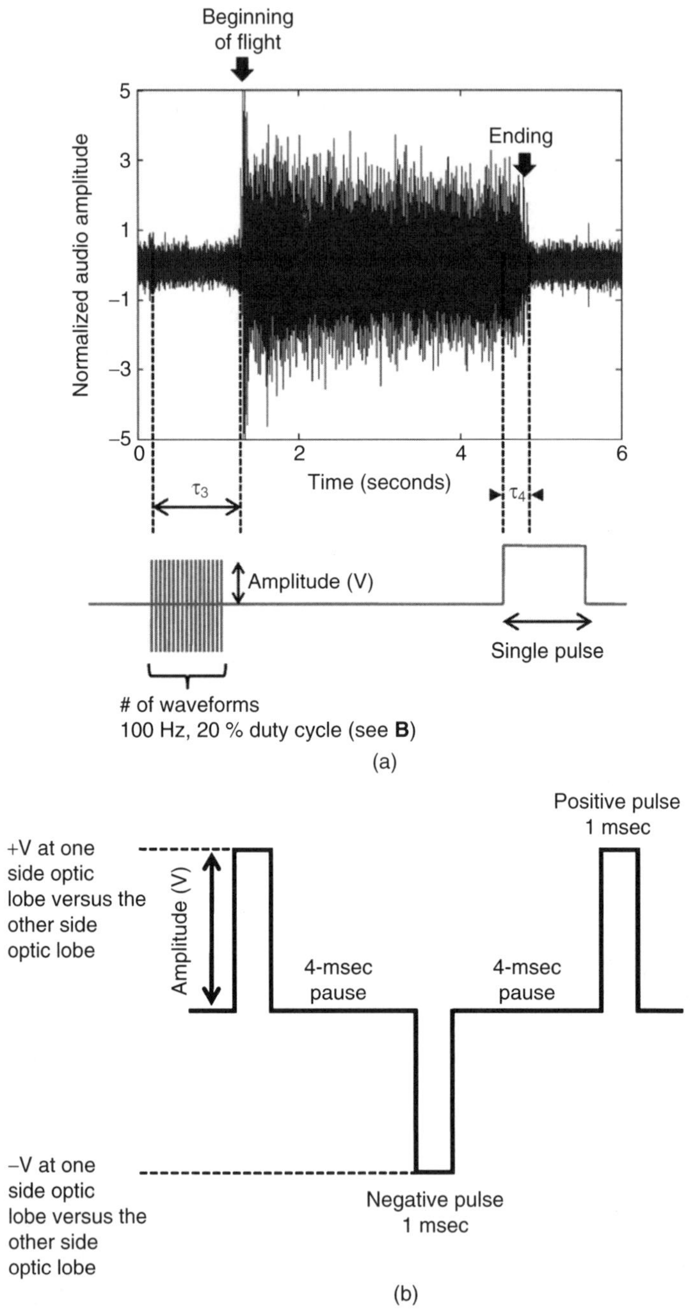
Beginning
of flight
Ending
Normalized audio amplitude
5
3
1
–1
–3
–5
0
2
4
6
Time (seconds)
τ3
τ4
Amplitude (V)
Single pulse
# of waveforms
100 Hz, 20 % duty cycle (see B)
(a)
Positive pulse
1 msec
+V at one
side optic
lobe versus the
other side
optic lobe
Amplitude (V)
4-msec
pause
4-msec
pause
–V at one
side optic
lobe versus the
other side
optic lobe
Negative pulse
1 msec
(b)

Figure 12.14. Initiation and cessation control of *Mecynorrhina torquata* beetle tethered flight. (A) Alternating positive and negative potential pulses (100 Hz, see [B] for the details of the waveform) applied between the left and right optic lobes initiated wing oscillations, while a single pulse ceased wing oscillations: (top) audio recording of tethered beetle; (bottom) applied potential to the one side optic lobe regarding the other side optic lobe. Delay, τ_3, is response time from beginning of the multiple pulse trains to beginning of the wing oscillation. Delay, τ_4, is response time from beginning of the single pulse to ending of wing oscillation. τ_3 and τ_4 for all the tested beetles are summarized in Sato et al. [69]. The sharp rise of audio amplitude at the beginning of oscillation is attributed to friction between the elytra and wings when the wings came out from the underneath of the elytra. The whole audio amplitudes were normalized by mean absolute value calculated for the middle period of the flight time (2.5–3.7 seconds). (B) Pulse trains applied between left and right optic lobes. Number of waveforms was swept from 1 to 100 in 1-waveform increments when testing for the number of waveforms required to trigger flight initiation [69].

the median response time from the first stimulation to flight initiation being 0.5 second (range 0.2–1.4 seconds; τ_3 in Fig. 12.4A). Median flight duration in response to stimulation was 46 seconds (range 33–2292 seconds). Stimulation voltage between 2 and 4 V did not affect the number of stimuli required to initiate flight, response time from stimulation to flight, or flight duration in *M. torquata* (Mann–Whitney *U*-tests; $p = 0.13, 0.46, 0.35$, respectively). We then demonstrated takeoff of beetles into the air (free-flight experiment) [69].

Once flight was initiated by our stimulation, the flight tended to persist without additional stimulation whether the beetle was either in the tethered or in free-flight condition. During normal flight, the beetle nervous system produces a pulse train with ~50-msec period to the basalar muscles [26, 27]. Artificially induced flight lasted far longer than 50 ms: median flight durations were 2.5 seconds (range 0.2–1793.1 seconds) for *C. texana*, and 45.5 seconds (range 0.7–2292.1 seconds) for *M. torquata*. Between given insects, flight bout duration was correlated with neither beetle mass nor stimulus amplitude. Furthermore, the beetle adopted a normal flight posture and continued flying in the air after the stimulus was turned off, indicating that the tonic neural signals required for flight maintenance continued after stimulus.

A single pulse applied between optic lobes stopped flight for *M. torquata* as shown in Figure 12.14. Ten insects were tested in tethered situation and each insect was repeated 10 times, that is 100 tests in total. All 10 insects tested were forced to stop flying by amplitudes of 6.0 V or less. The majority (77%) stopped with a 2–3-V amplitude. The median amplitude was 3.0 V (range 2–6 V). The majority (87%) showed a quite short response time, $\tau_4 < 100$ msec. This cessation can be seen in free flight [69].

12.2.2.2 Elevation Control. During flight, body pitch and wing oscillation frequency could be manipulated by modulating the wing oscillations with the

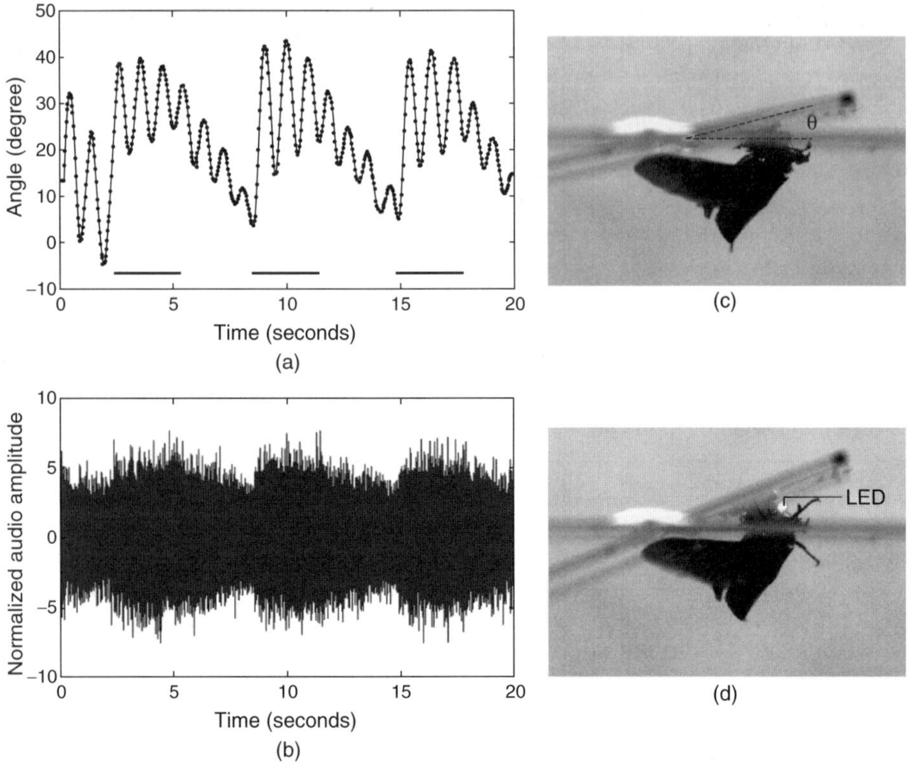

Figure 12.15. Elevation control of *Cotinis texana* beetle tethered on a custom pitching gimbal. Brain stimulus altered pitch of flying beetle. (A) Angle of attack of gimbal-mounted beetle during alternating periods of unstimulated and stimulated flights. Horizontal bars indicate duration of stimulus (3 seconds each); a 10-Hz, 3.0-V pulse train whose waveform is identical to that in Fig. 12.13A was applied during the indicated periods. (B) Audio recording corresponding to (A). The whole audio amplitudes were normalized by mean absolute value during unstimulated periods. Photographs of gimbal-mounted beetle during unstimulated (C) and stimulated (D) flights. A light-emitting diode (LED) mounted to the microcontroller acted as an indicator and blinked during stimulus.

neural stimulator. For *C. texana*, we observed that progressively shortening the time between positive and negative pulses led to a "throttling" of flight where the beetle's normal 76-Hz wing oscillation was strongly modulated by the 0.1–10-Hz applied stimulus as shown in Figure 12.12. A repeating program of 3-second, 10-Hz, 3.0-V pulse trains, followed by a 3-second pause (no stimulus) resulted in alternating periods of higher and lower pitch flight (see Fig. 12.15). The audio amplitude (likely reflecting stroke amplitudes) was enhanced by ca. 10% when the beetle was stimulated as shown in Figure 12.15B. High-speed video track of stimulated flight showed that wing oscillations during stimulated flight had 7% greater frequency than unstimulated normal fight [69]. For *M. torquata*, brain

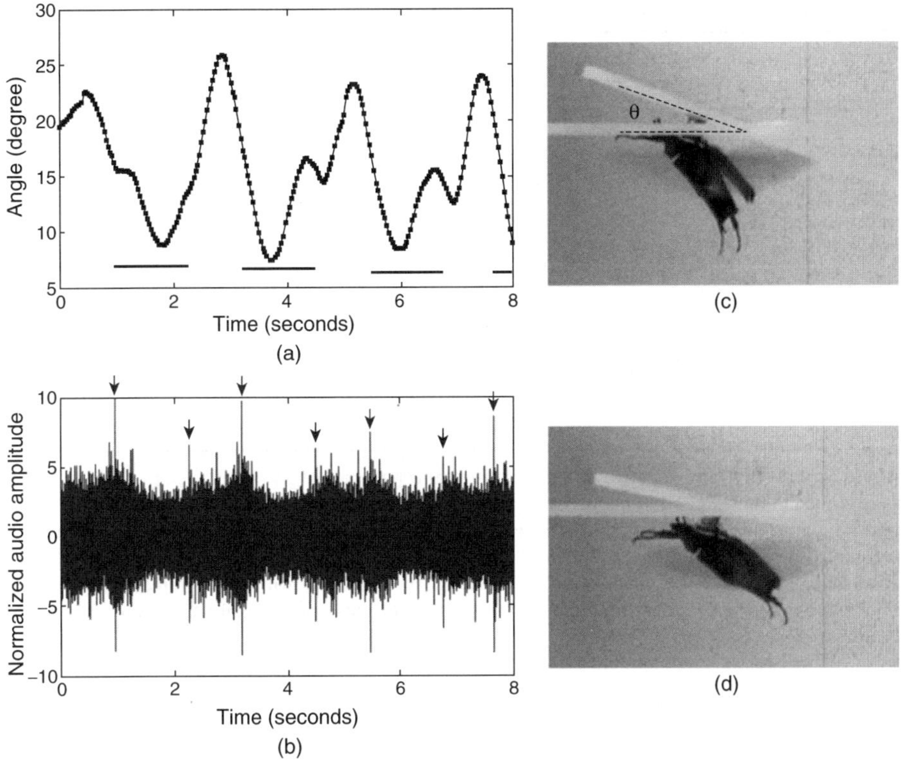

Figure 12.16. Elevation control of *Mecynorrhina torquata* beetle tethered on a custom pitching gimbal. Brain stimulus altered pitch of flying beetle (100-Hz, 2.0-V amplitude; see Fig. 12.13A for waveform). (A) Angle of attack of gimbal-mounted beetle during alternating periods of unstimulated and stimulated flights. Horizontal bars indicate duration of stimulus. (B) Audio recording corresponding to (A). The 1st, 3rd, 5th, and 7th arrows indicate the beginning of the stimulus to the brain, while the 2nd, 4th, and 6th arrows indicate the end. The sharp peaks at the arrows were attributed to the signal tone coming from the function generator to output the stimulus signal to the beetle's brain. The whole audio amplitudes were normalized by mean absolute value during unstimulated periods. Photographs of gimbal-mounted beetle during unstimulated (C) and stimulated (D) flights.

stimulus at 100 Hz, in the same manner as in *C. texana*, led to depression of flight. Set on a custom pitching gimbal, *M. torquata* could be repeatedly made to lower angle to horizon when stimulated as shown in Figure 12.16. Ten of the 11 tested beetles showed this tendency (see Table 12.2 for angle changes in individual insects). In some cases, the brain stimulus resulted in flight. The brain stimulus obviously weakened the *M. torquata* beetle's wing stroke amplitude (see wing blurs shown in Fig. 12.6C,D). In free flight, this corresponded to a controllable drop in altitude when stimulated as shown in Figure 12.17. Median drop in altitude caused by a 1-second stimulus was 60 cm.

TABLE 12.2. Gimbal Pitch Angle Change (*Mecynorrhina torquata*)

Insect	Number of Tests	Δθ (degree)	Number of Tests Resulting in Flight Cessation
1	29	−4.42	7
2	6	−0.69	2
3	21	−3.03	5
4	17	3.61	0
5	15	−0.52	1
6	11	−4.15	7
7	21	−1.51	5
8	8	−13.04	5
9	11	−0.33	7
10	28	−0.84	1
11	5	−0.39	1

Alternating positive and negative potential pulse trains (10 Hz, 2.0 V) were applied to the insect brain as described in the text and in Figure 12.16. Eleven beetles were implanted, and tested using a custom pitching gimbal. Δθ is the mean difference of gimbal pitch angle to horizon between unstimulated (θ_n) and stimulated (θ_s) flights: $\Delta\theta = \theta_n - \theta_s$. Negative value of Δθ indicates that the beetle climbed down when stimulated, and vice versa. Ten of 11 insects climbed down when stimulated: Only insect #4 climbed up. In some cases, the brain stimulus resulted in cessation of flight as shown in the fourth column.

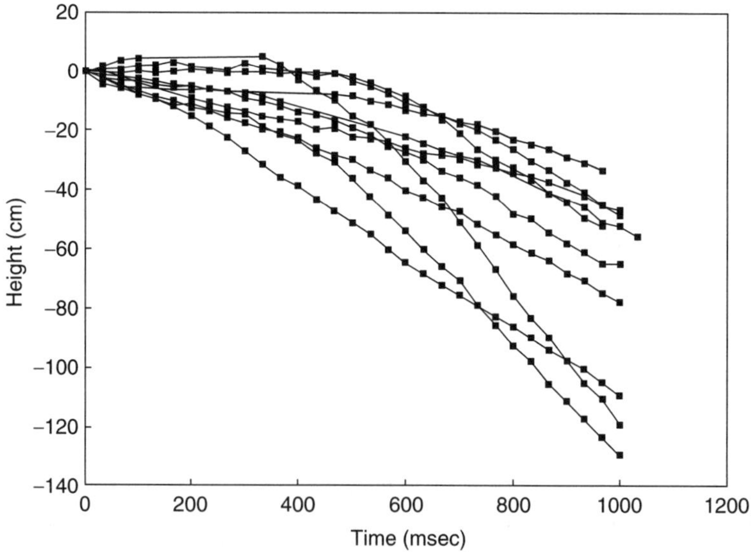

Figure 12.17. Elevation control of free-flying *Mecynorrhina torquata* beetle: temporal height change of a flying beetle (10 flight paths). Alternating positive and negative potential pulse trains at 100-Hz and 2.0-V amplitude to the brain caused the beetle to fly downward. The applied waveform was identical to that in Fig. 12.13A, but the frequency was different (100 Hz). The median height change was 60 cm (the range was from 33 to 129 cm).

12.2.2.3 Control of Turning. Turns were elicited by stimulus of the left and right basalar muscles with positive potential pulse trains. In *C. texana*, the basalar muscles normally contract and extend at 76 Hz when they are stimulated by ~8-Hz neural impulses from the beetle nervous system [26, 27]. It has been reported that the flight muscles in *Cotinis* produce maximum power when they are stimulated directly by electrical pulses at 100 Hz [27]. During flight, a turn was triggered by applying 2.0-V, 100-Hz positive potential pulse trains to the basalar muscle opposite to the intended turn direction as shown in Figure 12.18. A right turn, for example, was triggered by stimulating the left basalar muscle. In free-flying *M. torquata*, turning was elicited when either of the left or right basalar muscles was stimulated in the same manner as *C. texana* (Fig. 12.19). The success rates for left and right turn were 78% ($n = 42$) and 66% ($n = 68$), respectively. One second of left and right stimulation of free-flying beetles resulted in 1.7° and –9.0° median rolls to the ground and 20.0° and 32.4° median rotations parallel to the ground, respectively.

12.3 CONCLUSIONS

Advances in microtechnologies and chemical microinterfaces, the ever-decreasing feature size and power consumption of computational elements, and the advent of synthetic biology as an organized discipline are provoking fundamental questions: To what extent can multicellular development be used as a technology to make machines? [3] Can existing organisms be co-opted to new functions? What

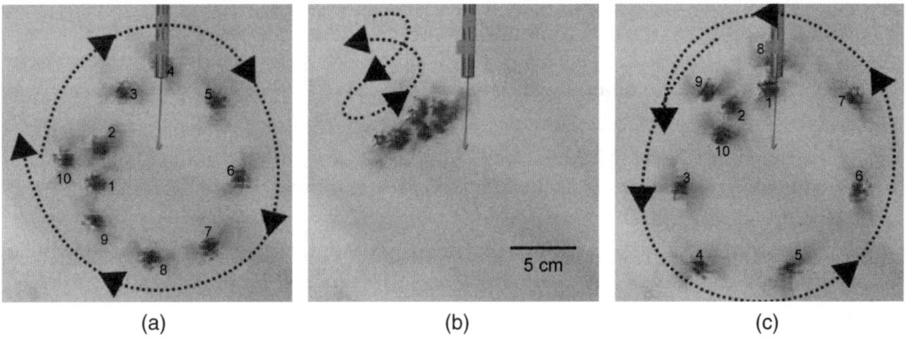

Figure 12.18. Turn control of *Cotinis texana* flight. A 100-Hz and 2.0-V positive potential (vs. counterelectrode at posterior pronotum) pulse train to the basalar muscle on one side of the beetle triggered a turn. Beetles mounted on a string (10 cm) were programmed with continuous sequences of left-pause-right-pause instructions; each instruction lasted 2 seconds. (A) Left basalar muscle stimulus generating a right turn, followed by (B) a pause during which the beetle zigged and zagged randomly, followed by (C) right basalar muscle stimulus generating a left turn. Each successive photograph consists of 10 frames; frames were taken every 0.2 second. Numbers in (A) and (C) signify the frame number.

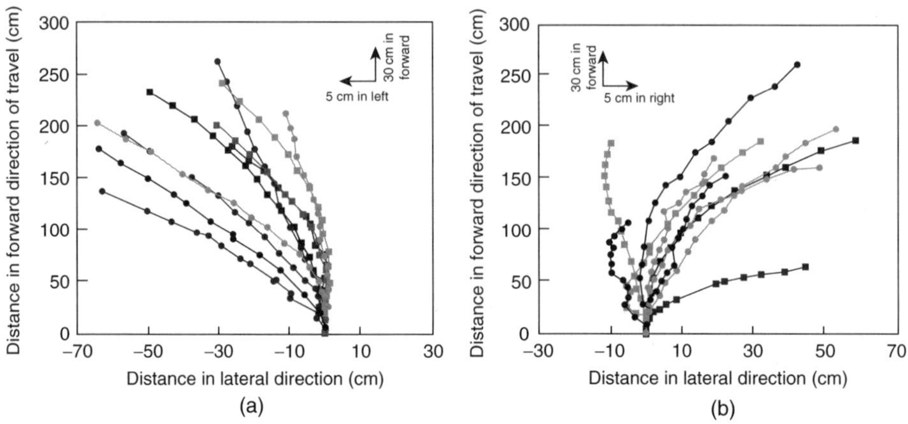

Figure 12.19. Turn control of free-flying *Mecynorrhina torquata* beetle. Pulse trains at 100-Hz and 1.3-V positive potential to the left or right basalar muscles triggered turns. Ten flight paths elicited by right (A) or left (B) basalar flight muscle stimulus for 0.5 second are shown. Each flight path is obtained after the three-dimensionally digitized flight path is projected on the XY plane (see text for the detailed method). Different shades of gray and shape plots show different beetles' flight paths.

is the ultimate plasticity of a particular developmental program? What are the ethical concerns in undertaking such an endeavor?

REFERENCES

[1] E. Andrianantoandro, S. Basu, D. K. Karig, and R. Weiss, "Synthetic biology: new engineering rules for an emerging discipline," *Mol. Syst. Biol.*, 2, epub, 2006.

[2] S. A. Benner and A. M. Sismour, *Nat. Rev. Genet.*, 6(7), pp. 533–543, 2005.

[3] R. F. Ismagilov and M. M. Maharbiz, "Can we build synthetic, multicellular systems by controlling developmental signaling in space and time?" *Curr. Opin. Chem. Biol.*, 11(6), pp. 604–611, 2007.

[4] C. Anderson, E. J. Clarkec, A. P. Arkin, and C. A. Voigt, "Environmentally controlled invasion of cancer cells by engineered bacteria," *J. Mol. Biol.*, 355(4), pp. 619–627, 2006.

[5] J. L. Fortman, S. Chhabra, A. Mukhopadhyay, H. Chou, T. S. Lee, E. Steen, and J. D. Keasling, "Biofuel alternatives to ethanol: pumping the microbial well," *Trends Biotechnol.*, 26(7), pp. 375–381, 2008.

[6] J. H. Park, T. Bansal, M. Pinelis, and M. M. Maharbiz, "Electrolytic patterning of dissolved oxygen microgradients during cell culture," *Lab Chip*, 6, pp. 611–622, 2006.

[7] T. Bansal and M. M. Maharbiz, "'Wet' AC actuated microfluidic micropore array for patterning diffusible gradients during cell culture," *Tenth International Conference on Miniaturized Systems for Chemistry and Life Sciences (MicroTAS)*, Tokyo, Japan, November 2006.

[8] T. K. Kim, M. Pinelis, and M. M. Maharbiz, "Generating steep, sheer-free gradients of small molecules for cell culture," *IEEE Biomed. Microdevices*, 11(1), pp. 65–73, 2009.

[9] M. Pinelis, T. Kim, and M. M. Maharbiz, "A high-yield method for generating mass-transfer gradients in elastomer microfluidics using impermeable capillaries," *IEEE Biomed. Microdevices*, 10(6), pp. 807-811, 2008.

[10] A. Michelsen, B. B. Andersen, W. H. Kirchner, and M. Lindauer, "Honey bees can be recruited by a mechanical model of a dancing bee," *Naturwiss.*, 76, pp. 277–280, 1989.

[11] S. Kishigami and Y. Mishina, "BMP signaling and early embryonic patterning," *Cytokine Growth Factor Rev.*, 16(3), pp. 265–278, 2005.

[12] L. Wolpert, *Principles of Development*, 2nd ed. Oxford, New York: Oxford University Press, 2002.

[13] W. C. Wu, L. Schenato, R. J. Wood, and R. S. Fearing, "Biomimetic sensor suite for flight control of a micromechanical flying insect: Design and experimental results," *Proc. IEEE Int. Conf. Rob. Autom.*, 1, pp. 1146–1151, 2003.

[14] R. J. Wood, "The first takeoff of a biologically inspired at-scale robotic insect," *IEEE Trans. Rob.*, 24, pp. 341–347, 2008.

[15] L. Schenato, W. C. Wu, and S. Sastry, "Attitude control for a micromechanical flying insect via sensor output feedback," *IEEE Trans. Rob. Autom.*, 20, pp. 93–106, 2004.

[16] S. A. Ansari, R. Zbikowski, and K. Knowles, "Aerodynamic modeling of insect-like flapping flight for micro air vehicles," *Prog. Aeosp. Sci.*, 42, pp. 129–172, 2006.

[17] D. J. Pines and F. Bohorquez, "Challenges facing future micro-air-vehicle development," *J. Aircr.*, 43, pp. 290–305, 2006.

[18] S. P. Sane, "The aerodynamics of insect flight," *J. Exp. Biol.*, 206, pp. 4191–4208, 2003.

[19] W. Shyy, M. Berg, and D. Ljungqvist, "Flapping and flexible wings for biological and micro air vehicles," *Prog. Aeosp. Sci.*, 35, pp. 455–505, 1999.

[20] P. Kloppenburg, S. M. Camazine, X. J. Sun, P. Randolph, and J. G. Hildebrand, "Organization of the antennal motor system in the sphinx moth *Manduca sexta*," *Cell Tissue Res.*, 287, pp. 425–433, 1997.

[21] S. P. Sane, A. Dieudonne, M. A. Willis, and T. L. Daniel, "Antennal mechanosensors mediate flight control in moths," *Science*, 315, pp. 863–866, 2007.

[22] N. Ando, I. Shimoyama, and R. Kanzaki, "A dual-channel FM transmitter for acquisition of flight muscle activities from the freely flying hawkmoth, *Agrius convolvuli*," *J. Neurosci. Methods*, 115, pp. 181–187, 2002.

[23] G. K. Taylor, "Mechanics and aerodynamics of insect flight control," *Biol. Rev.*, 76, pp. 449–471, 2001.

[24] R. Dudley, *The Biomechanics of Insect Flight: Form, Function, Evolution*. Princeton, NJ: Princeton University Press, 2000.

[25] M. H. Dickinson, "Insect flight," *Curr. Biol.*, 16, pp. R309–R314, 2006.

[26] R. K. Josephson, J. G. Malamud, and D. R. Stokes, "Asynchronous muscle: A primer," *J. Exp. Biol.*, 203, pp. 2713–2722, 2000.

[27] R. K. Josephson, J. G. Malamud, and D. R. Stokes, "Power output by an asynchronous flight muscle from a beetle," *J. Exp. Biol.*, 203, pp. 2667–2689, 2000.

[28] M. Burrows, *The Neurobiology of an Insect Brain*. New York: Oxford University Press, 1996.

[29] S. A. Budick, M. B. Reiser, and M. H. Dickinson, "The role of visual and mechanosensory cues in structuring forward flight in *Drosophila melanogaster*," *J. Exp. Biol.*, 210, pp. 4092–4103, 2007.

[30] M. S. Tu and M. H. Dickinson, "The control of wing kinematics by two steering muscles of the blowfly (*Calliphora vicina*)," *J. Comp. Physiol. A*, 178, pp. 813–830, 1996.

[31] D. J. Cohen, R. C. Morfino, and M. M. Maharbiz, "A modified consumer inkjet for spatiotemporal control of gene expression," *PLoS ONE*, 4(9), pp. e7086, 2009.

[32] D. Zicha, G. A. Dunne, and A. F. Brown, "A new direct-viewing chemotaxis chamber," *J. Cell Biol.*, 99, pp. 769–775, 1991.

[33] T. W. Engelmann, "Neue Methode zur Untersuchung der Sauerstoffausscheidung pfanzlicher and tierischer Organismen," *Pfluegers Arch. Gesamte Physiol. Menschen Tiere*, 25, pp. 285–292, 1881.

[34] S. Boyden, "The chemotactic effect of mixtures of antibody and antigen on polymorphonuclear leucocytes," *J. Exp. Med.*, 33, pp. 453–466, 1962.

[35] J. T. Bonner, *First Signals: The Evolution of Multicellular Development*. Princeton, NJ: Princeton University Press, 2000,

[36] L. N. Jeon, H. Baskaran, S. Dertinger, G. Whitesides, L. Van de Water, and M. Toner, "Neutrophil chemotaxis in linear and complex gradients of interleukin-8 formed in a microfabricated device," *Nat. Biotechnol.*, 20, pp. 826–830, 2002.

[37] T. M. Keenan and A. Folch, "Biomolecular gradients in cell culture systems," *Lab Chip*, 8, pp. 34–57, 2008.

[38] D. Chena, W. Dua, Y. Liua, W. Liua, A. Kuznetsovb, F. E. Mendezb, L. H. Philipsonb, and R. F. Ismagilov, "The chemistrode: A droplet-based microfluidic device for stimulation and recording with high temporal, spatial, and chemical resolution," *PNAS*, 105, pp. 16843–16848, 2008.

[39] T. Bansal, J. Lenhart, T. Kim, C. Duan, and M. M. Maharbiz, "Patterned delivery and expression of gene constructs into developing zebrafish embryos using microfabricated interfaces," *IEEE Biomed. Microdevices*, 11, pp. 633–641, 2009.

[40] G. Mayer and A. Heckel, "Biologically active molecules with a 'light switch'," *Angew. Chem. Int. Ed.*, 45, pp. 4900–4921, 2006.

[41] I. Shestopalov and J. Chen, "Chemical technologies for probing embryonic development," *Chem. Soc. Rev.*, 37, pp. 1294–1307, 2008.

[42] J. J. Chambers, M. R. Banghart, D. Trauner, and R. H. Kramer, "Light-induced depolarization of neurons using a modified shaker K+ channel and a molecular photoswitch," *J. Neurophysiol.*, 96, pp. 2792–2796, 2006.

[43] G. C. Ellis-Davies, "Caged compounds: Photorelease technology for control of cellular chemistry and physiology," *Nat. Methods*, 4, pp. 619–628, 2007.

[44] S. Basu, Y. Gerchman, C. Collins, F. Arnold, and R. Weiss, "A synthetic multicellular system for programmed pattern formation," *Nature*, 434, pp. 1130–1134, 2005.

[45] T. Brown, C. Chang, B. Heinze, P. Hollinger, J. Kittleson, K. MacDow, D. Reavis, J. Curry, and M. Riley, "Development of an inducible three colour bacterial water colour system," *IET Synth. Biol.*, 1, pp. 21–24, 2007.

[46] G. Gurtner, S. Werner, Y. Barrandon, and M. Longaker, "Wound repair and regeneration," *Nature*, 453, pp. 314–321, 2008.

[47] M. Lutolf and J. Hubbell, "Synthetic biomaterials as instructive extracellular microenvironments for morphogenesis in tissue engineering," *Nat. Biotechnol.*, 23, pp. 47–55, 2005.

[48] W. Gu, X. Zhu, N. Futai, B. S. Cho, and S. Takayama, "Computerized microfluidic cell culture using elastomeric channels and Braille displays," *PNAS*, 101, pp. 15861–15866, 2004.

[49] M. C. Peterman, J. Noolandi, M. S. Blumenkranz, and H. A. Fishman, "Localized chemical release from an artificial synapse chip," *PNAS*, 101, pp. 9951–9954, 2004.

[50] J. Merrin, S. Leibler, and J. Chuang, "Printing multistrain bacterial patterns with a piezoelectric inkjet printer," *PLoS ONE*, 2, e663, 2007.

[51] T. Boland, T. Xu, B. Damon, B. Manley, P. Kesari, S. Jalota, and S. Bhaduri, "Drop-on-demand printing of cells and materials for designer tissue constructs," *Mater. Sci. Eng.*, 27, pp. 372–376, 2007.

[52] R. E. Saunders, J. E. Gough, and B. Derby, "Delivery of human fibroblast cells by piezoelectric drop-on-demand inkjet printing," *Biomaterials*, 29, pp. 193–203, 2008.

[53] E. A. Roth, T. Xu, M. Das, C. Gregory, J. J. Hickman, and T. Boland, "Inkjet printing for high-throughput cell patterning," *Biomaterials*, 25, pp. 3707–3715, 2004.

[54] T. Boland, T. Xu, B. Damon, and X. Cui, "Application of inkjet printing to tissue engineering," *J. Biotechnol.*, 1, pp. 910–917, 2006.

[55] W. J. Rosoff, R. McAllister, and M. A. Esrick, "Generating controlled molecular gradients in 3D gels," *Biotechnol. Bioeng.*, 1, pp. 754–759, 2005.

[56] S. Ilkhanizadeh, A. I. Teixeira, and O. Hermanson, "Inkjet printing of macromolecules on hydrogels to steer neural stem cell differentiation," *Biomaterials*, 28, pp. 3936–3943, 2007.

[57] P. G. Campbell, E. D. Miller, G. W. Fisher, L. M. Walker, and L. E. Weiss, "Engineered spatial patterns of FGF-2 immobilized on fibrin direct cell organization," *Biomaterials*, 26, pp. 6762–6770, 2005.

[58] W. Saadi, S. W. Rhee, F. Lin, and B. Vahidi, "Generation of stable concentration gradients in 2D and 3D environments using a microfluidic ladder chamber," *Biomed. Microdevices*, 1, pp. 627–635, 2007.

[59] C. Lausted, T. Dahl, C. Warren, K. King, K. Smith, M. Johnson, R. Saleem, J. Aitchison, L. Hood, and S. Lasky, "POSaM: A fast, flexible, open-source, inkjet oligonucleotide synthesizer and microarrayer," *Genome Biol.*, 5, pp. R58–R58, 2004.

[60] M. Madou, M. Madou, J. Zoval, J. Zoval, G. Jia, G. Jia, H. Kido, J. Kim, and N. Kim, "Lab on a CD," *Annu. Rev. Biomed. Eng.*, 8, pp. 601–628, 2006.

[61] J. Ducree, S. Haeberle, S. Lutz, S. Pausch, F. von Stetten, and R. Zengerle, "The centrifugal microfluidic Bio-Disk platform," *J. Micromech. Microeng.*, 17, pp. 103–115, 2007.

[62] N. Yildirim and M. C. Mackey, "Feedback regulation in the lactose operon: a mathematical modeling study and comparison with experimental data," *Biophys. J.*, 84, pp. 2841–2851, 2003.

[63] E. M. Ozbudak, M. Thattai, H. N. Lim, B. I. Shraiman, and A. van Oudenaarden, "Multistability in the lactose utilization network of *Escherichia coli*," *Nature*, 427, pp. 737–740, 2004.

[64] A., Ahmadzadeh, A. Halasz, V. Kumar, S. Prajna, and A. Jadbabaie, "Analysis of the Lactose metabolism in E. coli using sum-of-squares decomposition," *Proceedings of the 44th IEEE Conference on Decision and Control*, pp. 879–884, 2005.

[65] J. D. Murray, *Mathematical Biology: II: Spatial Models and Biomedical Applications*. Berlin-Heidelberg: Springer-Verlag, 1993.

[66] P. Wong and S. Gladney, "Mathematical model of the lac operon: Inducer exclusion, catabolite repression, and diauxic growth on glucose and lactose," *Biotechnol. Prog.*, 13, pp. 132–143, 2008.

[67] A. L. Lehninger, D. L. Nelson, and M. M. Cox, *Principles of Biochemistry*, 4th ed. New York: W. H. Freeman, 2004,

[68] K. Amonlirdviman, N. A. Khare, D. R. P. Tree, W-S. Chen, J. D. Axelrod, and C. J. Tomlin, "Mathematical modeling of planar cell polarity to understand domineering nonautonomy," *Science*, 307, pp. 423–426, 2005.

[69] H. Sato, C. W. Berry, Y. Peeri, E. Baghoomian, B. E. Casey, G. Lavella, J. M. Vanden-Brooks, J. F. Harrison, and M. M. Maharbiz, "Remote radio control of insect flight," *Front. Integr. Neurosci.*, 3, pp. 24, 2009. doi:10.3389/neuro.07.024.2009.

[70] F. W. Darwin and J. W. S. Pringle, "The physiology of insect fibrillar muscle. I. Anatomy and innervation of the basalar muscle of lamellicorn beetles," *Proc. R. Soc. London Ser. B, Biol. Sci.*, 151, pp. 194–203, 1959.

13

TECHNOLOGIES FOR ARRAYED SINGLE-CELL BIOLOGY

Sarah C. McQuaide, James R. Etzkorn, and Babak A. Parviz

Conventional biology primarily deals with large ensembles of cells. In a typical experiment, thousands to millions of cells may be placed in a petri dish, challenged with a drug molecule, and then studied to determine the efficacy of the molecule. Such experiments average over a large number of cells and in many cases miss critically important data emerging from single cells or small subsets of cells. Collection of data at the single-cell level is imperative for developing a comprehensive understanding of biology and diseases that depend on small collections of cells in their initial stages such as cancer. In this chapter, we briefly introduce the concept of single-cell biology and discuss the type of tools needed to conduct the related experiments. In addition to the ability to manipulate single cells, these tools must have very high detection precision and be capable of conducting parallel experiments on a large number of single cells in order to generate biologically relevant and meaningful data sets. The recent advances in nanotechnology, microtechnology, and automation have had a transformative impact on this field. As a representative set of tools enabled by these technologies, we will discuss a nanoelectronic sensor for detection of biomolecules and a photonic method for measuring the oxygen consumption rate (OCR) of a single cell in the following sections.

13.1 THE IMPORTANCE OF STUDYING SINGLE CELLS

One of the primary goals of modern biology is to understand the mechanisms behind cell behavior, cell function, and the pathways that lead to disease. In the

CMOS Biomicrosystems: Where Electronics Meet Biology, First Edition. Edited by Krzysztof Iniewski.

past, cellular studies have primarily focused on bulk populations of cells and drawn conclusions about the behavior of individual cells from data collected from the entire population. However, as many biological systems are made of individual cells, information critical to the understanding of single-cellular behavior and cell-to-cell interactions is lost when an entire population of cells is studied. Cells of the same type under the same environmental conditions, for example, can exhibit different responses [1–3]. In the same way, cells of the same type can respond differently to the same stimulus or therapy. Consequently, studying the average response to a stimulus, as traditionally done in biological cell analysis, does not give a complete picture of individual cell behavior.

Obtaining the statistical distribution of responses from single cells is thus a very important topic for biologists, particularly for research pertaining to the three major causes of mortality in the United States: cancer, stroke, and heart disease [4]. By studying intracellular and intercellular parameters including OCR, pH, CO_2, and DNA composition, and relating this data to gene expression, much can be learned about how genotypic heterogeneity affects cell behavior. The following is a list of example cellular characteristics that would be better understood using single-cell analysis:

- Cell genetic heterogeneity
- Cell–cell communication
- Cell cycle
- Cell-surface interactions
- Cell differentiation
- Drug screening
- Concentration of critical metabolites and ions
- Patterns of cellular response to a given stimulus
- Cell synchronization

The understanding of the cell characteristics listed above will lead to better insight into the origins of disease and disease progression. We give a short explanation of disease progression here for background.

Inside each cell, housed in chromosomes, are genes that contain instructions for that cell. Each time a cell divides, it transfers its genetic code to the newly created cells. Disease can occur when genes are damaged or mutated. The majority of the time these damaged or mutated cells are destroyed by the organism's immune system as it instructs the cell to self-destruct, a process known as apoptosis. However, when a damaged cell survives, the incorrectly encoded genes are transferred through the cell division process and propagate. These erroneous genes may instruct the cell to act abnormally. If it were possible to detect such abnormalities in an individual cell or a small group of cells, disease diagnosis could happen at a much earlier stage, dramatically improving the potential effectiveness of treatments.

A look at the case of cancer is instructive as it is presumed it begins with a single cell. The genetic code of a cancer cell might have been damaged by a chemical carcinogen, radiation, or viral or bacterial infection, and starts to reproduce at a higher rate than healthy cells [5]. These cancerous cells, due to their increased activity, often consume the majority of the nutrients in their surrounding area and essentially suffocate the healthy cells nearby. Oftentimes, there is an observable difference in activity between a healthy cell and a cancerous cell in overdrive. Developing devices that can accurately measure such differences between normal and abnormal cells is a necessary step toward better drug discovery, disease diagnosis, and disease treatment for cancer.

In order to develop better methods of disease diagnosis and treatment, tools are needed to precisely measure single cells and their responses to stimuli such as a therapeutic drug. These measurements can be done invasively, for example, with a fluorescent indicator dye absorbed by the cell, or noninvasively, for example, by monitoring the immediate environment around a cell. One of the key physiological indicators of cell health that can be measured noninvasively on a single-cell level is respiration, a primary indicator of cell metabolism. Cellular metabolism directly corresponds to adenosine triphosphate (ATP) production, which is a reflection of the energy available for normal cell processes such as protein synthesis, mitosis, and the synthesis of DNA and RNA. In many damaged cells, cell metabolism—and thus cell respiration—is disturbed [6]. Another noninvasive cell response method is to measure a cell's external environment to directly detect changes in biological molecules such as proteins and DNA. This gives one the ability to verify, for example, whether a reaction has occurred and to precisely measure the response within the cell. The ability to detect proteins and DNA in single cells, combined with the tools for measurement of cell respiration rate and other extracellular parameters, shapes the overall picture of cell health and activity. Single-cell studies have become the next frontier in further understanding data that previously consisted of averages of hundreds or thousands of cells, providing a much more accurate picture of how cell heterogeneity is linked to cell response and disease.

An ideal platform for a single-cell analysis device would make possible real-time, repeatable, noninvasive, stimulus–response experiments. To accomplish this, one must be able to manipulate, isolate, and analyze individual cells, and for high-throughput purposes, accomplish this in a multicell format with densely packed sensing devices.

To this aim, the Microscale Life Sciences Center (MLSC) at the University of Washington, in collaboration with Arizona State University and the Fred Hutchison Cancer Research Center in Seattle, WA, is developing lab-on-a-chip technology for single-cell analysis [7]. The goal of the MLSC is to develop technology for multiparameter analysis of single cells, and to use this technology to understand heterogeneity in the life and death processes among cells. In this chapter, we address the MLSC's platform to manipulate, isolate, and analyze individual cells for electronic detection of molecules using amorphous silicon nanowire sensors and for the photonic detection of single-cell OCRs as a representative new tool under development for single-cell biology.

13.2 ELECTRONIC DETECTION OF MOLECULES IN THE NANOSCALE

One way of gathering comprehensive single-cell data is by examining molecular events of individual cells via direct measurements within the area immediately adjacent to the cell. A tool for this purpose requires the capacity to make precise measurements of the biomolecular environment of a volume on the order of the size of an individual cell while having minimal impact on cellular functions. This approach provides an understanding of the molecular events occurring in and around the cell, allowing for an investigation into the cellular decision-making process. In particular, measuring proteins and DNA in and around a single cell provides one with information that can be used to help determine the effectiveness of a drug therapy or to aid in the study of cancerous cell behavior.

As discussed previously, many cancer cells have a much higher metabolic rate than normal cells. This increase in activity can be verified by measuring the DNA content within the cell. In a normal human cell, chromosomes—composed of DNA—are organized in pairs; this is called a diploid cell. However, a cancerous cell can have an irregular number of pairs (aneuploidy); the ploidy number of a particular cell can be a strong indication of whether a cell is cancerous [8].

The protein cytokine also has the potential to be used as an indication of cancer. Cytokines are molecules secreted from a cell, usually in response to some stimulus, used to signal nearby cells and possibly change their behavior. These molecules are believed to be in higher concentration near cancerous cells, inflamed areas, and blood clots [9]. Monitoring the presence of cytokines around a cell can be another way to determine if the cell is diseased or otherwise unhealthy.

One tool that can meet the rigorous requirements for single-cell analysis is the amorphous silicon nanowire sensor [10], which can now be readily produced thanks to the advances in nanofabrication. The sensors can be made very sensitive, are very small and can fit in the volume related to a single cell, and can be arrayed and electronically interfaced with for parallel single-cell studies to collect statistical data sets. Our research team has worked on developing nanowire sensors for a number of targets including DNA, streptavidin, and H^+ ions at different concentrations in sample solutions.

13.2.1 Principle of Operation

The surface of the silicon nanowire was coated with receptors engineered to selectively bind to target molecules of interest. After the target molecule bonded with the selected receptor, a small net charge or electric dipole was created on the surface resulting in a change in the effective density of charge carriers immediately under the surface within the semiconductor. Due to the small size of the nanowire, roughly 50 nm × 50 nm in cross section, the variation in charge density near the surface of the nanowire was enough to detectably change the overall conductivity of the element (Fig. 13.1).

Measuring the sensor conductivity over time—before and after introducing the sample—results in an accurate method of determining varying concentrations

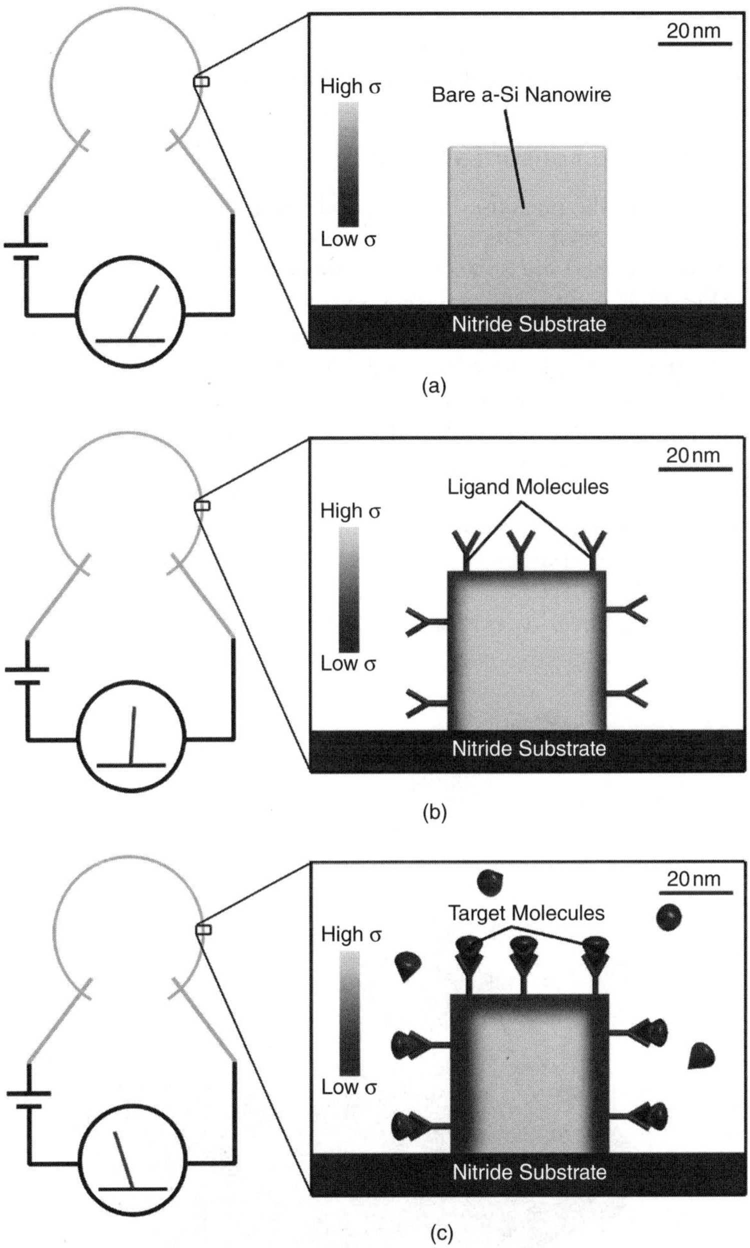

Figure 13.1. Schematic depiction of the operational principle of the nanowire sensor. (A) A voltage is applied across the semicircular sensor element, and the resultant current is measured. (B) Receptor molecules can be positioned on the surface of the semicircular element; the addition of the receptor may alter the conductivity. (C) Upon binding of the target molecules to the receptors on the sensor surface, the overall conductivity of the semicircular element changes. This change can be monitored with external electronics to determine the molecular binding events on the surface [10].

of the target molecule. The device had a small enough footprint and a high enough sensitivity that it could easily perform extracellular measurements and perhaps, in future renditions, gather intracellular data.

13.3.2 Nanowire Sensor Fabrication

Methods similar to the ones used in the production of integrated circuits can be used to make the sensors. The amorphous silicon nanowire sensors were fabricated on a boron doped silicon wafer in a 1-µm-radius semicircle with a thickness of 50 nm (Fig. 13.2). First, a 150-nm-thick low-stress silicon nitride dielectric layer was grown on to the entire wafer, followed by a 50-nm-thick layer of amorphous silicon. Electron-beam lithography with polymethyl-methacrylate (PMMA) resist was used to define the shape of the nanowire sensors and contact pads, 40 µm × 40 µm areas used for electrical connection between test equipment and the sensors. A chrome mask was then applied to the exposed amorphous silicon, followed by an acetone lift-off to remove the PMMA and attached chrome to allow for an SF_6 reactive ion etch (RIE) step-down to the silicon nitride layer. This resulted in 50-nm-tall amorphous silicon nanowires and contact pads. After the chrome was removed, a second electron-beam lithography step was performed, along with metal deposition and acetone lift-off, to apply a gold layer

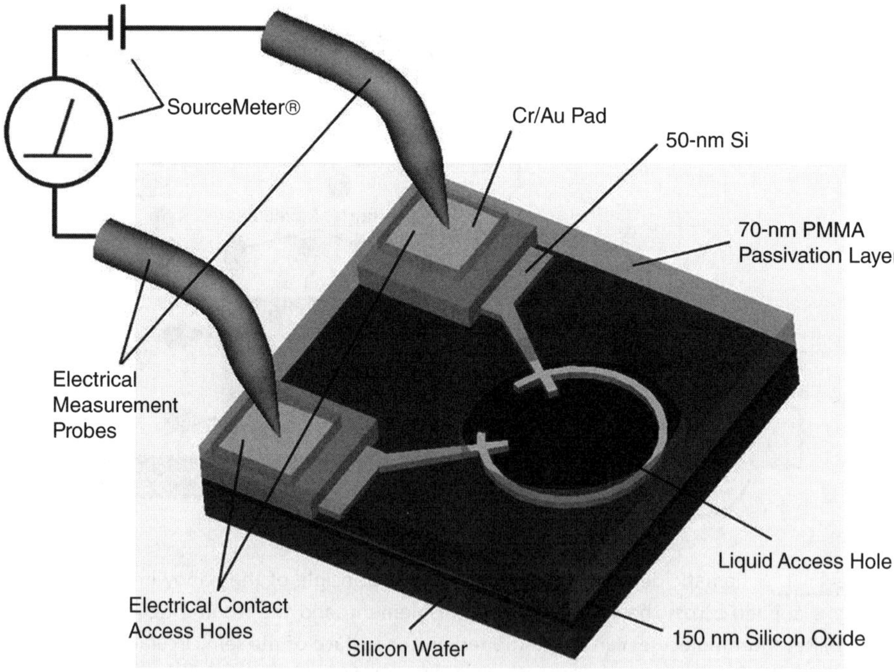

Figure 13.2. Illustration of a silicon nanowire sensor demonstrating the contact pads and sample well [10]. (Sourcemeter, Keithley, Cleveland, OH.)

onto the contact pads to assist in the electrical connection to the sensors. Finally, PMMA was patterned onto the sensors to isolate the contact pads and provide a circular well around the sensors to enclose the test solution.

13.2.3 Testing the Sensors

Preliminary testing of the nanowire sensors was performed with known concentrations of three different target molecules, which required three separate receptor layers. A monolayer of aminopropyltriethoxysilane (APTES) was used to detect H^+ ions, which have a direct correlation to the pH of the environment; a monolayer of mercaptopropyltrimethoxysilane (MPTMS) was used to bond acrydite DNA to detect its complementary DNA (cDNA) strand; and a layer of biotinamidocaproyl-labeled bovine serum albumin (BAC-BSA) was used to detect the protein streptavidin. The sensor remained fixed during testing while the sample solutions were directly added to the liquid access hole. Note that each nanowire sensor formed a native silicon dioxide layer that appears on the amorphous silicon when exposed to air, which electrically isolated the sensor from the receptor and test solution and prevented any short circuit that might result from conductive test solutions. Each receptor was tested separately on different nanowires to allow for setup time between experiments and to prevent any cross talk between target molecules.

The APTES-modified sensor was tested by measuring the current through the device in solutions of varying pH. APTES undergoes protonation, the process of adding or subtracting protons, when in the presence of a basic or acidic solution. Therefore, the current through the nanowire adjusted according to the pH of the solution due to the varying number of charge units attached to the sensor surface. Solutions with pH 2, 4, 5, 6, 7, and 9 were tested by adding hydrochloric acid or ammonia in varying concentrations to double-distilled H_2O. A droplet of solution was placed in the well surrounding the sensor followed by a current-voltage measurement made by connecting probes to the contact pads while maintaining a bias voltage under 100 mV. Between data points, the sensor was rinsed with deionized water and allowed to dry. Upon testing, a linear relationship between the pH and current was demonstrated with the current ranging from ~15 to 23 pA for the given pH values of 2–9 [10].

The MPTMS and acrydite DNA-modified sensor was tested by adding a solution containing single strands of DNA. In our pHs of interest, DNA molecules have a negatively charged backbone. Therefore, when a DNA strand attaches to the surface of the sensor, the density of charges flowing through the silicon due to an applied voltage is immediately impacted, in turn altering the overall conductivity of the nanowire sensor. To demonstrate this, an initial current measurement was performed with only the MPTMS and acrydite DNA as a reference for the subsequent measurements. Next, 10-μL solutions containing 200 nM concentrations of cDNA strands (DNA strands that contained the complement series of base pairs to the attached DNA strand) with zero, one, three, and five intentionally introduced sequence mismatch errors were individually measured with the sensor. An error was defined as C-G mismatch in base pairs, meaning a consecutive G-G or C-C pair, in the final DNA strand, effectively altering the binding energy of the two molecules. A phosphate buffered saline (PBS) rinse step was done between each test without allowing the element to dry between samples The resulting measurements confirmed that single-base-pair mismatches can be detected between two DNA strands [10]. These results are significant as such mismatches in the genome of a single cell may potentially result in large variations in the cell function and behavior.

Streptavidin, a protein of interest because of its attraction to biotin and thus a conduit for arrangement of biomolecules to a strong support, is another molecule whose presence could be detected by the nanowire sensor. BAC-BSA-modified nanowire sensors were used to detect the streptavidin. Biotin was used as a receptor for streptavidin because it was easily applied to the amorphous silicon surface and that the biotin–streptavidin bond is the strongest known protein–ligand interaction. A transient response of the nanowire sensor was measured with a 1-μL droplet of 500 nM streptavidin to demonstrate its effectiveness. After the streptavidin was introduced, the sensor took about 6 seconds to stabilize and displayed a current change of 1 pA. The time-lapse measurement lasted over 8 minutes, with an obvious drop in current once the sample was introduced [10].

13.2.4 Nanowire Sensor Application

The nanoscale sensor just described has a number of distinct advantages. The sensor itself was directly fabricated onto a silicon wafer, allowing the nanowires to be formed in an orderly fashion and placed in specific orientations and positions. This organized placement of the sensors allows for dense arrays that could potentially detect hundreds of different biomolecules in parallel, with each sensor modified with a different receptor. Furthermore, the large amount of data acquired by numerous sensors could be processed on the same chip that the measurements were made on since the nanoscale amorphous silicon structures are compatible with CMOS fabrication technology. Another advantage of the nanowire sensor is that it can directly detect the presence of a target molecule without the need to stain a cell with a specific dye, which introduces a foreign element to the cell that could disrupt its normal behavior and result in data inaccuracies.

Although these sensors have yet to be tested with live cells, their performance looks very promising. It has been demonstrated that the nanowire sensor can detect pH, DNA, and proteins; all of which are important to understanding cellular functions. Considering the small footprint of the nanoscale sensor and the relatively simple fabrication process, a chip with numerous single-cell analysis sites containing multiple sensors is feasible. Our group plans to incorporate these sensors with the oxygen-sensing method described in the following section, providing a multifunctional tool for single-cell analysis.

13.3 SINGLE-CELL RESPIRATION MEASUREMENTS

Our research center has developed a platform that measures single-cell OCRs, and in future iterations will measure additional parameters such as extracellular pH, calcium, carbon dioxide, and cytokine concentrations. We chose single-cell respiration as the first major parameter of interest because it provides a monitor of metabolism, a key physiological indicator of cell health. It is also one of the most challenging parameters to measure, but it paves the way for similar technology that will allow for the measurement of additional cellular parameters of interest.

Just as the respiration rate of an entire organism such as a human being can indicate its current state of health or stress, OCR is a fundamental indicator of single-cell health or stress. As early as 1927, the importance of single-cell metabolic rate was recognized and a system to measure oxygen consumption from a single paramecium both before and after feeding was devised [11]. The rate at which eukaryotic cells consume oxygen is directly related to their rates of ATP production, energy used in processes including the synthesis of proteins, cell division, and homeostasis [12]. The rate of oxygen consumption and thus ATP production can change drastically with cellular dysfunction. Understanding OCR variability can also lend insight into inflammatory disease progression including heart disease, stroke, diabetes, and cancer [13]. Oxygen uptake can also be a main parameter in the evaluation of cell stress response to various stimuli, toxicity determination, and general metabolic heterogeneity within a population [14–16].

In other applications, oxygen uptake can be a useful indicator of alternative energy sources [17]. Different biological communities utilize oxygen in ways that could be exploited for renewable energy sources, helping offset carbon emissions.

13.4 OXYGEN DETECTION

Detection of oxygen in microenvironments containing single cells is a difficult challenge. Because oxygen is a very small molecule, it can penetrate into and out of most materials including many plastics, making the vast majority of commercial cell culture equipment unsuitable as an accurate platform for oxygen consumption measurement. Some research groups are attempting to measure oxygen flux using these materials and compensating for leak rates, but these systems require complex statistical analysis to account for diffusion gradients.

For both bulk and single-cell analyses, Clark electrodes were used as an early method to study oxygen flux in the media surrounding a cell; however, they have the disadvantages of oxygen consumption by the probe itself, low sensitivities, and nondirect measurement of oxygen due to oxygen diffusion in and out of the nonsealed system. Scanning electrochemical microscopy (SECM) uses microelectrodes placed near the surface of a cell and then in gradual distance increments away from it to measure oxygen gradients [18]. Though useful for dynamic oxygen sensing, disadvantages of microelectrodes include sensor drift over time, sensitivity limits, and intensive user training for success in electrode alignment near a cell surface. In other oxygen measurement methods, nanobeads have been attached to cell surfaces to measure oxygen flux [19]. Though this method can be relatively high throughput, like all methods described here the cell environment is still open to oxygen diffusion, and the effects the attached nanobeads have on cells are unknown.

In order to directly measure the rate at which a cell is consuming a quantitative number of oxygen molecules, a platform must be used that isolates a single cell in a microenvironment made from materials impermeable to oxygen. A microwell array format further enhances this type of platform by allowing for spatial localization of multiple cells in a field of view, eliminating the need for complex cell-location software algorithms. A microwell array format also enables continuous monitoring of cells and multi-timepoint data, important for stimulus–response experiments. Glass is an ideal material for a microwell array, offering oxygen impermeability, optical transparency, and low autofluorescence.

Our approach to single-cell oxygen consumption measurement is to diffusionally isolate single cells in an array of glass microwells and monitor rates of oxygen consumption in each microwell using photonically interrogated microsphere sensors. An array of microwells can be viewed and monitored in a microscope field of view simultaneously, allowing for high-throughput measurements. Future versions of the platform will incorporate additional sensors for multiparameter single-cell measurements that will track the utilization of compounds (e.g., oxygen and carbon dioxide) or accumulation of compounds (e.g., cytokines) and could include the electronic nanoscale sensors discussed above.

13.5 OVERVIEW OF THE MINIATURE CELL INCUBATOR PLATFORM

The platform for single-cell measurements is the miniature cell incubator (MCI). The device consists of a micromachined borosilicate glass chip containing arrays of wells containing cells that sits inside a macrowell with a glass bottom, allowing the chip to be submerged in cell media while giving the optical microscope access from below (Fig. 13.3). An oxygen sensor consisting of a platinum compound is deposited inside the bottom circumference of each well; the phosphorescent lifetime of the sensor is inversely related to the oxygen concentration in that well. Each well has a glass rim, or lip, that sits flush against a smooth glass ceiling to create a seal during oxygen measurements. The glass ceiling is attached to the tip of a piston that is brought down with an automated actuator perpendicular to the chip surface; a load cell contained in the actuator provides the means for controlling the amount of force between the two glass surfaces. The voids between the lips of the wells allow for fluid movement when the glass ceiling is brought into contact with the wells. The MCI and piston actuator sit on a microscope

Figure 13.3. Miniature cell incubator (MCI) showing chip containing wells sitting inside the macrowell and the piston used for sealing. An actuator controls the piston's up/down movement to make a glass-on-glass contact to seal the wells containing cells.

stage; the entire microscope is housed in a Plexiglas box that allows for a constant temperature-controlled environment, while any gas mixture needed for incubation conditions enters the MCI through a port hole via a gas manifold.

To make an oxygen measurement, the glass lid is brought down by the piston actuator over the array of microwells at 10 lb of force and a glass–glass seal is made, isolating each cell from the next and preventing oxygen from entering or leaving the well. The phosphorescence of the porphyrin molecule sensor inside each well is then monitored and the oxygen concentration inside the well is measured and tracked over time. Automated software allows for the localization and monitoring of each well's sensor simultaneously so that independent data is acquired from all wells in the field of view of the microscope. Data is collected until a detectable OCR in the wells is observed, after which the glass lid is brought up and the cells are left to reoxygenate.

Because each well is diffusionally isolated from the next and no oxygen can leave or enter the wells, the platform directly measures the amount of oxygen molecules a single cell is consuming. The platform also has the novel capability of allowing the experimenter to repeat data collection on the same set of cells, allowing for validation and statistical data collection.

13.5.1 Microwell Chip Fabrication

Microwells were fabricated in arrays of 3×3 or 4×4 with nine arrays per chip. To begin the fabrication process, 3-in. borosilicate glass wafers (Erie Scientific, Portsmouth, NH) were soaked in piranha and SC-1 baths for 10 minutes each and rinsed. Then, 150 Å of Cr and 3000 Å of gold were thermally evaporated onto one side of the glass surface. Wafers were spin-coated with AZ1512 photoresist (Clariant Corp., Sommerville, NJ), soft-baked at 90°C for 1 minute, and UV-exposed through a transparency mask that created the pattern for the microwells. The wafers were developed in AZ300 MIF developer (Clariant Corp.) for 30 seconds and hard-baked at 110°C for 5 minutes. Exposed gold was etched in Gold Etch TFA (Creekside Technologies, Snohomish, WA) for 120 seconds and the chrome etched in Chrome Etch TFD (Transene Co., Danvers, MA) for 10 seconds. The remaining photoresist was washed away with acetone. Blue-Tack (Semiconductor Equipment Corp., Moorpark, CA) was then applied to the back-side of the wafers to prevent etching on that side. The exposed glass was etched in hydrofluoric acid to achieve the desired well depth. The lithography process and hydrofluoric acid etch were repeated with a second mask to create the seal-lips around each well. Remaining gold and chrome were removed with gold and chrome etch and wafers cleaned in acetone, isopropanol, and deionized water. A dicing saw was used to precisely cut the wafer into 1 cm × 1 cm chips, yielding 32 chips per 3-in. wafer. Before use, the chips were cleaned in fuming sulfuric acid for 30 minutes to destroy any organics remaining on the surface. Etch times were adjusted to tailor the microwell volume to a size appropriate to the cell of interest, and ranged from 50 to 500 pL. Figure 13.4 shows the microwell array and the chip arrangement relative to the microscope objective.

13.5.2 Oxygen Sensors and Sensor Deposition

Commercially available 1.0-μm platinum luminescent microspheres were used as oxygen sensors. When exposed to short-wavelength light from an excitation source, the platinum porphyrin molecule in the microspheres is excited to a triplet state. When the molecule relaxes back to the ground state, it emits light at a longer wavelength. Because the triplet state of the platinum porphyrin molecule is at a similar energy level as the triplet ground state of molecular oxygen, the presence of oxygen nearby absorbs the energy when the molecule relaxes to the ground state. When a large number of platinum porphyrin molecules for a given excitation pulse are present, a distribution of emitted light is captured and the phosphorescence emission lifetime can be measured. When there are numerous oxygen molecules present, a quenching of the Pt phosphorescence manifests itself in an observable shortening of the decay lifetime of the emitted light signal; conversely, when there are less oxygen molecules present, the emitted light has a longer decay lifetime. For our platform, a 405-nm excitation light source was pulsed at an array of wells, each containing sensor molecules, and a high-speed camera coupled with an image processing program measured the emission lifetime response of each well simultaneously through a microscope. Phosphorescent lifetime was translated into oxygen concentration using previously obtained calibration values.

Oxygen sensors were deposited into the microwells with wafer-level processing that allowed sensor deposition on tens to hundreds of wells quickly and easily. First, the microwell array chips were placed in an oxygen plasma etcher for 5

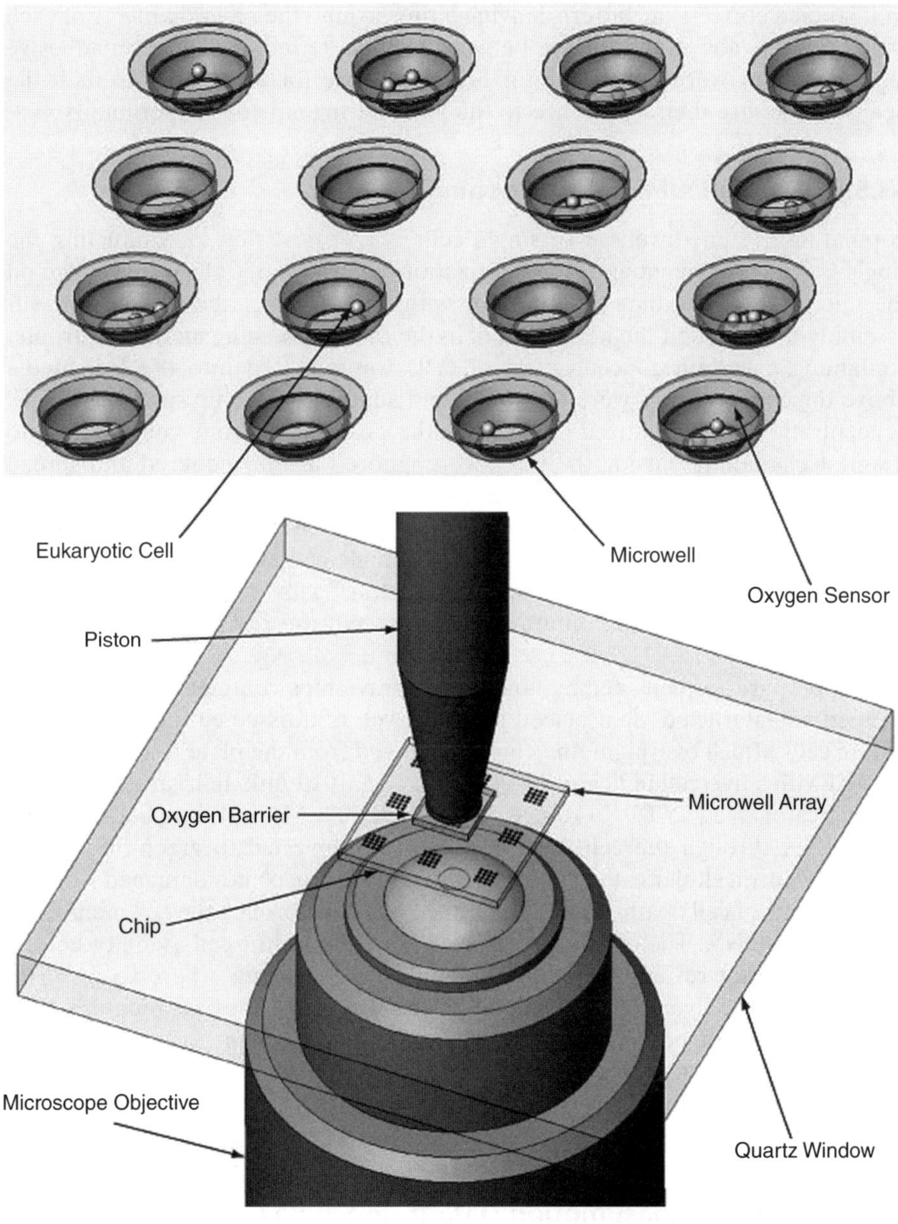

Figure 13.4. Illustration of the piston sealing one array of microwells.

minutes to render the glass surface hydrophilic. Then, 2 μL of beads in solution were then pipetted onto the microwell array surface and uniformly distributed using the pipette tip. The chips were then left to dry and subsequently put through another plasma etch, after which a second 2-μL layer of bead solution was deposited over the first. The beads, similar to other sediments in a solution, dry in a well-studied coffee-ring pattern, leaving a ring around the inner diameter of each well. Excess beads on the surface between wells were removed with an adhesive tape. The chips were then put on a 170°C hotplate for 10 minutes to melt the beads and assure their adherence to the glass during aqueous experiments.

13.5.3 Single-Cell Random Seeding Method

Populating the microwells with single cells is a critical step in conducting the single-cell measurement and characterization experiments. Cells were seeded on the microwell chip either randomly or with a trapping method based on self-assembly, described in the next section. In the random seeding method, a droplet containing a certain concentration of cells was pipetted into the cell media above the chip and cells were left to fall and adhere to the chip surface. The cell concentration was optimized to increase the possibility of one cell falling into a well. Cells falling outside the wells were ignored as they adhered and spread in the cavity space between well lips, while wells containing more than one cell (or zero cells) were used for data validation as they theoretically exhibited faster (or zero) OCRs as compared with single-cell wells. Random seeding with optimized concentrations resulted in a statistically predictable average of about 33% of microwells containing a single cell, the rest containing zero or multiple cells.

To prepare for cell seeding, microwell array chips containing sensors were cleaned and sterilized, then placed into one well of a tissue culture plate. After letting cells attach overnight, the chip was moved from the plate to the MCI and stained with a live-cell indicator dye (calcein AM, 10 μL/mL at 2.5 μM) and dead-cell indicator dye (SYTOX Orange, 2 μL/mL at 500 μM). In viable cells, Calcein AM passes through the cell membrane and is converted to green-fluorescent calcein by intracellular esterases. If a cell membrane becomes damaged—one of the first signs of cell death—SYTOX Orange will pass through the cell membrane and bind to DNA. The dyes are used as a way to monitor cell viability before, during, and after respiration measurements. In experiments where there was a concern that the dye might affect cellular behavior and thus invalidate results, dyes were only introduced to the media after an experiment was over and any fluorescent cell images were taken at that time. A representative image of a microwell array containing stained cells and oxygen sensors obtained with a fluorescent microscope is shown in Figure 13.5.

13.5.4 Oxygen Consumption Data from Single Cells Seeded Randomly

Our group has performed OCR measurements on a variety of cell types, including A549 human epithelial lung cancer cells, RAW264.7 murine macrophage cells (both from the American Type Culture Collection, Manassas, VA), and CP-D and GohTRT Barrett's esophagus precancerous cells provided by the Fred Hutch Cancer Research Center, Seattle, WA.

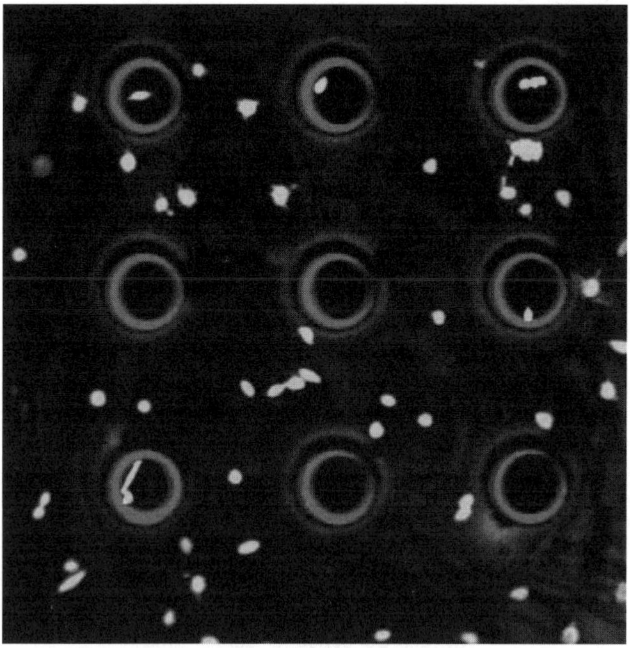

Figure 13.5. A 3 × 3 microwell array containing cells (white) and oxygen sensors (gray rings).

The studies have achieved repeatable stimulus–response OCR measurements on these three cell types, and are presently continuing to investigate various other cell types including bacteria. Further improvements to the single-cell measurement platform, including incorporation of more intra- and extracellular sensors, will further refine and develop the field of single-cell analysis, helping us in drawing conclusions about cell cycle, cell response to drug therapy, cell death pathways, cell response to changing nutrient conditions, and cell progression toward cancerous and inflammatory states (Fig. 13.6).

13.6 OXYGEN CONSUMPTION RATE MEASUREMENTS USING A SINGLE-CELL SELF-ASSEMBLY METHOD

As discussed previously, cells are spatially isolated by a passive random seeding method or an active trapping method. Random seeding resulted in a statistically predictable average of about 33% of the microwells capturing single cells, the rest containing multiple or no cells. Though random seeding has the advantage of speed and simplicity at the chip level, its drawback is an overall low single-cell placement yield. Before any conclusions can be made about a population of cells, there needs to be comprehensive data on a large number of cells, typically on the order of thousands of cells. Random seeding could, therefore, be an extremely time-consuming method of data acquisition.

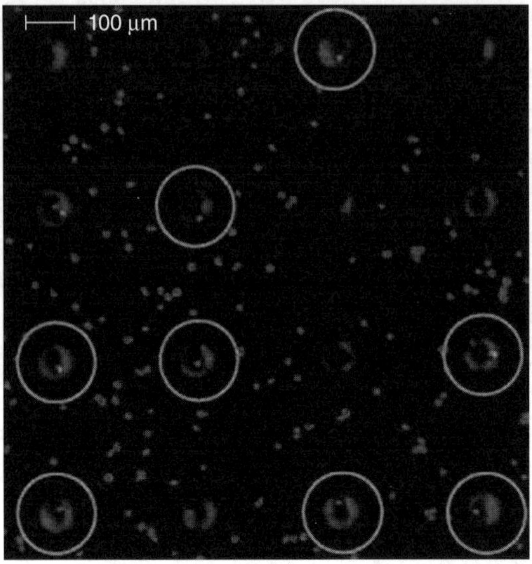
100 μm

(a)

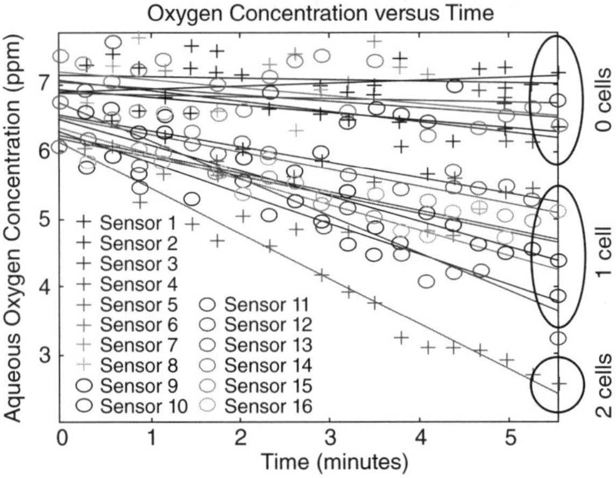
Oxygen Concentration versus Time
Aqueous Oxygen Concentration (ppm)
Time (minutes)
Sensor 1
Sensor 2
Sensor 3
Sensor 4
Sensor 5
Sensor 6
Sensor 7
Sensor 8
Sensor 9
Sensor 10
Sensor 11
Sensor 12
Sensor 13
Sensor 14
Sensor 15
Sensor 16
0 cells
1 cell
2 cells

(b)

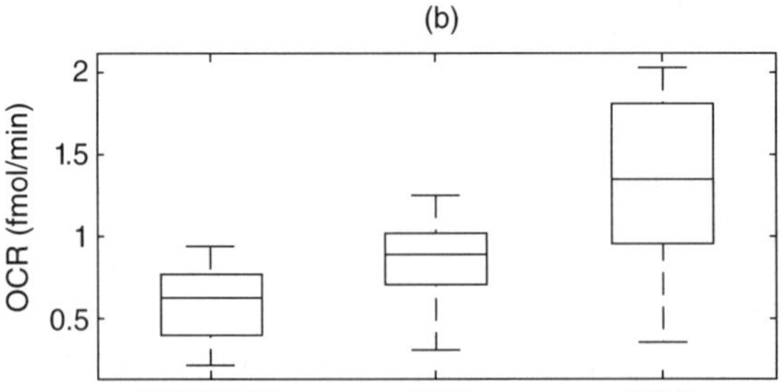
OCR (fmol/min)
RAW264.7
A549
CP-D

(c)

Figure 13.6. (A) Fluorescent image of Calcein AM-stained A549 human lung epithelial cells inside wells containing a platinum porphyrin-based oxygen sensor (light gray rings). Single cells are manually circled in white. Note that the cells in between the microwells are isolated and do not contribute to the changes observed inside the microwells. (B) Measurement results of oxygen concentration versus time in the sealed wells from (A). It can be seen that wells with zero cells had a net zero consumption rate, and wells with one cell had consumption rates lower than the well with two cells. (C) Oxygen consumption rate results of three cell types conducted at the single cell level: RAW264.7 mouse macrophages, A549 human epithelial cells, and Barrett's esophagus CP-D cells [20].

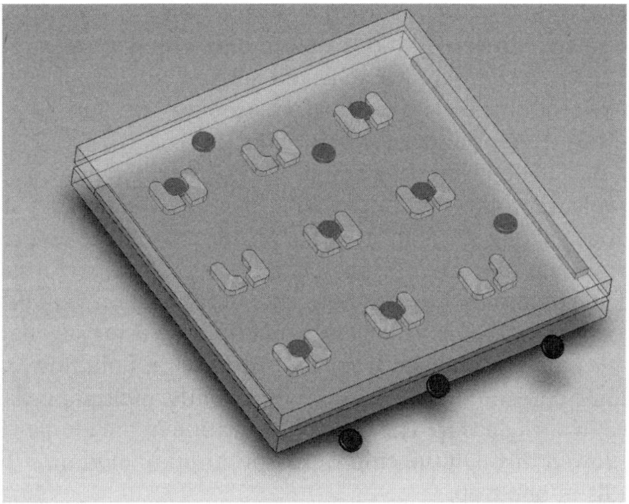

Figure 13.7. Illustration of the single-cell trapping process. SU-8 is patterned onto glass and later sealed with a second piece of glass. Cells are then flowed into the microchannel to be captured by the traps [21].

13.6.1 Single-Cell Self-Assembly

To improve single-cell localization yield, we developed a cell seeding method that involves trapping single cells with cuplike structures (Fig. 13.7). The structures were fabricated by photolithographically patterning SU-8-25 photoresist onto glass. Another piece of glass was then clamped on top of the SU-8 structures, creating a microchannel. Cells in the media were introduced at the entrance to the microchannel and were individually captured by the traps driven by fluid flow and gravitational forces. The traps were 35-μm tall and 50 μm in diameter, with a 24-μm "cup" diameter and an 8-μm gap at the bottom of each trap to allow fluid to exit. These dimensions could be easily altered to account for different cell sizes or other design parameters.

The traps were designed to encourage occupation by a single cell. When a trap was empty, fluid flowed to the left, to the right, and down the center of the trap through the fluid exit port at the bottom of the trap. However, when a trap was full, fluid only flowed to the left and right of the trap because the fluid exit port was blocked by the occupying cell. Any subsequent cells that came upon an occupied trap were carried by the flow around the occupied trap and continued to the next row.

Typically, this seeding method was performed with mammalian cells on a 12 mm × 12 mm glass chip containing nine arrays of nine traps, 81 traps in total. The arrays were spaced 2.5 mm apart in each direction. Once the cells were trapped, the chip and glass ceiling were placed in an incubator overnight at 37°C with 5% CO_2 to allow the cells to adhere to the bottom glass surface. After the cells attached, the top glass ceiling was removed to reveal attached single cells in predefined locations. The cells were then stained with the same dyes previously described and imaged to verify vitality before performing oxygen measurements. Using this trapping technique, we achieved single-cell trapping with a yield of 61%—a near twofold increase over random seeding—by flowing ~267,000 cells per milliliter at 3 mL per minute for 6 minutes [21].

Although the single-cell self-assembly method resulted in a better placement of individual cells than the random seeding method, the process still showed a distribution in occupying the trap sites. Therefore, there were empty traps at the end of a seeding run unless the cell assembly process was performed for a long time under ideal conditions. Furthermore, due to fabrication variability, the SU-8 traps were not always the same height, sometimes causing an inadequate seal between the glass ceiling and the traps that resulted in fluid flow paths above a trap that a single cell could not block. Consequently, multiple cells were sometimes captured in a single trap. The cell concentration in the media, flow rate, and total volume flowed through the chip were optimized to minimize the possibility of multiple cells in one trap.

13.6.2 SU-8 Toxicity Assessment

It is important to note that any single-cell analysis system is made of materials and structures that themselves must be tested to determine their effect on the biological system under consideration. As a representative case, we discuss the effect of SU-8 trap structures microfabricated and used to self-assemble single-cell arrays as mentioned above.

Before any data using SU-8 traps could be recognized as valid, the potential toxicity of SU-8 had to be addressed. An approach was taken involving two cell lines of interest, a Barrett's esophagus precancerous cell line (CP-DhTRT, Fred Hutchinson Cancer Research Center) and an epithelial lung cancer cell line (A549, American Type Culture Collection). Cells were grown and observed for 4 days under three different glass chip conditions. On the first day of the experiment, the cells were seeded onto borosilicate glass chips that contained (1) no SU-8, (2) a flat sheet of SU-8, and (3) patterned SU-8. The chips were then stained with the same cell vitality indicators as previously described and imaged on the

TABLE 13.1. Results of the Toxicity Experiment Showing No Obvious Toxicity to Cells of Interest Due to the Presence of SU-8 [21]

	Live	Dead	Live/Dead	% Dead
A549		TWO DAYS AFTER SEEDING		
No SU-8	8535	63	135.5	0.73
Flat SU-8	4412	79	55.8	1.76
Patterned SU-8	4001	137	29.2	3.31
		THREE DAYS AFTER SEEDING		
No SU-8	9448	44	214.7	0.46
Flat SU-8	13037	222	58.7	1.67
Patterned SU-8	7574	87	87.1	1.14
CP-DhTRT		TWO DAYS AFTER SEEDING		
No SU-8	4185	87	48.1	2.04
Flat SU-8	1353	25	54.1	1.81
Patterned SU-8	2884	42	68.7	1.44
		THREE DAYS AFTER SEEDING		
No SU-8	8247	46	179.2	0.55
Flat SU-8	2466	49	50.3	1.95
Patterned SU-8	3394	73	46.5	2.11

third and fourth day of the experiment, using new chips for each day to prevent any contamination due to prolonged exposure to the dyes. The data was processed with an in-house cell counting software to determine the total number of live and dead cells (Table 13.1). Neither cell line exhibited a major difference between the three chip conditions, which led to the conclusion that SU-8 was nontoxic to these cell lines.

13.6.3 Oxygen Measurement of Cells Trapped Using Self-Assembly Method

The glass chip containing the traps and attached cells was placed inside the MCI described previously, allowing the chip to be submerged in media while giving optical microscope access from below. Just as with the random seeding method, a glass lid was brought down by the piston actuator over the array of traps to form a glass–glass seal. However, in this case, the glass ceiling was not flat but contained a 3 × 3 array of inverted microwells that aligned with the traps (Fig. 13.8). Each well's glass rim sat flush against the glass chip, enclosing the traps to create a seal during oxygen measurements. As with the wells used during passive cell seeding, a platinum porphyrin-based phosphorescent oxygen sensor was deposited inside the bottom circumference of each inverted well. As described previously, the phosphorescent lifetime of the sensor is inversely related to the oxygen concentration in that well. To begin an experiment, the glass-tipped

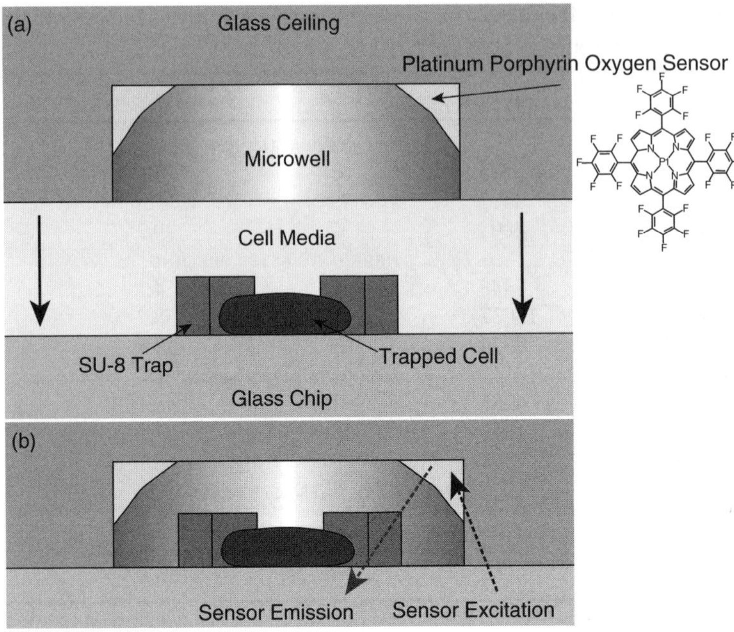

Figure 13.8. Illustration of the inverted wells containing the oxygen sensor sealing a single cell trapped by the SU-8 structure. (A) Once the ceiling is aligned, it was brought down over the trapped cell and a 10-lb force was applied to isolate the cell. (B) After an appropriate seal was achieved, the emission lifetime of the excited sensor was measured and correlated to the cell's oxygen consumption rate [21].

piston was brought down with an actuator perpendicular to the chip surface and a load cell contained in the actuator allowed for the piston to apply a consistent amount of force between the two glass surfaces. The rest of the experiment was carried out in a similar fashion as for randomly seeded cells as described previously.

13.6.4 Single-Cell Oxygen Consumption Rate Measurement Results

A549 human lung epithelial cells were used to demonstrate the single-cell self-assembly trapping method followed by the photonic oxygen measurements system previously described. The cells were captured and allowed to attach over night. Figure 13.9 displays the drawdown rate of three single cells from three separate experiments. The data were added to the same graph for clarity and to display consistency between measurements.

The image processing software developed at the MLSC outputs a graph of oxygen consumed in parts per million plotted against time. The actual oxygen consumed in fmol/min was calculated using the well volume. The calculated

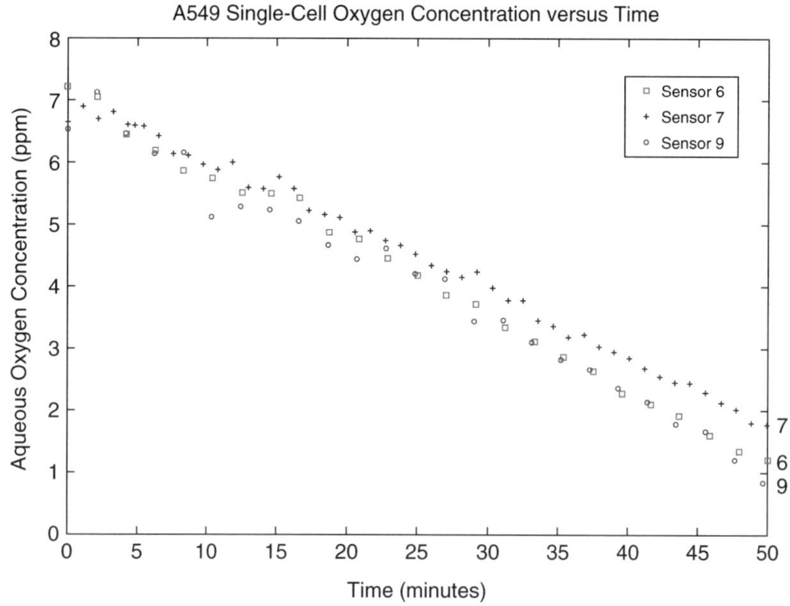

Figure 13.9. Output graph of three separate drawdowns of single A549 cells. Drawdown rates of 0.83, 0.71, and 0.82 fmol/min for three different sensors were measured [21].

OCRs for the three isolated live A549 cells were 0.83, 0.71, and 0.82 fmol/min, comparable with the results published using the random seeding method of 0.91 ± 0.39 fmol/min for A549 cells [20].

13.7 CONCLUSION

Recent advances in nanotechnology, microfabrication, and automation have enabled the opening of new vistas in biological research. One of the newly enabled research paradigms is single-cell biology in which the biochemical environment of a single cell, and arrays of single cells, is studied in detail to determine molecular pathways and, eventually, the nature of cell–cell interactions. Such single-cell studies reveal previously inaccessible data to researchers in biology and medicine and place very stringent requirements on the tools that are used to carry out the experiments. The tools must have sensitivities at levels comparable with the number and concentration of molecules found in a single cell. In addition, these tools must be able to experiment on a large number of single cells to generate biologically meaningful data. Preferably, these tools can be produced at a low cost and made available widely to the research community.

As examples of how electronics and photonic techniques may be deployed in this area, we discussed two examples involving a nanoelectronic molecular detector designed to sense hydrogen ions, short DNA molecules, and proteins, and a photonic system capable of measuring the OCR of a single cell and of performing a large number of experiments in parallel.

We believe that as the new tools from electronics, photonics, nanofabrication, microtechnology, computing, and automation are transitioned increasingly more to the biological and medical domain, more sophisticated single-cell analysis systems will become available. Lastly, we note the truly interdisciplinary nature of conducting these experiments that involve electrical engineers, mechanical engineers, material scientists, chemists, biologists, and physicians. This is another trend that is likely to become more prominent in the coming years as complex and large-scale experimental work is carried out by multidisciplinary teams focused on solving biological and biomedical research problems.

ACKNOWLEDGMENTS

The research covered in this chapter was conducted within the MLSC at the University of Washington and supported by the National Human Genome Research Institute at the National Institutes of Health. A large number of researchers have contributed to this work. We especially thank professors D. Meldrum, M. Lidstrom, L. Burgess, A. Jen, B. Reid, and B. Cookson, and Dr. M. Hall for their invaluable contributions to building the single-cell analysis platform.

REFERENCES

[1] L. Trumper et al., "Single-cell analysis of Hodgkin and Reed-Sternberg cells: Molecular heterogeneity of gene expression and p53 mutations," *Blood*, 81(11), pp. 3097–3115, 1993.

[2] R. Bucy et al., "Heterogeneity of single cell cytokine gene expression in clonal T cell populations," *J. Exp. Med.*, 180(4), pp. 1251–1262, 1994.

[3] M. T. Barrett et al., "Evolution of neoplastic cell lineages in Barrett oesophagus," *Nat. Genet.*, 22(1), pp. 106–109, 1999.

[4] M. Heron et al., "Deaths: Final data for 2006," *Natl. Vital. Stat. Rep.*, 57(14), pp. 1–134, 2009.

[5] S. Preston-Martin et al., "Increased cell division as a cause of human cancer," *Cancer Res.*, 50(23), pp. 7415–7421, 1990.

[6] D. C. Wallace, "A mitochondrial paradigm of metabolic and degenerative diseases, aging, and cancer: A dawn for evolutionary medicine," *Annu. Rev. Genet.*, 39, pp. 359–407, 2005.

[7] M. E. Lidstrom and D. R. Meldrum, "Life-on-a-chip," *Nat. Rev. Microbiol.*, 1(2), pp. 158–164, 2003.

[8] L. H. Hartwell and M. B. Kastan, "Cell cycle control and cancer," *Science*, 266(5192), pp. 1821–1828, 1994.

[9] F. Balkwill and A. Mantovani, "Inflammation and cancer: Back to Virchow?" *Lancet*, 357(9255), pp. 539–545, 2001.

[10] J. Lund, R. Mehta, and B. A. Parviz, "Label-free direct electronic detection of biomolecules with amorphous silicon nanostructures," *Nanomedicine*, 2(4), pp. 230–238, 2006.

[11] R. B. Howland and A. Bernstein, "A method for determining the oxygen consumption of a single cell," *J. Gen. Physiol.*, 14(3), pp. 339–348, 1931.

[12] J. Hynes et al., "Fluorescence based oxygen uptake analysis in the study of metabolic responses to apoptosis induction," *J. Immunol. Methods*, 306(1–2), pp. 193–201, 2005.

[13] S. L. Fink and B. T. Cookson, "Apoptosis, pyroptosis, and necrosis: Mechanistic description of dead and dying eukaryotic cells," *Infect. Immun.*, 73(4), pp. 1907–1916, 2005.

[14] J. J. Casciari, S. V. Sotirchos, and R. M. Sutherland, "Variations in tumor cell growth rates and metabolism with oxygen concentration, glucose concentration, and extracellular pH," *J. Cell Physiol.*, 151(2), pp. 386–394, 1992.

[15] P. Jorjani and S. S. Ozturk, "Effects of cell density and temperature on oxygen consumption rate for different mammalian cell lines," *Biotechnol. Bioeng.*, 64(3), pp. 349–356, 1999.

[16] J. Shen et al., "Oxygen consumption rates and oxygen concentration in molt-4 cells and their mtDNA depleted (rho0) mutants," *Biophys. J.*, 84(2 Pt 1), pp. 1291–1298, 2003.

[17] P. A. Del Giorgio and P. J. I. Williams, *Respiration in Aquatic Ecosystems*. New York: Oxford University Press, 2005.

[18] H. Shiku et al., "Oxygen consumption of single bovine embryos probed by scanning electrochemical microscopy," *Anal. Chem.*, 73(15), pp. 3751–3758, 2001.

[19] Y. Kuang and D. R. Walt, "Detecting oxygen consumption in the proximity of *Saccharomyces cerevisiae* cells using self-assembled fluorescent nanosensors," *Biotechnol. Bioeng.*, 96(2), pp. 318–325, 2007.

[20] T. W. Molter et al., "A microwell array device capable of measuring single-cell oxygen consumption rates," *Sens. Actuators B Chem.*, 135(2), pp. 678–686, 2009.

[21] J. R. Etzkorn et al., "Forming self-assembled cell arrays and measuring the oxygen consumption rate of a single live cell," *Solid-State Sensors, Actuators and Microsystems (IEEE Transducers)*. 2009. p. 2374–2377.

14

APPLICATION OF BACTERIAL FLAGELLAR MOTORS IN MICROFLUIDIC SYSTEMS

Steve Tung, Jin-Woo Kim, and Ryan Pooran

14.1 INTRODUCTION

In the current era of micro- and nanotechnology, microfluidics occupies an important position both in terms of the fundamental science and practical applications due to the simple fact that the majority of living microorganisms is either surrounded by fluids or requires fluids for survival. Microfluidics, according to the most commonly accepted definition, is the study of "transport phenomena and fluid-based devices at the microscopic length scales" [1]. The primary reason for separating microfluidics from the conventional discipline of fluid mechanics, which has been going strong for well over 100 years, is because the assumptions of continuum and no-slip boundary conditions, two of the major cornerstones in deriving the governing equations for incompressible flows, begin to break down as the characteristic flow scale reaches the micrometer range. In this range, factors such as surface tension and other interfacial phenomena that are more or less omitted in the treatment of macroscale fluid flows become increasingly important due to rising surface-to-volume ratio.

The development of microelectromechanical and the more recent nanoelectromechanical systems (MEMS and NEMS, respectively) fabrication techniques has greatly accelerated the transformation of microfluidics from laboratory-based studies to commercially viable applications. Using these techniques,

CMOS Biomicrosystems: Where Electronics Meet Biology, First Edition. Edited by Krzysztof Iniewski.

conventional flow-control devices in the macroscale can be shrunk down to the micro- and nanometer scales where individual microorganisms inhabit. One prominent example in the application of MEMS techniques to microfluidics is the lab-on-a-chip device or the increasingly popular μTAS, or the micrototal analysis system [2]. The fundamental idea and ultimate goal of a μTAS is to be able to execute a complete set of biological or biomedical diagnostics on a small microliter- or picoliter-sized sample in one single shot using a massively integrated and preferably handheld, battery-powered device. Such a system will have tremendous advantages over the conventional laboratory-based systems in terms of speed, sample size, ease of use, and the always important cost per test. The main component of a μTAS is an interconnected system of microchannels through which the sample medium is directed into different microwells and reservoirs either for additional processing or for analysis. To make medium transport possible, the microchannels are typically connected to various fluid propelling and manipulating devices such as micropumps, micromixers, and microseparators. These devices can be fabricated either separate from the microchannels (the hybrid design) or on a common substrate (the monolithic design).

Of the various μTAS components, the micropump is perhaps the most researched element because of the obvious reason that if there is no flow pumping, there is no flow sensing. The current MEMS micropumps can be roughly divided into two main categories: the nonmechanical and the mechanical designs [3]. The nonmechanical micropumps drive the microflow by converting nonmechanical power into kinetic energy for the fluid [4]. Their pumping mechanisms are usually based on electrokinetic, magnetohydrodynamic, or electrohydrodynamic principles. One distinct drawback of the nonmechanical pumps is the requirement of imposing an electrical bias to the fluid in the microchannel. Since the biological medium in the μTAS is typically ionic, the electrical bias can have a negative effect on the electrical and magnetic properties of the medium. In contrast, the mechanical micropumps use mechanical actuators to drive the fluid, and any electrical bias, when needed, is only applied to power the actuators and not the fluid. They usually contain moving parts, which are essentially miniaturized versions of the macroactuators in the conventional pumps. There are two types of micromechanical pumps: the displacement and the dynamic pumps. In the displacement pumps, fluid flow is generated either by imposing a positive pressure gradient in the direction where the flow is intended or by creating a moving boundary where the fluid is translated through viscous drag. Examples of the displacement mechanical pumps include check valve pumps, peristaltic pumps, and rotary pumps. In the dynamic pumps, work is continuously performed on the fluid mass through nonconventional means such as ultrasound and centrifugation to initiate fluid velocity [5].

In designing a μTAS, the selection of one pumping design over the others can depend on many factors: the flow rate requirement, the level of back pressure, the fluid being pumped, and the availability of particular microfabrication facilities. The nonmechanical pumps, with no moving parts, are typically easy to fabricate and are well suited for small chip size and highly integrated designs. Most of them consist of custom-designed microelectrodes incorporated into a microchannel system. The electrodes and channels can be fabricated on one common substrate or on different substrates and then integrated into one device by using a wide variety of chip/wafer bonding techniques. Since the actuation mechanism of the nonmechanical pumps does not generate a high fluid pressure, there is generally no danger of fluid leakage through the bond interface. The downside is the pump's poor ability in handling back pressure and a small flow rate (about 10–100 μL/min). Additionally, as indicated earlier, the electrical and magnetic properties of the medium can have a significant impact on the effectiveness of the pumping mechanism.

The mechanical pumps, with their moving actuators, are more difficult to fabricate and also integrate with other components in the μTAS; but they produce a large flow rate (about 10 μL/min–10 mL/min) and can handle back pressures much better than the nonmechanical pumps. Additionally, the performance of the mechanical pumps usually does not depend on the electrical properties of the medium. This is attributed to the fact that the pumping action is generated through the physical movement of a solid interface between an actuator and the fluid, where the only material property involved is the viscosity of the fluid. The pump's actuator can be either external or integrated monolithically with the pumping system. The external actuators can be driven by piezoelectric, pneumatic, and shape memory alloy (SMA) principles. Such actuation mechanisms have the advantage of a large driving force but suffer from a large footprint and thus are restricted by the level of miniaturization achievable. The integrated actuators are usually micromachined devices controlled through electrostatic, thermopneumatic, or electromagnetic means. The electromagnetic actuators have a fast response time but are not capable of producing large forces. The thermopneumatic actuators can generate large pressures but require a large electrical power and a good thermal management for operations. In addition, mechanical micropumps usually create a large pressure inside the pump chamber that requires excellent sealing for good performance and reliability.

14.2 FLAGELLAR MOTOR MICROPUMP

The majority of the mechanical micropumps require a large pressure drop inside their pumping chambers in order to generate a desirable flow rate. This creates leakage problems that require an excellent sealing scheme for the pump to achieve good performance and reliability. An alternative mechanical design that does not require high pressures but is still capable of large flow rates is the viscous pump. Viscous pumping is based on the classical Couette flow principles [6]. In this design, fluid is translated by moving an immersed solid wall where the no-slip boundary condition dictates that the fluid next to the wall must move at the same velocity as the wall. The velocity profile generated by a Couette flow can be determined by solving the Navier–Stokes equation for incompressible flows. Assuming the flow is trapped between two "infinitely" long (in the

x-direction) and wide plates (in the z-direction), the Navier–Stokes equation is reduced to

$$0 = \mu \frac{d^2u}{dy^2}, \tag{14.1}$$

where u is the velocity in the streamwise (x) direction. Solving this equation while applying the appropriate boundary conditions of one stationary and one moving wall results in the simple equation of

$$u = \frac{Uy}{h}, \tag{14.2}$$

where U is the velocity of the moving plate and h is the distance between the stationary and moving plates. Integrating the velocity profile in Equation 14.2 results in the volumetric flow rate Q (per unit width) between the two plates:

$$Q = \int_0^h u dy = \frac{Uh}{2}. \tag{14.3}$$

Based on Equation 14.3, to maintain a flow rate of 100 μL/min (water) in a 10 μm by 300 μm (height and width) microchannel, a plate velocity U of 2.6 m/s will be required.

In reality, designing and fabricating a microchannel with a mobile sidewall can be somewhat difficult for the existing microfabrication tool bag. Instead, a more feasible design for the viscous micropump can be similar to the schematics displayed in Figure 14.1. Here, a rotating cylinder is placed off center, close to the bottom sidewall of the microchannel. As the cylinder rotates in the clockwise direction, the spacing difference creates a viscous drag differential between the area above and below the cylinder, which in turn generates an asymmetric flow around the cylinder: the flow rate above the cylinder to the right is larger than the flow rate below to the left. In essence, a net flow to the right is generated whenever the cylinder rotates in the clockwise direction. Sen et al. [7] studied the effects of various parameters on the performance of a microviscous pump with a similar design. Based on their results, the velocity profile in a viscous pump was parabolic

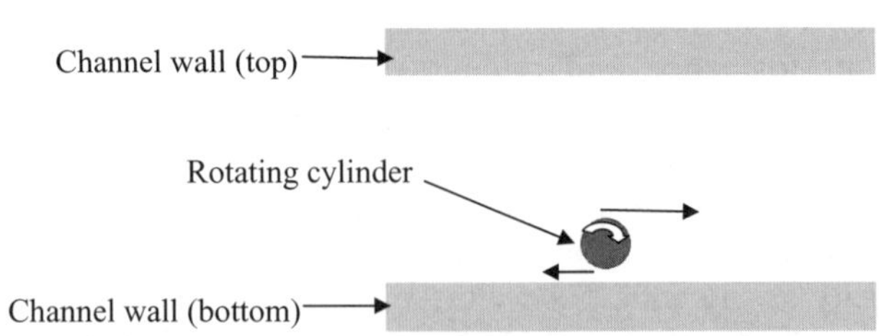

Figure 14.1. Schematic of a microviscous pump.

in nature and the average velocity was about two-thirds of the peak value in the profile. Using numerical simulations, DeCourtye et al. [8] analyzed a similar configuration and found that as the width of the channel increased, the average velocity increased to a maximum level before leveling off asymptotically.

To convert the schematics in Figure 14.1 into an actual microviscous pump requires the implementation of a MEMS actuator capable of 360° rotation, preferably without the need for a complex actuation mechanism. Throughout the last decade, research institutes such as the Sandia National Laboratory have produced a number of actuators capable of 360° rotation. The Sandia actuators are fabricated by a five-layer polycrystalline surface micromachining process SUMMiT V™ [9]. They are capable of high rotational rates with a reasonable level of mechanical torque. However, the SUMMiT V process is extremely complex and contains proprietary processes for producing low thin-film intrinsic stress. Additionally, the process is extremely sensitive to surface topography requiring chemical-mechanical polishing (CMP) to achieve the necessary surface smoothness. In addition to the complexities involved in the fabrication process, there are other drawbacks in utilizing these actuators such as the high voltage required for actuation and difficulties in operating in a fluidic environment. Finally, since these rotors spin at extremely high speeds, friction can very easily lead to device failure. To combat this problem, researchers sometimes have had to resort to exotic mechanisms for minimizing friction by providing sufficient lubrication between the moving and stationary parts of the actuator [10].

In the world of microbiology, mechanical work is routinely performed through the use of molecular motors such as myosin, F_1-ATPase, and the bacterial flagellar motor [11]. While myosin is a linear motor, F_1-ATPase and the flagellar motor are rotary motors. Molecular motors are important not only because they are responsible for a wide range of biological functions but also because they perform those functions at such a high efficiency level that is difficult to duplicate using man-made systems. A comparison of F_1-ATPase and the bacterial flagellar motor is shown in Table 14.1. Notice that both motors have an efficiency of over 80%.

The F_1-ATPase motor is a multisubunit enzyme that synthesizes adenosine triphosphate (ATP), the central energy storage molecule of living organisms [12].

TABLE 14.1. Comparison of F_1-ATPase and Bacterial Flagellar Motor

Property	F_1-ATPase	Bacterial Flagellar
Motor size	8 nm	45 nm
Propeller size	750 nm	3–5 μm
Driving force	Adenosine triphosphate	Proton gradient
Maximum free rotation rate	130 Hz	300 Hz
Tethered rotation rate	4.5 Hz	10 Hz
Maximum torque	40 pN-nm	4000 pN-nm
Maximum power output	10^{-18} W	10^{-15} W
Efficiency	80%	100%

Protons passing through the F_1-ATPase induce conformal changes in the enzyme causing the rotation of a small complex of subunits. Flagellar motors are found in many species of bacteria including *Escherichia coli* and *Salmonella typhimurium*. They rotate the flagellar filaments on the cell bodies to propel the bacteria forward in search of nutrients (Fig. 14.2). Each motile *E. coli* cell has an average of three to five helical flagellar filaments [13]. Each filament, 20 nm in diameter and 10-μm long, can rotate up to 100 revolutions per second (rps). Under normal conditions, rotation of the flagellar filament is bidirectional. However, mutations that affect the chemotactic signaling pathway can lock the motors in one direction of rotation. KAF95 is a mutant strain of *E. coli*. It carries two mutations that (1) force the flagellar motor to rotate only in the counterclockwise direction and (2) allow the flagellar filament to attach spontaneously to a negatively charged surface [14]. A KAF95 cell has a cylindrical shaped body with hemispheric ends. Depending on the growth conditions, it is typically 3-μm long and 1 μm in diameter. When a single flagellar filament on the cell body is permitted to tether to a flat substrate surface, the flagellar motor attached to the filament turns the cell body like a merry-go-round at a rotational speed of about 10 rps in the clockwise direction (Fig. 14.3).

The highly efficient nature of the F_1-ATPase and *E. coli* flagellar motors suggests the possibility of utilizing the molecular motors as rotating actuators in a

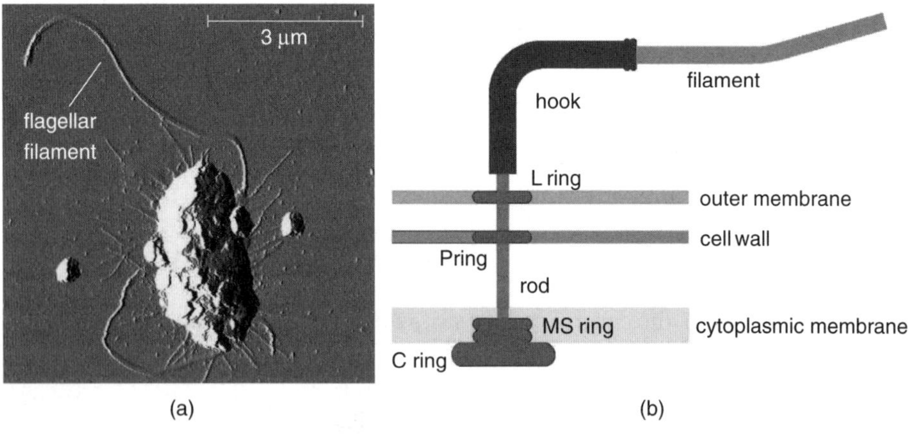

Figure 14.2. (a) Atomic force microscope (AFM) scan of an *E. coli* cell. (b) Structural components of a flagellar motor.

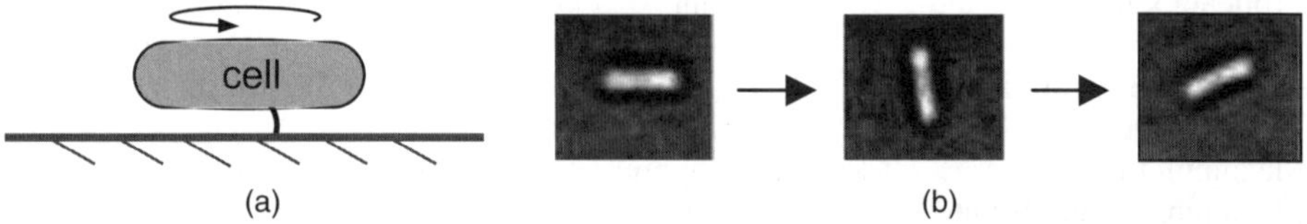

Figure 14.3. (a) Cell tethering through a short filament. (b) Cell body rotation (top view).

microviscous pump. Between the F_1-ATPase and flagellar motors, the latter appears to be better suited for the pumping application due to several reasons. First, the actuator portion of the bacterial flagellar motor is the body of the cell and is already "built-in" the system. The F_1-ATPase motor does not come with a natural actuator; an artificial nanorod must be added to the rotor, but the typical success rate of attaching the nano rod while maintaining the motoring functionality is extremely low [15]. Second, as shown in Table 14.1, the torque output of the flagellar motor is about 100 times higher than the F_1-ATPase motor and the mechanical power output is three orders of magnitude larger.

Other advantages in using the *E. coli* flagellar motors include the fact that *E. coli* has been widely studied over the years and a wealth of knowledge is known about its motility system, biophysics, genetic makeup, and biochemical processes. The KAF95 strain is nonpathogenic and requires no special handling precautions. The nutritional needs of the strain are simple and well understood, and a large number of cells can be obtained rapidly (doubling times of 30–90 minutes) in a simple growth medium without sophisticated culture techniques.

To evaluate the feasibility of utilizing *E. coli* flagellar motors as actuators in a microviscous pump, two fluid dynamics simulations were performed to determine the flow-generating capability of such device [16]. In the first simulation, the bacterial cell was modeled in ANSYS/FLOTRAN as a stationary cylinder with a length of 3 μm and a diameter of 0.7 μm. It was tethered 0.2 μm above the sidewall and 1 μm from the bottom of a microchannel. The (top) width of the channel was from 5 to 20 μm. To simulate a rotating cylinder, a three-dimensional velocity field was assigned as a boundary condition to the wall of the cylinder. The velocity field was equivalent to the cell rotating at 10 rps. The two-dimensional contour lines in Figure 14.4a represent the resultant longitudinal (along the channel) velocity distribution on a cross-sectional plane normal to the flow direction. As expected, the most intense velocity fields are found at the two ends of the cell, where the fluid flows in opposite directions as indicated by the shades of the contours. Viscous damping at the lower end of the cell results in a higher average velocity at the upper end, resulting in a net flow in the longitudinal direction. Figure 14.4b shows the volumetric flow rate determined by integrating the longitudinal velocity distributions over the cross section. The maximum flow rate is about 0.12 nL/min. In the second simulation, the relationship between the location of the cell in the channel and the resulting flow rate is studied. The result indicates that the highest flow rate is achieved when the cells form a linear array near the sidewall of the microchannel and rotate in phase with each other. Based on the simulation results, a conceptual drawing of the optimized design for the flagellar motor driven microviscous pump is shown in Figure 14.5.

14.3 EXPERIMENTAL STUDIES

In the past few years, the concept of a hybrid micropump consisting of "living" flagellar motors as the primary actuation source has been actively explored in

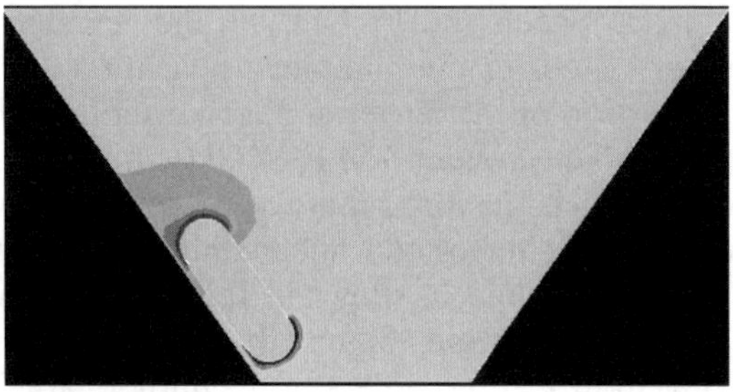

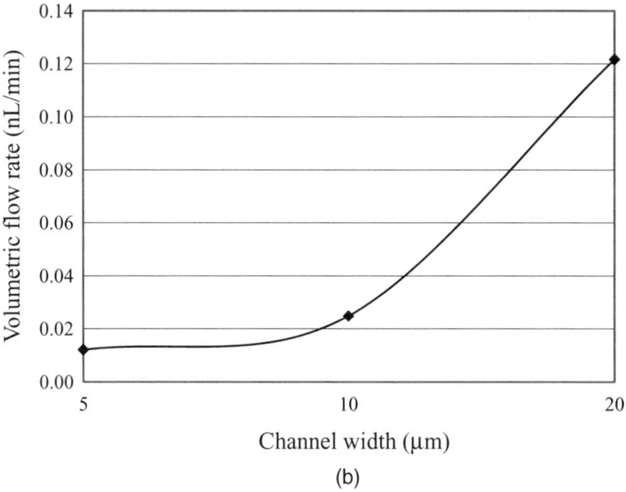

Figure 14.4. (a) Longitudinal velocity contours. Velocity is positive (out of the page) at the top of the cell and negative (into the page) at the bottom. (b) Volumetric flow rate.

our laboratories. The focus of our study is the development of the necessary micro-/nanoengineering tools needed to successfully integrate flagellar motors with microchannel devices without compromising their biological motility. These tools, as will be discussed in details in the following, are developed from a combination of microbiological and MEMS processing protocols. Frequently, compromises must be made between the best microchannel design achievable and the optimized environmental conditions for maximizing the longevity and durability of the cells.

14.3.1 Cell Preparation Protocol

The KAF95 cells, in its original form, are not suitable for integration with microchannels. They must be processed following a specific protocol in order to be tethered onto the surface of a microfabricated device in a controlled manner. The basic cell preparation protocol consists of three main parts: cell growth, cell harvesting, and cell shearing.

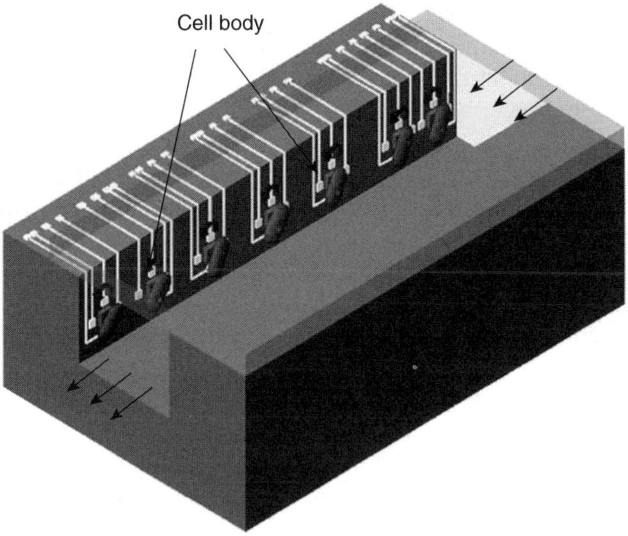

Figure 14.5. Schematic of a microviscous pump actuated by tethered *E. coli* cells.

14.3.1.1 Cell Growth. The KAF95 bacterial cells were grown from a stock solution supplied by Dr. Karen Fahrner of Harvard University. Originally, the stock solution was stored at –80°C to prevent any activity within the cell culture. When the stock solution was thawed, a sample was extracted by using a loop and was streaked onto a Luria–Bertani (LB) agar plate (1% tryptone, 0.5% yeast extract, 0.5% NaCl, pH 7.0) supplemented with 100 mg/mL of ampicillin. After streaking, the cell culture was incubated for 10–12 hours at 37°C to allow for the growth of colonies onto the agar plate.

Once the cell colonies were formed on the agar plate, they were transferred to 10 mL of fresh liquid tryptone broth (1% trypton, 0.5% NaCl, pH 7.0) and incubated for 10–12 hours at 30°C in a rotary shaker at a shaking speed of 200 rpm. After incubation, the cell culture (2.5% inoculum) was transferred to a fresh tryptone broth supplemented with 100 mg/mL of ampicillin and incubated for an additional 4 hours at 30°C.

14.3.1.2 Cell Harvesting. After the 4-hour incubation, the cell culture was ready to be harvested. This was carried out by pouring the 100 mL cell solution into a 500-mL centrifugation bottle and centrifuging at 3400 rpm for 7 minutes at room temperature. During centrifugation, the cell culture settled to the bottom of the centrifugation bottle leaving the supernatant liquid with impurities to be removed and discarded. The cell culture obtained from centrifugation was resuspended in a 25 mL motility medium (10 mM potassium phosphate with 0.1 mM EDTA). Afterward, the cell sample was placed on a Genie 2 vortex system (VWR, West Chester, PA) and shaken vigorously until the cell culture was saturated in the motility medium solution. Once the cell solution was saturated, it was transferred to a 50-mL centrifuge tube and a series of three centrifugations and resuspensions in motility medium was performed. Following the third centrifugation, the supernatant was removed and the cell suspension was transferred to 1 mL of fresh buffer.

14.3.1.3 Cell Shearing. This cell processing step is not a standard step in the general bacterial culturing protocol. It must be carefully controlled in order to achieve a high "tethering efficiency" when combining the flagellar motors with the microchannels. As mentioned earlier, each *E. coli* bacterial cell has an average of six flagellar filaments, each about 10-μm long. To achieve successful tethering, two conditions need to be met. The first condition is that tethering occurs via only one flagellar filament. This is essential because if the cell body is tethered through two or more filaments rotation of the cell body will not occur after tethering. The second condition is that the flagellar filament must be shortened. If the filament is too long, only a low percentage of cells will tether and even if tethering occurs, wobbling instead of smooth rotation will occur.

To satisfy these two conditions, the cell culture is taken through a process known as shearing [17]. In this process a 1 mL cell culture is passed back and forth through a polyethylene tube (0.58-mm inner diameter [ID]; 12-cm long) between two syringes (26 gauge) to shorten the flagellar filaments. To complete the cell preparation protocol, the sheared cell culture is taken through a series of three centrifugations and resuspensions in 25 mL of motility medium to remove the sheared flagellar filaments. After the final wash and removal of the supernatant, the cell culture is resuspended in 1 mL of motility medium and serial dilutions of 10 and 100 times are obtained. At this point, the cell culture is ready for tethering.

A systematic experiment has been carried out to determine the optimum number of passages required to achieve a satisfactory tethering result. In this experiment, five different cell cultures were prepared, each having been sheared a different number of times ranging from 0 to 100 times in increments of 20. Each of the sheared samples was then inserted into five different glass microchannels and left for approximately 20 minutes to allow sufficient time for tethering. The microchannels were then washed with motility buffer to remove untethered cells. In each microchannel, the number of successfully tethered cells was recorded at three different locations. A tethering event is considered successful if the cell body can be seen rotating 360° in a smooth manner. Tethering is considered unsuccessful when either no rotation or only slight movement is observed.

Results of the shearing experiment are displayed in Figure 14.6. As the number of passages increases from 0 to 80, the number of successfully tethered bacteria also increases. However, when the passage number reaches 100, the number of successfully tethered bacteria begins to decrease, indicating that the optimum passage number should be between 80 and 100. Shearing a dense cell suspension tends to cause the flagellar filaments to tangle and fracture. From the results of the shearing experiment, it can be concluded that a passage number between 80 and 100 results in the optimized filament condition, both in terms of

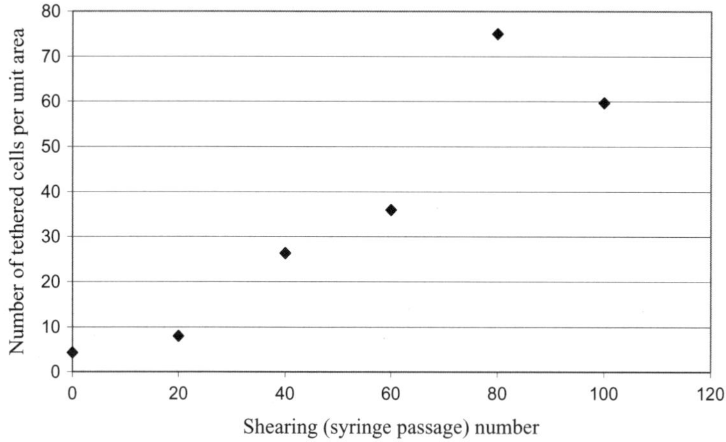

Figure 14.6. Effect of filament shearing on cell tethering efficiency.

length and number, for tethering. When the passage number exceeds 100, it is likely that too many filaments are stripped from the cell bodies and the remaining filaments are too short for successful tethering. It is also possible that excessive shearing can damage the structure of the cells.

14.3.2 Microchannels for Cell Tethering

A properly designed microchannel is critical to a flagellar motor micropump as it provides the necessary fluidic environment for cell tethering and the subsequent rotation. Two different types of microchannels have been proven for the task: an all-glass design and a hybrid design consisting of a polydimethylsiloxane (PDMS) microchannel bonded to a glass capping chip. Prior to our study, cell tethering has only been demonstrated on a flat glass substrate. As will be shown, in addition to glass, cell tethering is also possible on a polymer substrate. Details of the fabrication schemes for both the glass and PDMS microchannels have been extensively documented in the past [16, 18]. Here, a brief description of the fabrication scheme will be provided.

The fabrication process of the all-glass microchannels is as followed. A soda lime glass wafer was first thoroughly cleaned in a piranha solution to remove organic contaminants on the wafer surface. Next, a 200-Å thick layer of chrome and a 2000-Å thick layer of gold were evaporated onto the glass wafer to act as a masking material for wet etching. The metal layers were patterned by photolithography to expose the glass surface where microchannels are desired. Following this step, the wafer was immersed in a liquid etching mixture composed of 100 mL buffered oxide etch (BOE), 120 mL HCl, and 120 mL water. The inclusion of HCl in the etching mixture improved the surface smoothness of the etched area and prevented the formation of sodium dendrites. An unexpected result of the HCl addition was that the etch rate increased by a factor of 10, which had the benefit of avoiding lift-off of the metal masking layer during the etching process. Following etching, the glass wafer was diced into individual microchannel chips and their metal masking layers were stripped using commercial etchant solutions. Thermal fusion bonding was used to add a matching glass capping chip to each microchannel chip. The capping chip contains predrilled through holes designed to allow controlled fluid flow into the microchannel. The fusion bonding process began with cleaning the microchannel and capping chips with a piranha solution. They were then brought into intimate contact to form a prebond. The prebonded chips were placed in a vacuum oven for 30 minutes at 150°C to remove any trapped moisture. Afterward, they were then placed in a muffle furnace and the temperature was ramped at a rate of 5°C/min to the glass melting point of 635°C. The prebonded chips were held at this temperature for 60 minutes to allow the two chips to fuse permanently. Following the bonding process, microfluidic connectors were bonded to the inlet and outlet holes of the capping chip to complete the fabrication process.

The PDMS microchannel was fabricated using soft lithography with the assistance of a SU-8 microcasting mold. The fabrication process for the mold and the subsequent casting of the PDMS is as follows. In the mold fabrication process, liquid SU-8 2100 (MicroChem Corp., Newton, MA) photoresist was first spin-coated onto a cleaned silicon wafer at a speed of 3000 rpm, yielding a film thickness of 100 μm. The wafer was then placed on a level surface for 30–45 minutes to allow for planarization. Afterwards, the wafer was prebaked and allowed to cool slowly to room temperature. The slow cooling step was critical because rapid cooling typically leads to cracking and delamination of the photoresist. After cooling, the photoresist was exposed to UV light for 20 seconds at 900 W in a Karl Suss mask aligner. The exposed photoresist was developed in a SU-8 developer (MicroChem Corp., Garching, Germany) after a postexposure bake. Based on profilometer scans, the accuracy of the height of the SU-8 mold was within ±5 μm.

Casting of the PDMS microchannel began with mixing the liquid PDMS elastomer base with a curing agent at a ratio of 1:10 by volume (Sylgard 184 Silicone Elastomer Kit, Dow Corning, Midland, MI). The mixture was then left in a vacuum chamber for a short period of time to remove dissolved gas. Next, the mixture was poured over the silicon wafer that consisted of the SU-8 mold. The liquid PDMS conformed to the contour of the SU-8 structure and replicated the features of the mold. The molded PDMS was cured for 90 minutes at 80°C. After curing, the solidified PDMS film with molded microchannels was peeled away from the mold.

Similar to the glass microchannel, a predrilled glass chip was used to cap off the PDMS microchannel. However, instead of fusion boding, the glass chip was attached by exposing both the glass and PDMS surfaces to a short oxygen plasma treatment, which has the effect of creating active silanol (Si-OH) groups that allow the two surfaces to be joined together irreversibly with a large bond strength. Fabrication of the PDMS microchannel is completed by attaching microfluidic connects to the inlet and outlet holes on the glass chip. Figure 14.7 demonstrates a completed PDMS microchannel device ready for cell tethering.

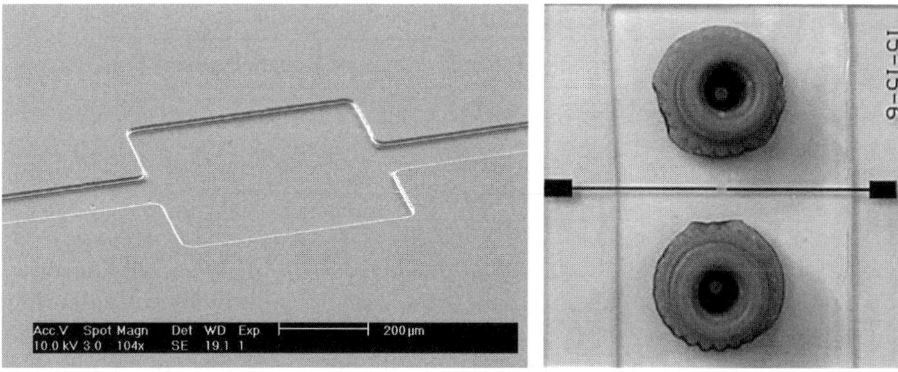

(a) (b)

Figure 14.7. PDMS microfluidic system. (a) SEM picture of the PDMS microchannel. (b) A completed PDMS microchannel device with microfluidic connectors attached.

14.3.3 Cell Observation and Control

The experimental setup was developed to inject the processed cell culture into the microchannel with high precision and analyze the subsequent behavior of the cells. The setup consists of a Leica DM/IL inverted phase contrast microscope (Leica, Wetzlar, Germany), an Advanced Strobe Electronics microstroboscope (Harwood Heights, IL), a Wavetek 19 function generator (Aeroflex GmbH, Ismaning, Germany), a Harvard PHD2000™ syringe pump (Harvard Apparatus, Holliston, MA), a QImaging QiCAM™ cooled CCD camera (QImaging, Surrey, BC, Canada), and a Dell Optiplex™ GX260 PC (Dell, Round Rock, TX). When necessary, a Zeiss Axiophot fluorescent microscope (Carl Zeiss Microimaging GmbH, Gottingen, Germany) was also used to obtain stationary close-up pictures of individual cells. The syringe pump was used to deliver a controlled liquid flow into the microchannel system, when needed. The CCD camera and PC were used to record the behavior of cells through the phase contrast microscope. The PC was also used as a processing station for analyzing the frame-by-frame information recorded by the camera. The microstroboscope was used to measure the rotational rate of tethered cells. The function generator controlled the strobe frequency. To measure the rotational rate, a suitable starting strobe frequency was used at the beginning to flash the cells. It was then gradually increased until the cells appeared to be stationary under the strobe light. At this point, the strobe frequency is in sync with the rotational frequency of the cells.

14.3.4 Tethering Efficiency and Rotation Rate Distribution

Integration of the flagellar motors with the microchannel involves the loading of processed KAF95 cell cultures into the microchannel where the cells are allowed to tether for 20–30 minutes before being "cleaned" by a motility buffer to remove untethered cells [19]. Not all cells introduced to the channel will tether and only a percentage of the tethered cells will rotate eventually. A statistical study was

TABLE 14.2. Cell Tethering on Glass-Type Substrates

Surface	% Tethered	Surface Roughness (μm)
Glass cover slide	66.7	0.14
Quartz	64.7	0.2
Soda lime	0.0	3

carried out to determine the tethering efficiency of the KAF95 cells on three common substrate materials that have been used previously for cell tethering: smooth soda lime glass (of a cover slide), etched soda lime glass, and etched quartz. Here, the tethering efficiency is defined as the percentage of the tethered cells that rotate smoothly. The cell concentration used in the study was about 8×10^7 cells/mL, while the capacity of the microchannels was approximately 16 μL. When the microchannel is fully loaded with the cell suspension, there should be about 1.3×10^6 cells inside the channel.

Disregarding the substrate material used, approximately 30% of the cells introduced into the microchannel actually adhere onto the substrate. However, the surface quality of the substrate will affect the ability of the tethered cells to rotate. As demonstrated in Table 14.2, the glass cover slide as well as the quartz substrate has a similar tethering frequency of 65%. In contrast, the etched soda lime glass has almost zero tethering. Based on this result, it is clear that the tethering efficiency is correlated with the surface roughness of the substrate. When the average roughness is higher than the tethering height of the cell, cell rotation is prevented. This explains why rotation is difficult to achieve on etched soda lime glass, which has a surface roughness more than one order of magnitude higher than etched quartz glass and glass cover slide. Further confirmation was obtained when an improved etching solution produced a much smoother etch surface on the soda lime glass and cell tethering on the surface achieved a similar tethering efficiency as the quartz and glass surfaces.

The tethering experiment also yielded the rotation rate distribution of the tethered cells. Figure 14.8 demonstrates that distribution measured from a sample of 75 tethered cells on a glass cover slide. The smooth curve represents the normal distribution and serves as a reference for the experimental distribution. The mean rotation rate of the distribution is approximately 6.5 Hz, with a variation ranging from 1.5 to 12 Hz.

14.3.5 Life Expectancy in a Microchannel

Life expectancy of the tethered *E. coli* cells can obviously have a significant impact on the overall lifespan of the flagellar motor micropump [20]. As such, experiments were performed to determine the average survival time of tethered cells in a microchannel environment and possible means for controlling or enhancing it. For the purpose of this study, the survival time was defined as the length of time it took for the percentage of live rotating cells to be reduced to 45% of the original population. Two types of experiments were conducted. In the

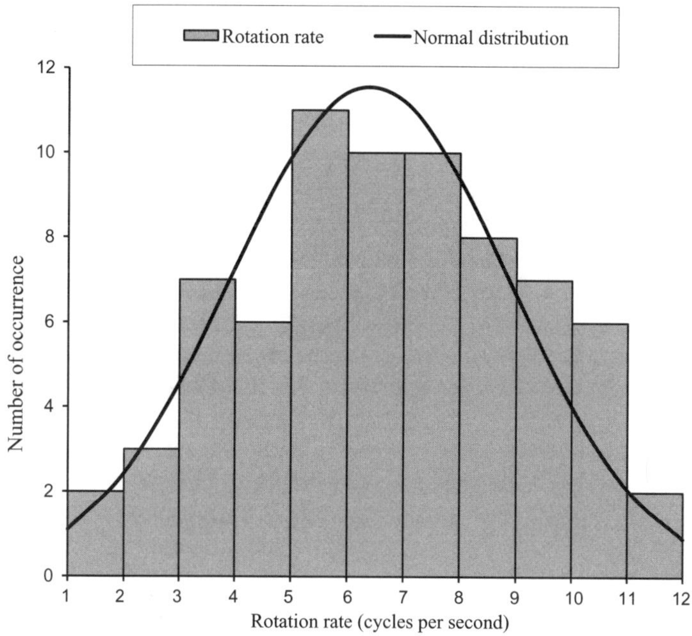

Figure 14.8. Rotation rate distribution based on a sample of 75 tethered cells.

first experiment, cells were allowed to tether onto the surface of a PDMS microchannel and the number of rotating cells was counted. The same cells were then recounted every 4 hours and the new number was recorded. The total experimental time was about 170 hours (1 week). In the second experiment, a similar procedure was performed, except that the tethered cells were replenished with fresh motility buffer at every 12-hour mark. Without the introduction of fresh motility buffer, the number of rotating cells decreases exponentially with time. Based on the 45% cutoff mark definition, the average survival time of tethered cells is about 20 hours. After 80 hours, the number of rotating cells stabilizes to about 10% of the original number. When the experiment was terminated at the 170-hour mark, about 5% of the original cells still remain active.

The addition of fresh motility buffer to the rotating cells improved the survival time of the cells. In fact at the 45% cutoff mark, cells with fresh motility buffer outlived those without by almost 2.5 times. Furthermore, fresh motility medium stabilizes the number of rotating cells over several hours once it is introduced, as indicated by plateaus in the cell rotation rate that occurs immediately after the addition of the fresh medium. There is also evidence that the addition of fresh medium can potentially revive cells that have stopped rotating. This is supported by the slight increase in the percent of rotating cells when fresh medium is introduced.

In a separate experiment that used a similar procedure as the one described above, the aging effect of a group of rotating cells was monitored through their

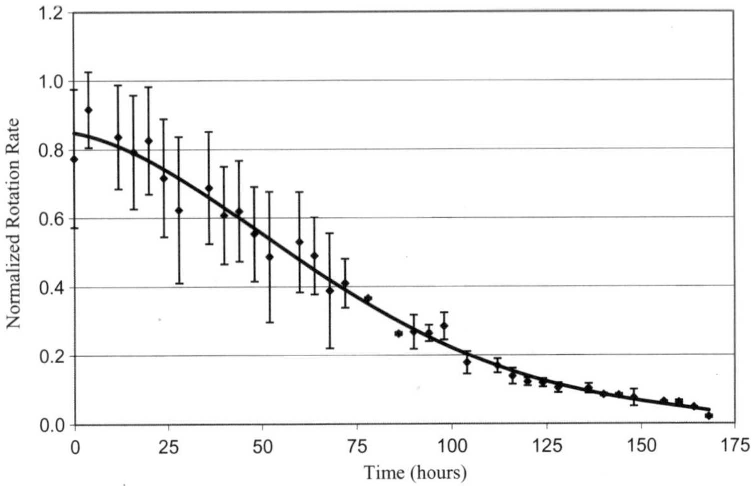

Figure 14.9. Rotation rate decay of tethered cells.

rotational rate. Figure 14.9 shows the result based on a sample size of 100 cells. In this figure, the raw rotational rate data for each cell was normalized by the maximum rotation rate of that cell and the normalized data was then averaged at every given time. In the aging experiment, the motility buffer was not replenished. As can be observed, the normalized rotational rate decreases fairly linearly with time from 0 to 100 hours. Afterward, the decay is mostly exponential. The average time for the rotational rate to reach 45% of the original level is approximately 60 hours. A number of tethered cells remained active after 168 hours albeit rotating at a very small percentage of the original rotation rate.

14.3.6 Hydrodynamic Loading

The rotation rate of a tethered cell is a function of the torque capability of its flagellar motor and the fluidic shear stress in the space between the substrate and the rotating cell body [21]. Assuming the cell is rotating in the counterclockwise direction, a fluid flow moving from left to right across the cell will impose a drag force on the cell body. The level of the drag force is linearly proportional to the surface shear stress τ_w on the substrate where

$$\tau_w = \frac{6\mu}{Wh^2}Q. \tag{14.4}$$

Here, μ is the viscosity of the fluid, Q is the flow rate, W is the width of the microchannel, and h is the depth. The surface shear stress imposes a counter-torque T on the flagellar motor that can be estimated by

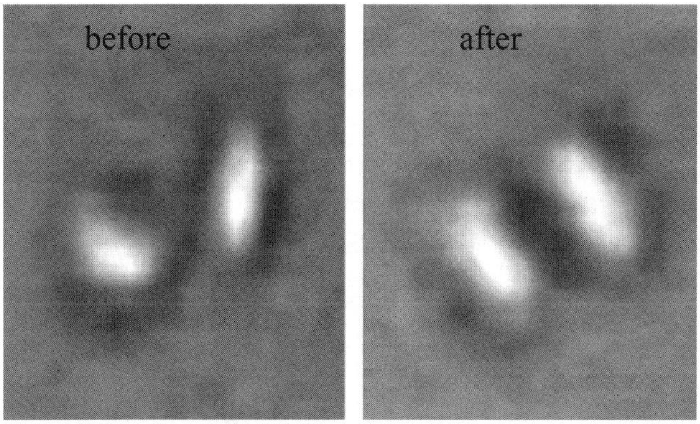

Figure 14.10. Effect of hydrodynamic loading on a pair of tethered cells.

$$T = \int_0^l \tau_w l_c d_c dl. \tag{14.5}$$

Since T is proportional to τ_w, it is also proportional to Q. Based on this, by using different levels of Q, different levels of countertorque or hydrodynamic loading can be imposed on the tethered motor. Figure 14.10 demonstrates the effect of hydrodynamic loading on the rotational behavior of a pair of neighboring cells in a microchannel. Without a background flow in the channel, the cells rotate out of phase with each other. When a flow is initiated, the cells begin to experience additional shear stress and their rotational speed declines accordingly. When the flow rate reaches a critical level, both cells cease rotation and orient themselves at a 45° angle with respect to the impinging flow. When the micro channel flow is turned off, the stopped cells resume their rotation, and at this time in-phase with each another.

The hydrodynamic loading results indicate that a microchannel flow rate of about 1000 μL/min will freeze the rotating cells. Based on Equations 14.4 and 14.5, the corresponding hydrodynamic T imposed on the cells is 3.9 nN-nm. This value is very close to the previously reported 4 nN-nm torque capability of the flagellar motor, indicating that the tethered cells cease rotation when the hydrodynamic loading equalizes the motor torque.

The hydrodynamic loading experiment also reveals the possibility of using a microflow-based method to synchronize the rotational behavior of the tethered cells. This concept is demonstrated in Figure 14.11. From 0 to 0.2 second when hydrodynamic loading is not applied, the phase angles of a pair of rotating cells are not synchronized. From 0.25 to 0.45 second when hydrodynamic loading is applied, the same two cells are locked in at a constant phase angle. At 0.5 second when hydrodynamic loading is removed, their phase angles become synchronized. Using this mechanism, it is possible to synchronize the rotation of a group of tethered cells, which can be important in optimizing the performance of the flagellar motor micro pump. Overall, hydrodynamic loading is a very robust

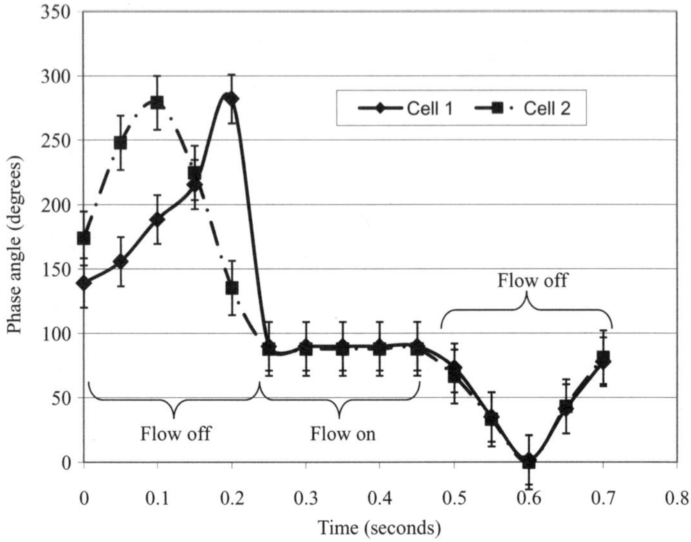

Figure 14.11. Phase angles of tethered cells before (flow off), during (flow on), and after hydrodynamic loading (flow off).

noncontact cell manipulation method that does not require extensive equipment needed in other techniques such as optical trapping and electrorotation. At this point, there is no indication of any irreversible ill effect on the cells that were manipulated by hydrodynamic loading. Those cells typically resume their rotation once hydrodynamic loading is removed.

14.3.7 Flagellar Filament Adhesion Force

As the microchannel flow rate is raised beyond the 1000 μL/min mark, the stopped cells begin to rotate toward the direction of the flow. Complete alignment occurred at 9000 μL/min. Between 9000 and 15,000 μL/min, no observable change in the cell orientation can be detected. When the flow rate is higher than 15,000 μL/min, the cells begin to disappear from the substrate. Using Equation 14.4, the corresponding shearing force is determined to be approximately 2.4 pN. Not all tethered cells disappear at the same shearing force. In fact, for the ~200 cells monitored during the experiment, cell shearing took place in a stepwise manner. Most cells (~150) were sheared off the substrate between 6.7 and 17 pN. At the lower limit of 2.4 pN, only ~10 cells were sheared off. Similarly, at the upper limit of 38.3 pN, only ~5 cells were sheared off. The average shearing force required to remove a tethered cell is approximately 14.5 pN. This value can be considered the average adhesion force of the flagellar filament.

To determine the nature of the filament adhesion force, results from other forms of microbiological binding systems can be used as a guideline. Liang et al. [22] investigated the adhesion forces of piliated *E. coli* to mannose surfaces. Pili are hollow, hairlike protein structures that originate from the cytoplasmic membrane of the bacteria. They allow the bacterial cells to attach to each other and also to foreign surfaces. Unlike flagellar filaments, pili are not attached to rotating motors. They also tend to be much shorter. Liang et al.'s study found that the adhesion force of pili to a mannose surface was between 3.5 and 18 pN. Another microbiological binding system that the filament adhesion can be compared with is the streptavidin–biotin ligand–receptor interaction, which has been extensively applied in the purification and sensing of biomolecules. The bonding force in the streptavidin–biotin system covers a wide range, from a few to hundreds of picoNewtons [23]. Our results indicate that the flagellar filament adhesion force falls in the same range as that of the pili–surface and streptavidin–biotin systems.

14.3.8 Cell Response to External Stimuli

Just like free-swimming bacteria, the tethered cells in a microchannel can be extremely sensitive to the presence of external stimuli. The energy source for the flagellar motor is a proton flux moving through the cytoplasmic membrane of the cell. It is therefore not surprising that the environmental pH would affect the rotation rate of the tethered cells. With the assistance of a microstroboscope, experiments were performed to determine the relationship between environmental pH and cell rotation for a group of about 50 tethered cells in a microchannel. Motility mediums (25 mL each) with different acidity conditions were prepared by adding either hydrochloric acid or potassium hydroxide to the medium while keeping the medium viscosity constant. The pH level tested ranged from 5.5 to 8.5 in increments of 0.5. Changing the pH value was performed by injecting the microchannel with 1 mL of the new medium. As soon as a new medium was added, the rotation rates of the tethered cells were measured and recorded. The same procedure was repeated five times for each pH value to obtain a statistically converged result. The result shows that the rotation rate is the highest at pH 7. At pH 5.5 and 8.5 (the experimental pH extremes), the rotation rate is about 50 and 70% of the pH 7.0 value, respectively. This finding corresponds very well with the study by Meister and Berg [24], which demonstrated that the torque generated by tethered *Streptococcus* cells dropped by around 80% from the pH 7.0 level when the extracellular pH was either lowered to 5.5 or raised to 8.0.

Another stimulus that the tethered cells are sensitive to is UV exposure. This phenomenon is known as photodamage [25]. While the exact cause of this damage is currently unknown, three possible mechanisms have been proposed: transient local heating, optical absorption, and the creation of toxic chemical species within the cell [26]. For *E. coli*, the creation of a toxic excited state singlet oxygen due to UV light is well documented. In our experiment, the rotation rate of multiple tethered cells was measured at 10-second intervals during exposure to UV light at the 100-W intensity level commonly used in fluorescent microscopy. Results shown in Figure 14.12 indicate that for an exposure time of up to 20 seconds, cell rotation rate is not affected. As the cumulative time exceeds 20 seconds, a linear decline in the rotation rate is detected for up to 60 seconds. Beyond the 60-second

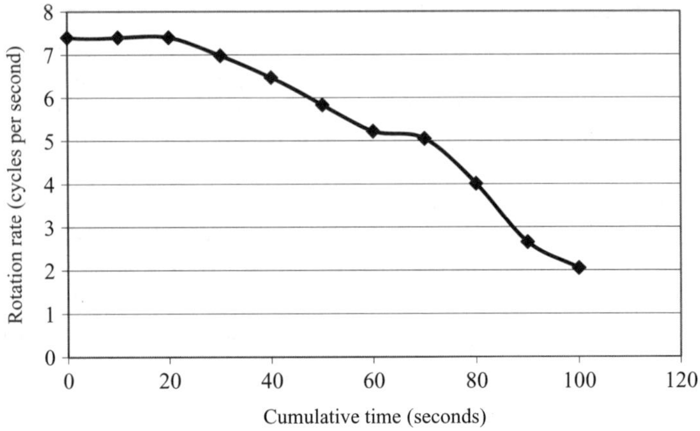

Figure 14.12. UV effect on rotation rate of tethered cell.

mark, decline in the cell rotation rate accelerates, and at the 100-second mark, the rotation rate has dropped to about 25% of the original value. This result indicates that photodamage to cell rotation is a gradual, instead of a catastrophic, process. There are two linear stages in the decline of the rotation rate. The first one takes place between 20 and 60 seconds and the second one between 70 and 100 seconds. The decline rate of the second stage is twice of that of the first, indicating an accelerated decay in the capability of the flagellar motors during the second stage.

Similar result has also been observed by Neuman et al. [26]. However, they found a constant decay slope and speculated that the toxic species had a short lifetime. Results of our study indicate that the effect of photodamage is accumulative instead. A possible explanation for the discrepancy is that the UV intensity in our study was 1000 times higher than the one in Neuman et al.'s (100 W vs. 100 mW). A high UV intensity can cause toxic species to be created at such as high rate that the cells are ill equipped to handle, thus causing the accumulation of the harmful species inside the cells.

14.3.9 Preferential Tethering

The ability to transform tethered cells into functioning microfluidic devices lies largely in the ability to dictate the locations where cell tethering is allowed to take place in the microchannel. For the flagellar motor micropump, the KAF95 cells must be tethered "specifically" along a single line close to the sidewall in order for the pump to generate a measurable flow. There are two possible approaches for achieving this goal: by patterning the microchannel with a thin-film material that the cells have strong affinity to (preferential tethering) or by using a selective removal process similar to the lift-off process used in metal thin-film patterning in MEMS fabrication (microsieve tethering). Between the

TABLE 14.3. Cell Tethering on Various Materials

Material	Tethered	Rotate	Surface Characteristic
Spin-on-glass	Yes	Yes	Negatively charged
Polyimide	Yes	Yes	Hydrophobic
PDMS	Yes	Yes	Hydrophobic
Benzocyclobutene (BCB)	Yes	Yes	Hydrophobic
Polyethylene glycol	Yes	Yes	Hydrophilic
Silanized glass	Yes	Yes	Positively charged
Poly-L-lysine treated glass	Yes	Yes	Positively charged

two approaches, preferential tethering is preferable since the processing protocol is more "cell friendly" and relies mostly on the natural self-assembling behavior of the cells. In our flagellar motor pump project, preferential tethering was examined extensively utilizing glass, a negatively charged material, as the tethering material in conjunction with a broad range of previously considered nontethering substrate surfaces. The glass tethering sites were fabricated from spin-on-glass (SOG). SOG is similar to bulk glass in terms of chemical and mechanical properties except that it can be spin-coated and patterned photolithographically. In our study, for each tethering/nontethering material combination, SOG tethering islands were patterned on a nontethering substrate and then integrated into a PDMS microchannel system. Processed cell culture was then introduced into the channel and allowed to tether on the material surface as previously described. Table 14.3 displays the preferential tethering results conducted for different combinations of tethering and nontethering materials.

Perhaps the most significant result of our experiment is that flagellar filament binding occurs on all material combinations regardless of the charge properties and hydrophobicity of the materials. This suggests that flagellar filament binding to a substrate material is nonspecific in nature. Consequently, it will be impossible to accomplish specific tethering through preferential tethering alone, at least with the nontethering materials examined. Additional experiments were conducted to measure the filament adhesion force in order to determine the feasibility of using hydrodynamic loading to selectively remove the tethered cells on the nontethering materials. Unfortunately, the experiment revealed that the filament adhesion force on various materials was approximately the same. In essence, if a sufficiently high hydrodynamic loading is applied, tethered cells on both the tethering and nontethering materials will be sheared off.

When the preferential tethering results are combined with the filament adhesion force results described earlier, it can be concluded that the interaction between flagellar filaments and the tethering materials tested is van der Waals in nature. The surface charge characteristics and hydrophobicity of the tethering material have no significant impact on filament binding. In our experiments, negatively charged glass, positively charged poly-L-lysine, hydrophilic poly(ethylene glycol) (PEG), and hydrophobic PDMS were used. None of them led to a different tethering efficiency or filament adhesion force. The consistency observed in the tethering behavior from surface to surface is evident that hydrogen bond, hydrophobic interaction, and ion–dipole interaction could not have played a significant role in the bonding mechanism.

14.3.10 Microsieve for Specific Tethering

A microsieve is a thin, flexible PDMS membrane with arrays of patterned through holes [27, 28]. To achieve specific tethering using a micro sieve, the sieve is first attached to a flat substrate and cells are then allowed to tether on the sieve surface and inside the sieve holes. When tethering is complete, the sieve is lifted off the substrate, leaving cells tethered specifically at locations where the sieve holes previously occupy. Cells tethered on the sieve surface are removed along with the sieve. The sieve was fabricated by spin coating a 30-μm-thick PDMS layer over a mold of either silicon (Fig. 14.13a) or SU-8 posts. The silicon posts were fabricated by deep reactive ion etch (DRIE), while the SU-8 post was patterned on a silicon wafer using UV lithography. After the PDMS film was cured, it was peeled away from the posts resulting in a PDMS sieve with arrays of holes that correspond to the diameter of the posts (Fig. 14.13b). In some instances, the spin-coated PDMS layer was thicker than the height of the silicon posts and the sieve holes were partially or wholly covered by a thin layer of PDMS after the sieve was lifted off from the mold. When this occurred, the PDMS sieve was placed in a reactive ion etcher (RIE) and the blocking PDMS layer was removed by dry etching using O_2 and CF_4. The RIE-processed PDMS surface was typically rougher than the unprocessed surface but did not affect cell tethering significantly.

The PDMS sieve can be used to pattern tethered cells in two approaches. In the first approach, the sieve is attached to a glass cover slide and two to three drops of the cell suspension are placed on top of the sieve. After 15–20 minutes of cell tethering time, the glass cover slide is immersed in a motility medium and shaken gently to wash away the untethered cells. Following this, the cover slide is removed from the medium and the sieve is lifted off. This method is effective

(a) (b)

Figure 14.13. Cell patterning with PDMS microsieve. (a) Silicon mold. (b) PDMS microsieve.

in confining tethered cells at designated locations on the cover slide but ineffective in maintaining a fluidic environment necessary for cell motility.

In the second approach, the sieve is first integrated with a PDMS microchannel. Cell culture is then injected into the channel and allowed to tether for the same amount of time. Untethered cells are removed by a flow of motility medium through the channel. With the microchannel, tethered cells can be specifically patterned while remaining in a fluidic medium. The rotation rate of the specifically tethered cells was measured using the microstroboscope and compared with that of randomly tethered cells on a glass surface. The result indicated that the application of the PDMS sieve to achieve specific cell tethering did not affect the motility of tethered cells unfavorably.

Microsieves with different hole dimensions were experimented. The average number of tethered cells in a 50-μm sieve is about 50. For the 30-μm sieve, using an identical cell concentration, that number is reduced to 25. Further reduction in the number of tethered cells can be achieved by lowering the cell concentration. When the 30-μm sieve is used along with a cell concentration of $10^{-1.5}$, arrays of 6–15 tethered cells can be consistently achieved, as displayed in Figure 14.14. When a lower cell concentration of 10^{-2} is used instead, single-cell tethering is achieved in two out of an array of nine sieve holes.

Figure 14.15 shows the relationship between the number of tethered cells at each tethering location and the sieve hole size for three different cell concentrations. Based on the figure, it is tempting to conclude that single-cell tethering is indeed achievable when a small sieve hole is used with a low cell concentration. In reality, this is hardly the case because as the cell concentration and sieve hole size are reduced, the probability for a single cell to successfully enter a sieve hole is also reduced. In the microsieve experiments, the area ratio of sieve holes to PDMS substrate was approximately 1/200 for the 30-μm sieve. This implies that when the cells were deposited on the sieve, assuming an even distribution, most of the cells would land on the sieve surface as opposed to inside the sieve holes.

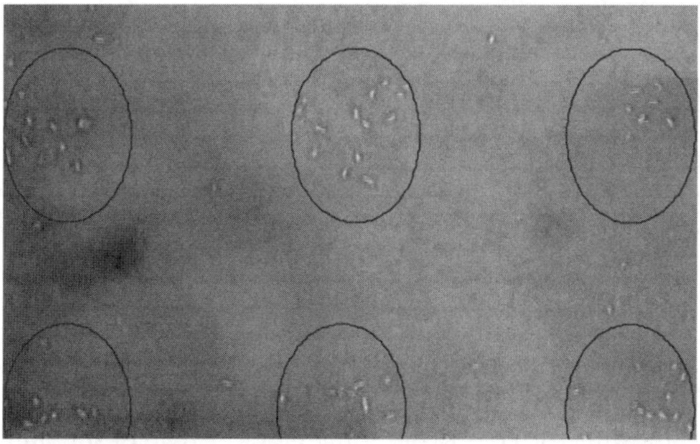

Figure 14.14. Preferential cell tethering using a 30-μm microsieve.

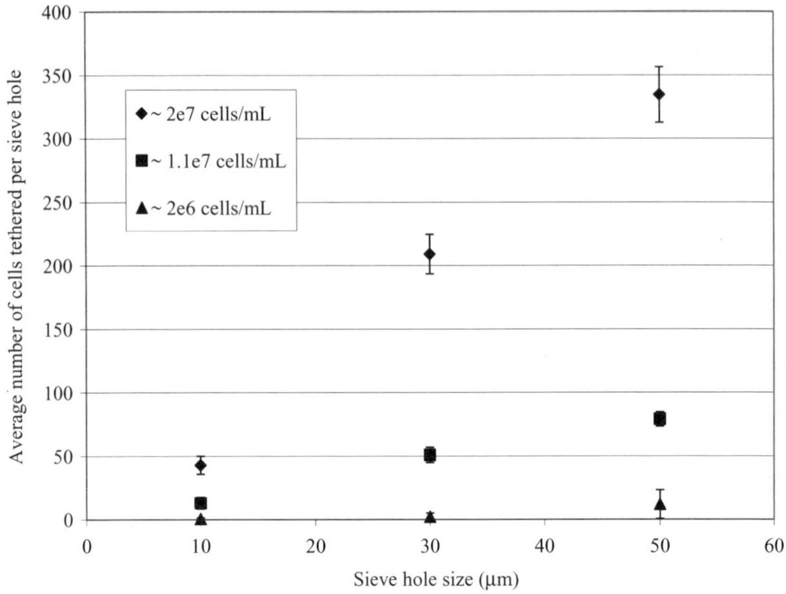

Figure 14.15. Effect of sieve hole size on preferential tethering.

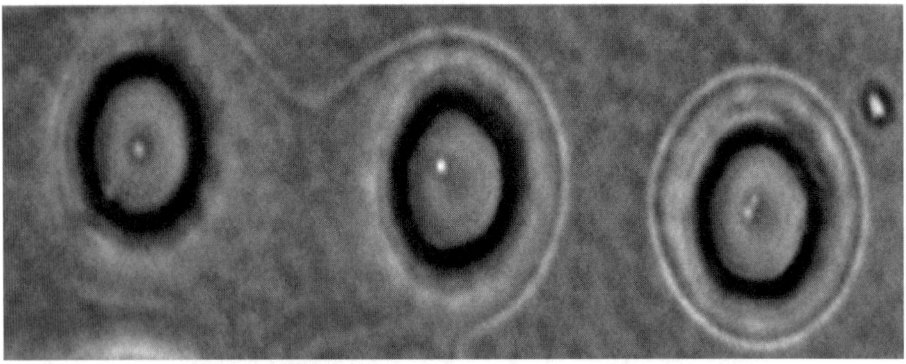

Figure 14.16. Single-cell tethering using a 10-μm sieve and the dip-pen technique.

Our experiments indicated that single-cell confinement was achieved in about 20% of the sieve holes. When this level of cell confinement is coupled with the 30% tethering efficiency, the probability of achieving an individual rotating cell is only about 6%.

The success rate of single-cell tethering can be significantly improved by utilizing the dip-pen technology. In this method, a micropipette tip can be used to transfer a low-concentration (10^{-2}) cell culture directly to the top of the sieve holes, thereby allowing the cells to enter the holes instead of spreading out across the sieve. This method for cell patterning provided immediate results in our experiments, as seen in Figure 14.16, which shows a line of tethered cells confined by the sieve holes numbering one, one, and two. The probability of single-cell tethering in a sieve hole using the dip-pen technology is estimated to be about 30% as opposed to 6% without the dip pen.

The PDMS microsieve along with the dip-pen method has by far provided the most consistent and reliable technique in achieving single-cell tethering. However, further improvement is still required in order to raise the success rate needed for actual flagellar motor device fabrication. One possibility is to develop a high-precision cell delivery method so that the micropipette can deliver single cells to individual sieve holes. This method must also be accompanied with improved cell processing protocols where cell motility and its tethering behavior can be more accurately predicted and controlled.

14.4 CONCLUDING REMARKS

This chapter is intended as an introduction to the concept of utilizing bacterial flagellar motors as microactuators in microfluidic systems. Most of the information provided in the chapter is related to the unique biological and microfabrication techniques developed specifically to tether KAF95 *E. coli* cells in a microchannel system while maintaining, and sometimes prolonging, their motile behavior. It is our hope that researchers interested in flagellar motor devices will utilize this information and expand the study into areas beyond microfluidics. Flagellar motor is truly one of the most amazing, and in some cases most debated, microorganisms in nature. It has such a large power-to-size ratio that should be useful to all micromachines that require an efficient mechanical power supply. However, just like any biological entities, flagellar motors can only function in a biologically friendly environment, which is difficult to create in a man-made microsystem. As a result, future success in flagellar motor devices will depend largely on innovative designs that maximize the potential of flagellar motors without severely compromising the engineering performance.

ACKNOWLEDGMENTS

The primary funding for our work is provided by the Electrical, Communications and Cyber Systems (ECCS) and Civil, Mechanical and Manufacturing Innovation (CMMI) Divisions of the National Science Foundation (ECS-0201004, ECS-0401196, and CMS-0508435). Partial funding is also supplied by the Arkansas Biosciences Institute. The authors would like to thank Dr. Karen Fahrner of Harvard University for generously supplying the KAF95 cell strain.

REFERENCES

[1] P. Gravesen, J. Branebjerg, and O. S. Jensen, "Microfluidics—A review," *J. Micromech. Microeng.*, 3, pp. 168–182, 1993.

[2] A. Manz, N. Graber, and H. M. Widmer, "Miniaturized total chemical analysis systems: A novel concept for chemical sensing," *Sens. Actuators B*, 1, pp. 244–248, 1990.

[3] N. Nguyen, X. Huang, and T. Chuan, "MEMS-micropumps: A review," *J. Fluids Eng.*, 124, pp. 384–392, 2002.

[4] S. Zeng, C. H. Chen, J. C. Mikkelsen Jr., and J. G. Santiago, "Fabrication and characterization of electroosmotic micropumps," *Sens. Actuators B*, 79, pp. 107–114, 2001.

[5] R. M. Moroney, R. M. White, and R. T. Howe, "Ultrasonically induced microtransport," *Proceedings of MEMS, 4th International Workshop on Micro Electromechanical Systems*, Nara, Japan, pp. 277–282, 1991.

[6] F. M. White, *Viscous Fluid Flow*, 3rd ed., Columbus, OH: McGraw-Hill, 2006.

[7] M. Sen, D. Wajerski, and M. Gad-el-Hak, "A novel pump for MEMS applications," *J. Fluids Eng.*, 118, pp. 624–627, 1996.

[8] D. Decourtye, M. Sen, and M. Gad-el-Hak, "Analysis of viscous micropumps and microturbines," *Int. J. Comput. Fluid Dynam.*, 10, pp. 13–25, 1998.

[9] L. M. Phinney, J. D. Kuppers, and R. C. Clemens, "Thermal conductivity measurements SUMMiT™ V polycrystalline silicon," Sandia Report, SAND2006-7112, 2006.

[10] K. Strawhecker, D. B. Asay, J. McKinney, and S. H. Kim, "Reduction of adhesion and friction of silicon oxide surface in the presence of n-propanol vapor in the gas phase," *Tribol. Lett.*, 19(1), pp. 17–21, 2005.

[11] P. Nelson, *Biological Physics*. New York: W. H. Freeman, 2004.

[12] E. Muneyuki, T. Watanabe-Nakayama, T. Suzuki, M. Yoshida, T. Nishizaka, and H. Noji, "Single molecule energetics of F_1-ATPase motor," *Biophys. J.*, 92(5), pp. 1806–1812, 2007.

[13] H. C. Berg, "Constraints on models for the flagellar rotary motor," *Philos. Trans. R. Soc. Lond., B, Biol. Sci.*, 355, pp. 491–501, 2000.

[14] K. A. Fahrner, "*Studies of bacterial flagellar motors and filaments*," PhD dissertation, Harvard University, Cambridge, MA, 1995.

[15] R. K. Soong, G. D. Bachand, H. P. Neves, A. G. Olkhovets, H. G. Craighead, and C. D. Montemagno, "Powering an inorganic nanodevice with a biomolecular motor," *Science*, 290(5496), pp. 1555–1558, 2000.

[16] S. Tung, J. W. Kim, A. Malshe, C. C. Lee, and R. Pooran, "A cellular motor driven microfluidic system," *Transducers'03–12th International Conference on Solid-State Sensors, Actuators and Microsystems*, Boston, MA, June 8–12, 2003.

[17] X. Chen and H. C. Berg, "Torque-speed relationship of the flagellar rotary motor of *Escherichia coli*," *Biophys. J.*, 78, pp. 1036–1041, 2000.

[18] R. Pooran, M. Al-Fandi, S. Tung, J.-W. Kim, N. Kotagi, and J.-S. Lee, "Bacterial flagellar motors as microfluidic actuators," *International Mechanical Engineering Congress and RD&D Expo*, IMECE2004-61224, Anaheim, CA, November 13–19, 2004.

[19] S. Tung and J.-W. Kim, "Microscale hybrid devices powered by biological flagellar motors," *IEEE Trans. Autom. Sci. Eng.*, 3(3), pp. 260–263, 2006.

[20] M. Al-Fandi, J. W. Kim, A. Malshe, S. Tung, J. Jenkins, and R. Pooran, "Chemosensitivity and reliability of flagellar rotary motor in MEMS microfluidic actuation system," *Sens. Actuators B, Chem.*, 114(1), pp. 229–238, 2006.

[21] S. Tung, J.-W. Kim, and R. Pooran, "Rotational control of tethered bacterial flagellar motor," *Proceedings of the IEEE NANO Conference*, Dallas, TX, August 2008.

[22] M. N. Liang, S. P. Smith, S. J. Metallo, I. S. Choi, M. Prentiss, and G. M. Whitesides, "Measuring the force involved in polyvalent adhesion of uropathogenic *Escherichia coli* to mannose—Presenting surfaces," *Proc. Natl. Acad. Sci. U. S. A.*, 97(24), pp. 13092–13096, 2000.

[23] J. Wong, A. Chilkoti, and V. T. Moy, "Direct force measurements of the streptavidin—Biotin interaction," *Biomol. Eng.*, 16, pp. 45–55, 1999.

[24] M. Meister and H. C. Berg, "The stall torque of the bacterial flagella motor," *Biophys. J.*, 52, pp. 413–419, 1987.

[25] K. Konig, Y. Tadir, P. Patrizio, M. W. Berns, and B. J. Tromberg, "Effects of ultraviolet exposure and near infrared laser tweezers on human spermatozoa," *Hum. Reprod.*, 11, pp. 2162–2164, 1996.

[26] K. C. Neuman, E. H. Chadd, G. F. Liou, K. Bergman, and S. M. Block, "Characterization of photodamage to *Escherichia coli* in optical traps," *Biophys. J.*, 77, pp. 2856–2863, 1999.

[27] R. Pooran, M. Al-Fandi, S. Tung, and J.-W. Kim, "Patterning of *Escherichia coli* flagellar motors in a microfluidic system," *International Conference on Bio-Nano-Informatics Fusion*, Marina Del Ray, CA, July 2005.

[28] S. Tung and J.-W. Kim, "Putting *E. coli* to good use," *IEEE Nanotechnol.*, 2(1), pp. 4–8, 2008.

15

GENE INJECTION AND MANIPULATION USING CMOS-BASED TECHNOLOGIES

Arati Sridharan and Jit Muthuswamy

15.1 INTRODUCTION

The aim of this chapter is to explore avenues by which CMOS technologies could play an important role in accelerating advances in genetic engineering. Direct gene manipulation in living organisms involves (1) understanding the general control parameters for a particular gene, (2) synthesizing a nucleic acid-based target that controllably manipulates specific gene expression to give desired results, and (3) injecting the synthetic target construct via various gene injection methods into living cells. The former two steps have been already extensively studied, and the fundamental basis of biological information processing is summarized in the next section. Additionally, CMOS technologies already have a major impact in understanding genetic circuits and other gene interactions via the development of deoxyribonucleic acid (DNA)-based gene arrays. The third step involving delivery of desired nucleic acid payloads into living cells and tissues is a significant problem since the cell membrane is a formidable barrier to circumvent. Various methods of gene injection that either take advantage of the inherent cell transport properties or alter the physical state of cell membrane have been developed to deliver these synthetic gene constructs. Techniques such as engineered viruses, electroporation, sonoporation, and other mechanical membrane perforations are currently being developed.

In this chapter, our purpose is to allow the reader to critically assess the complexities of injecting synthetic gene constructs or nucleic acids into living cells

CMOS Biomicrosystems: Where Electronics Meet Biology, First Edition. Edited by Krzysztof Iniewski.

using CMOS-based architectures. In Section 15.1, we briefly review the fundamental roles of genes in biological information processing. In Section 15.2, we lay out the current physical methods of gene injection and explore the potential of integration with CMOS technologies. Promising methods for gene transfection, such as electroporation methods, that can be incorporated readily into CMOS-based microscale platforms will be highlighted. In Section 15.3, we assess the general complexities of interfacing biological cells and tissues with current CMOS technology for gene injection. Key technical challenges that need to be overcome for the success of hybrid CMOS platforms for gene injection are biocompatibility issues, ensuring adequate cell adhesion at the electrode interface, handling and upkeep of living cells, media and other biological molecules, integration of biological assays and measurements, long-term stability, and so on. Engineering challenges also include the multiplexing of analog systems, microfluidic chambers, and relevant microarrays with microelectromechanical systems (MEMS) technologies with the digital architecture. CMOS-based bioelectronics can play a significant role in exciting future applications in lab-on-chip monitoring and diagnostic devices, synthetic biology applications, high-throughput assays for the drug development industry, and biological computing.

15.1.1 Fundamentals of Biological Information Processing

From a biological perspective, the fundamental information processing occurs at the genetic level [1]. Genes in most life-forms are composed of DNA components. DNA is composed of four nucleotides: adenine (A), thymidine (T), cytosine (C), and guanine (G). When encoded, DNA forms a double-stranded helix composed of complementary sequences of base pairs (A-T) and (G-C). The central dogma of molecular biology suggests that at a high level, information flow within a cell occurs from DNA to RNA to protein manufacture as shown in Figure 15.1. Essentially, the genetic information encoded in DNA (which serves as the master copy) is *transcribed* into RNA (temporary template copy), which in turn is *translated* into constructing proteins that perform as "nanomachines" within the cells. The minimum information unit required for encoding a single amino acid (basic unit of a protein) is three nucleotides, which is ~0.8–1.0 nm in length estimated from the average distance between the base pairs on the DNA strand.

The cracking of the genetic code has led to the genome sequencing of various life-forms ranging from bacteria to human. To date, 180 organisms have been fully sequenced and most of the genomic sequence data is freely available on the National Institutes of Health (NIH) and other government websites [2, 3]. It is generally understood that almost every single cell within the body contains the same genomic content. Differential expression of genes within the genome leads to the formation of the wide diversity of cells within the same life-form. The primary research focus of most biologists has been to understand how the same genomic content of each cell can be manipulated to give different protein expression levels.

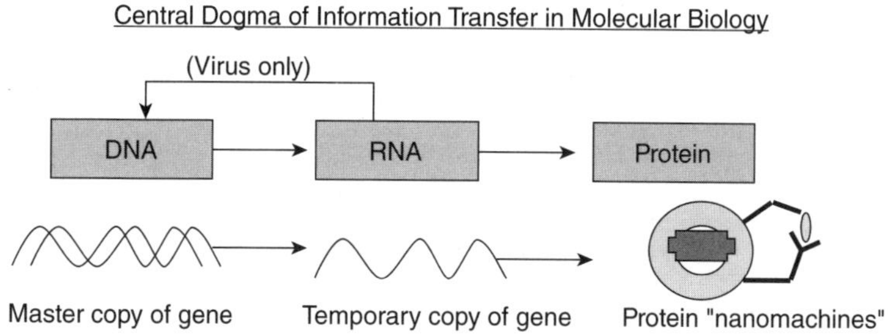

Figure 15.1. High-level diagram of biological information flow. RNA represents the temporary, short-term copy of the required genetic code from the larger DNA genomic "library." The information on the RNA is like a "blueprint" for constructing a particular portion of a protein. Within the cell, *ribosomes* (protein factories) use the RNA blueprint to build proteins with the desired amino acid sequence. Note that only certain viruses that encode for the *reverse transcriptase* gene can make RNA into DNA and incorporate themselves into the host genome.

Modulation of gene expression has been shown to be a complex signal processing sequence that has many controls (Fig. 15.2). Typically in mammalian systems, genes are turned on via the promoter region in the DNA, where specific sequences, for example, TATA, signal the start of a specific gene. Small molecules or proteins that act as promoters bind this region to signal to the RNA polymerase (enzyme that makes the RNA strands) to actively transcribe the gene of interest. Either DNA strand may code for a gene. The strand that codes for the gene is known as the *sense* strand and its complement is known as the *antisense*.

Protein expression levels are extremely dependent on feedback control. Enhancers and repressors may bind to specific regions within the gene to modulate the protein output. The actual gene that will be transcribed and eventually translated into a protein may be coded along the DNA strand in various sections known as *exons*. Regions known as *introns* do not code for the protein and are spliced out in the final RNA strand (messenger RNA [mRNA]) that will used to make proteins. The function of introns is not known. It is speculated that these are mostly junk DNA that have accumulated in the genome since life began. Specific genetic codes (TAA, TAG, TGA) signal the end of the gene and stop transcription. Aberrant transcription of the genetic code or improper modulation of the feedback loop to control protein expression leads to many disease states such as cancer.

Similar to the expression of the gene in the DNA strand, mRNA that is used to make protein is also regulated. These mRNA strands often contain special sequences such as the nuclear export sequence (NES), which signals transport out of the nucleus. Sometimes, the sequences are very specific to the manner of transport out of the nucleus. For instance, the G-quartet or RGG peptide motifs specifically signal fragile X mental retardation protein (FMRP) complex to bind and transport the selected mRNA to a particular location inside the cell [4]. Other regulatory mechanisms include modification of the mRNA strand

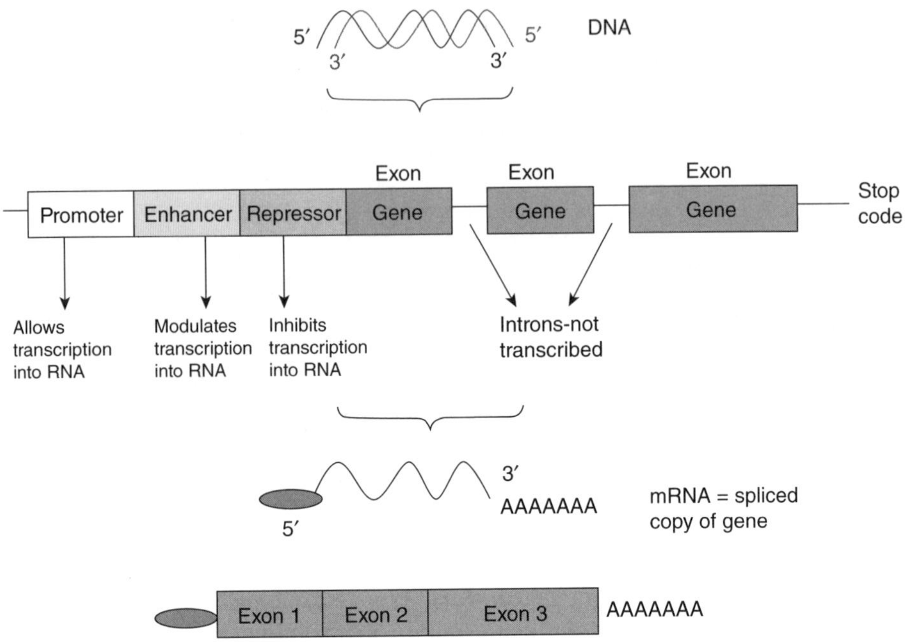

Figure 15.2. Genetic control mechanisms in mammalian systems. The expression of genes is controlled at every step. Access to the genomic library at the DNA level is modulated typically by protein complexes that may promote, enhance, or repress the transcription of a particular gene. Gene expression is also modulated by controlling the maturation of the final *messenger RNA* (mRNA) transcript. Splicing of the mRNA transcript is also highly regulated. The final mRNA transcript requires a "cap" at the 5′ end and a poly-A tail at the 3′ end for transport from the nucleus to the cytoplasm.

itself. The final mRNA copy is modified with a 5′ cap and a poly-A-3′ tail. This is a requirement for signaling transport out of the nucleus to the ribosomes located in the cytoplasm where proteins are made. Without the 5′cap or poly-A tail, the mRNA strand will be degraded and protein expression will not occur. Another control mechanism is the endogenous microRNA (miRNA) pathway. Here, recently discovered small RNA strands (~20–23 nucleotides) bind as the antisense of the selected mRNA, forming a double-stranded mRNA complex at the 5′ end. This is then tagged for degradation by the cellular machinery [5]. The endogenous microRNA pathways have been found to modulate numerous mRNA transcripts. Antisense technology, such as interference RNA (RNAi) whose variations include miRNA and short hairpin RNA (shRNA), is currently used to suppress gene expression either transiently or permanently within a cell [6, 7]. This unique suppression technology is speculated to occur as an evolutionary response to a virus attack, which typically have double-stranded RNA. Excellent reviews of antisense technology are given in [8, 9].

Biologically engineered, synthetic genetic circuits that modulate gene expression are currently being designed to deductively understand cellular biological

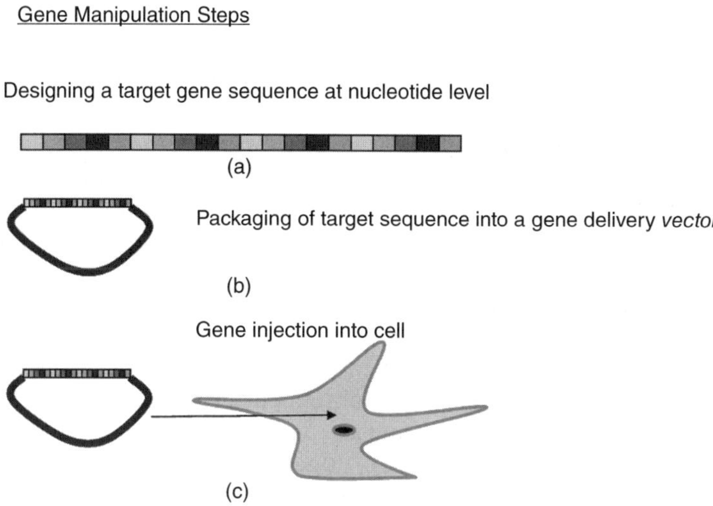

Figure 15.3. Steps in genetic manipulation of a living cell. (a) Sequences have to be appropriately designed to precisely target specific genes. (b) Part of the design process includes choosing an appropriate packaging vector to deliver the desired target sequence. (c) The packaged genetic construct is then allowed to associate with the target cell or tissue and injected into the cell via various methods such as viral delivery, chemical transfection methods, electroporation, and sonoporation.

signal processing mechanisms [10–12]. The effects of artificial combinations of various types of promoter, suppressor, and enhancer genes are currently being studied in cells not only to better understand the design rules for engineering applications but also for better understanding of expression and processing of biological information. Therefore, synthetic biology in generating biohybrid or bioinspired logic circuits is expected to play a significant role in the future.

15.1.2 Controlled Genetic Manipulation

Considering the programmed functioning of genes and their behavior within cells, external manipulation to modulate genetic expression is the main focus of the field of genetic engineering. To control gene expression, the key steps are to *design* an appropriate sequence (either DNA or RNA), and then to *deliver* the target sequence into a living cell or tissue by penetrating the cell membrane (Fig 15.3).

To manipulate genes in cells, desired genes may either be expressed or suppressed. Specific gene sequences may be obtained from existing databases, ligated into specific genetic vectors, and transfected into desired cells. (Transfection is the process of gene injection into cells.) Artificial hybrid genes may also be created by combining sequences, where multiple genes are fused together or coexpressed with the same promoter. A common example is the fusion of the green fluorescent protein (GFP) reporter gene with a desired gene of interest

[13]. The expression of the desired gene of interest is subsequently evaluated biophotonically using GFP fluorescence properties. Another example might be to introduce variations within a gene sequence, transfect, and study its phenotypic effect on a cell.

Alternatively, the inherent cellular gene expression may be modulated by targeting the various enhancers, promoters, and suppressors controlling the gene. Therefore, the target sequence may be anywhere between a few to several thousands of kilobases. In this strategy, small gene sequences can be designed to be incorporated into specific locations in the desired gene sequence within the cellular genome. Incorporation into the desired gene sequence then disrupts the transcription of the desired gene, effectively altering the control mechanisms for that gene. Using this technique, genes can be constitutively turned on or off or modulated to increase or decrease its subsequent protein production. From an engineering perspective, specific sequences for precise targeting of a gene are preferred in order to prevent off-target effects where unrelated genes with similar sequences may be inadvertently affected [14]. Designing a target sequence requires careful comparison and verification across various gene databases. A popular website where sequences can be compared is the *basic local alignment search tool* (BLAST), which is managed by the NIH and the National Center for Biotechnology Information (NCBI) [15].

Gene expression may also be controlled at the mRNA level using antisense oligonucleotides. Typically, mRNA for a gene of interest can be suppressed using RNAi or its variants. RNAi molecules are comparatively smaller (~20–23 nucleotides in length) and have been shown to achieve a high degree of specificity in suppressing their target. Nevertheless, the actual oligonucleotide sequence needs to be carefully designed to avoid cross-reactions. To study essential genes where suppression or overexpression may lead to cell death, a decision of transient versus permanent incorporation of the desired gene construct needs to be made. Additionally, using small oligonucleotides to control gene expression is advantageous in subsequent gene delivery steps. Also, the use of small sequences enables easier manipulation of multiple genes in a cell. Smaller sequences, in general, also have a higher transfection efficiency and generate a lower immune response *in vivo*. In summary, the key aspects of sequence design are

- Precise sequence specificity for desired gene of interest
- Short sequence length for gene delivery
- Addition of appropriate selection markers and reporters within the sequence to identify transfected cells

15.1.3 Gene Transfection

Once designed, the target sequence needs to be injected into the living cell. The cell has been shown to take up naked DNA/RNA constructs if small enough at relatively low transfection efficiencies [16]. However, as the size of genetic construct increases, the lipid bilayer that constitutes the cell membrane serves as an insurmountable barrier that blocks charged particles. Large molecules such as proteins and engineered nucleic acid payloads (DNA, RNA, etc.) cannot easily cross the cell membrane in their natural state due to their size and inherent electrostatic charge. The cell has distinct ion channels and other transport mechanisms that regulate in a precise and specific manner the types of charged molecules that can cross the bilayer. In order to overcome this barrier, desired genetic sequences need to be suitably packaged in gene delivery vectors [17, 18]. These packaging vectors allow close association of the genetic construct with the membrane, which in turn is either injected in via physical means (electrical force, mechanical penetration) or via natural cellular transport mechanisms (chemical, viral vectors). Common packaging vectors are typically derived from either natural gene transfer mechanisms or synthetic liposomes. Plasmids, for instance, are genetic constructs (DNA based) that are transferred between certain species of bacteria (e.g., *Escherichia coli*). The genetic sequence of the plasmids can be altered to contain a combination of engineered sequences and its natural sequences. These designed genetic constructs can be injected into mammalian cells and expressed to produce desired genes, promoters, enhancers, and so on. Other types of packaging vectors include using altered sequences of lambda phages, adenovirus, herpes simplex virus, and retroviruses. Many of these viral packaging vectors have sequences that allow for direct incorporation into genome for permanent gene expression. Synthetic liposomes are generally made of phospholipids with a net cationic charge that can easily associate and merge with the cell membrane when in close proximity. Nucleic acid payloads within the liposomes are delivered into the cytoplasm of the cell upon merging with the membrane [19].

The choice of genetic vector is heavily dependent on the amount of genetic information that needs to transfected or injected into cells (Table 15.1). Large sequences, such as those required for genomic libraries, typically utilize bacterial or yeast artificial chromosomes (BACs or YACs, respectively).

BACs and YACs have been used to store sequences during the Human Genome Project. For typical genetic experiments (<20 kb), plasmids, viruses, cationic lipids, and even naked DNA/RNA sequences are generally used. For small sequences (<15 kb), naked DNA/RNA sequences and plasmids may be delivered into the cell using physical methods of membrane penetration. Commercial cationic lipids such as lipofectamine, transfectin, polyethyleneimine, and numerous other cationic molecules that have been developed (1) enable close association of the genetic construct with the cell membrane and (2) permit gene delivery via permeabilization of the membrane [20, 21]. Cationic lipids are positively charged amphiphilic molecules that are electrostatically attracted to the negatively charged lipid bilayer. Transfection efficiencies typically are in the range of 40–80%. Alternatively, calcium chloride precipitation methods may also be used. Key impediments in using chemical methods are toxicity to cells, lack of cell specificity in the delivery, varying transfection efficiencies based on cell type, and relatively high cost, especially for patented lipids.

TABLE 15.1. Packaging Sequence Limitations for Genetic Vectors

Common Gene Packaging Vectors	Maximum Sequence that Can be Packaged (kilobases [kb])
Cationic liposomes/lipids	0–8[a]
Adenovirus/adeno-associated virus	1–8
Retrovirus	1–7.5
Herpes simplex virus (HSV)	1–20
Plasmids	10–15
Lambda phage	20–25
Cosmids	30–50
Bacterial artificial chromosomes (BACs)	150–350
Yeast artificial chromosomes (YACs)	100–3000

[a] Theoretically, the maximum packaging sequence is unlimited for liposomes; however, due to toxicity issues and lower transfection efficiencies, the typical range used in literature is indicated.

Gene delivery using viruses is far more efficient due to natural biological evolution. Viruses typically utilize the natural cellular transport mechanisms to deliver their nucleic payloads [22]. The viral nucleic acid material (which can be either DNA or RNA) is typically packaged in a proteinaceous outer coat that "docks" on the cell membrane. The viral genetic material is then transferred into the cell via various cell mechanisms such as endocytosis. A highly mechanical and complex method of viral gene delivery is seen in the T4 bacteriophage that attacks *E. coli* (Fig 15.4).

These phages have a complex body that acts as a "molecular syringe" to deliver its genetic material [23]. The phage docks onto the outer surface of the bacterium, and subsequently, it mechanically injects and forces its genetic material into the cell. This is one of the finest examples of nature's finesse in gene delivery mechanisms since it achieves a highly controlled and precise gene design and delivery to a cell. Due to controlled delivery methods, the preference of current gene therapy methodology is to utilize viruses. Key advantages to using the viral delivery methods are high transfection efficiencies, and potential ability to either incorporate into the genome for a "permanent" gene transfection or remain outside of the genome for a more "transient" effect. Some cellular specificity may be achieved using cell type-specific viruses. However, selective transfection of specific cells within the same cell types is not generally possible. Key disadvantages are low cell viability and high risk potential to infect other cells especially in an *in vivo* environment.

15.2 PHYSICAL STRATEGIES FOR GENE INJECTION

Even if viral or chemical methods are used, a key step in gene injection is to bring the packaged nucleic payloads in close proximity to the surface of the cell membrane for efficient transfection. Both chemical and viral methods lack the ability

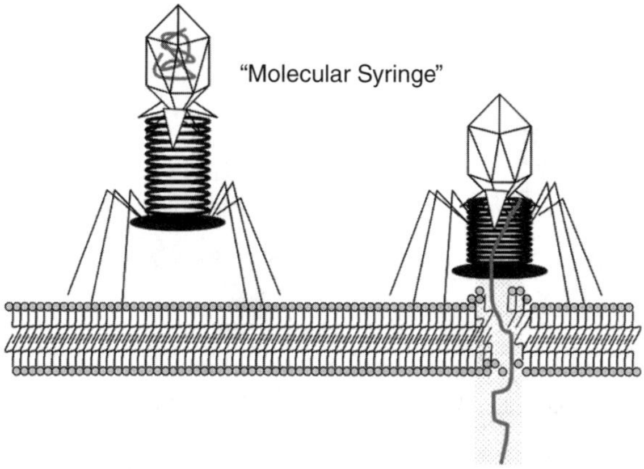

Figure 15.4. Example of a natural gene delivery mechanism. Depicted is a T4 bacteriophage that infects the *Escherichia coli* bacterium. Its complex delivery mechanism involves docking onto the surface of the bacterium and mechanically injecting its genetic material directly into the cell.

to target specific cells. In addition, prolonged exposure to chemically based cationic treatments is often toxic to the cells and leads to decreased cell viability. For both chemical and viral methods, therefore, spatiotemporal control of gene delivery is hard to achieve. To enhance transfection efficiencies and improve upon spatiotemporal control, physical means of membrane penetration, such as using electric fields, magnetic fields, ultrasonic energies, and mechanical forces, have been extensively studied [24]. The utilization of physical forces to attain efficient gene delivery could also potentially allow for integration with CMOS technologies. To date, direct gene injection using CMOS-based technologies to manipulate the genetic content of a cell has not yet been demonstrated. Development of such a technology platform must be capable of directly manipulating and modulating gene expression at the cellular level. The main advantages of using CMOS technologies in gene injection and manipulation systems are in

- Reducing the device footprint, lowering power consumption, and dramatically increasing throughput
- Allowing for spatiotemporal control of the physical gene injection mechanism at a single-cell or a localized subcellular level
- Integrating analytical, logic-based measurands within the gene delivery platform for rapid and accurate assessment of transfection and subsequent changes in the cell phenotype

The ultimate goal of integrating gene injection systems with CMOS technologies is to drive forward a consumer-friendly lab-on-chip platform with real applications such as rapid and accurate gene diagnostics, therapies, and also in discovery of novel molecules in the drug development industry. Broadly, current *physical*

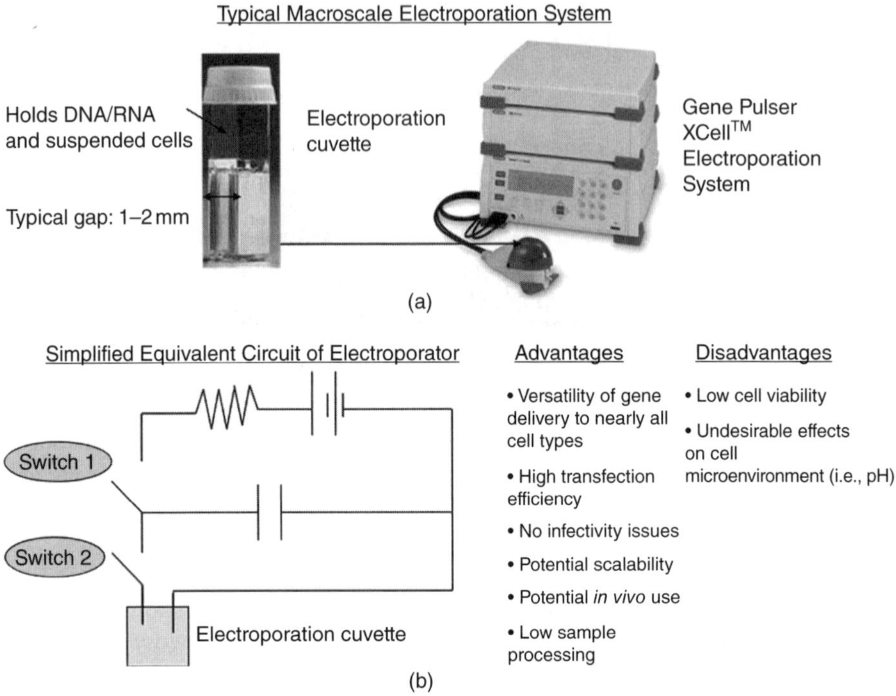

Figure 15.5. A macroscale electroporation device. (a) A typical electroporation cuvette with two metallic, conductive plates spaced 1–2-mm apart that can be placed in an electroporation chamber (Gene Pulser XCell, Bio-Rad Laboratories, Hercules, CA) that is capable of delivering high voltages. Copyright © 2010. (b) An equivalent electrical circuit diagram for most electroporators is shown. When the "switch 1" is closed, the capacitor is charged up. When "switch 2" is closed, a high-voltage pulse is delivered to the electroporation cuvette. Typical electroporation parameters for most cells range from 10,000 to 100,000 V/cm. For mammalian cells, it ranges from 100 to 1000 V/cm.

methods of gene delivery fall into two main categories—electroporation-based approaches and mechanical approaches for gene injection. Electroporation utilizes electric fields to manipulate the lipid bilayer of the cell membrane to form small perforations for molecules to pass through. A typical macroscale electroporation cuvette and device is shown in Figure 15.5. Electroporation cuvettes have large metallic plates that are placed ~1–5-mm apart.

Cells are suspended in electroporation buffers, which are hypo-osmolar, having low conductivity (<10 mS/cm), along with desired nucleic acid materials in these cuvettes. Hypo-osmolar solutions allow cells to swell and enhance gene delivery upon voltage application. High-salt solutions (high conductivity) are not desirable due to electrical arcing issues. High voltages ranging from 100 to 100,000 V/cm are delivered to the metal electrodes. An equivalent electrical circuit diagram of a typical electroporator is shown in Figure 15.5b. Essentially, a capacitor is charged for a set time till desired voltage is reached and discharged

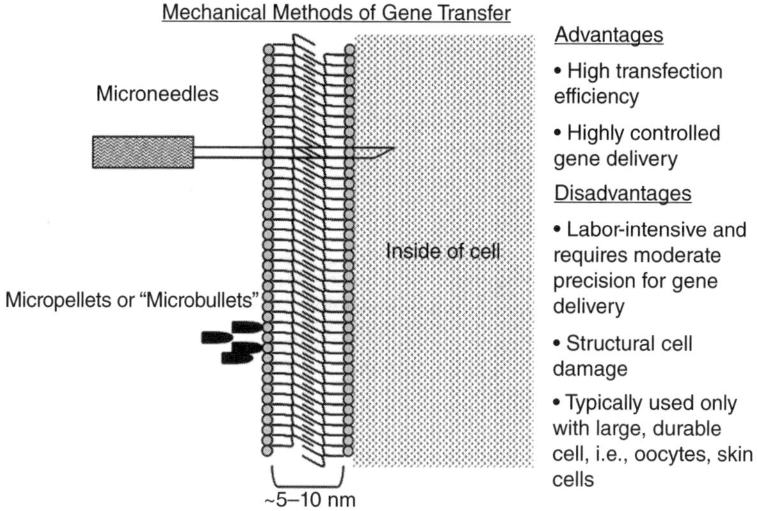

(a)

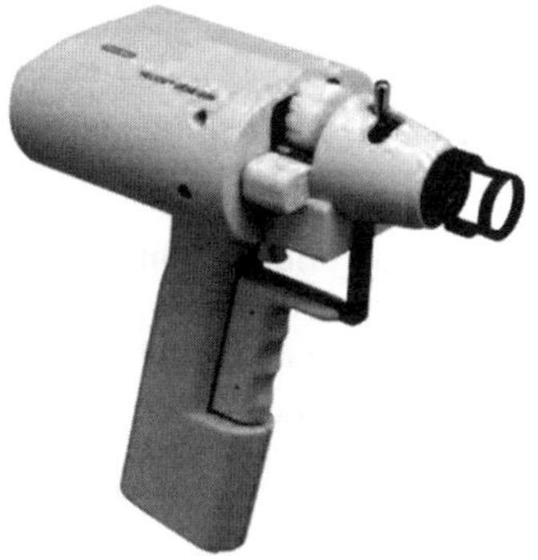

(b)

Figure 15.6. Mechanical methods of membrane penetration. (a) Schematic shows that cell membranes may be penetrated with micro- or nanoparticles or with microneedles. (b) A picture of the Helios™ Biolistic Gene Gun (Bio-Rad Laboratories) for nucleic acid delivery (http://www.biorad.com).

to the electroporation cuvette. A significant problem is low cell viability due to the high voltages used.

Mechanical approaches to gene injection rely on a variety of micro- and nanoparticles and microbubbles to penetrate the cell membrane (Fig 15.6).

Mechanical methods use a wide range of physical energies to penetrate the membrane. These include using (1) shooting microparticles (biolistics) or microfluidic droplets (jet injection) carrying nucleic acid payloads at high velocities into cells, (2) creating acoustic microcavitation bubbles to disrupt the cell membrane (sonoporation), (3) attaching nucleic material to magnetic particles and bombarding them into cells using magnetic fields (magnetoporation), or (4) driving microneedles into the cells and directly injecting genetic material into the

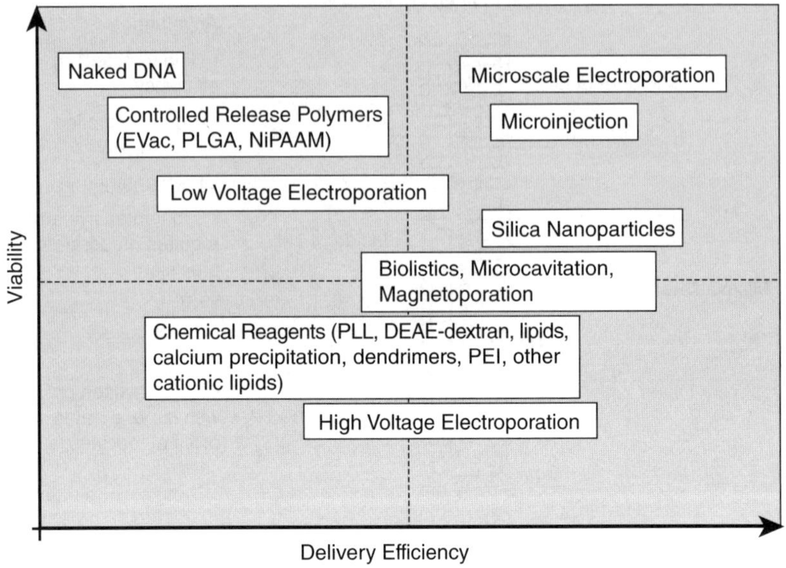

Figure 15.7. Comparison of current genetic delivery techniques. *Source*: Adapted from Lee, 2008. Copyright © 2009 RSC (Royal Society of Chemistry, *Integrative Biology*).

target. Currently, a majority of these gene injection techniques are used in macroscale applications, requiring high energies and sacrificing spatial resolution. For electroporation, sonoporation, and magnetoporation, cells typically have to be detached from their natural adherent state into suspension cells for transfection. The use of microneedles help achieve high spatial resolution, but the process of injecting individual cells is cumbersome and time-consuming and requires large cells that are readily observable by microscopy.

A significant engineering trade-off that must be addressed in gene injection delivery platforms is optimizing between transfection efficiency and cell viability. In Figure 15.7, various methods of transfection are compared. Exposing cells to naked DNA or controlled-release polymers typically have low transfection yields. To increase transfection efficiencies, low-voltage electroporation or biolistics may be used [25–27].

However, cell viability is significantly compromised with these techniques. Bombardment with microparticles, for instance, could damage the structural integrity of the cell upon transfection and decrease cell viability. Using nanoparticles, such as silica or carbon nanotubes, seem to slightly enhance cell viability [28, 29]. Gene injection using chemical reagents or high-voltage electroporation result in the lowest cell viabilities and have a wide range of reported transfection efficiencies. Chemical reagents typically fall within 30–80% efficiencies, while macroscale, high-voltage electroporation have 20–60% efficiencies, depending on cell type and other environmental conditions [30]. It should be noted that cell viability and transfection efficiencies are heavily dependent on

- Size of nucleic acid payload
- Packaging vector
- Cell type and size
- Cell stage of differentiation
- Other environmental issues such as ionic strength and osmolarity

The state-of-the-art microscale technologies offer a promising approach to achieve high cell viability with high transfection efficiencies. Microscale electroporation and microinjection techniques not only reduce the foot print of device but also minimize several problems that are typically experienced in macroscale applications. Transfection efficiencies greater than 80% have been achieved with high cell viability using microscale electroporation methods [31]. Additional advantages of using microscale technologies include low environmental disturbance (i.e., stable pH) for cells and low probability for sample contamination. Also, adherent cells may be transfected without significant sample processing. The integration of microscale gene delivery technologies with a CMOS-based platform is therefore expected to lead to simultaneous, spatiotemporal control of gene transfection and rapid, posttransfection analysis. In order to inject genetic material across a highly complex lipid bilayer, one of the challenges that CMOS technologies face is the integration of different physical modalities on a single platform. The force per unit length needed to penetrate a cell membrane has been estimated to be ~4 dynes/cm (40 μN/cm) [32]. (There is, however, high variability in this value due to heterogeneity in the material composition of the lipid bilayer itself and other cellular protein-based structural reinforcements.) Using CMOS architectures, current physical gene injection methods may be coupled with micropower systems that are capable of generating desired electrical, magnetic, and other physical energies to induce transfection. Incorporation of MEMS technologies with CMOS may further allow micromotors to be integrated for controllable propulsion of objects or needles coated with nucleic acid material into cells. In addition, spatiotemporal control with integrated digital logic circuits for rapid interrogation of transfections may be envisioned. Potential strategies to integrate CMOS technologies with electroporation and selected mechanical methods of gene injection will be examined in subsequent sections.

15.2.1 Microscale Electroporation

Gene transfection using pulsed electric fields is one of the most efficient and rapid delivery techniques [33]. Application of electric fields to cells is relatively easy and allows for quick gene manipulation in cells with minimal equipment. The electroporative technique is fairly simple and can be easily scaled down to microelectrodes. Due to highly controllable application of electric fields at the microscale, precise spatiotemporal control for cell positioning and localized gene delivery is easily achieved. Studies have shown that efficient electroporation of molecules can be achieved for small cell populations (6–10 cells) [34–36], single cells [37–39], or even subcellular features of a neuronal axon [40]. Transfection efficiencies for microscale electroporation range from 50 to 80% while maintaining high cell viability [35, 36]. Even difficult-to-transfect cells such as neurons are efficiently transfected with high cell viability [34]. A major advantage of this gene delivery technique is that it is applicable to nearly all types of living cells without significant customization. Microscale electroporation has the additional advantage of having a lower power consumption compared with its macroscale counterpart. The simplicity of the technique, relative scalability of the electrodes, and high transfection efficiencies while maintaining high cell viability makes microscale electroporation the most promising candidate for integration with CMOS architecture.

In order to achieve microscale electroporation and eventual integration with current CMOS systems, miniaturized electroporation chambers where electric fields are directed toward living cells need to be constructed. Electroporation chambers are in essence electrochemical cells, containing a ground and a working electrode in a conductive, ionic, liquid medium. Living cells may either be suspended in the liquid or adhered to an electrode in a microscaled device. A simplified, equivalent electrical circuit of the electrochemical cell with adherent cells on the electrodes is depicted in Figure 15.8.

Each electrode is represented by an electrode–electrolyte interfacial impedance and a Helmholtz-based capacitive double layer. Living cells are typically

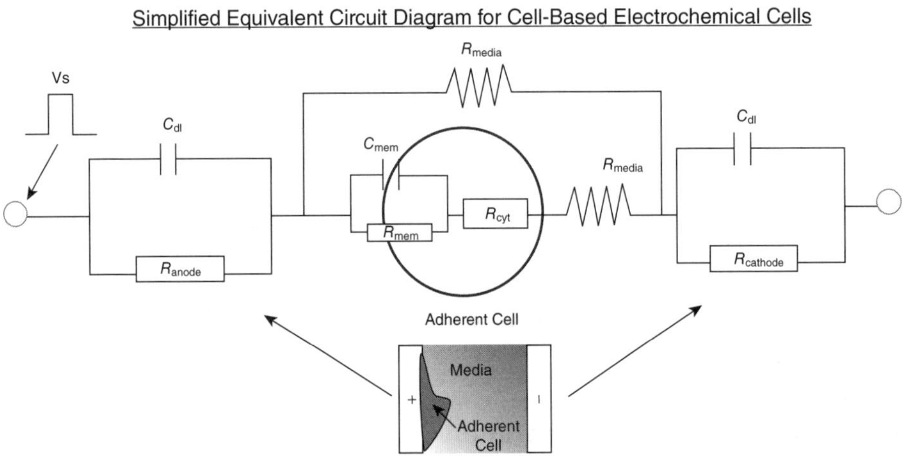

Figure 15.8. Simplified equivalent circuit diagram for cell-based electrochemical cells. "C_{dl}" represents the Helmholtz double layer that forms at the electrode–electrolyte interface. "R_{anode}" and "$R_{cathode}$" represent the equivalent electrode–electrolyte interfacial charge transfer impedance. "R_{media}" represents the bulk electrolyte resistance. Here an adherent cell is juxtaposed to an electrode as a simple system with a lipid bilayer membrane and an ionic cytoplasm with resistance "R_{cyt}." The lipid bilayer (dark gray) is represented with a membrane-associated capacitance ("C_{mem}") and conductance for ionic movement ("R_{mem}"). Voltage ("Vs") is pulsed at the electrode closest to the cells.

modeled with a capacitive membrane boundary and resistive cytoplasm, which contains ions. Additionally, the resistance of the membrane (R_{mem}) due to pore formation and subsequent flow of charged particles across the membrane is also included. In its native state, the contribution of the R_{mem} parameter is expected to be minimal. In Figure 15.8, the simplified circuit reflects the presence of an adherent cell on an electrode and assumes that no media is present between the cell and the electrode. This model is an example of the conditions that might be present in the chamber and is expected to be different for suspended cells and weakly adhered cells. CMOS-based electroporation devices should be designed such that the leakage current is minimal in order for the bulk of the charge to be injected entirely across the cell–device interface for electropermeating the membrane.

In order for CMOS-based technology to perform electroporation, a high electric field needs to be generated at the biological cell–electrode interface. In the resting state, all mammalian, living cells have a transmembrane potential ranging from –40 to –90 mV. During microscale electroporation, a voltage pulse is typically applied to an electrode, which perturbs the cell equilibrium potential [41]. The impedance of membrane layer (R_{mem} in Fig. 15.8) is expected to significantly decrease due to pore formation and the movement of charged molecules through the pores. During the voltage pulse, the transmembrane potential is expected to increase to >1 V in localized areas in the cell membrane that are typically closest to the electrodes [42, 43]. The biophysics of pore formation and the mechanisms of electroporation are poorly understood [44]. Electroporation is known to occur at least by two different mechanisms—"reversible" and "irreversible" electroporation [45]. In irreversible electroporation, the molecules in the cell membrane lipid bilayer are totally ionized. This leads to a total material breakdown analogous to a dielectric breakdown of an insulator. In this type of electroporation, the integrity of the membrane is compromised, leading to permanent pore formation by membrane damage. The continuous influx due to the membrane breach leads to cell death by lysis. Irreversible electroporation has applications in the development of novel cancer treatment devices, where microscaled electrodes can selectively lyse cancerous cells. In reversible electroporation, the cell membrane is permeated and forms transient pores when an electric field is applied. When the field is removed, the pores reseal after a time delay. The most prevalent understanding of the reversible electroporation mechanism is shown in Figure 15.9.

Reversible pores are theorized to form due to the compressive force exerted by the electric field on the membrane bilayer. When the compressive force that is normal to the membrane exceeds the line tension along the membrane, the lipids at the localized area move apart and rearrange themselves to create small pores. An alternative theory suggests that pore formation is more indirect. Molecules are suggested to rearrange themselves in localized areas of the membrane due ionic imbalances and voltage-induced polarization at the interface caused by the external electric field (Maxwell–Wagner polarization). In reality, both mechanisms are likely to contribute in generating reversible pores. Typical

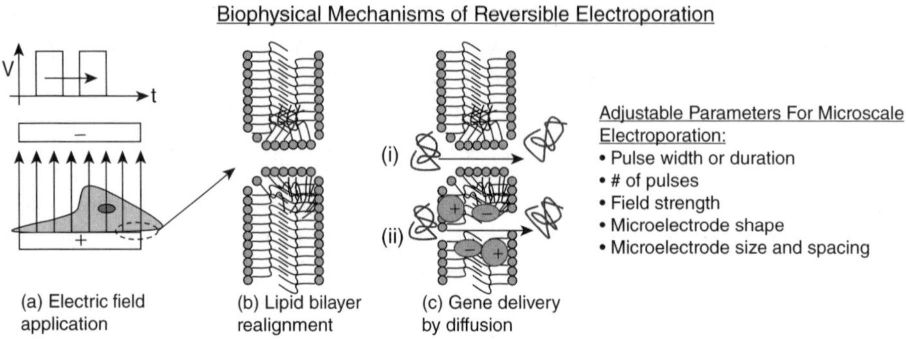

Figure 15.9. Biophysics of reversible electroporation. (a) A voltage pulse with high field strengths is applied to cells. (b) The electric field imposes a compressive force on the cell membrane that disrupts the tension forces that maintain membrane integrity. At a certain threshold in localized areas where the compressive force exceeds tension, the lipid molecules realign to form pores. (c) Nucleic acid delivery is expected to occur via diffusion through the pores. In mechanism (i), pores form only due to the realignment of lipid molecules. In mechanism (ii), when the lipid molecules displace due to the field-compressive forces, small hydrophilic ions help to form and maintain a hydrophilic channel. Mechanism (i) is more widely accepted in literature.

pore sizes are estimated to be ~5 nm in diameter. Once formed, the stability of the pore needs to be maintained to allow for transporting the desired vector. Current views suggest two mechanisms that stabilize the pore as depicted in Figure 15.9c. In the first theory that is most commonly cited, the hydrophilic heads of the phospholipids are thought to fold in to create a hydrophilic channel conducive to ion movement. This action is expected to maintain the energy balance where the exposure of the hydrophobic tails of the membrane phospholipids to the aqueous media is minimized. In the less common, second theory, ions are expected to line the perforated bilayer to allow passage of hydrophilic molecules such as DNA, RNA, and other nucleic acid materials. Once the pore is formed, the gene delivery mechanism by itself is expected to occur by diffusion and electrophoretic methods. Passage of up to 750 kD (1.2-kb DNA or relevant nucleic material) is expected to occur by diffusion, while larger nucleic acid materials such as plasmids are expected to migrate via electrophoretic (differential charge-induced flow) mechanisms [44, 46–48]. The *keys* to the gene delivery are

- Length of time that pores are open
- Degree of association between the nucleic material and the pore

Experimental studies show that small fluorescent dye molecules and oligonucleotides (siRNA, antisense DNA) enter the cell upon electroporation almost within a few minutes [48]. Injection of larger nucleic material, such as GFP plasmids, has a more complicated mechanism. Upon application of the voltage pulse and subsequent pore formation, the plasmids associate with the pores almost instantaneously. However, once associated with the pores, traversing the membrane into the cytoplasm takes an additional 30–60 minutes [44, 47, 49]. The internalization of the plasmid has been suggested to occur by natural endocytotic mechanisms.

Therefore, pore formation is fundamental to the success of electroporation-based gene injection. Application of the optimal field strength and adequate charge transfer at the cell–electrode interface are essential. The shape of voltage pulses that may be applied are exponential, sinusoidal, or square (which is most popularly used). Typical pulse durations last between 1 and 10 ms. Field strengths for mammalian cells range from 100 to 600 V/cm depending on cell type, size, shape, membrane curvature, and other factors. (Higher field strengths are used for bacterial and plant cells.) The typical number of pulses used in literature is widely varied. A suggested heuristic is to apply a short pulse followed by a long pulse. An initial short pulse is suggested to allow desired nucleic acids to associate closely with the membrane via electrophoretic mechanisms. This is then followed by a longer pulse to electroporate or "electropermeate" the bilayer. This type of regimen has been suggested to allow for a more efficient and optimal gene delivery technique [50]. Alternative application using nanopulses have also been suggested to improve gene delivery techniques for mammalian cells [51]. Since mammalian cells contain internal compartments, such as mitochondria and nucleus, the genetic vector has the additional step of traversing across the cytoplasm to the nucleus in order to have effective gene transcription. Application of short nanosecond pulses with duration smaller than the charging time constant of the main cell membrane lipid bilayer has been shown to help with the internal electroporation of the nucleus and other compartments and achieve more efficient gene delivery. Controlled delivery of variable frequencies and pulse amplitudes to electrodes can be easily done in current CMOS configurations. Floating-gate transistors within the CMOS architecture can be customized with programmable circuits and feedback loops using electronic erasable programmable read-only memory (EEPROM) to deliver electrical energy based on cell type and required environmental condition. Typical feedback information will most probably entail transient thermal and current changes at the cell–electrode interface.

Typical CMOS architecture is currently optimized for low-power and low-noise level applications. However, for microscale electroporation, studies suggest that desired current densities are in the microampere range [35]. A potential strategy for delivery of high levels of charge is to use Darlington pairs. Additional strategies that CMOS systems may adopt are to lower the power consumption using microelectrode arrays (MEAs) for microscale electroporation. A typical *macroscale* electroporator with 1–2-mm electrode distance requires ~50–100 V to generate sufficient field strength for mammalian cell electropermeation. MEAs offer a distinct advantage by reducing power consumption by decreasing the interelectrode distance to the microscale. By doing so, the required operational

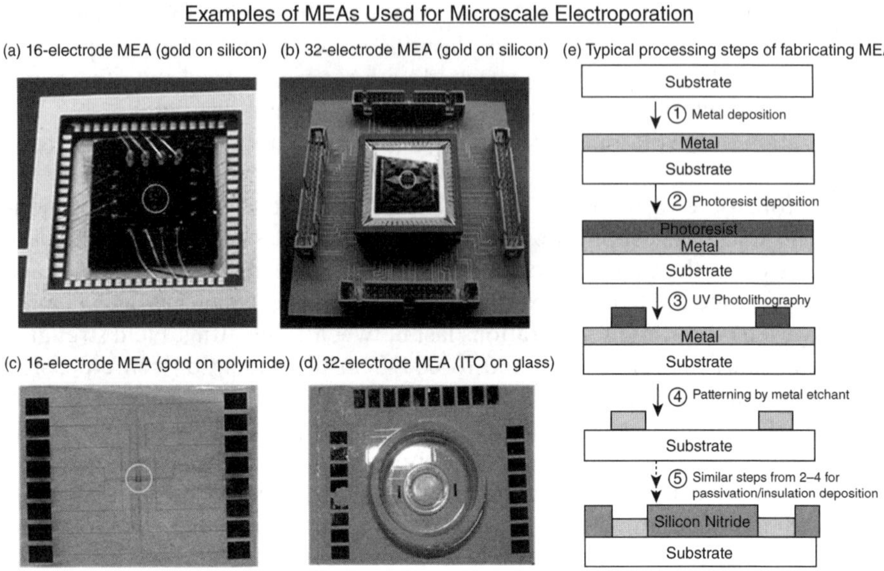

Figure 15.10. Planar MEAs used in microscale electroporation. Circles represent areas where living cells will be grown. MEAs are manufactured using standard fabrication methodologies. Typical MEAs shown here have either16 microelectrodes (a,c) or 32 microelectrodes (b,d). Individual microelectrode sizes vary from 100 μm (a) to 50 μm (b–d) in diameter. ITO microelectrodes inside the wells are transparent in the visible spectrum (d) and can be used for optical measurements compared with gold microelectrodes (a–c), which are opaque. Panels (a) and (b) show a dual in-line package (DIP)-type packaging, while panels (c) and (d) show a more benchtop-friendly alternative with miniature wells and easy, external clip-on bond pads for electrical manipulation with a standard, computer-controlled software. Alternative materials that may be used are doped polysilicon (not shown), which can be directly incorporated in the CMOS process. ITO or gold may also be incorporated but with significant postprocessing fabrication steps. (e) Typical processing steps to fabricate a planar MEA.

voltages also drop significantly due to scaling effects. Microscale electroporation methods that use MEAs require only 1–10 V for adequate reversible cell permeation. In addition, MEAs offer distinct spatial and temporal control over an area for selective transfection of cells. The fabrication of MEAs use fairly standardized photolithographic patterning techniques (Fig. 15.10). Typical electrode size range anywhere from 10 μm to greater than 100 μm, with diverse shapes such as squares and circles.

The microelectrode shape plays a significant role in determining the spatial electric field distribution as determined by finite element modeling of the spatial distribution of electric field strengths using commercial software such as ANSYS or COMSOL packages. Square-shaped microelectrodes, for instance, have areas of higher electric field intensities at the corners of the microelectrode compared with round ones (edge effect). The goal is to attain uniform fields over the area where living cells are in contact for predictable electroporation. The number of microelectrodes used in arrays varies widely and is only limited by photolithographic capabilities. Microscale electroporative living cell-based arrays tend to use anywhere from 8 to 64 electrodes. CMOS-based biosensors that use MEAs can range from 400 (Nanogen, San Diego, CA) [52] to over 16,000 sites [53]. MEAs are easily fabricated using standard semiconductor processing techniques and can be easily incorporated into CMOS architecture. Doped polysilicon, which is standard in CMOS processing, has been shown to be fairly stable under electrochemical and biological conditions [54–56]. Gold, platinum, silver, and other metals typically used in MEAs and electrochemical systems may require some additional postprocessing steps for deposition. Nevertheless, since MEAs are the interfacial contact to liquids and cells, special attention is required to ensure biocompatibility with cells and general protection from corrosion. Therefore, the key parameters to using MEAs for microscale electroporation are

- Minimization of edge effects (uniform electric fields)
- Maximizing charge transfer across cells
- Staying within the electrochemical "water window" of electrode materials

MEA platforms are fairly versatile and may serve as electroporators, biosensors (i.e., impedance sensing, current/voltage sensing), and for manipulating cell movement (dielectrophoresis). A MEA platform integrated with CMOS technologies will be a powerful tool in cell biology and medicine. A recent study used a two metal-layer fabrication method that is compatible with CMOS processes to manipulate cell movement using dielectrophoretic principles. [57]. (In dielectrophoresis, AC electric fields are used to transport cells toward or away from a given electrode.) Since multiple metal layers are possible in CMOS architecture, a wide array of sensing and control schemes can be built into a CMOS platform. Other relevant environmental conditions, such as fluid flow, temperature, and gas exchange, could be controlled using sensors in feedback loops and very large scale integration (VLSI) circuitry. Therefore, microscale electroporation using MEAs offers distinct advantages, such as

- Selectively spatial and temporal addressability
- Rapid gene delivery
- Lower power consumption
- Easy integration into standard CMOS architecture
- Versatile "mixed-mode" applications

Future CMOS-based systems are expected to have a hybrid CMOS architecture that caters for high-charge delivery for electroporative purposes and integrated low-noise, low-power biosensor circuitry for monitoring and manipulating the cells *in situ*.

15.2.2 Microinjection

Mechanical gene delivery approaches using microneedles and micro- and nano-projectiles are also promising in terms of integration with current CMOS systems. Both of these methods use either a tethered (needle) or mobile (particle) object to physically penetrate the cell membrane at high forces and directly inject genetic material into the cell. Tethered objects, such as microneedles, offer considerable spatial control with high precision for gene delivery in a cell culture, tissue, or *in vivo* microenvironments. In addition, physical penetration, especially by microneedles, offers high transfection rates at high cell viabilities and is currently the "gold" standard for cell transfection and cell selectivity. Reported transfection efficiencies using the microneedle approach are between 90 and 100%, with survival rates from 80 to 100%. Traditional microneedle approaches use sharpened, glass capillaries that are manually targeted by a skilled technician to a cell or embryo under the microscope. The smallest, commercially available needles are 30-gauge (~160-μm nominal inner diameter) to 34-gauge (~83-μm nominal inner diameter) needles [58]. These needles are manually aligned with the target cell at an angle. The cells are then mechanically pierced to deliver the desired genetic material. This method is a highly cumbersome and laborious since it requires

- Precise, visual coordination between the cell and the needle
- Optimal penetration force customized to cell type and condition
- Optimal injection rate, since pumping too fast would burst the cell

Typical skilled technicians can inject large cells such as ova or frog eggs at ~15 cells/min and small cells such as endothelial and fibroblast cells at ~5 cells/min [59, 60]. Advances to automate the process has resulted in some success using CMOS or CCD cameras to align the cells on the surface with a microneedle using precision, computer-controlled x-y-z motorized microscope stages. In one study, approximately 100 drosophila embryos could be injected with RNAi using silicon nitride hollow microneedles and automated microcontrollers in <7 minutes [61]. In another similar study, motorized, computer-controlled delivery was achieved for adherent cells at ~25 cells/min, with a transfection success rate of 82% and a viability of 95% [60]. The manual delivery using microinjection in the same study resulted in a viability of ~40%. The decrease in the manual injection study was primarily due to cell damage stemming from improper penetration force and poorly controlled injection rates. It should be noted that even the computer-controlled microneedle penetration would cause some physical and structural damage to the punctured cell. Therefore, to achieve optimal gene delivery via microinjection, the key parameters to consider are controllable cell membrane penetration force and picoliter fluid delivery. Strategies to control and minimize the penetration force include using (1) a rotary drilling method, (2) a repetitive retraction technique with the microneedle, and (3) a piezo-controlled vibration method. All three methods allow for gentle penetration of a highly deformable membrane. For precise liquid delivery, external computer-controlled fluid injectors are used, where 10–100 pL of fluid delivery can be achieved without cell bursting. In these current microinjection systems, CMOS technology is mainly used in the microcontroller and feedback circuits and CMOS-based cameras for optical alignment of cell and microneedle. Despite high selectivity and control over the gene injection process, limitations of this process are

- Slow turnover of transfected cells due to precision alignment requirements
- Not readily scalable to high-throughput applications
- Lack of portability
- High instrumentation complexity

Additionally, since current CMOS imaging systems have a minimum pixel size of ~7 μm, which is of the same size as typical living cells such as neurons, selecting small cells or subcellular gene delivery targets is difficult. Recent advances toward portability have utilized a microfluidic approach to select suspended cells and direct them toward needles [62]. However, optical selection requirements and integration of CMOS architecture with appropriate control circuits for microfluidic valves and micropumps into a single lab-on-chip environment for gene delivery has not yet been achieved.

To overcome low turnover rates for transfecting cells and for targeting small cells, high-density microneedles using CMOS fabrication techniques have been developed. High-density microneedles have been demonstrated to deliver DNA vaccines, antisense oligonucleotides, and plasmids in skin cells, tumors, plant tissue, and whole animals such as rats [63]. Typical substrate materials for microneedles, such as silicon, polysilicon, glass, silicon nitride, stainless steel, and organic polymers, are produced in closely packed arrays. Some examples of organic polymers are polylactic acid (PLA), polyglycolic acid, poly(lactic-co-glycolic acid) (PLGA), amylopectin, and carboxymethylcellulose [64]. These closely packed, sharp needle arrays are used to puncture living tissues and cells for gene or drug delivery applications. Fabrication is highly comparable with CMOS-compatible MEMS devices. Microneedles can be fabricated either in "*in-plane*" or "*out-of-plane*" configurations. In-plane microneedles have the needle lumen aligned in parallel to the substrate. The needles are typically fabricated in a single one-dimensional (1D) array. Additional postprocessing steps may be required for making multi-dimensional arrays. The main advantage with in-plane microneedles is that the length of the needle can be varied from tens of micrometers to several millimeters. The variability in length allows access to hard-to-reach depths within a tumor or multilayered tissues for gene or drug delivery. Polysilicon is extensively used in manufacturing MEMS devices and may be used to fabricate in-plane microneedles. In-plane needles of complex shapes such as serrated tips or pyramidal points have been fabricated [65]. For instance, complex features that mimic the mosquito proboscis have been made for painless, diabetic blood testing applications. A key disadvantage to this type of in-plane microneedles is the possibility of residual stresses in long cantilevers that can make these structures curl up if not designed appropriately. A "rolling" method by Prinz et al. has also been suggested to generate in-plane microneedles [66]. In this method, InGaAs/GaAs/SiGeSi films are deposited on a substrate and selected sacrificial layers are etched. The dislodged films due to incongruent surface stresses roll themselves into nanotubes. The hollow, cylindrical tubes are typically 300 nm in diameter and resistant to buckling and breakage. They have been demonstrated to deliver DNA solutions for transfection into thick-walled onion cells multiple times.

Out-of-plane microneedles, in comparison with their in-plane counterparts, have various advantages in fabrication and mechanical strength. Out-of-plane microneedles can be described as having needle lumens that are aligned normal to the substrate. Even though they require some customization in standard CMOS manufacturing, these microneedles can be bulk machined by a single mask process using standard photolithographic techniques. Additionally, micromolding techniques can be used for mass production, reducing manufacturing costs. A common master mold technique using SU-8 epoxy to generate various polymer and metal microneedles has been demonstrated to be highly reproducible over several molding iterations [67]. A significant advantage is that highly dense, two-dimensional (2D) microneedle arrays can be easily manufactured for simultaneous delivery over large tissue areas. Typical microneedle density can vary from a few hundred to >100,000/cm^2. In one study, a confluent monolayer of small, adherent cells was simultaneously transfected with an array of 160,000 microneedles/cm^2 [68]. Therefore, small cells or subcellular features (<50 μm) can be targeted with ease. Similar to in-plane microneedles, out-of-plane microneedles may be manufactured using various substrates, such as stainless steel, silicon, and polymers. Another advantage is that the thickness of the microneedle can be easily adjusted depending on application to required strengths to resist buckling. However, the length of an out-of-plane microneedle is restricted to ~500–1000 μm due to etchant limitations. Therefore, the penetration depth for delivering genetic material to tissue is limited and may require other methods for gene delivery.

Fabrication of out-of-plane microneedles is a relatively straightforward process that requires a combination of isotropic and anisotropic etchants. Multiple sizes and complex shapes can be manufactured as seen in Figure 15.11. Solid or hollow types of needles with sharp, blunt, asymmetric, symmetric, and other combinations have been demonstrated widely in literature. An interesting application of out-of-plane microneedles are painless, transdermal delivery of drugs and genetic material, such as hGh, insulin, and DNA vaccines. Macroflux® Corp. (now Zosano Pharma) has commercialized this "microprojection" technology to deliver hormones and drugs painlessly [69]. Another application is the delivery of drugs and RNAi inhibitors in cardiovascular stent technologies. Drug-coated and genetic construct (RNAi)-coated microneedles incorporated with stents has demonstrated that atherosclerotic plaques in arteries are not regenerated after their removal with angioplasty [70]. A typical process flow that was developed by Boris Stoeber and utilizes CMOS fabrication methodologies for generating symmetric, hollow microneedles is shown in Figure 15.11. The two-mask process flow devel-

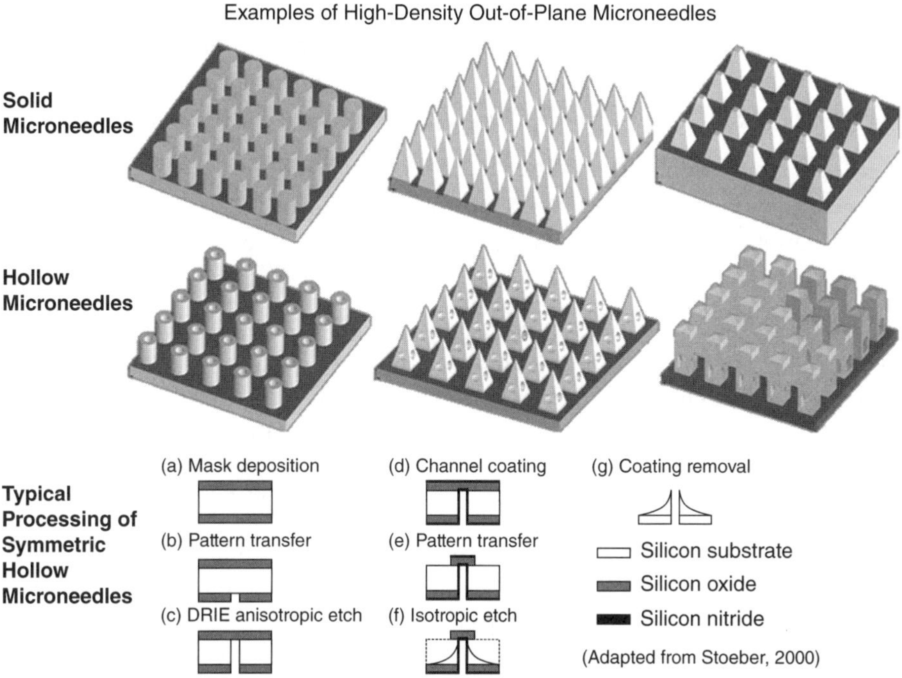

Figure 15.11. High-density out-of-plane microneedles. CAD recreations of various types of high-density microneedles fabricated in the past are shown. The top row shows solid needles that are blunt (cylindrical), sharp, or pyramidal. Their design is reflective of their application. Gene and drug delivery is typically achieved via coatings of the needle array and subsequent diffusion upon tissue or cell penetration. For liquid delivery, hollow microneedle designs, as seen in the middle row, may be used. Holes on the penetrating face (as seen in the cylindrical tubes) may be impeded by fouling and clogging. Alternative hollow microinjection delivery designs have ports on the side. In the bottom row, a typical two-mask process flow for generating symmetric hollow microneedles using a combination of anisotropic and isotropic etching techniques is shown. *Source*: Adapted from Stoeber, 2000.

oped by Boris Stoeber uses the deep reactive-ion etch (DRIE)-Bosch process (an anisotropic dry etch method) for making a hollow channel, while using an isotropic etch to make the outer cylindrical and symmetric shape of the microneedle [71].

Typical UV-based photolithographic methods may be used for the pattern transfer steps. Microchannels may also be made using the Lebouitz process, where holes are made for etchants to flow [63]. In these hollow microneedles, genetic delivery may be delivered in liquid medium. Controlled delivery may be achieved using CMOS control circuits that manipulate openings and valves for microfluidic channels and networks for selective delivery to individual microneedles. Although not yet demonstrated, spatially and temporally selective microinjection-based gene injection systems integrated with CMOS control circuitry are plausible.

The key to gene delivery in microneedle approaches lies in its design. The shape of the microneedle is extremely critical to proper delivery of the drugs and genetic material. Three types of gene delivery mechanisms are possible using high-density microneedle technology:

- External coatings of genetic material on the needle
- Microinjection delivery via hollow microfluidic channels
- Iontophoretic delivery via electrically conductive needles (microelectrodes)

The iontophoresis of DNA vaccines via skin using microneedles has shown to be a highly effective method *in vivo*. (Iontophoresis refers to diffusions of ions, drugs, and other charged particles such as DNA that is controlled by external electric fields.) Typical iontophoretic microneedles may be generated from solid, metallic microneedles or solid microneedles that have an outer metallic coating deposited via electrochemical means. In general, solid microneedles tend to have higher mechanical strength than hollow ones. Gene delivery is typically reliant on diffusion mechanisms and is not very spatially controllable, but may be highly effective systemic gene delivery applications. However, microinjection via hollow tubes is preferable for localized, spatial, and temporal selectivity of target cells. For instance, single cell resolution that may be needed for pulsatile delivery of picoliter quantities of gene expression inhibitors (i.e., RNAi for cancer and tumor cells) can be achieved using CMOS-controlled individual micropumps and microfluidic networks. The design of hollow microneedles also needs to consider the location of the fluid port. In order to prevent biological fouling and clogging of the fluid ports when inserted, the microfluidic ports may be located on the side of the needle. Alternatively, the microneedle may be designed asymmetrically, with a sharp taper for penetration and fluidic ports on the opposite end. Injection rates should also be considered. Optimal cytoplasmic fluid delivery of 0.1–0.3 pL per cell may be achieved without bursting the cell [72, 73]. If the target is the nucleus, even smaller volumes (0.01–0.05 pL) need to be delivered. Typical hollow port designs that have been used in transdermal and plant cell wall puncturing are shown in Figure 15.11 [63, 74–76].

The penetration capabilities of the microneedle also need to be carefully addressed in its design. Too little force may lead to improper or failed gene injection and delivery, while too much force may cause significant mechanical injury to cells, causing cell death. However, this design parameter is highly dependent on the cell or tissue type. Table 15.2 depicts the typical penetration forces needed for various tissue and cell types [77, 78].

The typical skin insertion forces exerted by transdermal devices ranges from 0.5 to 3 N. The biological effects of silicon microneedle remnants, for instance, have shown to cause chronic inflammation and cellular structural integrity. Methods to circumvent this issue might be (1) application of minimal force and (2) utilization of biodegradable polymers as microneedle substrate material. Considering that a typical cell has a highly deformable cell membrane, the design of a microneedle also needs to account for its sharpness. It has been suggested

TABLE 15.2. Typical Penetration Force Requirements for Cells and Tissue

Skin	0.1–3 N
Plant cell wall	$<1\,\mu N$
Blood or fibroblast cell	$<10\,nN$

that the radius of curvature at the tip should be $<0.1\,\mu m$ for preventing deformation of the membrane that provide mechanical stimulation and long-term, chronic injury to the target cell. In addition, the penetration force decreases with sharpness. For instance, in one study, silicon microneedles with a pyramidal shape exerted 0.7–20 nN to penetrate the cell membrane [78]. However, needles constructed with tips with the sharpness of multiwalled carbon nanotubes exerted only 0.1–0.2 nN to penetrate a culture of pleural mesothelial cells. The lower penetration force will certainly help with biocompatibility issues, especially those that relate to injury to the cytoskeletal structure that supports the lipid membrane. Other mechanical considerations are buckling and insertion rate. Therefore, the mechanical design of microneedles requires

- Small radius of curvature
- High aspect ratio
- High elastic modulus

An alternative to tethered needles to deliver genetic material is using high-velocity projectiles like micro- and nanoparticles. Small particles are typically loaded with desired genetic materials by electrostatically coating the external layer with cationic lipids or polymers. These positively charged molecules attract and adhere to nucleic material, such as RNAi or plasmids. It should be noted that toxicity effects from these cationic lipids and polymers such as polyethylene imine (PEI) and other related molecules (polyethylene oxide [PEO], imine derivatives) is a serious issue in using this method of gene delivery. Nanocrystals and nanoparticles therefore serve as the nucleic acid loading mechanism. Multiple shapes of the micro- or nanoparticles have been used in literature, ranging from spherical, cylindrical to spearlike structures [28, 79, 80]. These bioloaded particles are then propelled at high forces that puncture living cell and nuclear membranes. CMOS technologies can likely play a role in assisting the propulsion mechanisms of these nanoparticles for gene injection.

CMOS technologies can be used to spatially and temporally control the propulsion forces using electric or magnetic fields. Dielectrophoretic methods, for instance, capitalize on the polarized movement of ions or charged particles such as cells in solution in the presence of an electric field. Manipulating particles using electric fields is still a semi-invasive process, requiring electrode contact with the medium. The use of magnetic forces to manipulate particle movement and transfect cells (magnetofection) is highly desired since it is relatively noninvasive. Unlike electroporation, the energy requirements for the magnetic field to permeate the membrane are high, resulting in bulky instrumentation. To circumvent the large magnetic forces required to move molecules, ultrasmall superparamagnetic iron oxide and ferrite particles are typically used to bombard the living cells at high velocity. The principle by which these nano- and microparticles can be moved is the Lorentz force, where the force generated is directly proportional to the particle net charge, velocity, and the magnetic field strength. Experiments with magnetic sources suggest that at comparatively low magnetic strengths, nickel- and carbon-based nanotubes coated with nucleic material could efficiently transfect a cell [28, 29, 81]. The synergistic combination of conventional viral gene delivery methods (adenovirus) with coated magnetic beads show significant improvement in transfection as compared with viral methods alone [82]. Additionally, spatial discrimination of cells within a culture using magnets has been demonstrated. In one study, inductive current passed via microcoil to an overlapping grid of gold wires induced enough dielectrophoretic force to move individual cells in a culture [83]. The magnet microcoils were on the order of several hundred microns in diameter. Smaller microcoils (~100 µm in diameter) have been demonstrated in other MEMS applications [84]. An alternate magnetically based spatial selectivity uses a magnetic array of cylinders that are ~6-mm high beneath a 96-well plate containing transfecting plasmid and ferrite nanoparticles. A magnetic flux density of 100 mTesla was adequate to successfully transfect living cells over 12 wells in a 96-well plate [81]. The main advantages of magnetofection if it is developed at a microscale are

- Noninvasive incorporation into platforms
- Adequate transfection efficiencies
- Incorporation of spatial and temporal selectivity using inductive coupling methods in current CMOS architecture

The disadvantage is the lack of microscale resolution in transfection. In the McBain study, the minimum distance between the magnetic cylinders for isolated fields is 12 mm [81]. It is also potentially toxic due to the lack of dissolution or degradation of the nucleic acid load-carrying particle. So, once the particle in lodged in a cell, it remains intact in the cell. This may create inadvertent mechanical stimuli, false positives, and cell death. In comparison with needles, magnetoporation may be easier to implement in CMOS architecture rather than microneedles. While magnetofection is likely to result in higher cell viability compared with transfection, it is still not a high-throughput method for selective transfection.

Key items that mechanical approaches to gene injection need to address are

- Selective and efficient targeting in 2D/three-dimensional (3D) cell cultures and tissue (needle has advantage)
- CMOS manufacturing complexity (projectile delivery has advantage)
- Semi-invasive gene delivery (projectile delivery has advantage)
- Biocompatibility

Therefore, CMOS-driven lab-on-chip integration of mechanical gene delivery systems might be complex due to the extensive postprocessing that is necessary for assembly. CMOS-controlled MEMS integration of valves, microfluidics, and control circuits for precise microscale movement dynamics of needles and projectiles are necessary for this gene delivery system. In combination with electroporation, microelectrode needle may achieve 3D cell transfections with higher efficiencies. However, integration of CMOS technologies with mechanical gene delivery approaches will be more difficult to achieve compared with integration with electroporative approaches.

15.2.3 Microcavitation

Similar to magnetofection, microcavitation transfection methods apply a noninvasive, external power source (acoustic pressure generated by ultrasound) to perforate cells. The acoustic pressure caused by ultrasound on liquid by itself is not energetic enough to manipulate pore formation in membranes. Instead, similar to magnetofection, which uses impermeable, solid micro- and nanoparticles to bombard the lipid bilayer, ultrasound methods generate hollow, gas-encapsulated microbubbles that bombard and perforate the bilayer. Ultrasound is typically defined as frequencies above 20 kHz. The roles that ultrasound plays in microcavitation are (1) generation of microbubbles and (2) generation of optimal pressure waves to burst the microbubbles at the cell interface. High efficiency in transfection requires close association of a stable microbubble with the nucleic acid sequence. To successfully transfect cells, DNA or other nucleic acid material is (1) loaded inside the stable microbubble using chemical engineering techniques and (2) incorporated in the microbubble air–liquid interface either by manipulating the surface tension of the gas/liquid interface or by adhering electrostatically to the outside of the microbubble by changing the surface charges with certain molecules [85, 86]. The key event that makes this method effective is the volatile bursting of the microbubbles as the pressure wave meets the membrane–liquid interface. This destructive phenomenon described by the violent collapse of the microbubble at the membrane interface is known as *microcavitation*. Typical pore sizes that are generated using this phenomenon are 35–75 nm in diameter in the membrane. As the microbubble is destroyed, its nucleic acid payload is automatically deposited by diffusive means into the cell. The biophysical mechanism of sonoporation and microcavitation is shown in Figure 15.12. Alternative methods to generate cavitation is to use high-power lasers to focus their energies onto distinct regions within a cell culture and transfect them [87, 88].

Although successful transfection has been seen at frequencies as low as 20 kHz (4% tranfection success, 40% viability), optimal transfections for mammalian cells are seen in the 1–3-MHz range. Interestingly, DNA uptake has also been reported at extremely high diagnostic frequencies (>40 MHz). Typical operational parameters for ultrasound-induced transfection (also known as sonoporation) are as follows [89]:

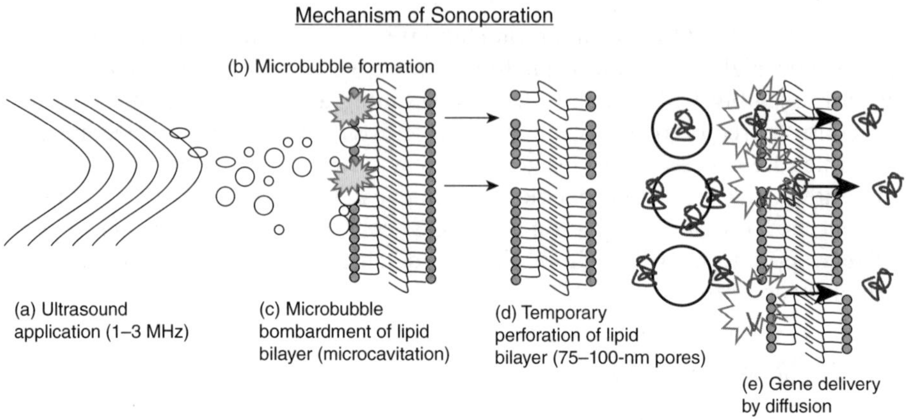

Figure 15.12. Biophysical principle of microcavitation. (a) The mechanism typically begins with the application of ultrasound (1–3 MHz) or laser induction, which generates microbubbles (b). (c) When the frequency of the acoustic pressure wave is in resonance with the bubble interface (center frequency), the microbubbles burst in a phenomenon known as microcavitation. (d) The intense release of energy generates large perforations in the membrane. (e) DNA or other genetic material may be loaded (1) inside, (2) embedded, or (3) electrostatically adhered to the outside of the microbubble.

- 1–3 MHz
- 0.5–2 W/cm^2 acoustic power
- 10-second–30-minute insonation times (depending on cell type and environmental conditions)

For plant or bacterial cells, which has thick cell walls, transfection has been observed at higher frequencies (200–400-MHz range). Therefore, *key* parameters to manipulate in sonoporation devices are as follows [90]:

- Microbubble cavitation size (bubble radius)
- Amount of genetic material loading (genetic payload)
- Acoustic application frequency (kHz→mHz)
- Frequency pulse length (insonation time)
- Mechanical index should not exceed 1.9 (FDA guidelines)

The mechanical index, which is defined by the U.S. Food and Drug Administration (FDA) as the quantity of mechanical work per ultrasound cycle, measures the point where ultrasound exposure will produce an adverse biological effect by a nonthermal mechanism. For classic diagnostic and imaging applications, the ratio of the peak acoustic pressure to the square root of the incident frequency should not exceed 1.9.

The microbubble size is one of the most important parameters to control and can be varied by proportional increases in frequency. Smaller bubbles tend to be more stable, while larger bubbles hold more payload and generate more energy upon cavitation, causing larger pores. Low acoustic pressure causes more generated microbubbles to be in close proximity to the cell for appropriate contact. In general, the fate of microbubbles is

- Localized sites of intense heat
- Localized sites of intense light emission (sonoluminescence)
- Free radical generation
- Shock wave generation

To better control microbubble stability and loading ability, it may be more easy to use commercially available microbubble products such as Optison™ (GE Healthcare, Piscataway, NJ) and use ultrasound and lasers just for generating pressure waves and, eventually, cavitation. These commercial products typically stabilize the surface tension between the liquid, microbubble surface, and internal gas interfaces for long-lasting microbubbles. In the case of Optison, the microbubble interface is albumin coated and artificial organic gases are internalized in the bubble. Typical bubble sizes are 0.5–6 μm in diameter [91]. Key disadvantages of microcavitation methods are

- Potential cell death due to free radical formation
- Heat stress
- Severe mechanical damage to the cell membrane
- Inflammation

Precise spatial discrimination is more difficult to achieve because the acoustic pressure wave may be generated throughout a given area. Controlled and focused external lasers and ultrasound transducers may be used to generate localized pressure waves. However, the needed spatial resolution still may prove elusive for targeting small cells and subcellular features. For lab-on-chip applications, microsized ultrasound transducers may be utilized. Typical surface acoustic wave (SAW) devices and piezoelectric devices use quartz or other crystalline structures for sensitive measurement. However, these materials are costly and incorporation into the CMOS architecture may require extensive postprocessing fabrication. To better integrate CMOS technology in ultrasound generators, piezoelectric lead–zirconate–titanate (PZT) materials may be directly integrated into microchambers to generate microbubbles. There are various CMOS-compatible processes to incorporate PZT-based capacitors. Typical fabrication steps utilize sol-gel deposition with wet etching methods. PZT films can be generated and patterned easily by wet etch techniques [92]. The devices made were able to image at 20–50 MHz for diagnostic applications. Alternative CMOS incorporation of PZT with thin film deposition techniques was demonstrated in biosensor applications that utilized magnetically actuated microcoils (200 μm in diameter). These microcoils exerted a Lorentz force on the PZT films, which in turn was used to generate acoustic pressure [93]. Similar films may be optimized and integrated into lab-on-chip devices for generating lower center frequencies for transfection purposes. CMOS-based ultrasound was used to manipulate nonmammalian cells at 380 MHz in a lab-on-chip application [94]. Additional CMOS circuitry to control frequencies from 1 to 10 MHz, with desired acoustic power (~2–3), and variable pulse cycle modulation and insonation times may be easily envisioned for customized transfection protocols. Therefore, it is feasible to integrate CMOS technology with ultrasound-based microcavitation. It is also not expected to be as intensive in fabrication methodology, lowering manufacturing and processing cost. However, in terms of spatial resolution, control of microbubble cavitation may prove elusive at the microscale level unless alternate technologies using wave interference techniques to dissipate the acoustic pressure waves at undesired locations are incorporated. Perhaps the microfluidic chambers with cells could be designed for discrete localization of a generated pressure wave.

15.3 CHALLENGES IN BIOCHIP–CMOS INTEGRATION

Currently, the gene manipulation process involves thousands of cells and depends largely on a statistical assessment of changes in the phenotype. Typical genetic studies track one or two genes under multiple conditions using a 96-/384-well format. To assess the functional effect of gene suppression or overexpression, fluorescence or chemiluminescence techniques are often used to quantify protein expression. For the successful application of any bio–CMOS hybrid system for gene injection, the system will need to be capable of performing each of the steps involved in gene manipulation in a single platform. The platform will need to simultaneously or sequentially inject multiple genetic constructs in living cells and tissue with high throughput in order to quickly assess the basic gene function, its interactions with other genes, and its combinatorial effect on the cell phenotype. Currently, gene studies use microarray technology to profile the levels of different genes expressed in a cell [95]. In this approach, short strands of DNA sequences are immobilized or tethered in an addressable manner at designated microelectrodes. The site-specific binding of DNA oligomers is typically monitored by measuring the microelectrode–electrolyte interfacial impedance. Immobilization of DNA oligomers usually involves the use of biotin–streptavidin or other common covalent linkages. In certain single nucleotide polymorphism (SNP) detection gene chips, nucleic acid-based test samples are directed to designated areas by surface electrodes [96]. To assess genetic expression in a certain cell type or tissue, cDNA (expressed DNA) of the same cells grown under a particular environmental stressor is compared with the gene library from the gene chip. Current gene chips require lengthy cDNA sample preparation process before simultaneous genetic expression levels of multiple genes can be analyzed on the gene chip. Cells are grown under desired conditions and then mRNA is harvested from them using standard cell lysis techniques. The harvested mRNA is then processed using reverse transcription methods to obtain cDNAs.

Affymetrix Gene Chips Illumina Gene Chips

Figure 15.13. Examples of gene chips. Affymetrix uses optical assessment techniques to test for gene interactions using genomic DNA that has been spot-addressed. Illumina uses cDNA sampling techniques in a 96-well format. Preprocessing of cells is similar for both gene chips. Copyright © 2010 (http://www.affymetrix.com); Copyright © 2010 (http://www.illumina.com).

Fluorescently tagged cDNA from the sample is then hybridized with the gene library and an optical mode of assessment is used to analyze the general gene expression levels. Numerous gene libraries for different species are now commercially available. Commercially available gene chips include those from Affymetrix (Santa Clara, CA) and Illumina (San Diego, CA), shown in Figure 15.13. Microarray formats vary between Affymetrix and Illumina. Affymetrix uses oligomers that are ~25 nucleotides that are spatially addressed in ~200-µm-diameter wells in a planar microarray format. Illumina uses a novel microdroplet technology (~3-µm droplets), reducing the DNA probing area significantly by creating a high-density microarray. A typical gene chip costs ~\$350–\$1000, with analysis equipment >\$100,000.

The role of CMOS technology in gene chips is related mainly to impedance and capacitance sensors to test for (1) DNA oligomer localization, (2) hybridization, and (3) on-chip signal amplification. In addition, CMOS-based external imagers collect the relative gene expression data for a given stressor simultaneously for all genes in a species genome using fluorescence as a photonic marker. The high-throughput nature of the whole-genome assay helps assess multiple gene expression interactions and pathways simultaneously [97]. However, the above information is still assessed on a large population rather than small groups or even individual cells. For example, the differences in genetic expression between individual cell–cell interactions as seen among neurons in a network are not readily assessed in the current gene chip. In addition, the real-time, functional response of stressors on the living cell cannot be easily assessed using current

CMOS-based gene chips. Therefore, there is a critical need for scalable technologies that will not only enable gene injection and manipulation in *living cells* but also make critical assessments of functional responses to different stressors on living cell cultures.

CMOS-based field effect transistors (FETs) and ion-selective FETs (ISFETs) are novel biosensors that represent the next step in generating scalable CMOS technologies. In ISFETs, the ionic electrolyte interface interacts to change the properties of the gate to permit current flow. Portable biosensors using CMOS integrated with living cells have been demonstrated for "canary"-type sensor technologies [98]. Silicon dioxide is typically used as a gate material to measure pH and carbon dioxide in the liquid medium. Alternatively, temperature measurements using thermocouple-like configurations (i.e., p–n junctions) have been devised. Even low-noise electrical signals from neurons have been recorded using CMOS-based FETs [53]. In these unique studies, the neuron forms a connection between the source and the drain. To obtain low-noise electrical recordings, the FET array takes advantage of a small, electrolyte-filled space between the cell and the gate. As neurons become electrically excited, relative changes in the drain-source current are correlated to changes in the local field potential at the gate, similar to an ISFET concept. Therefore, *key* technologies that can be adapted to future bio–CMOS hybrid system for gene injection by electroporation are

- Sensor technology for measuring cell adherence, such as quartz crystal microbalance (QCM), impedance sensing, and current/voltage sensors
- Environmental feedback controls for maintaining cell cultures, such as pH, temperature, and gas (CO_2 and O_2)
- Fabrication methodology that renders the semiconductor interface inert to liquid media and is nontoxic to living cells, such as passivation layers (silicon dioxide, silicon nitride, and aluminum oxide)

The use of passivation layers is essential since the majority of CMOS-based materials are incompatible to liquids due to corrosion issues. While substrates such as silicon are the standard in the semiconductor industry, integration of CMOS technologies with biological medium can be better achieved in the future with alternative flexible and transparent (for optical measurements) substrates such as polyimide, parylene, and indium tin oxide (ITO) on polyethylene.

Some of the desired features of a future, bio–CMOS gene manipulation chip would involve the following:

- Electronic and photonic addressing systems for spatial and temporal control of gene manipulation
- Impedance and other integrated biosensors for real-time feedback on the viability of immobilized biomolecules and/or cells
- On-chip signal processing and readout to enable quantitative assessments at every step before and after gene injection
- Communication circuitry to communicate with external computers and other control devices

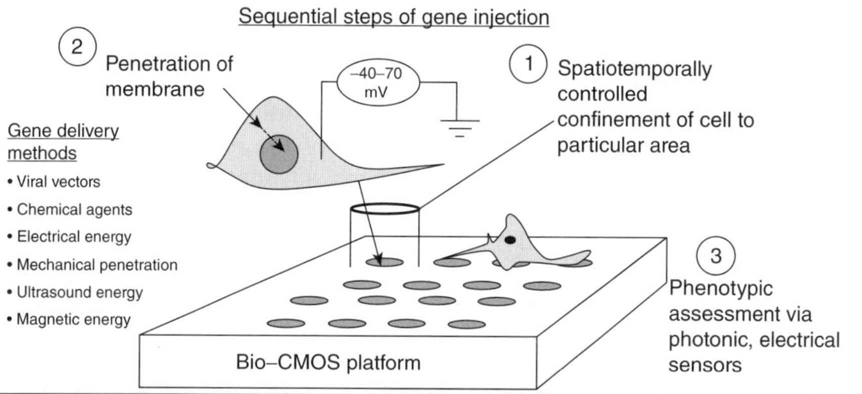

Gene delivery considerations	General cell considerations	Interfacial considerations	Engineering considerations
• Size of genetic construct • Transient or permament gene suppression/ expression • Cell viability • Dividing/nondividing cell	• Resting cell potential • Isotonic environment • Cell type—suspension/adherent • Chamber atmosphere—isothermic, % O_2, % CO_2 • Cell morphology • Lipid bilayer characteristics	• Surface topology • Surface charge • Adhesion parameters • Electrochemically inert platform • Low immunogenicity (especially *in vivo*) • Proximity to preferred gene delivery method	• Isolation of internal electrical circuitry from aqueous cell enviroment • Integration of MEMS/microfluidics • Integration of reaction chambers and cell housing • Digitally controlled gene delivery • On-chip power source • Optial transparency for photonic assessment

Figure 15.14. Broad components of a typical bio–CMOS chip for gene injection and assessment of changes in phenotype. The three broad components are (1) cell selection or isolation for spatially targeted gene injection, (2) mechanism for penetration of cell membrane, and (3) assessment of cell phenotype. Some of the key factors that will impact the design and performance of the system are listed in the table. A total lab-on-chip platform for gene injection will need to address all these concerns.

- Biocompatibility at the cell–device interface
- On-chip power source for controlled gene injection

Incorporation of living cells into a gene chip with CMOS architecture involves additional feedback control mechanisms to maintain cell viability. For real-time assessment, the bio-CMOS gene injection chip will need to have a mixed-mode architecture, with an analog component that probes and genetically manipulates living cells under digital control mechanisms (Fig. 15.14). The broad components of such a bio–CMOS gene injection chip along with all the key factors that will need to be addressed are illustrated in Figure 15.14. The mode of gene injection can be viral, chemical, electrical, mechanical, and so on, using methods outlined in the previous section. The design of the gene vector itself needs to be optimized for size, cell viability, and cell type.

General examples of difficult-to-transfect cells are primary cells, stem cells, and neurons. These cells have complex environmental requirements to keep them viable. Therefore, a general design guideline is that the desired gene vector should not interfere with cell homeostasis. For appropriate maintenance of cell homeostasis, the environmental conditions should not be perturbed at length. The cell–device interface should have a high biocompatibility in order to obtain credible gene manipulation results. Since the cell–device interface is typically an ionic liquid/conductive solid interface, the substrate of the bio–CMOS chip and the microelectrodes (if any) should be electrochemically inert. In addition, to control the bioelectrochemical interface, digital control circuitry and power delivery method needs to be integrated on the same chip. Hermetic sealing and isolation of the electronics from the cell housing is necessary to isolate the wet and dry components of the system. Therefore, the future development of CMOS-based gene injection systems would enable the following:

- Spatiotemporal control of gene manipulation in a small population of cells
- Microscale precision for targeting specific single cells or subcellular structures
- Simultaneous phenotype assessment of multigene knockouts or application of multiple stressors
- High transfection efficiency and ability to maintain viable cells
- High throughput

15.3.1 Biocompatibility at the Bioelectronic Interface

15.3.1.1 Factors that Impact Cell Viability. The challenge for creating such a CMOS-based system that is capable of gene injection requires solving several biological issues at the liquid–cell–device interface. From a living cell's perspective at the interface, the underlying electronics and CMOS architecture must be designed to prevent *cell death* while being injected with nucleic material. The modes of cell death may be immediate or may occur several hours or even days after the initial insult. Cells may die for a variety of reasons as a result of the gene injection process. Too much membrane permeation from high electric field strength, ultrasonic frequency, or excess particle bombardment or other mechanical injuries may result in an uncontrolled flow of the external liquid into cell. This osmotic imbalance may result in an immediate cell lysis, where the integrity of the cell membrane is compromised and the cell literally bursts. This type of cell death is also known as *necrosis*. Alternatively, during the gene injection process, toxic materials may be generated as electrochemical by-products (free radicals, hydrogen peroxide, soluble, corrosive metal products, etc.). Additionally, the vector materials used to deliver genes may interfere with key metabolic processes within the cell. If toxic material is taken in by the cell, it may directly impair mitochondrial function impacting energy production. This in turn results in a form of cell death known as *apoptosis*, where cells initiate an autodestruction sequence of cellular events and self-destruct. Metabolic toxicity need not be immediate and may affect indirect or alternative, redundant energy-producing pathways. (A typical living cell is highly adaptable and has diverse modes of creating energy to sustain itself under a variety of environments. But it also makes the cell vulnerable to different sources of toxicity.) Apoptotic signaling cascade may take hours to days in external cell culture environments, and weeks to years in *in vivo* environments, depending on how acute the stress is [99, 100].

A third manner by which apoptotic cell death may occur is due to changes in the interfacial surface charge or topology during the gene injection process. External environmental conditions (not just oxygen and other dissolved gas and nutrient levels in the medium) play a significant role in keeping cells alive. Cells, especially adherent cells, live in complex tissues that contact various surfaces and multiple cell types. Surface topology and charge play a significant role in giving cells appropriate survival signals, without which cells will commit suicide by apoptosis. (In malignant tumor cells, this signaling pathway is typically disabled, allowing cancerous cells to spread and invade different tissues and multiple organs.) In a gene injection chip, the application of high energies may result in a long-term change in surface potential and charge. As a result, cells may not adhere properly or receive inappropriate signals and undergo apoptosis [101]. Corrosion on the surface or microscale transistor features may result in changes in surface roughness. Surface roughness of ~2 μm is known to affect cell morphology in certain cell types. This would be a serious issue to consider in single-platform, microneedle-based gene delivery systems for adherent cells. Growth on sharp points may result in altered morphologies and generate artifacts in general cell behavior. Therefore, placing microneedles on a different platform or making them movable may need to be implemented to prevent cell death. Additionally, the design and deposition of materials to make transistors in the CMOS architecture need to consider surface topology. In general, an optimally rough surface tends to favor cell adhesion, while extremely smooth surfaces do not allow cell adhesion. The degree of ideal surface roughness is highly dependent on cell type or stage of cell differentiation. Surface charge also matters. An example is neurons, which tend to favor positively charged, polylysine-coated surfaces. Different cell types may also require additional complex protein coatings with specific peptide sequences or may need additional cell feeder layers to assist in cell survival. These preferential cell adhesion mechanisms may also be used as guidelines to engineer and pattern CMOS-based device surfaces. Using surface topology and surface charge properties, cells can be manipulated to grow over selected microelectrodes to achieve spatial and temporal control over gene delivery.

15.3.1.2 Corrosion of Interfaces. Corrosion of the CMOS process material may be a significant issue in future bio–CMOS platforms. CMOS-based architecture with three layers have various surfaces that are biocompatible, such as the silicon substrate layer (both n-type and p-type) and passivation layer, which is mainly silicon dioxide (Fig. 15.15). Further, the surface chemistry of glass is easily modifiable to accommodate necessary surface coatings to promote cell adhesion. The gate electrodes are typically made of doped polysilicon that have been demonstrated to be biocompatible [55, 56]. Metal layers such as aluminum

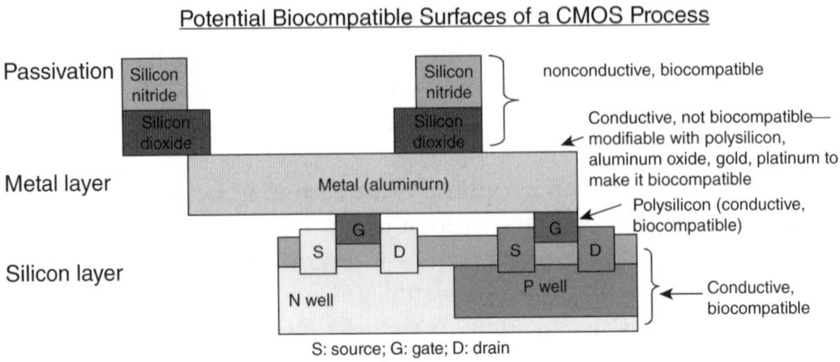

Figure 15.15. Biocompatible surfaces in a CMOS device. The general layers of a typical CMOS-based architecture are shown, with silicon, metal, and passivation layers. Biocompatible layers are silicon, polysilicon, silicon nitride, and silicon dioxide.

are the least biocompatible. Aluminum is generally prone to corrosion in an electrochemical environment. The dissolution of aluminum to aluminum hydroxide is a spontaneous reaction, with an electrochemical potential of ~2.3V [102]. The application of high electrical energy may further speed up the electrochemical reaction. A potential solution is coat corrosion-prone layers with electrochemically inert metals such as gold or platinum. However, a potential drawback is that gold and platinum deposition is not part of the standard CMOS process. In SNP detection-based CMOS gene chips, platinum is preprocessed and deposited separately by sputtering and holes for exposed electrodes are etched after assembly. Alternatively, an electrochemical deposition protocol may be used to deposit these metals as a postprocessing step for a CMOS-based device. A typical bioengineering solution to prevent corrosion of metals (i.e., titanium, aluminum) with body fluids is to add extremely thin passivation layers on top of the metal. Titanium, for example, forms a thin oxide layer when in contact with aqueous solutions. The oxide layers serves as a barrier to corrosion. A similar solution for aluminum metal may be applied, where a thin aluminum oxide (alumina) layer may be generated. Fromhertz et al. showed that a thin, nanoporous alumina layer prevents corrosion and yet allows for liquid–electrode contact. The nanoporous alumina layer is generated by anodizing the aluminum layer for prescribed amounts of time. Controlled and optimized anodizing voltage application may be used to engineer desired surface topologies and charges for appropriate cell adhesion. Alternatively, since the amount of applied electrical energy is directly proportional to the dissolution rate, designing MEAs that operate under low power consumption circuits may be necessary for polysilicon-based systems. Potential strategies for using polysilicon microelectrodes in electroporation strategies are as follows:

- Keeping microelectrode area low and minimizing edge effects (for low voltages)

- Reducing stimulation time (<1 msec)
- Increasing resistance of the media (for low solution currents)
- Lowering interfacial (cell–electrode) resistance (for efficient electroporation and maximum charge transfer across cellular membrane)
- Staying within the water window (stable electrochemical ionization limits) of an electrochemical system

The above outlined strategies are valid for any future CMOS-based gene injection platform with a liquid/device conductive interface to prevent or minimize corrosion.

15.3.1.3 Maintaining Cell Homeostasis. To achieve a total lab-on-chip gene injection platform, maintaining cell homeostasis is an extremely important criterion. Aside from surface topology, keeping appropriate nutrient and gas levels in the liquid medium is essential to having good cell health. Typical external cell culture requirements use a diffusion-based mechanism for gas exchange and intermittently exchange with fresh media to remove cellular waste products, refresh nutrients, and maintain pH levels (generally at 7.2–7.4). General guidelines in macroscale cultures require 5–10% CO_2 content and 1–2 mM glutamate, along with other essential ions and nutrients. It is essential that during the gene injection process, the nutrients levels are unperturbed and the pH levels are maintained. Acidic conditions, for instance, could promote apoptosis. Cells also require an isothermal environment. If the energy delivered causes a thermal increase of ~5°C, heat-shock proteins in response to the stress will be activated and cause potential artifacts in the gene manipulation experiments. Excessive heat production could result in rapid denaturation of proteins, often resulting in fixing or "gluing" the cell to the surface permanently. Since extremely small culture volumes are being used, the ability to heat media solutions from 37°C (mammalian cells) by ~5°C would require ~2 J for a 100 μL solution, assuming the medium has a similar heat capacity to water. This means that, in the case of electroporation, the total applied power over the length of the electroporation process should not exceed 2 J to prevent a heat-shock response. Considering all these parameters, frequent and extensive monitoring of general cell health is critical to prevent unintended consequences as a result of gene injection. Cells need to be monitored for morphology, behavior, and biocompatibility. Conventional methods typically require a human-based optical assessment of general cell well-being. However, at the microscale and at a high-throughput level, manual assessment might be error prone and tedious. CMOS-based digital circuitry could assess medium conditions such as temperature, pH, and nutrient level. Cell lysis may be detected by measuring membrane capacitance as a marker of acute toxicity. Morphology markers for general cell well-being are harder to assess electronically. Probing for the number of cell adhesion sites using impedance or mass balance methods might be a solution. Nevertheless, one of the critical requirements of a bio–CMOS platform would be to maintain cell homeostasis and prevent irreversible cell injury.

TABLE 15.3. Typical Cell Sizes

Cell Type	Size
Ostrich egg cell	150 mm
Most plant cells	10–100 μm
Most animal cells	10–30 μm
Longest human neuron	1.5 m × 10 μm
Most bacteria	1–10 μm

15.3.2 Scalability of a Bio–CMOS Hybrid System for Gene Injection

The MEA provides a convenient, scalable platform that provides spatial control and selectivity in gene injection via its array configuration. The number of discrete microelectrodes on the sensing surface also provides the ultimate resolution for the system, which theoretically is limited only by the resolution of the photolithography process. In a recent study, a CMOS-based biosensor array for neurons was made with a 16,384 FET-based microelectrode array, with each transistor/microelectrode having a ~7-μm feature size [103]. Therefore, microscaled sensor arrays with appropriate signal amplification are readily achieved with CMOS-based systems. The two primary determinants for the level of spatial resolution required in gene injection platforms are (1) target cell size or cell population and (2) degree of spatiotemporal discrimination for gene injection. As seen in Table 15.3, typical cells have a wide range of sizes. Egg cells are typically very large, ranging from few millimeters to several hundred millimeters. Typically, for large cells, microinjection is the most efficient form of nucleic acid delivery. In fact, most automated microinjection technologies are targeted toward individual inoculation of egg cells and embryos within them. However, smaller cells such as plant (10–100 μm) or mammalian cells (~10 μm) are more difficult to target with microinjection technologies since they require more precision. Current MEA fabrication technologies are well within current limits for targeting even single mammalian cells. Typical gene injection microelectrodes may be designed to be as large as 100 μm or as small as 1 μm for potential single or small populations of bacterial transformations. Therefore, a significant challenge is to customize the microelectrode size to the desired target size.

Microelectrode sizes may further be decreased in size beyond the single-cell range for targeting subcellular structures such as axons or nucleus in neurons. Additionally, cells are highly deformable and can easily change shape to accommodate to its surroundings. As seen in Figure 15.16, a silicon-based MEA with adherent fibroblast cells growing on it is shown to stretch and spread along the surface. In order to maintain the cell volume, individual cells can spread into extremely thin configurations along the surface in response to environmental cues. Fibroblast cells, for example, can spread to become a wide structure that stretches for greater than 100–200 μm in length. Therefore, smaller microelec-

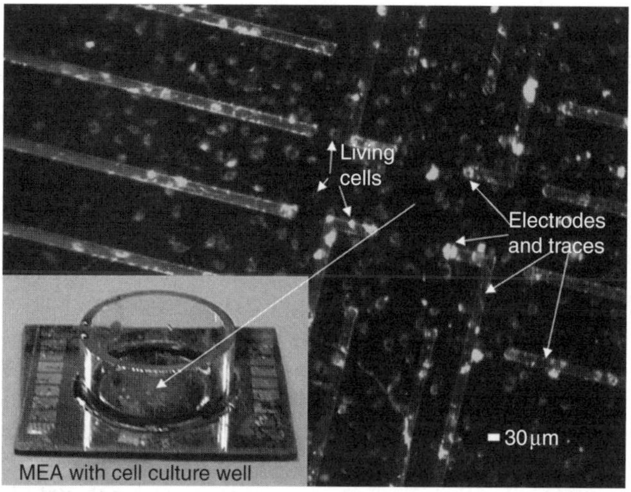

Figure 15.16. Bright-field image of NIH3T3 cells grown on silicon-based 4×4 MEA. Electrodes are 30 μm in diameter, with equivalent dimension for electrode traces.

trode sizes can be used to preferentially target distinct areas in a cell. Additionally, closely spaced electrodes may be used to probe for "soft spots" on the lipid bilayer membrane. The cell membrane is highly heterogeneous in composition, having distinct regions in the bilayer that are more fluid than others. There also exist regions within the cell membrane that are more conducive to endocytosis and internalization mechanism. Typically, these regions have a high concentration of "lipid rafts" or unique lipid aggregates in the membrane where cell signal transduction commences. Moreover, spreading and stretched cells may have cytoskeletal reinforcements under areas of strain in the lipid bilayer, making certain membrane regions less penetrable for gene injection than others. Therefore, a higher density of electrodes may help increase the efficiency of transfection by optimally targeting cells at vulnerable points on the cell membrane. Localizing gene delivery to specific points in turn may help increase cell viability during and after gene delivery. Also, cells on artificial surfaces tend to migrate. Having a higher density of electrodes may help better target position shifting cells. Using built-in impedance-based sensors, it is possible to detect the cell position and target cells on certain electrodes.

The MEA platform could be potentially a cross-cutting platform for all microscale gene injection technologies including magnetoporation, sonoporation, microneedle injection, and electroporation technologies. However, spatial resolutions of magnetoporation- and sonoporation-based approaches are inferior to that of microneedle- and electroporation-based approaches. For instance, microcoils for magnetoporation and PZT-based ultrasonicators are still on the order of several hundred microns in size. Therefore, these gene delivery methods might be more suitable for targeting small to larger populations of cells. High-density microneedle-based gene injection, electroporation, or their combination (conductive microneedles, also known as 3D electroporation) offer the best resolution for spatially controlled injection and delivery of nucleic acids to cells. A key parameter to optimize for electroporation-based methods is the distance between the electrodes. The goal is to ensure that no irreversible damage is caused to the membrane or essential cell functions. By scaling MEAs to desired sizes, the expected voltage requirements for electroporative gene delivery would also steadily decrease. In essence, lower voltages are needed to maintain the same field strengths as the distance between any two electrodes decreases. So, transistor-transistor logic (TTL) stimulation (±5 V) would supply the necessary voltage for electroporation. A hypothetical field strength of 500 V/cm would require just 5 V with an interelectrode distance of 100 μm. However, as the electrode size decreases, the signal-to-noise ratio (SNR) of recorded electrical activity from electrically active cells is expected to decrease due to higher noise levels. In addition, with decreasing electrode area, the electrode impedance would increase and charge transfer across the interface would decrease. For gene injection, charge transfer across the cell–electrode interface needs to be high. Therefore, an engineering trade-off between reducing electrode size for higher spatial resolution and maintaining optimal charge injection at the interface needs to occur. In addition, TTL-based stimulation may not provide sufficient currents. Some alternatives to ponder are to incorporate current amplifiers or use a Darlington pair to rapidly inject current into the system.

15.3.3 Programmability of Individual Microelectrodes for Gene Injection

High-density MEA systems also have the challenge of addressing individual microelectrodes. Smaller distance between microelectrodes, while advantageous for reducing operational voltages, increases the likelihood of cross talk across two needles or electronic coupling between any two adjacent electrodes. To maintain discrete control over the microelectrodes, the interelectrode distance is another parameter that needs to be optimized. COMSOL- or ANSYS-based finite element software may be used to simulate the effective electric fields generated by a set of microelectrodes and assess optimal electrode distances that are customized to the application. Individual addressability and programmability of microelectrodes is a highly desired element of MEA design. However, a key issue, especially in planar MEAs, is that the extensive area required for the interconnecting traces from the actual MEA to external bond pads is often larger than the array area itself. As seen in Figure 15.17, the array area where cells would be in contact is ~10–12 times smaller than its interconnecting traces. If each individual microelectrode were connected individually in, for example, 100 × 100 MEA, 100,000 traces need to be accommodated. This complicates MEA design and the purpose of miniaturization. Additionally, this may require large interelectrode separation distances, increasing operational voltages. CMOS-based digital controllers may provide a potential solution to MEA addressability using row

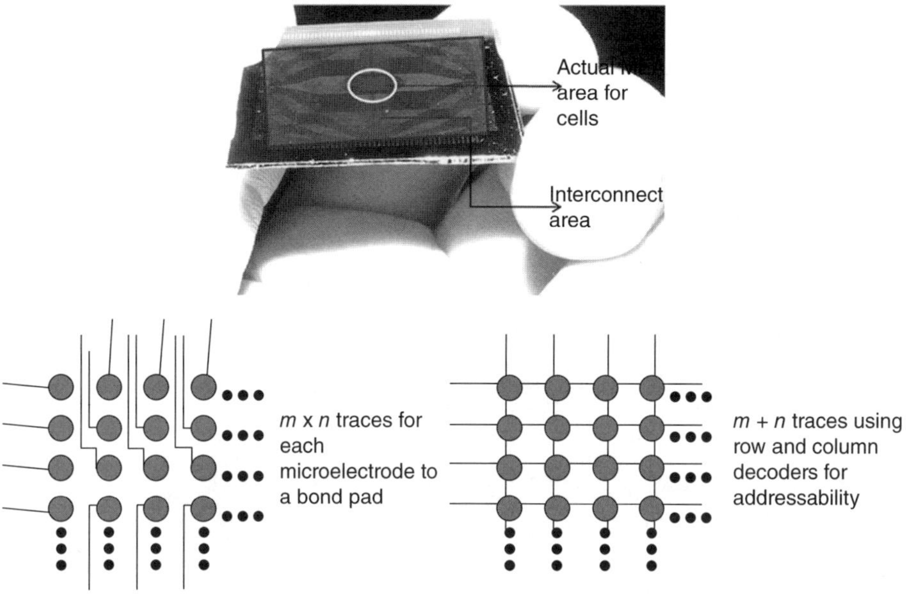

Figure 15.17. Programmability and addressability of a CMOS-based MEA. Electrodes are 30 μm in diameter, with equivalent dimension for electrode traces. (Top) The white circle denotes actual electrode area where cells might be located. The gray rectangle indicates the interconnections and traces from each microelectrode to an outer bond pad. The space needed to accommodate the interconnect area is illustrated here to exceed the actual electrode area by at least 10 times. (Bottom left) To individually connect each microelectrode in an $m \times n$ array would require $m \times n$ traces. (Bottom right) However, using simple digital logic-based controls, the number of traces can be reduced to $m + n$ wires.

and column decoders. Using an AND logic operator, individual microelectrodes can be addressed by selecting a desired row and column. This effectively reduces the number of traces to 200 for a 100×100 array. The significant reduction in traces allows for further miniaturization. If a parallel or variable mode of operation of these addressable electrodes is desired, more complex field-programmable gate arrays (FPGAs) or analog-based field programmable analog array (FPAA) (which uses a switched capacitor system for control) may be incorporated into the biochip. Alternatively, the use of optical interconnects may help miniaturize. Additional digital circuitry may also provide the temporal control of the MEA. Similar circuitry may be used to manipulate cell movement using dielectrophoretic principles. All together, CMOS-based logic systems may be used to selectively provide gene delivery to desired cells.

15.3.4 Integration of Multiple Systems

A significant hurdle in developing the CMOS-based gene injection platform is the integration of multiple functions that may need the presence of optics, fluidics,

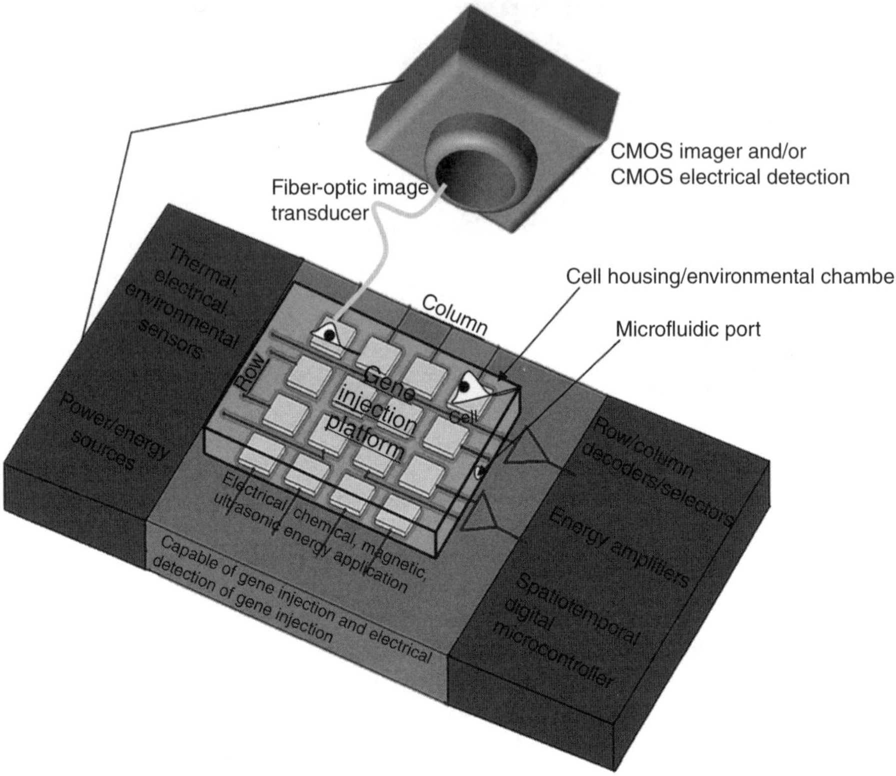

Figure 15.18. Typical CMOS-based biochip for gene injection. The comprehensive CMOS-based gene delivery platform would have at least two distinct modules—a housing for living cells (wet) and the CMOS circuitry (dry) controlling the platform. The CMOS circuitry needs to be hermetically sealed from the wet components. CMOS-based current amplifiers, energy sources, microcontrollers for MEAS, and biosensors are all part of this multifunction biochip. The housing holds the desired cells for gene injection along with a microfluidic chamber for media renewal. A third module that may be integrated with the digital control system is the CMOS-based imaging system. Photonic assessment is currently the only way to determine gene delivery success *in situ*.

and MEMS on the same platform. Such a bio–CMOS platform would require that the wet and the dry components of the system are hermetically packaged. A potential bio–CMOS device is depicted in Figure 15.18. The chamber housing the cells needs to be sealed properly to prevent leakage from liquid media.

It also needs to be transparent to allow for optical assessment of cells. Microfluidic ports and networks for media and nutrient renewal will need to be incorporated in a postprocessing step. One of the disadvantages of a standard process CMOS-based biochip is the lack of transparency of the substrate, especially when all current approaches to assess gene injection are based on photonic markers. Potential solutions might be to develop CMOS circuitry on transparent substrates, such as polyimide, cellulose derived products, and parylene to name a few. Alternatively, fluorescent semiconductor-based quantum dots or other luminescencing markers may be used to determine the efficacy of gene injection. Reflective methods may also be attempted. CMOS imagers to collect photonic measurements may be incorporated with the digital module or attached separately. Current CMOS imagers are of high quality and pixel density, with built-in buffering systems and signal processing to enhance resolution. Typical CMOS-based imagers that use silicon-based photodiodes to capture light have medium efficiencies and are poor at resolving point sources of light, especially at low light levels in very small-scale devices. At low light levels, high thermal noise tends to persist. However, with adequate light amplification techniques and electronic biasing, efficiencies can be easily improved.

If MEMS is incorporated into a CMOS biochip system, a challenge is fabrication. While CMOS circuit designs and some MEMS share similar processing steps, significant incompatibilities in temperatures exist. Alternative approaches might be to produce the MEMS structure (for instance, a cantilever or microneedle) and CMOS circuitry in separate platforms and to use flip-chip to bond the two in a postprocessing step. The integration of MEMS and CMOS circuitry on a single platform still needs to resolve significant fabrication issues for future gene delivery systems.

With this multisystem integration, living cells may be monitored, gene injected, and assessed for gene expression or inhibition.

15.4 CONCLUSIONS AND FUTURE OUTLOOK

Gene delivery is a complex, engineering endeavor that can be achieved at the single-cell level. Current trends in gene delivery utilize viral and chemical transfection methods to deliver desired nucleic acid material to cells. However, these traditional methods lack precision and typical cell viability is poor. The integration of electronics and CMOS-based technologies with physical gene transfection methods will pave the way for a lab-on-chip device. Additionally, using standard microelectrode and other microscale technologies, genetic material can be delivered with high spatial and temporal discrimination. CMOS technologies will be critical in the development of high-throughput gene injection and manipulation systems that are needed in genetic engineering studies as well as in the drug development industry.

Of the many gene injection techniques, microscale transfection methods such as microinjection and microelectroporation have been demonstrated to have the highest cell viability and transfection success. The ability to deliver with high efficiencies at discrete locations allows for novel interactive cell culture studies (i.e., neuromorphic networks or tissues with multiple cell lines). Microscale technologies also give a high-throughput screening technology for simultaneous assessment of multiple genes. Physical methods of gene delivery have the best chance of incorporation with CMOS technologies since conventional electrical, magnetic, or acoustic power sources can be integrated into potential gene delivery systems. Of these, electric field-derived gene delivery offers the easiest route to developing a comprehensive living cell gene transfection system using standard CMOS processing. However, there are still significant engineering challenges to developing CMOS-based gene injection technologies, such as biocompatibility, electrochemical corrosion, and mechanical integration with sensor technologies, among others. Nevertheless, the promise of a CMOS-based microscale gene delivery system remains significant for both *in vivo* and *in vitro* applications. This is especially demonstrated by the recent trends in transdermal microneedle technology for drug and gene delivery used for DNA vaccines. It is expected that future CMOS-based gene delivery systems will utilize a combination of power sources to transfect cells. Noninvasive approaches, such as acoustic, magnetic, or even RF-mediated methods, have some appeal but are still in the developmental stages for the required cellular precision control for targeting. However, semi-invasive methods such as electroporation offer an easier manufacturing route since the principle of transfection is simple and requires standard manufacturing practices and operational voltages. Future designs for gene delivery would be expected to take a mixed-mode approach. It would probably incorporate multiple gene transfection technologies, such as combining smart polymer chemicals and optimally designed viral vectors with microneedles coated with metal for electroporative purposes. Direct genetic manipulation of biological systems has already revolutionized the biotechnology and biomedical fields, creating new avenues for gene therapy, rapid lab-on-chip-based diagnostics, forensic analysis, and agriculture. With the promise of novel treatments in gene therapy, and rapid diagnosis using miniaturized, portable bioelectronic devices, CMOS integration of gene delivery systems is but a stepping stone to larger possibilities.

REFERENCES

[1] B. Alberts, A. Johnson, J. Lewis, M. Raff, K. Roberts, and P. Walter, *Molecular Biology of the Cell*, 4th ed., New York: Garland Science, 2002.

[2] "Genome: information by genome sequence." Available: http://www.ncbi.nlm.nih.gov/sites/entrez?db=genome (accessed February 11, 2011).

[3] "Nucleotide." Available: http://www.ncbi.nlm.nih.gov/nuccore (accessed February 11, 2011).

[4] R. B. Denman, "Deja vu all over again: FMRP binds U-rich target mRNAs," *Biochem. Biophys. Res. Commun.*, 310, pp. 1–7, 2003.

[5] P. Nelson, M. Kiriakidou, A. Sharma, E. Maniataki, and Z. Mourelatos, "The microRNA world: small is mighty," *Trends Biochem. Sci.*, 28, pp. 534–540, 2003.

[6] A. Dallas and A. V. Vlassov, "RNAi: a novel antisense technology and its therapeutic potential," *Med. Sci. Monit.*, 12, pp. RA67–RA74, 2006.

[7] S. Brantl, "Antisense-RNA regulation and RNA interference," *Biochim. Biophys. Acta*, 1575, pp. 15–25, 2002.

[8] Y. Dorsett and T. Tuschl, "siRNAs: Applications in functional genomics and potential as therapeutics," *Nature*, 3, pp. 318–329, 2004.

[9] S. T. Crooke, "Antisense strategies," *Curr. Mol. Med.*, 4, pp. 465–487, 2004.

[10] M. L. Simpson, "Rewiring the cell: Synthetic biology moves towards higher functional complexity," *Trends Biochem. Sci.*, 22, pp. 555–557, 2004.

[11] M. Tigges, T. T. Marquez-Lago, J. Stelling, and M. Fussenegger, "A tunable synthetic mammalian oscillator," *Nature*, 457, pp. 309–312, 2009.

[12] J. Hasty, D. McMillen, and J. J. Collins, "Engineered gene circuits," *Nature*, 420, pp. 224–230, 2002.

[13] J. M. Tavare, L. M. Fletcher, and G. I. Welsh, "Using green fluorescent protein to study intracellular signalling," *J. Endocrinol.*, 170, pp. 297–306, 2001.

[14] A. L. Jackson and P. S. Linsley, "Recognizing and avoiding siRNA off-target effects for target identification and therapeutic application," *Nat. Rev. Drug Discov.*, 9, pp. 57–66, 2010.

[15] "BLAST: Basic local alignment search tool." Available: http://blast.ncbi.nlm.nih.gov/Blast.cgi (accessed February 11, 2011).

[16] J. A. Wolff and V. Budker, "The mechanism of naked DNA uptake and expression," *Adv. Genet.*, 54, pp. 3–20, 2005.

[17] D. T. Curiel, "Tailoring vectors through DNA shuffling," *Nat. Gen.*, 25, pp. 366–367, 2000.

[18] S. van Drunen and L.-V. den Hurk, "Novel methods for the non-invasive administration of DNA therapeutics and vaccines," *Curr. Drug Deliv.*, 3, pp. 3–15, 2006.

[19] M. Johnson-Saliba and D. A. Jans, "Gene therapy: Optimising DNA delivery to the nucleus," *Curr. Drug Targets*, 2, pp. 371–399, 2001.

[20] M. A. Mintzer and E. E. Simanek, "Nonviral vectors for gene delivery," *Chem. Rev.*, 109, pp. 259–302, 2009.

[21] V. Russ and E. Wagner, "Cell and tissue targeting of nucleic acids for cancer gene therapy," *Pharm. Res.*, 24, pp. 1047–1057, 2007.

[22] A. E. Smith, "Viral vectors in gene therapy," *Annu. Rev. Microbiol.*, 49, pp. 807–838, 1995.

[23] P. G. Leiman, S. Kanamaru, V. V. Mesyanzhinov, F. Arisaka, and M. G. Rossman, "Structure and morphogenesis of bacteriophage T4," *Cell. Mol. Life Sci.*, 60, pp. 2356–2370, 2003.

[24] S. Mehier-Humbert and R. H. Guy, "Physical methods for gene transfer: Improving the kinetics of gene delivery into cells," *Adv. Drug Deliv. Rev.*, 57, pp. 733–753, 2005.

[25] J. E. Biewenga, O. H. J. Destree, and L. H. Schrama, "Plasmid-mediated gene transfer in neurons using the biolistic technique," *J. Neurosci. Methods*, 71, pp. 67–75, 1997.

[26] J. Villemejane, L. Mir, and M., "Physical methods of nucleic acid transfer: General concepts and applications," *Br. J. Pharmacol.*, 157, pp. 207–219, 2009.

[27] H. Wellman, B. Kaltschmidt, and C. Kaltschmidt, "Optimized protocol for biolistic transfection of brain slices and dissociated cultured neurons with a hand-held gene gun," *J. Neurosci. Methods*, 92, pp. 55–64, 1999.

[28] D. Cai, J. M. Mataraza, Z.-H. Qin, Z. Huang, J. Huang, T. C. Chiles, D. Carnahan, K. Kempa, and Z. Ren, "Highly efficient molecular delivery into mammalian cells using carbon nanotube spearing," *Nat. Methods*, 2, pp. 449–454, 2005.

[29] D. Cai, C. A. Doughty, T. B. Potocky, F. J. Dufort, Z. Huang, D. Blair, K. Kempa, Z. F. Ren, and T. C. Chiles, "Carbon nanotube-mediated delivery of nucleic acids does not result in non-specific activation of B lymphocytes," *Nanotechnology*, 18, pp. 365101–365111, 2007.

[30] W. G. Lee, U. Demirci, and A. Khademhosseini, "Microscale electroporation: Challenges and perspectives for clinical applications," *Integr. Biol.*, 1, pp. 242–251, 2008.

[31] F. Yamauchi, K. Kato, and H. Iwata, "Spatially and temporally controlled gene transfer by electroporation into adherent cells on plasmid DNA-loaded electrodes," *Nucleic Acids Res.*, 32, pp. e187, 2004.

[32] J. Akinlaja and F. Sachs, "The breakdown of cell membranes by electrical and mechanical stress," *Biophys. J.*, 75, pp. 247–254, 1998.

[33] N. Odani, K. Ito, and H. Nakamura, "Electroporation as an efficient methods of gene transfer," *Dev. Growth Differ.*, 50, pp. 443–448, 2008.

[34] T. Jain and J. Muthuswamy, "Microelectrode array (MEA) platform for targeted neuronal transfection and recording," *IEEE Trans. Biomed. Eng.*, 55, pp. 827–832, 2008.

[35] T. Jain and J. Muthuswamy, "Bio-chip for spatially controlled transfection of nucleic acid payloads into cells in a culture," *Lab Chip*, 7, pp. 1004–1011, 2007.

[36] T. Jain and J. Muthuswamy, "Microsystem for transfection of exogenous molecules with spatio-temporal control into adherent cells," *Biosens. Bioelectron.*, 22, pp. 863–870, 2007.

[37] M. Khine, A. Lau, C. Ionescu-Zanetti, J. Seo, and L. P. Lee, "A single cell electroporation chip," *Lab Chip*, 5, pp. 38–43, 2005.

[38] B. Judkewitz, M. Rizzi, K. Kitamura, and M. Hausser, "Targeted single-cell electroporation of mammalian neurons *in vivo*," *Nat. Protoc.*, 4, pp. 862–869, 2009.

[39] Y. Huang and B. Rubinsky, "Microfabricated electroporation chip for single cell permeabilization," *Sens. Actuators A*, 89, pp. 242–249, 2001.

[40] W. C. Chang and D. W. Sretavan, "Single cell and neural process experimentation using laterally applied electrical fields between pairs of closely apposed microelectrodes with vertical sidewalls," *Biosens. Bioelectron.*, 24, pp. 3600–3607, 2009.

[41] K. DeBruin and W. Krassowska, "Modeling electroporation in a single cell. I. Effects of field strength and rest potential," *Biophys. J.*, 77, pp. 1213–1224, 1999.

[42] R. Davalos, Y. Huang, and B. Rubinsky, "Electroporation: Bio-electrochemical mass transfer at the nano scale," *Microscale Therm. Eng.*, 4, pp. 147–159, 2000.

[43] K. DeBruin and W. Krassowska, "Modeling electroporation in a single cell. II. Effects of ionic concentrations," *Biophys. J.*, 77, pp. 1225–1233, 1999.

[44] J.-M. Escoffre, T. Portet, L. Wasungu, J. Teissie, D. Dean, and M.-P. Rols, "What is (still not) known of the mechanism by which electroporation mediates gene transfer and expression in cells and tissues," *Mol. Biotechnol.*, 41, pp. 286–295, 2009.

[45] J. C. Weaver and Y. A. Chizmadzhev, "Theory of electroporation: A review," *Bioelectrochem. Bioenerg.*, 41, pp. 135–160, 1996.

[46] J. C. Weaver, "Electroporation of biological membranes from multicellular to nanoscales," *IEEE Trans. Dielectr. Electr. Insul.*, 10, pp. 754–768, 2003.

[47] T. D. Xie and T. Y. Tsong, "Study of mechanisms of electric field-induced DNA transfection, V. effects of DNA topology on surface binding, cell uptake, expression, and

integration into host chromosomes of DNA in the mammalian cell," *Biophys. J.*, 65, pp. 184–1689, 1993.

[48] S. I. Sukharev, V. A. Klenchin, S. M. Serov, L. V. Chernomordik, and Y. A. Chizmadzhev, "Electroporation and electrophoretic DNA transfer into cells: The effect of DNA interaction with electropores," *Biophys. J.*, 163, pp. 1320–1327, 1992.

[49] C. Favard, D. S. Dean, and M.-P. Rols, "Electrotransfer as a non viral method of gene delivery," *Curr. Gene Ther.*, 7, pp. 67–77, 2007.

[50] Y. Akaneya, B. Jiang, and T. Tsumoto, "RNAi-induced gene silencing by local electroporation in targeting brain region," *J. Neurophysiol.*, 93, pp. 594–602, 2005.

[51] R. Sundararajan, "Nanosecond electroporation: another look," *Mol. Biotechnol.*, 41, pp. 69–82, 2009.

[52] D. Hodko, P. Swanson, D. Dehlinger, B. Sullivan, and M. J. Heller, "CMOS electronic microarrays in diagnostics and nanotechnology," in *CMOS Biotechnology*, H. Lee, D. Ham, and R. M. Westervelt, Eds. New York: Springer US, 2007, pp. 179–206.

[53] M. Hutzler, A. Lambacher, B. Eversmann, M. Jenkner, T. Thewes, and P. Fromherz, "High-resolution multi-transistor array recording of electrical field potentials in cultured brain slices," *J. Neurophysiol.*, 96, pp. 1638–1645, 2006.

[54] N. Jackson, "Neural recordings in the brain using novel moveable MEMS microelectrode arrays," Ph.D. dissertation, Arizona State University, 2009.

[55] B. H. Stark, M. R. , and K. Najafi, "Improving corrosion-resistance of polysilicon using boron doping and self-induced galvanic bias," *IEEE Trans. Adv. Packag.*, 26, pp. 295–301, 2003.

[56] R. Saha and J. Muthuswamy, "Structure-property relationships in the optimization of polysilicon thin films for electrical recording/stimulation of single neurons," *Biomed. Microdevices*, 9, pp. 345–360, 2007.

[57] B. M. Taff and J. Voldman, "A scalable addressable positive-dielectrophoretic cell-sorting array," *Anal. Chem.*, 77, pp. 7976–7983, 2005.

[58] P. Aggarwal and C. R. Johnston, "Human blood rheology in MEMS-based microneedles," *Proc. SPIE*, 5651, pp. 185–195, 2005.

[59] W. H. Wang and X. Y. Liu, "High-throughput automated injection of individual biological cells," *IEEE Trans. Autom. Science Eng.*, 6, pp. 209–219, 2009.

[60] W. Wang, Y. Sun, M. Zhang, R. Anderson, L. Langille, and W. Chan, "A system for high-speed microinjection of adherent cells," *Rev. Sci. Instrum.*, 79, pp. 104302, 2008.

[61] S. Zappe, M. Fish, M. P. Scott, and O. Solgaard, "Automated MEMS-based *Drosophila* embryo injection system for high-throughput RNAi screens," *Lab Chip*, 6, pp. 1012–1019, 2006.

[62] A. Adamo and K. F. Jensen, "Microfluidic based single cell microinjection," *Lab Chip*, 8, pp. 1258–1261, 2008.

[63] D. V. McAllister M. G. Allen, and M. R. Prausnitz, "Microfabricated microneedles for gene and drug delivery," *Annu. Rev. Biomed. Eng.*, 02, pp. 289–313, 2000.

[64] K. M. Ainslie and T. A. Desai, "Microfabricated implants for applications in therapeutic delivery, tissue engineering, and biosensing," *Lab Chip*, 8, pp. 1864–1878, 2008.

[65] M. L. Reed and W.-K. Lye, "Microsystems for drug and gene delivery," *Proc. IEEE*, 92, pp. 56–75, 2004.

[66] A. V. Prinz, V. Y. Prinz, and V. A. Seleznev, "Semiconductor micro- and nanoneedles for microinjections and ink-jet printing," *Microelectron. Eng.*, 67–68, pp. 782–788, 2003.

[67] R. Luttge, E. J. W. Berenschot, M. J. de Boer, D. M. Altpeter, E. X. Vrouwe, A. van den Berg, and M. Elwenspoek, "Integrated lithographic molding of microneedle-based devices," *J. Microelectromech. Syst.*, 16, pp. 872–884, 2007.

[68] D. V. McAllister, P. M. Wang, S. P. Davis, J.-H. Park, P. J. Canatella, M. G. Allen, and M. R. Prausnitz, "Microfabricated needles for transdermal delivery of macromolecules and nanoparticles: Fabrication methods and transport studies," *Proc. Natl. Acad. Sci. U.S.A.*, 100, pp. 13755–13760, 2003.

[69] W. Lin, M. Cormier, A. Samiee, A. Griffin, B. Johnson, C.-L. Teng, G. E. Hardee, and P. E. Daddona, "Transdermal delivery of antisense oligonucleotides with microprojection patch (macroflux) technology," *Pharm. Res.*, 18, pp. 1789–1793, 2001.

[70] H. Takahashi, D. Letourneur, and D. W. Grainger, "Delivery of large biopharmaceuticals from cardiovascular stents: A review," *Biomacromolecules*, 8, pp. 3281–3293, 2007.

[71] B. Stoeber and D. Liepmann, "Two-dimensional arrays of out-of-plane needles," *Proceedings of the 2000 ASME International Mechanical Engineering Congress and Exposition, Micro-Electro-Mechanical Systems (MEMS)*, Orlando, FL, 2000, pp. 355–359.

[72] F. Chabri, K. Bouris, T. Jones, D. Barrow, A. Hann, C. Allender, K. , and J. Birchall, "Microfabricated silicon microneedles for nonviral cutaneous gene delivery," *Br. J. Dermatol.*, 150, pp. 869–877, 2004.

[73] P. Bucher, A. Fischer, P. L. Luisi, T. Oberholzer, and P. Walde, "Giant vesicles as biochemical compartments: The use of microinjection techniques," *Langmuir*, 14, pp. 2712–2721, 1998.

[74] S. P. Davis, B. J. Landis, Z. H. Adams, M. G. Allen, and M. R. Prausnitz, "Insertion of microneedles into skin: Measurement and prediction of insertion force and needle fracture," *J. Biomech.*, 37, pp. 1155–1163, 2004.

[75] T. Shibata, S. Yamanaka, N. Kato, T. Kawashima, M. Nomura, T. Mineta, and E. Makino, "Fabrication of micromanipulator array for cell patterning," *Microelectron. Eng.*, 86, pp. 1439–1442, 2009.

[76] P. W. Lucas, I. M. Turner, N. J. Dominy, and N. Yamashita, "Mechanical defences to herbivory," *Ann. Bot.*, 86, pp. 913–920, 2000.

[77] S. A. Safran, N. Gov, A. Nicolas, U. S. Schwarz, and T. Tlusty, "Physics of cell elasticity, shape, and adhesion," *Physica A*, 352, pp. 171–201, 2005.

[78] I. U. Vakarelski, S. C. Brown, K. Higashitani, and B. M. Moudgil, "Penetration of living cell membranes with fortified carbon nanotube tips," *Langmuir*, 23, pp. 10893–10896, 2007.

[79] U. O. Hafeli, J. S. Riffle, L. H.-S. A. Carmichael-Baranauskas, F. Mark, J. P. Dailey, and D. Bardenstein, "Cell uptake and *in vitro* toxicity of magnetic nanoparticles suitable for drug delivery," *Mol. Pharm.*, 6, pp. 1417–1428, 2009.

[80] Y. Namiki, T. Namiki, J. Yoshida, Y. Ishii, A. Tsubota, S. Koido, K. Nariai, M. Mitsunaga, S. Yanagisawa, H. Kashiwagi, Y. Mabashi, Y. Yumoto, S. Hoshina, K. Fujise, and N. Tada, "A novel magnetic crystal-lipid nanostructure for magnetically-guided *in vivo* gene delivery," *Nat. Nanotechnol.*, 4, pp. 598–606, 2009.

[81] S. C. McBain, U. Griesenbach, S. Xenariou, A. Keramane, C. D. Batich, E. W. F. W. Alton, and J. Dobson, "Magnetic nanoparticles as gene delivery agents: Enhanced transfection in the presence of oscillating magnet arrays," *Nanotechnology*, 19, pp. 405102–405107, 2008.

[82] K. Kamei, Y. Mukai, H. Kojima, T. Yoshikawa, M. Yoshikawa, G. Kiyohara, T. A. Yamamoto, Y. Yoshioka, N. Okada, S. Seino, and S. Nakagawa, "Direct cell entry of gold/iron-oxide magnetic nanoparticles in adenovirus mediated gene delivery," *Biomaterials*, 30, pp. 1809–1814, 2009.

[83] H. Lee, A. M. Purdon, and R. M. Westervelt, "Manipulation of biological cells using a microelectromagnet matrix," *Appl. Phys. Lett.*, 85, pp. 1063–1065, 2004.

[84] K. Kratt, V. Badilita, T. Burger, J. G. Korvink, and U. Wallrabe, "A fully MEMS-compatible process for 3D high aspect ratio micro coils obtained with an automatic wire bonder," *J. Micromech. Microeng.*, 20, pp. 015021–015031, 2010.

[85] K. Ferrera, R. Pollard, and M. Borden, "Ultrasound microbubble contrast agents: fundamentals and application to gene and drug delivery," *Annu. Rev. Biomed. Eng.*, 9, pp. 415–447, 2007.

[86] C. R. Mayer, N. A. Geis, H. A. Katus, and R. Bekeredjian, "Ultrasound targeted microbubble destruction for drug and gene delivery," *Expert Opin. Drug Deliv.*, 5, pp. 1121–1138, 2008.

[87] S. L. Gac, E. Zwaan, A. van den Berg, and C.-D. Ohl, "Sonoporation of suspension cells with a single cavitation bubble in a microfluidic confinement," *Lab Chip*, 7, pp. 1666–1672, 2007.

[88] Y. Hosokawa, S. Iguchi, R. Yasukuni, Y. Hiraki, C. Shukunami, and H. Masuhara, "Gene delivery process in a single animal cell after femtosecond laser microinjection," *Appl. Surf. Sci.*, 255, pp. 9880–9884, 2009.

[89] D. L. Miller, S. V. Pislaru, and J. F. Greenleaf, "Sonoporation: mechanical DNA delivery by ultrasonic cavitation," *Somat. Cell Mol. Genet.*, 27, pp. 115–134, 2002.

[90] C. Pichon, K. Kaddur, P. Midoux, F. Tranquart, and A. Bouakaz, "Recent advances in gene delivery with ultrasound and microbubbles," *J. Exp. Nanosci.*, 3, pp. 17–40, 2008.

[91] S. Ohta, K. Suzuki, Y. Ogino, S. Miyagawa, A. Murashima, D. Matsumaru, and G. Yamada, "Gene transduction by sonoporation," *Dev. Growth Differ.*, 50, pp. 517–520, 2008.

[92] T. Bland, J. Ohni, A. Patel, P. Kirbby, M. Robinson, and J. Kerr, "Integration of sol-gel PZT with silicon-on-sapphire CMOS circuitry," *Microelectron. Eng.*, 29, pp. 29–32, 1995.

[93] I. G. Mina, H. Kim, I. Kim, S. K. Park, K. Choi, T. N. Jackson, R. L. , and S. Troiler-McKinstry, "High frequency piezoelectric MEMS ultrasound transducers," *IEEE Trans. Ultrason. Ferroelectr. Freq. Control*, 54, pp. 2422–2430, 2007.

[94] T. C. Marentis, B. Kusler, G. G. Yaralioglu, S. Liu, E. O. Haeggstrom, and B. T. Khuri-Yakub, "Microfluidic sonicator for real-time disruption of eukaryotic cells and bacterial spores for DNA analysis," *Ultrasound Med. Biol.*, 31, pp. 1265–1277, 2005.

[95] A. Schulze and J. Downward, "Navigating gene expression using microarrays–A technology review," *Nat. Cell. Biol.*, 3, pp. E190–E195, 2001.

[96] M. J. Heller, "DNA microarray technology: devices, systems, and applications," *Annu. Rev. Biomed. Eng.*, 4, pp. 129–153, 2002.

[97] J. B. Heng, A. Akisimentiev, C. Ho, V. Dimitrov, T. Sorsch, J. Miner, W. Mansfield, K. Schulten, and G. Timp,"Beyond the gene chip," *Bell Labs Tech. J.*, 10, pp. 5–22, 2005.

[98] K. H. Gilchrist, V. N. Barker, L. E. Fletcher, B. D. DeBusschere, P. Ghanouni, L. Giovangrandi, and G. T. A. Kovacs, "General purpose, field-portable cell-based biosensor platform," *Biosens. Bioelectron.*, 16, pp. 557–564, 2001.

[99] S. Elmore, "Apoptosis: A review of programmed cell death," *Toxical. Pathol.*, 35, pp. 495–516, 2007.

[100] N. G. Hattangady and M. S. Rajadhyaksha, "A brief review of *in vitro* models of diabetic neuropathy," *Int. J. Diabetes Dev. Ctries.*, 29, pp. 143–149, 2009.

[101] H.-A. Pan, Y.-C. Hung, C.-W. Su, S.-M. Tai, C.-H. Chen, F.-H. Ko, and G. S. Huang, "A nanodot array modulates cell adhesion and induces an apoptosis-like abnormality in NIH 3T3 cells," *Nanoscale Res. Lett.*, 4, pp. 903–912, 2009.

[102] A. J. Bard and L. R. Faulkner, *Electrochemical Methods: Fundamentals and Applications*, 2nd ed. New York: Wiley.

[103] F. Heer, W. Franks, A. Blau, S. Tachini, C. Ziegler, A. Hierlemann, and H. Baltes, "CMOS microelectrode array for the monitoring of electrogenic cells," *Biosens. Bioelectron.*, 20, pp. 358–366, 2004.

16

LOW-COST DIAGNOSTICS: RF DESIGNER'S APPROACH*

Nan Sun, Yong Liu, and Donhee Ham

16.1 INTRODUCTION

Silicon radio frequency (RF) integrated circuits (ICs) have been widely used in modern wireless communication systems in the past few decades. In this chapter, we present a different application of silicon RF IC, that is, sensing biological objects.

In an early stage of disease development, biomolecules characteristic to the disease, such as viruses and cancer marker proteins, emerge. The ability to sense these biomolecules would facilitate early disease detection. Researchers from many areas of science and engineering are developing a variety of biosensors, aiming at increased sensitivity or low-cost diagnostics [1]. Our RF biosensor represents a "circuit designer's way" to pursue highly sensitive biosensing in a portable form.

Figure 16.1 shows our RF biosensor, central to whose operation is the silicon RF transceiver IC. The underpinning physical phenomenon of our sensing method is nuclear magnetic resonance (NMR), the resonant interaction between RF

* Portions reprinted with permission from N. Sun, Y. Liu, H. Lee, R. Weissleder, and D. Ham, "CMOS RF biosensor utilizing nuclear magnetic resonance," *IEEE J. Solid-State Circuits*, 44, pp. 1629–1643, 2009. © IEEE.

CMOS Biomicrosystems: Where Electronics Meet Biology, First Edition. Edited by Krzysztof Iniewski.

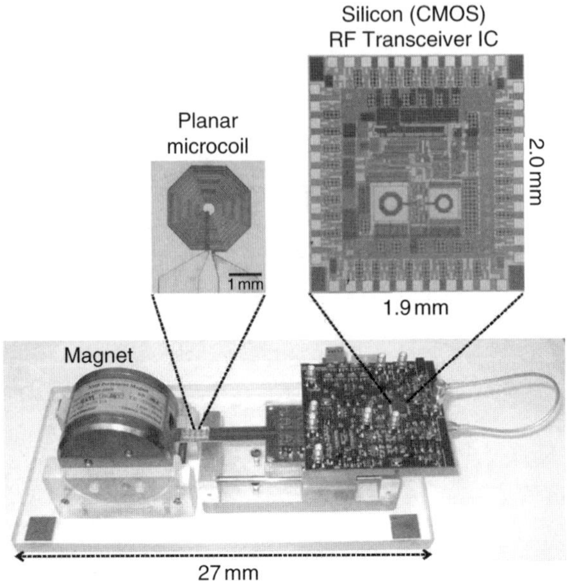

Figure 16.1. CMOS RF biosensor (this work) utilizing NMR. The entire system weighs 2 kg, where the commercial magnet dominates the weight. Prior to this work, we built an intermediate miniature NMR biosensor [4] where we used the same magnet and in-house fabricated microcoil with discrete electronics. (From N. Sun, Y. Liu, H. Lee, R. Weissleder, and D. Ham, "CMOS RF biosensor utilizing nuclear magnetic resonance," *IEEE J. Solid-State Circuits*, 44, pp. 1629–1643, 2009.)

magnetic fields and protons in water, which is altered by the presence of target biological objects. This NMR-based biosensing was developed in 2002 [2] and has been used within a state-of-the-art commercial benchtop NMR system [3], which, however, is bulky and heavy (120 kg). The main contribution of our work is to drastically shrink the *entire* NMR system by developing the RF transceiver IC, hence enabling NMR-based biosensing in the portable form of Figure 16.1. Occupying only 2.5 L and weighing only 2 kg, our system is 60 times lighter, 40 times smaller, yet 60 times more mass sensitive than the state-of-the-art benchtop NMR system: Our system is actually the smallest complete NMR system ever built. As a biosensor, it detected 20 fmol, 1.4 ng of avidin (protein) in a 5 μL sample volume. This detection threshold can be even further improved, for it is currently limited not by our transceiver sensitivity, but by the specific bioassay used [4–5]. Our system might offer a new way to pursue early disease detection in a low-cost, portable platform.

This chapter describes the entire work (design of, and experimentation with, the RF IC) in Sections 16.4–16.7. We also add one perspective section, Section 16.3 to explain how our RF IC facilitated dramatic miniaturization. All of this would be best understood with some familiarity with NMR, so we start by reviewing the NMR basics.

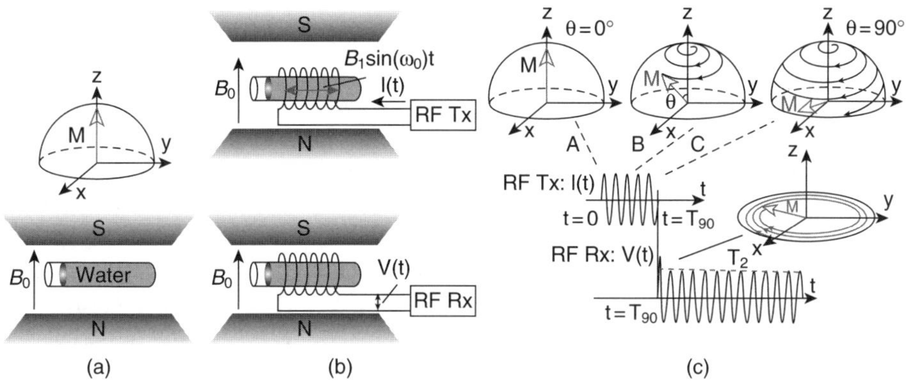

Figure 16.2. (a) Water in a magnetic field B_0. (b) Excitation (top) and reception (bottom) phase in NMR. (c) Motion of magnetic moment M in NMR. (From N. Sun, Y. Liu, H. Lee, R. Weissleder, and D. Ham, "CMOS RF biosensor utilizing nuclear magnetic resonance," *IEEE J. Solid-State Circuits*, 44, pp. 1629–1643, 2009.)

16.2 REVIEW OF NUCLEAR MAGNETIC RESONANCE

Although NMR is a well-established subject [6], we provide this review to quickly introduce the basic concepts of NMR relevant to this work, so that readers less familiar with NMR can rapidly access the main contribution portion of this chapter.

16.2.1 Nuclear Magnetic Resonance

Atomic nuclei with spin act like tiny bar magnets. These nuclear magnets are the main actors in the NMR phenomenon. We explain NMR using protons, the simplest nuclei. Proton NMR is observed with as common a substance as water, for it contains numerous hydrogen atoms whose nuclei are protons.

Let water be placed in a static magnetic field B_0 (Fig. 16.2a). After several seconds, the system reaches an equilibrium, where the proton magnets preferentially line up along the direction of B_0, just like a compass needle lines up along the Earth's field. Macroscopically, in the equilibrium, the water exhibits a magnetic moment M along the direction of B_0, which we take as the z-axis. This equilibrium state is the lowest energy state the system assumes.

Now if one wraps a coil around the water and sends an RF current into the coil, one can produce an RF magnetic field $B_1 \sin \omega_0 t$ across the water (Fig. 16.2b, top), where B_1 and ω_0 are the amplitude and angular frequency of the RF field. If ω_0 is not the *Larmor frequency*, defined as

$$\omega_L = 2\pi f_L = \gamma B_0, \tag{16.1}$$

where γ is a proton-specific constant yielding $f_L = 21.3\,\text{MHz}$ for $B_0 = 0.5\,\text{T}$, the proton magnets stay indifferent to the RF field, and M remains along the z-axis. But for $\omega_0 = \omega_L$, proton magnets absorb energy from the RF field, raising the potential energy of M by misaligning M from B_0 with an increasing angle θ between them. In other words, M is tipped away from the z-axis. As shown at three time instants **A**, **B**, and **C** in Figure 16.2c, this downward motion of M is superposed with a much faster precession of M at the Larmor frequency ω_L about the z-axis. Overall, M exhibits a spiral downward motion when absorbing energy from the RF field. This absorption of RF energy at ω_L by the proton magnets is termed nuclear magnetic resonance (NMR). During energy absorption, the time taken for M to enter the xy-plane ($\theta = 90°$; **C**, Fig. 16.2c) from its alignment along the z-axis ($\theta = 0°$; **A**) is

$$T_{90^\circ} = \frac{\pi}{2\gamma B_1}. \tag{16.2}$$

Typically $B_0 >> B_1$, which is why the downward motion of M (growth of θ) is much slower than its precession at $\omega_L = \gamma B_0$.

How does one figure out if ω_0 was tuned to ω_L and NMR took place? One can cease the RF transmission at $t = T_{90}$, connect the coil to an RF receiver, and observe the coil output after $t = T_{90}$ (Fig. 16.2b, bottom) If $\omega_0 \neq \omega_L$, NMR did not occur and M stayed along the z-axis. Thus, nothing registers at the receiver after $t = T_{90}$. If $\omega_0 = \omega_L$, NMR took place during the RF field transmission ($0 < t < T_{90}$) and M is in the xy-plane at $t = T_{90}$ (Fig. 16.2c). After $t = T_{90}$, there is no more RF field for protons to absorb energy from, and hence, M will stay in the xy-plane, maintaining $\theta = 90°$ with a constant potential energy.[1] M, however, continues to precess with ω_L about the z-axis in the xy plane. Therefore, the coil experiences a periodic magnetic flux variation that induces an RF voltage $V(t)$ of frequency ω_L across the coil. $V(t)$ registers at the RF receiver. Note that the RF excitation is stopped at $t = T_{90}$ because M in the xy-plane induces the largest signal across the coil.

$V(t)$ is actually an exponentially damped sinusoid (Fig. 16.2c) with characteristic decay time called T_2. The damping arises because the precession of an individual proton magnet is randomly perturbed by tiny magnetic fields produced by nearby proton magnets (proton–proton interaction). Due to the perturbation, each proton precesses at ω_L but with its own phase noise. Consequently, precessions of a large number of individual protons, which were initially in phase, grow out of phase over time in the xy-plane. As phase coherence is lost, the vector sum M of individual proton magnetic moments exponentially decays, yielding the

[1] As mentioned, after several seconds, M will lose energy to the environment, aligning along B_0 to reach equilibrium. The time considered here in the main text, however, is far shorter than the equilibrium restoration time.

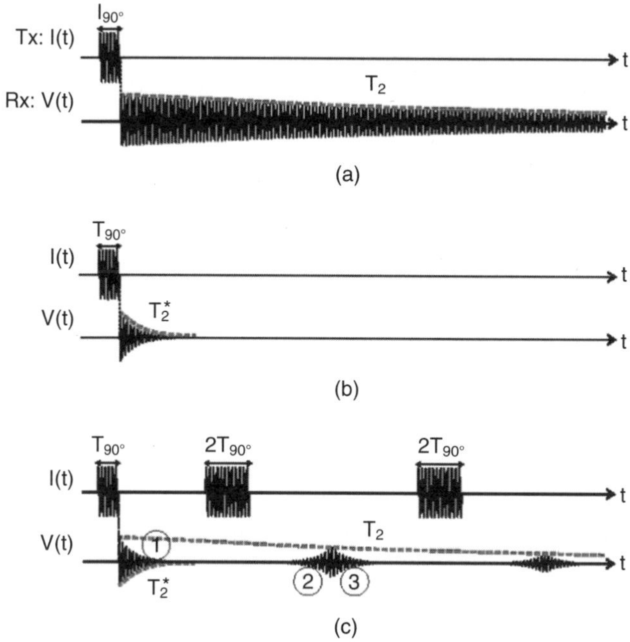

Figure 16.3. (a) T_2 damping due to the basic proton-proton interaction. (b) T_2^* damping due to the static magnetic field inhomogeneity. (c) Echo technique. (From N. Sun, Y. Liu, H. Lee, R. Weissleder, and D. Ham, "CMOS RF biosensor utilizing nuclear magnetic resonance," *IEEE J. Solid-State Circuits*, 44, pp. 1629–1643, 2009.)

damped signal. For water, $T_2 \sim 1$ second and the quality factor $Q = \omega_L T_2$ of the damped sinusoid with $f_L = 21\,\text{MHz}$ is about 10^8.

Through the NMR experiments described above, one obtains the values for ω_L and T_2, which carry information on the compositions and properties of the substances in study. In this way, NMR has found applications in fields such as chemical spectroscopy, magnetic resonance imaging (MRI), and biosensing.

16.2.2 Echo Technique

We have so far assumed a uniform static magnetic field B_0 across the water. In this case, all protons precess at the same frequency $\omega_L = \gamma B_0$ with phase noise caused by proton–proton interactions, yielding the T_2-damped sinusoid $V(t)$ at the receiver (Fig. 16.3a). In reality, any magnet suffers a spatial variation of the field magnitude around the intended value B_0. When such field inhomogeneity is present, $V(t)$ is damped much faster with decay time T_2^*, overriding the T_2 damping (Fig. 16.3b). The faster damping arises because protons at different positions precess at different Larmor frequencies due to field inhomogeneity, making them rapidly grow out of phase. To estimate T_2^*; assume that the field inhomogeneity is 50 parts per million (ppm) around $B_0 = 0.5\text{T}$. If we denote this fractional variation as Δ, then the precession frequencies range from $\gamma(B_0 - \Delta B_0/2)$ to $\gamma(B_0 + \Delta B_0/2)$, and hence, the damped sinusoid of Figure 16.3b has a bandwidth of $\gamma\Delta B_0$ around γB_0. Thus $Q = \omega_L T_2^* \sim (\gamma B_0)/(\gamma \Delta B_0) = 2\times 10^4$, or $T_2^* \sim 200\,\mu\text{sec} \ll T_2 \sim 1\,\text{second}$: Even with 50 ppm inhomogeneity, the T_2^* damping completely masks the T_2 damping.

Since field inhomogeneity is deterministic, one can effectively remove the T_2^* damping. Such a technique is important, as T_2, which carries useful information, would be immeasurable otherwise. Figure 16.3c illustrates one widely used technique [6]. The T_{90} RF pulse is followed by repeated RF transmissions, each of duration $2T_{90}$. After the initial T_{90} pulse, the proton magnets brought down to the *xy*-plane precess at different Larmor frequencies: Those at a stronger field position precess faster and ahead of those at a weaker field position, resulting in the T_2^* damping (①, Fig. 16.3c). Now the $2T_{90}$ pulse performs the same function as the T_{90} pulse, but twice, changing θ by 180°. Thus, after the $2T_{90^\circ}$ pulse, the proton magnets return to the *xy*-plane, but with the faster precessing protons now lagging the slower ones (this inversion in relative positions may be difficult to grasp without detailed explanation, but let us accept it and move on, as the goal here is to become familiar with basic characteristics of the resulting signal, $V(t)$). The former will catch up the latter, regaining phase coherence, hence showing the growing sinusoid (②, Fig. 16.3c). After all protons acquire the same phase, the faster precessing protons again get ahead of the slower ones, once again resulting in dephasing and the T_2^* damping (③, Fig. 16.3c). As the $2T_{90^\circ}$ pulse repeats, multiple such "echoes" register at the receiver. Since T_2 damping arising from random proton–proton interaction cannot be removed, the envelope of the echoes slowly decays with T_2, allowing its measurement despite field inhomogeneity.

16.2.3 Nuclear Magnetic Resonance-Based Biosensing

Consider detecting certain proteins (e.g., receptors of circulating cancer cells, or soluble cancer biomarker proteins such as VEGF, PSA, CEA, and AFP) in a blood sample. Nanoscale magnetic particles [7] coated with antibodies that specifically bind to the target proteins are suspended in the sample (Fig. 16.4). If the proteins exist, the magnetic particles self-assemble into clusters around the proteins [2]. This self-assembly can be detected by performing proton NMR: A significant portion of blood is water, so we may regard the blood sample as water containing many protons for our purpose.

Figure 16.4 illustrates the echo signal $V(t)$ from precessing *protons in water* in three different cases. In Figure 16.4a, water with no magnetic particles yields T_2 due to the basic proton–proton interactions. Figure 16.4b shows the case of water containing monodispersed magnetic particles in the absence of target proteins. The magnetic particles incessantly move around due to inevitable Brownian motion and produce fluctuating local magnetic fields, perturbing precessions of

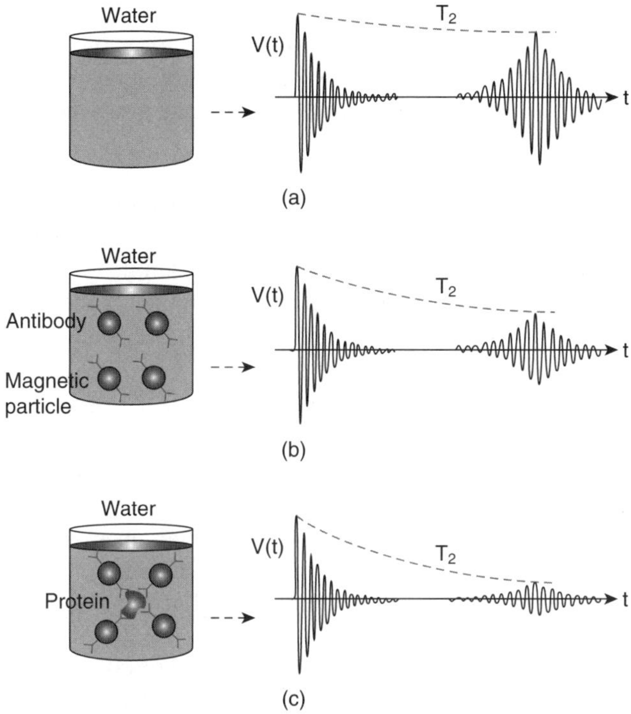

Figure 16.4. NMR-based biosensing [2]. (From N. Sun, Y. Liu, H. Lee, R. Weissleder, and D. Ham, "CMOS RF biosensor utilizing nuclear magnetic resonance," *IEEE J. Solid-State Circuits*, 44, pp. 1629–1643, 2009.)

proton magnets and causing more phase noise than is expected from the basic proton–proton interactions. Thus, phase coherence is lost more rapidly, reducing T_2. In Figure 16.4c, due to the presence of target proteins, the magnetic particles self-assemble into larger magnets, which are even more efficient in increasing phase noise in the proton magnets' precession, further accelerating dephasing, yielding the smallest T_2. Thus, by monitoring T_2, one can detect target objects [2].

16.3 CMOS RADIO FREQUENCY TRANSCEIVER INTEGRATED CIRCUIT DESIGN: CHALLENGES

The NMR-based biosensing requires an NMR system that can faithfully measure T_2. To this end, the 120-kg commercial benchtop system [3] (Section 16.1) has been used. The main focus of this work is the drastic miniaturization of the NMR system, enabling NMR-based biosensing in the portable form of Figure 16.1. Here we describe challenges that lie in such miniaturization, and how we met them through our RF IC design.

As seen in Figure 16.2, any NMR system consists of a magnet, a coil, and an RF transceiver. By far the largest component is the magnet (e.g., the 120-kg weight of the benchtop system is largely due to a large magnet). Therefore, any effort to shrink an NMR system should start with using a small magnet. An NMR system with a small magnet, however, suffers from a reduced signal $V(t)$ at the input of the receiver.

To see this, we note that $V(t)$, induced in the coil due to the periodic variation of magnetic flux caused by the precession of M, is proportional to the rate of change of flux, ω_L, and the total magnetization $M \times$ (sample volume): The volume factors in here because M is measured per unit volume. Since $M \propto B_0$ at room temperature[2] and $\omega_L = \gamma B_0$, we have

$$V(t) \propto B_0^2 \times [\text{Sample volume}]. \tag{16.3}$$

This equation explains why a small magnet yields a small $V(t)$. First, in general, a small magnet tends to produce a weak field B_0, and $V(t)$ is reduced with quadratic dependence on B_0. Second, a smaller magnet exhibits a more pronounced field inhomogeneity. In the presence of field inhomogeneity, protons at different positions of the sample have different Larmor frequencies, and thus, they are excited at different frequencies. Although the $T_{90°}$ RF pulse has other frequency components than $\omega_L = \gamma B_0$ due to its finite duration of $T_{90°}$, it still has a limited bandwidth ($\sim 1/T_{90°}$). Therefore, the $T_{90°}$ pulse can efficiently excite only protons in the region where their Larmor frequencies fall within the bandwidth of the $T_{90°}$ pulse. If the field inhomogeneity is increased with a smaller magnet, the spatial extent the $T_{90°}$ pulse can excite is reduced, effectively reducing the sample volume in Equation 16.3, thus decreasing $V(t)$, regardless of the physical sample size. One may increase the bandwidth of the excitation pulse by decreasing the signal duration $T_{90°}$, but this requires a larger B_1 (see Eq. 16.2) and thus a larger transmitted signal power, which is electronically limited.

In summary, with a small magnet, $V(t)$ is substantially reduced and its detection requires a highly sensitive RF transceiver. For facile NMR experiments avoiding the stringent requirement for the RF transceiver, people have used large magnets, leading to the existing bulky NMR instrumentation.

To achieve drastic miniaturization, we took the opposite approach: We use a very small magnet the size of an adult fist (Fig. 16.1), and to detect the resulting, substantially reduced signal $V(t)$ (amplitude: 130 nV; available power: 0.5 fW), we design a highly sensitive RF transceiver. By opting to use a very small magnet for miniaturization, we shifted the entire system design burden onto the RF transceiver design, which we successfully carried out. Furthermore, we integrated the RF transceiver onto a CMOS IC. As a very small, low-cost magnet is used,

[2] This well-established relation is derived from the Boltzmann's law, which we will use without showing the derivation.

the circuit integration makes sense: in a conventional system where a large, expensive magnet dominates the system size and cost, integration of the RF transceiver would hardly reduce either the system size or the cost.

Our design also uses a planar microcoil on a glass substrate (Fig. 16.1), having in mind mass production of the coil via standard microfabrication techniques. As compared with commonly used solenoidal coils, the planar coil's geometry does not allow effective induction of $V(t)$ from precessing protons, which was partly the reason for the weak signal amplitude of 130 nV. Once again, we overcame this undesirable effect by the design of the highly sensitive RF transceiver.

16.4 CMOS RADIO FREQUENCY TRANSCEIVER INTEGRATED CIRCUIT DESIGN: ARCHITECTURE

Figure 16.5 shows the architecture of our NMR RF transceiver IC, which we describe now. As our magnet has $B_0 = 0.5$ T, the RF design is based on the Larmor frequency of 21.3 MHz.

16.4.1 Receiver Path

Let us first focus on the receiver path in Figure 16.5. The signal appearing at node ② is a weak echo signal whose available power is 0.5 fW. The bandwidth of the echo signal is determined by the faster T_2^* damping due to field inhomogeneity, instead of the slower T_2 damping due to proton–proton interaction. As discussed in Section 16.2.2, the bandwidth corresponding to T_2^* damping is $\gamma\Delta B_0$, where Δ signifies the fractional field inhomogeneity. As we have $\Delta = 50$ ppm, the echo signal in the frequency domain (Fig. 16.5) is centered at the Larmor frequency 21.3 MHz, with a bandwidth of $\gamma\Delta B_0/(2\pi) \approx 1.1$ kHz.

The echo signal is amplified by a low-noise amplifier (LNA), and then the frequency down-converted using mixers and a local oscillator. Choosing the proper frequency for the local oscillator is a critical task in dealing with the weak echo signal. Let us denote the oscillator frequency as $\omega_L + \delta$. If δ is set to zero, the down-converted signal at node ③ is swamped by $1/f$ noise. On the other hand, if δ is too large, rejecting noise outside the signal bandwidth of 1.1 kHz after the frequency down-conversion is difficult, as it requires a high quality factor for the band-pass filter. Since the signal is very weak, rejecting out-of-band noise in order not to corrupt the signal-to-noise ratio (SNR) is a critical task. We select $\delta/(2\pi) = 3$ kHz, with which the down-converted signal is not centered at *dc*, but very close to *dc* (Fig. 16.5), facilitating out-of-band noise rejection using a band-pass filter with a moderate quality factor.

As in any heterodyning receiver, image noise would add 3 dB to the receiver noise figure, unless suppressed. We incorporate an image rejection algorithm in the digital domain after the down-converted signal is digitized using analog-to-digital converters (ADCs) (Fig. 16.5). A Hilbert transformer shifts the phase of

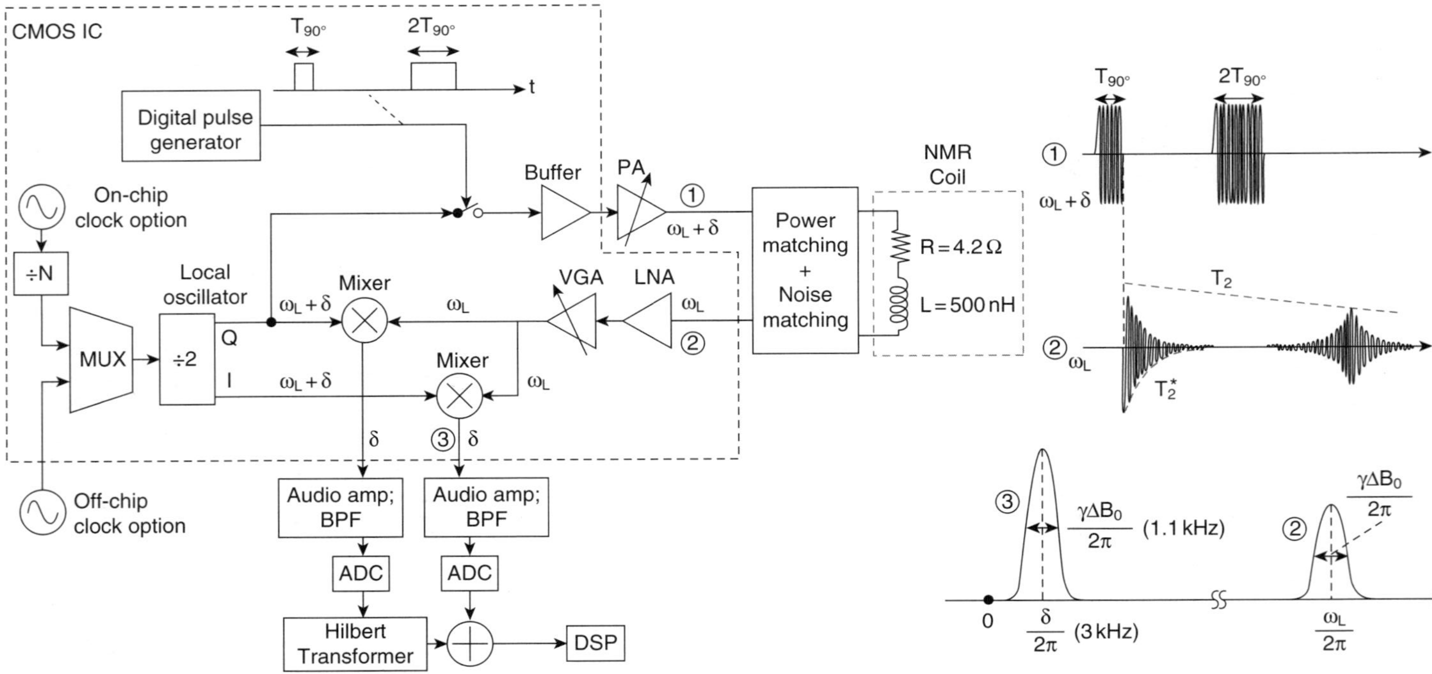

Figure 16.5. The architecture of our NMR RF transceiver IC. (From N. Sun, Y. Liu, H. Lee, R. Weissleder, and D. Ham, "CMOS RF biosensor utilizing nuclear magnetic resonance," *IEEE J. Solid-State Circuits*, 44, pp. 1629–1643, 2009.)

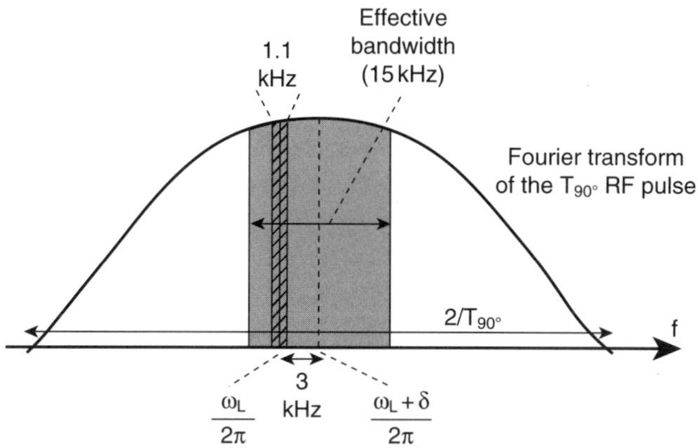

Figure 16.6. Fourier transform of the $T_{90°}$ pulse with $\omega_L + \delta$ excitation frequency. (From N. Sun, Y. Liu, H. Lee, R. Weissleder, and D. Ham, "CMOS RF biosensor utilizing nuclear magnetic resonance," *IEEE J. Solid-State Circuits*, 44, pp. 1629–1643, 2009.)

the signal down-converted by the quadrature-phase local oscillator by –90°. Addition of the Hilbert transformer output to the signal down-converted by the in-phase local oscillator nulls image noise.

16.4.2 Transmitter Path

Consider the transmitter path in Figure 16.5. To produce the RF magnetic field in the coil for excitation of proton magnets, we use the same local oscillator used for the frequency down-conversion in the receiver. The frequency of the local oscillator, $\omega_L + \delta$, is not the Larmor frequency ω_L, but this arrangement is still effective in causing NMR, as we will explain shortly. The RF signal produced by the local oscillator is gated by the digital pulse generator in order to generate the $T_{90°}$ pulse followed by repeated $2T_{90°}$ pulses (Fig. 16.3c), which is to obtain the echo signal at the receiver.

How can the $T_{90°}$ pulse with the off-tuned frequency $\omega_L + \delta$ excite proton magnets in the sample? On the one hand, the bandwidth of the $T_{90°}$ pulse may be taken as 30% (15 kHz) of $2/T_{90°}$ (shaded region in Fig. 16.6) centered about $(\omega_L + \delta)/(2\pi)$. On the other hand, across our 5 µL sample, the static field inhomogeneity Δ is 50 ppm, and thus, the Larmor (excitation) frequencies of various positions of the sample range within the bandwidth of $\gamma\Delta B_0/(2\pi) = 1.1$ kHz (hashed region in Fig. 16.6) centered about $\omega_L/(2\pi) = \gamma B_0/(2\pi)$. Combining, the entire sample is excited, as the range of the Larmor frequencies of the sample lies well within the bandwidth of the $T_{90°}$ pulse. Of course, this scheme is possible because our sample is very small and the corresponding field inhomogeneity (and the resultant Larmor frequency variation) is relatively small.

We could use a separate oscillator for the transmitter, but in such a scenario, we would have to tune the frequencies of both transmitter and receiver oscillators separately in the process of finding the frequency for NMR. In contrast, in our strategy sharing the same oscillator between the transmitter and receiver, we can just tune this single oscillator to find the right frequency for NMR, until the maximum signal appears at the output of the band-pass filter of Figure 16.5.

16.5 CMOS RADIO FREQUENCY TRANSCEIVER INTEGRATED CIRCUIT DESIGN: FRONT-END DESIGN

The planar microcoil (Fig. 16.1) is modeled as an inductor, $L = 500\,\text{nH}$, in series with parasitic resistance, $R = 4.2\,\Omega$, (Fig. 16.5). Its quality is $Q = \omega_L L/R = 16$. Although the coil is inherently an inductor while a radio antenna assumes a 50-Ω impedance, the coil is analogous to the antenna in that it emits power and picks up signals. Just like in radios, the matching of the transceiver to the coil and the low-noise design of the receiver amplifier are crucial, especially in our case where $V(t)$ is diminutive. Here we describe the front-end design.

16.5.1 Power Matching

The RF excitation pulses are sent to the coil through a power amplifier (PA). For maximum power transfer, the PA whose output impedance is R_{PA} (real number, e.g., $50\,\Omega$) is to be impedance matched to the coil. Two capacitors C_1 and C_2 arranged next to the coil in Figure 16.7a form an impedance matching network widely used in NMR system design. With suitably chosen C_1 and C_2, the real part of Z_1 becomes R_{PA} while Z_1's inductive reactance is resonated out by C_2, leading to $Z_2 = R_{\text{PA}}$.

16.5.2 Noise Matching via Passive Amplification

In the receiving stage of NMR experiments, the coil is connected to an LNA to amplify the signal $V(t)$ across the coil induced by precessing proton magnets. To minimize the noise figure, the LNA should be connected to the coil cautiously. In the widely practiced discrete design of the NMR RF transceiver, an off-the-shelf discrete LNA with a 50-Ω input impedance is impedance matched to the coil using the same C_1–C_2 network discussed above. See Figure 16.7b, where V_{rms} in the coil's model is the root mean square (rms) value of $V(t)$ and $\overline{v_n^2}/\Delta f$ is the power spectral density of the thermal noise generated by the coil's resistance R. This predominantly used strategy will work satisfactorily, as long as the LNA's noise figure is small at the source impedance of $50\,\Omega$.

Here we will seek a different LNA–coil connection option to obtain minimum noise figure, without demanding a priori either the 50-Ω input impedance for the LNA or the impedance matching condition. To this end, consider a general passive network between the coil and the LNA (Fig. 16.8a). The passive network has a voltage transfer function $H(\omega)$, which includes the effect of LNA's input impedance, Z_{LNA}. Define $\alpha \equiv |H(\omega_L)|$. If noise in the passive network is ignored,

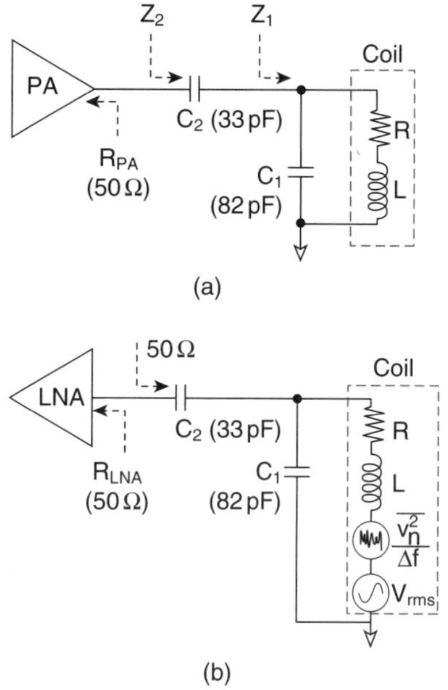

Figure 16.7. 50-Ω impedance matching: (a) PA–coil; (b) LNA–coil. (From N. Sun, Y. Liu, H. Lee, R. Weissleder, and D. Ham, "CMOS RF biosensor utilizing nuclear magnetic resonance," *IEEE J. Solid-State Circuits*, 44, pp. 1629–1643, 2009.)

or equivalently, if the passive network has negligible loss (this crucial assumption will be justified shortly), the passive network will "amplify" both the voltage signal and noise of the coil by the same factor α, while maintaining the SNR: At the output of the passive network, the signal is αV_{rms} and the noise power spectral density is $\alpha^2 \overline{v_n^2}/\Delta f$.

If α is large enough to put $\alpha^2 \overline{v_n^2}/\Delta f$ far beyond the LNA's input-referred noise $\overline{v_{a,n}^2}/\Delta f$ (Fig. 16.8b, left), the SNR is not degraded appreciably by the LNA, and the noise figure is small. In contrast, if α is not large enough and $\alpha^2 \overline{v_n^2}/\Delta f$ is comparable with, or smaller than, $\overline{v_{a,n}^2}/\Delta f$ (Fig. 16.8b, right), the SNR is degraded considerably by the LNA, resulting in a high-noise figure. Therefore, noise figure minimization entails two tasks: maximization of α and minimization of the LNA's input-referred noise, $\overline{v_{a,n}^2}/\Delta f$. We defer the discussion of the second task until Section 16.5.4, and here discuss how to maximize α.

We can immediately note that Z_{LNA} should be infinite to maximize α, as any finite Z_{LNA} will only lower the voltage at the output of a given passive network. Now with $Z_{LNA} = \infty$, we design a passive network that returns the maximum α. Various topologies may be worked out, and here we use the C_1–C_2 network: Figure 16.8c. The aim is to choose C_1 and C_2 that maximize α. C_2 should be infinite (short circuit), as any finite impedance in the C_2 path will lower the voltage at the passive network's output. With $C_2 = \infty$, we choose C_1 that maximizes α, which is given as a function of C_1:

$$\alpha(C_1) = \frac{1}{\sqrt{(1-\omega_L RC_1 Q)^2 + (\omega_L RC_1)^2}}, \tag{16.4}$$

where $Q = 16$ is the coil's Q. $\alpha(C_1)$ becomes maximum

$$\alpha_{\max} = \sqrt{Q^2+1} \approx Q \tag{16.5}$$

when

$$C_1 = \frac{1}{\omega_L R} \cdot \frac{Q}{Q^2+1} \approx \frac{1}{\omega_L RQ} = \frac{1}{\omega_L^2 L}, \tag{16.6}$$

that is, when C_1 resonates with L at ω_L.

We have found that for a fixed LNA input-referred noise, a shunt capacitor C_1 that resonates with the coil inductor at ω_L maximally amplifies the input signal and noise with a gain of Q, minimizing the noise figure. We assumed no noise (no loss) in the capacitor. This is to be justified, as any significant noise directly translates to noise figure, defeating the purpose of our technique. The assumption is indeed valid, for the quality factor of the capacitor is on the order of several hundreds, while the coil's Q is 16. In this connection, the LC resonator in the minimum noise figure configuration may be understood as a passive preamplifier with a gain of 16 but with almost zero noise figure, which is unattainable with active amplifiers. Forming the LC resonator as a preamplifier naturally exploits the intrinsic inductor L already available in the coil.

Note that the minimum noise figure configuration does not involve impedance matching: The impedance looking into the passive network is Q^2R, while $Z_{LNA} = \infty$ (Fig. 16.8c). Not the power, but the voltage is maximally delivered to the LNA, in achieving the minimum noise figure.

Numerical examples may help appreciate this technique. The maximum amplitude of $V(t)$ is 130 nV, thus $V_{rms} = 92$ nV. The noise generated in the coil is $\overline{v_n^2}/\Delta f = 4kTR = (0.26\ \mathrm{nV}/\sqrt{\mathrm{Hz}})^2$, and the input-referred noise of our LNA is $\overline{v_{a,n}^2}/\Delta f = (1.3\ \mathrm{nV}/\sqrt{\mathrm{Hz}})^2$ (Section 16.5.4). From these values, we created Table 16.1, where the passively amplified rms signal power $\alpha^2 V_{rms}^2$, passively amplified coil noise over the signal bandwidth (BW = 1.1 kHz), $\alpha^2\overline{v_n^2} = \alpha^2\overline{v_n^2}/\Delta f \times \mathrm{BW}$, and LNA's input-referred noise over the bandwidth, $\overline{v_{a,n}^2} = \overline{v_{a,n}^2}/\Delta f \times BW$, are given for three values of α. For each α, it also shows the noise figure, which was calculated using $\mathrm{NF} = 10\log[(\alpha^2\overline{v_n^2} + \overline{v_{a,n}^2})/(\alpha^2\overline{v_n^2})]$.

From Table 16.1, we note the following:

- $\alpha = Q = 16$: In this optimum case (Fig. 16.8c), the passively amplified coil noise dominates the LNA's input-referred noise, yielding a noise figure of only 0.4 dB.

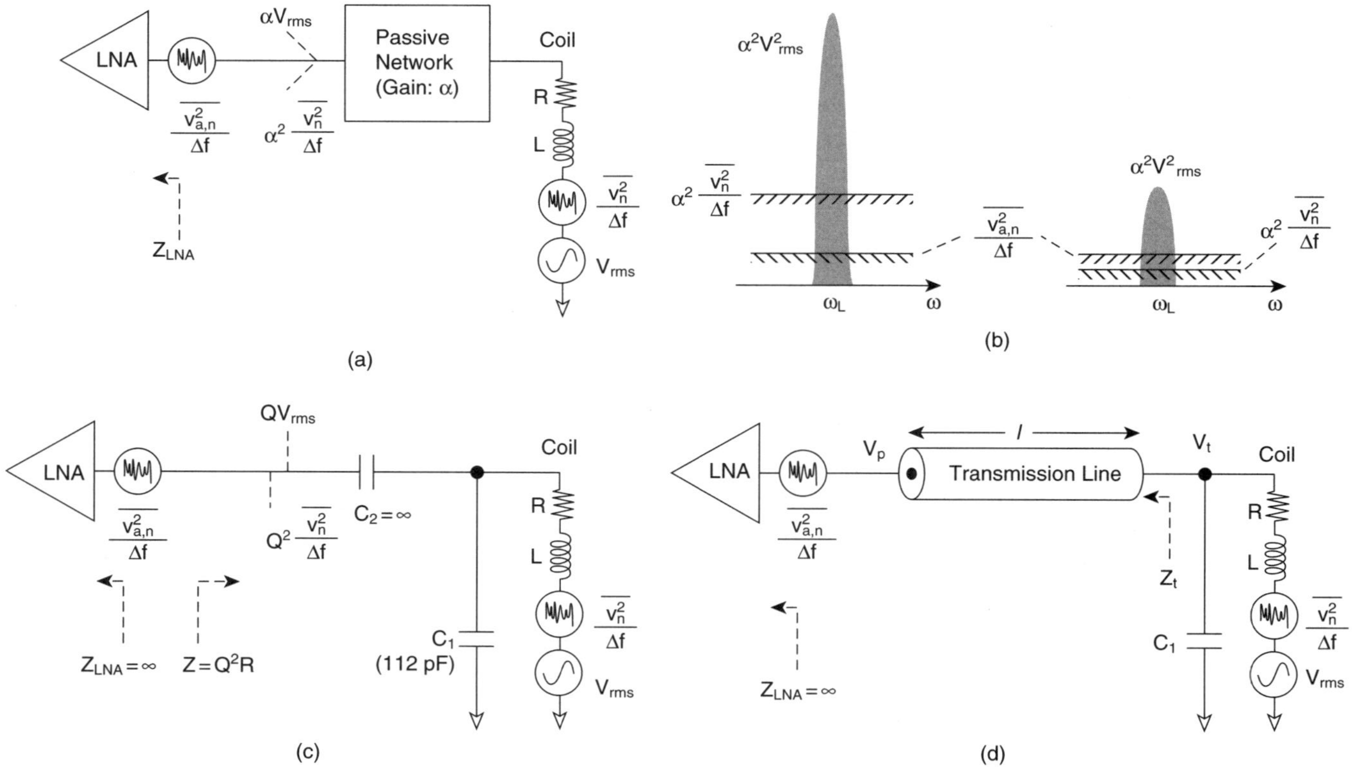

Figure 16.8. (a) General passive network between a coil and an LNA. (b) Passively amplified signal and coil noise, and LNA's input referred noise in frequency domain for a large (left) and small (right) α . (c) Maximum passive amplification. (d) Maximum passive amplification in the presence of a transmission line. (From N. Sun, Y. Liu, H. Lee, R. Weissleder, and D. Ham, "CMOS RF biosensor utilizing nuclear magnetic resonance," *IEEE J. Solid-State Circuits*, 44, pp. 1629–1643, 2009.)

TABLE 16.1 LNA Noise Figures for Various Values of α

α	16	1	3.4
$\alpha^2 V_{rms}^2$	$(1471\,nV)^2$	$(92\,nV)^2$	$(313\,nV)^2$
$\alpha^2 \overline{v_n^2}$	$(139\,nV)^2$	$(8.8\,nV)^2$	$(29.7\,nV)^2$
$\overline{v_{a,n}^2}$	$(43\ nV)^2$	$(43\,nV)^2$	$(43\,nV)^2$
Noise figure	0.4 dB	14 dB	4.9 dB

- $\alpha = 1$: This is the case where the coil is directly connected to the LNA with $Z_{LNA} = \infty$ with no passive network. The LNA's input-referred noise dominates the passively amplified coil noise. Consequently, a large noise figure of 14 dB results.
- $\alpha = 3.4$: If one uses the configuration of Figure 16.7b with C_1 and C_2 yielding 50 Ω but with an LNA with $Z_{LNA} = \infty$, $\alpha = 3.4$. This is a scenario where the C_1–C_2 network already designed for impedance matching for the PA is directly connected to the LNA with $Z_{LNA} = \infty$. The noise figure is 4.9 dB.

As seen, the noise figure varies substantially with α, as it determines the relative difference between the passively amplified coil noise and LNA's input-referred noise. The point of our technique is to maximize the passively amplified coil noise so that it is far larger than LNA's input-referred noise.

16.5.3 Generalization of Passive Amplification

As NMR transceivers have been realized mostly at the discrete level, the 50-Ω matching with abundantly available 50-Ω off-the-shelf LNAs (Fig. 16.7b) has been the dominant choice for the LNA–coil connection. Regarding the resonance noise matching with the infinite-impedance LNA (Fig. 16.8c), we suspected that this simple yet powerful technique might have been used during the 60 years of NMR, but there are very limited publications on the technique, and it is difficult to know the actual degree of its awareness and usage. It appears that the technique has been rarely conceived because custom design of an LNA (to produce non-50-Ω impedance) is seldom exercised in the conventional discrete NMR system design. We found only two papers [8, 9] reporting the technique, where it was empirically demonstrated but with no explanation of its underlying machinery. We also found a patent [10] that discusses it but wrongly disqualifies it as the optimum option. In this connection, our foregoing elucidation of resonance noise matching may help its further recognition and appreciation, and clarify its concept.

The reader may be concerned that resonance noise matching may not work in case the high-impedance LNA is connected to a coil through a 50-Ω coaxial cable (Fig. 16.8d) due to impedance mismatches at both ends of the line (this is a legitimate concern in that even at 21 MHz, a line length of, say 20 cm, introduces a wave effect). This concern may be another reason for the dominance of the 50-Ω matching: when a 50-Ω cable is used, "everything 50 Ω" offers the most convenient design choice. Contrary to the concern, however, resonance noise matching works well with a more general passive network containing a transmission line.

To show this, let us work out the resonance noise matching in Figure 16.8d based on our articulation of it from the previous subsection. The parameters of the transmission line are characteristic impedance: Z_0 ; wave velocity: v; length: l; and wave number: $\beta = \omega/v$. We seek to find C_1 that maximizes α for a fixed l. Assume $\beta l \ll l$, although we can solve the problem generally. V_p and V_t in Figure 16.8d are related through $V_p = V_t(\omega)/\cos(\beta l) \approx V_t$. The impedance Z_t is $Z_t = -jZ_0 \cot(\beta l) \approx -jZ_0/\beta l = vZ_0/(j\omega l)$ since $Z_{\mathrm{LNA}} = \infty$. Therefore, the transmission line behaves like a capacitor $C_t = l(vZ_0)^{-1}$ at the coil end. Thus, to maximize α, L should resonate with $C_1 \parallel C_t$ at ω_L:

$$C_1 = \frac{1}{L\omega_L^2} - \frac{l}{vZ_0}. \tag{16.7}$$

The resulting maximum α is again $Q = 16$. This design is valid, as the line has a negligible measured loss (less than 0.05 dB for a 30-cm cable).

16.5.4 Design of the Low-Noise Amplifier

As discussed in Section 16.5.2, to minimize noise figure, α should be maximized and LNA's input-referred noise should be minimized. We have just discussed the former task. We now describe our LNA design to perform the latter. As $Z_{\mathrm{LNA}} = \infty$ is a prerequisite for maximum α, a common source topology is chosen for the LNA. It has a capacitive input impedance, which may be regarded as infinite at low enough frequencies.

Figure 16.9 shows the schematic of our LNA. Since the design frequency, 21.3 MHz, is smaller than the $1/f$ noise corner (~50 MHz) of NMOS transistors in the 0.18-μm technology used, we use PMOS devices with the smaller $1/f$ noise corner (~1 MHz) [11] as input transistors $\mathbf{M}_1$ and $\mathbf{M}_2$. PMOS transistors bring an additional benefit, the mitigation of the coupling of the substrate noise produced by the digital pulse generation circuits into the LNA input: The substrate noise would otherwise interfere with the weak signal at the LNA input, which is 4 μV even after maximal passive amplification. On the same token, we must minimize the coupling of the strong local oscillator signal into the LNA input: It occurs first through the mixer, then through the LNA's input–output coupling capacitors. To minimize the latter coupling, cascode transistors $\mathbf{M}_3$ and $\mathbf{M}_4$ are used. Finally, to suppress the common mode noise, the amplifier is designed in a differential form and a common-mode feedback circuit (CMFB) is employed. The capacitor C_3 and the resistor R_3 are for frequency compensation.

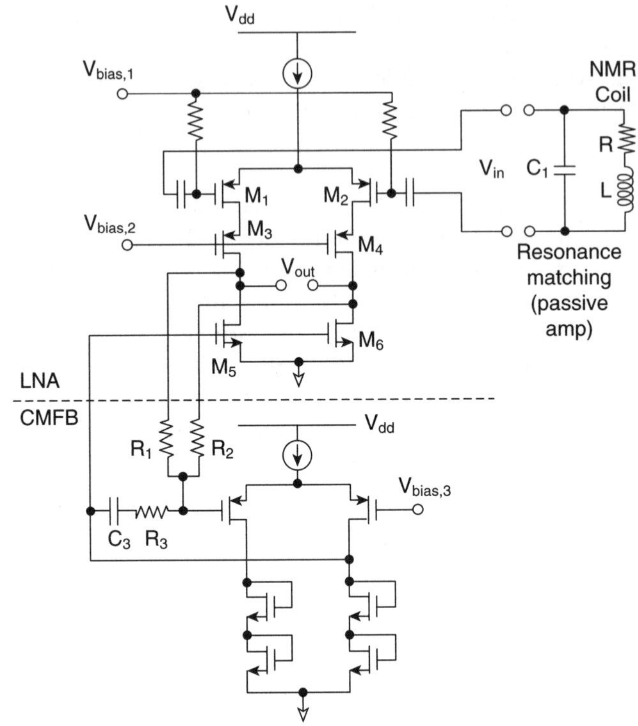

Figure 16.9. LNA schematic. (From N. Sun, Y. Liu, H. Lee, R. Weissleder, and D. Ham, "CMOS RF biosensor utilizing nuclear magnetic resonance," *IEEE J. Solid-State Circuits*, 44, pp. 1629–1643, 2009.)

With the arrangement above, the total noise of the LNA is dominated by the channel thermal noise of the transistors. The input-referred noise of the LNA calculated through the standard small-signal analysis is given by

$$\overline{v_{a,n}^2} \approx 2\overline{v_{n,M_1}^2} + 2\overline{v_{n,M_5}^2} \cdot \frac{g_{m5}^2}{g_{m1}^2} + \frac{2\overline{v_{n,M_3}^2}}{g_{m1}^2 r_{ds1}^2}, \tag{16.8}$$

where $\overline{v_{n,M_i}^2}$ ($i = 1,3$ and 5) is the transistor $\mathbf{M}_i$'s gate-source referred channel thermal noise. We have ignored the noise of the CMFB, and assumed perfectly matched differential pairs, to obtain the general guideline for how to size the transistors. The third term is negligible as $g_{m1}r_{ds1} \gg 1$. To minimize the second term, active loads $\mathbf{M}_5$ and $\mathbf{M}_6$ are made much smaller than input transistors $\mathbf{M}_1$ and $\mathbf{M}_2$: this leads to $g_{m5}/g_{m1} \ll 1$ and the total input-referred noise is dominated by the first term, $\overline{v_{n,M1}^2} \approx 4kT\gamma g_{m1}^{-1}$ where g_{m1} is the transconductance of the input transistors. To reduce this first term, a large tail current (4 mA) and wide input transistors (900 μm) are used to maximize g_{m1}. From SPICE simulation, the input-referred noise of the LNA at 21.3 MHz is $1.3\,\text{nV}/\sqrt{\text{Hz}}$ and its voltage gain is 41 dB. Its input

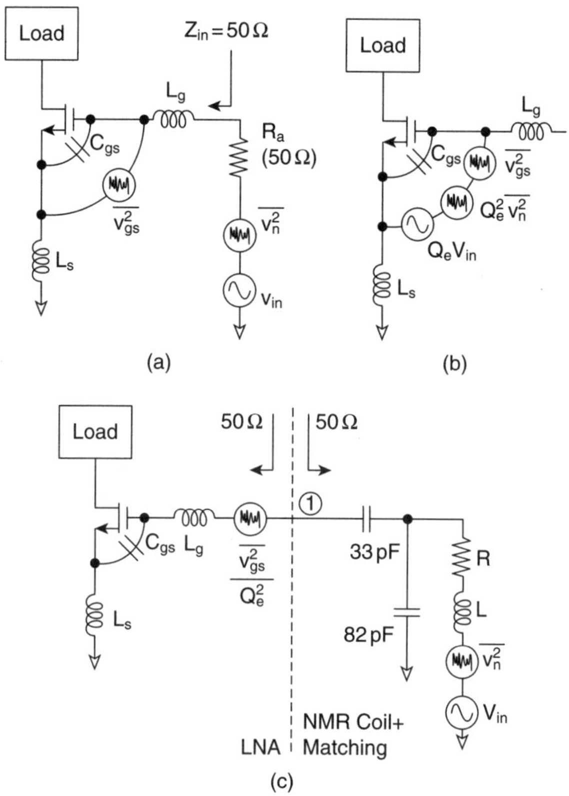

Figure 16.10. (a) LNA with inductive source degeneration. (b) Passive amplification in this LNA. (c) Use of this LNA in Fig. 16.7b. (From N. Sun, Y. Liu, H. Lee, R. Weissleder, and D. Ham, "CMOS RF biosensor utilizing nuclear magnetic resonance," *IEEE J. Solid-State Circuits*, 44, pp. 1629–1643, 2009.)

impedance, even with the wide input transistors, is still large enough at 21.3 MHz to guarantee $Z_{\text{LNA}} = \infty$.

16.5.5 Comparison with Inductively Source-Degenerated LNA

The passive amplification to minimize the noise figure is also employed in the inductively source-degenerated LNA (Fig. 16.10a) widely used for wireless RF design [12, 13], although configured differently. As many designers are familiar with this LNA, here we compare it with our LNA.

In the inductively source-degenerated LNA of Figure 16.10a designed for a wireless receiver, V_{in} and $\overline{v_n^2}$ are the input signal and noise from an antenna, while $\overline{v_{gs}^2}$ is the gate-source referred channel thermal noise of the transistor. $R_a = 50\ \Omega$ is the antenna impedance. Inductors L_s and L_g and capacitor C_{gs} arranged around the transistor are chosen such that they resonate at the design frequency, and also present 50 Ω to the antenna: in Figure 16.10a, $Z_{in} = L_s g_m / C_{gs} = 50\ \Omega$. This 50-Ω matching is required in wireless RF design, often reinforced by the band selection filter positioned between the antenna and the LNA. With such inductor and capacitor values, the LC network passively amplifies the input signal and noise, returning their amplified versions across the gate and source of the transistor (Fig. 16.10b) [12, 13]. The gain is $Q_e \equiv \omega(L_g + L_s)/(2R_a)$, the effective quality factor of the LC tank including the antenna resistance (the factor 2 is because $Z_{in} = R_a$). If Q_e is large enough to put $Q_e^2 \cdot \overline{v_n^2}$ far beyond the transistor noise $\overline{v_{gs}^2}$, the noise figure of the LNA will be small, as the SNR is minimally affected by the transistor noise. This is clearly the same passive amplification idea.

Passive amplification is realized differently between our LNA (Figs. 16.8c, or 16.9) and the inductively source-degenerated LNA (Fig. 16.10). In the wireless RF design where 50-Ω matching is required, the source-degenerated LNA is well suited, as it simultaneously offers passive amplification and 50-Ω matching. In contrast, NMR system design does not demand 50-Ω matching. This, in conjunction with that the NMR coil has an inherent inductance, makes the implementation of passive amplification for NMR systems much simpler: The "free" inductor of the coil is used as an integral part of the LC passive amplifier, without adding extra inductors.

Note that the inductively source-degenerated LNA with a 50-Ω impedance can be used for the widely used 50-Ω matched front-end of the NMR receiver (Fig. 16.7b): such a design is shown in Figure 16.10c. Here although α is only 1.7, the noise figure is small because the LNA's input-referred noise at node (1) is significantly reduced, as it is the transistor's gate-source referred channel thermal noise $\overline{v_{gs}^2}$ divided by Q_e^2: This is another way of looking at passive amplification in the source-degenerated LNA. This design, however, does not exploit the free inductance of the coil. It instead transforms the coil impedance to 50 Ω, then performs passive amplification, introducing extra inductors. Our design (Figs. 16.8c, or 16.9) is thus a simpler choice as far as NMR system design is concerned.

16.5.6 Simultaneous Noise and Power Matching

We have seen that receiver noise matching (Fig. 16.8c) and transmitter power matching (Fig. 16.7a) require different passive networks. We may arrange the two networks separately, and to switch the coil to the LNA through the noise matching network during the receiving mode, and to the PA through the power matching network during the transmission mode (Fig. 16.11a). This strategy, however, suffers from the loss associated with the switch. In the receiving mode, the turn-on loss of the switch will directly translate to noise figure.

To avoid the use of a switch in the receiving mode, we devised a new, differential matching network in Figure 16.11b. In the transmission mode, the switches next to the PA are turned on, and the PA is connected to the coil through the inductors (L_a) and capacitors (C_a, C_b, C_c) (Fig. 16.11c). In the receiving mode, the switches are off, effectively making the inductors disappear, and the LNA is connected to the coil, whose impedance is modified by the capacitors (C_a, C_b, C_c)

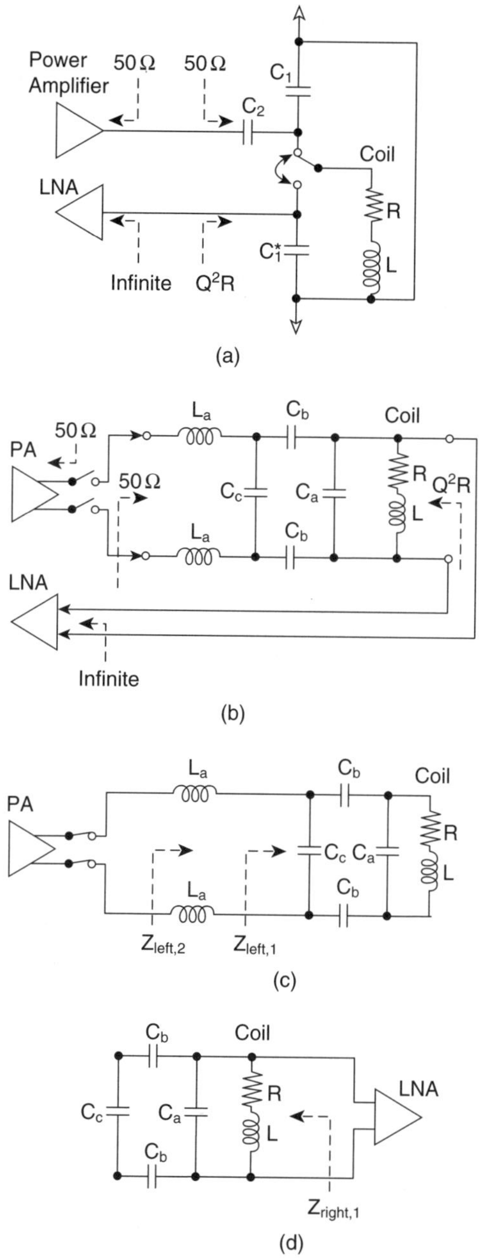

Figure 16.11. Noise/power matching with (a) and without (b) a switch in the receiver path; (c) Transmitting mode. (d) Receiving mode. (From N. Sun, Y. Liu, H. Lee, R. Weissleder, and D. Ham, "CMOS RF biosensor utilizing nuclear magnetic resonance," *IEEE J. Solid-State Circuits*, 44, pp. 1629–1643, 2009.)

(Fig. 16.11d). The first point of this circuit is that the LNA in the receiving mode and the PA in the transmitting mode see two different impedances, that is, $Z_{\text{left},2} \neq Z_{\text{right},1}$ in Figure 16.11. Thus, one can choose capacitors and inductors appropriately to provide power matching for the PA and noise matching for the LNA. Note that the difference between $Z_{\text{left},2}$ and $Z_{\text{right},1}$ arises not merely from the fact that the transmitting path has additional inductors, but more importantly from the fact that the impedance of the network of capacitors and the coil seen from the left is completely different from that seen from the right, that is, $Z_{\text{left},1} \neq Z_{\text{right},1}$ in Figure 16.11. The second point is that neither lossy switches nor lossy inductors are present in the receiving mode (Fig. 16.11d), and hence, the matching network hardly degrades the receiver noise figure.

We now consider how to select capacitor and inductor values quantitatively. For simplicity, we introduce the following notation: $X_L \equiv w_L L$, $X_a \equiv -1/(w_L C_a)$, $X_b \equiv -1/(w_L C_b)$, and $X_c \equiv -1/(w_L C_c)$. $Z_{\text{right},1}$ in Figure 16.11d is given by

$$\begin{aligned} Z_{\text{right},1} &= (R + jX_L) \| jX_a \| (2jX_b + jX_c) \\ &= (R + jX_L) \| j\frac{X_a(2X_b + X_c)}{X_a + 2X_b + X_c}, \end{aligned} \tag{16.9}$$

that is, when seen from the coil, the capacitors in Figure 16.11d behave like one effective capacitor, whose impedance is to the right of the $\|$ symbol. The noise matching condition (see Eq. 16.6) for $Q \gg 1$ is that the effective capacitor resonates with L at ω_L. This condition may be written as

$$X_L + \frac{X_a(2X_b + X_c)}{X_a + 2X_b + X_c} = 0. \tag{16.10}$$

In the transmitting mode, $Z_{\text{left},1}$ of Figure 16.11c is expressed as

$$\begin{aligned} Z_{\text{left},1} &= jX_c \| [2jX_b + jX_a \| (R + jX_L)] \\ &= \frac{(X_a X_c)^2}{R(X_a + 2X_b + X_c)^2} + j\frac{(X_a + 2X_b)X_c}{X_a + 2X_b + X_c}. \end{aligned} \tag{16.11}$$

For power matching, the real part of $Z_{\text{left},1}$ should equal the output resistance R_{PA} of the PA, while its capacitive reactance is to be resonated out by the inductors L_a:

$$2\omega_L L_a + \frac{(X_a + 2X_b)X_c}{X_a + 2X_b + X_c} = 0 \tag{16.12}$$

$$\frac{(X_a X_c)^2}{R(X_a + 2X_b + X_c)^2} = R_{\text{PA}}. \tag{16.13}$$

If three conditions Equations 16.10, 16.12, and 16.13 are simultaneously met, noise and power matching are simultaneously obtained. Since there are four unknown design parameters (C_a, C_b, C_c, and L_a), we have a large degree of freedom in choosing them.

16.6 CMOS RADIO FREQUENCY TRANSCEIVER INTEGRATED CIRCUIT DESIGN: OTHER CIRCUIT CONSIDERATIONS

16.6.1 Receiver Linearity

Although the planar microcoil is the ultimate part of our system, it is desirable for the system to be able to accommodate a variety coils for varying experimental needs. The coil's Q varies widely depending on its geometry and construction: for example, while $Q = 16$ for the planar microcoil, $Q = 200$ for a solenoidal microcoil we built. The signal grows with the coil Q, especially with the resonance noise matching. To provide more amplification for a smaller signal, but to ensure that a larger signal does not enter the nonlinear regime of the mixer,[3] we designed a variable gain amplifier (VGA) following the LNA (Fig. 16.5). The receiver linearity is important, as we seek to accurately extract T_2.

The VGA is depicted in the middle of Figure 16.12. It is a differential common-source amplifier with two tunable loads shown within the two dashed boxes. The VGA's gain is altered by tuning the load impedance. Each load has three branches in parallel. Consider the branch built around transistor $\mathbf{M}_1$. As the control voltage V_c is increased from zero, the operation regime of $\mathbf{M}_1$ undergoes transitions from turn-off to pinch-off to triode, decreasing its drain-source resistance, $r_{ds} = \partial V_{ds}/\partial I_{ds}$. Therefore, the first branch serves as a voltage-controlled load. The second branch behaves essentially the same way, with decreasing r_{ds} of $\mathbf{M}_2$ with increasing V_c, but this time, the gate of $\mathbf{M}_2$ sees a voltage lower than V_c due to $\mathbf{M}_3$, thus the impedance-versus-V_c curve for the second branch is that for the first branch, but translated along the V_c axis. The third branch involving three transistors, $\mathbf{M}_4$, $\mathbf{M}_5$, and $\mathbf{M}_6$, also exhibits the same impedance-versus-V_c curve, but even more shifted along the V_c axis. By combining the three branches in parallel, the load impedance is tuned more smoothly over a wider range of V_c, as compared with the case any single branch is used alone. The voltage gain tuning range of the VGA is from 0.8 to 22. The VGA's noise hardly contributes to the receiver noise figure, as the LNA's gain is large (41 dB). Therefore, little noise minimization effort is required for the VGA.

16.6.2 Receiver-Transmitter (RX-TX) Switching Scheme

As discussed in Section 16.5.6, to avoid the use of lossy switches in front of the LNA, in our design the LNA stays connected to the coil during the $T_{90°}$ and $2T_{90°}$

[3] The input to our mixer can be as large as 10 mV within 5% gain reduction.

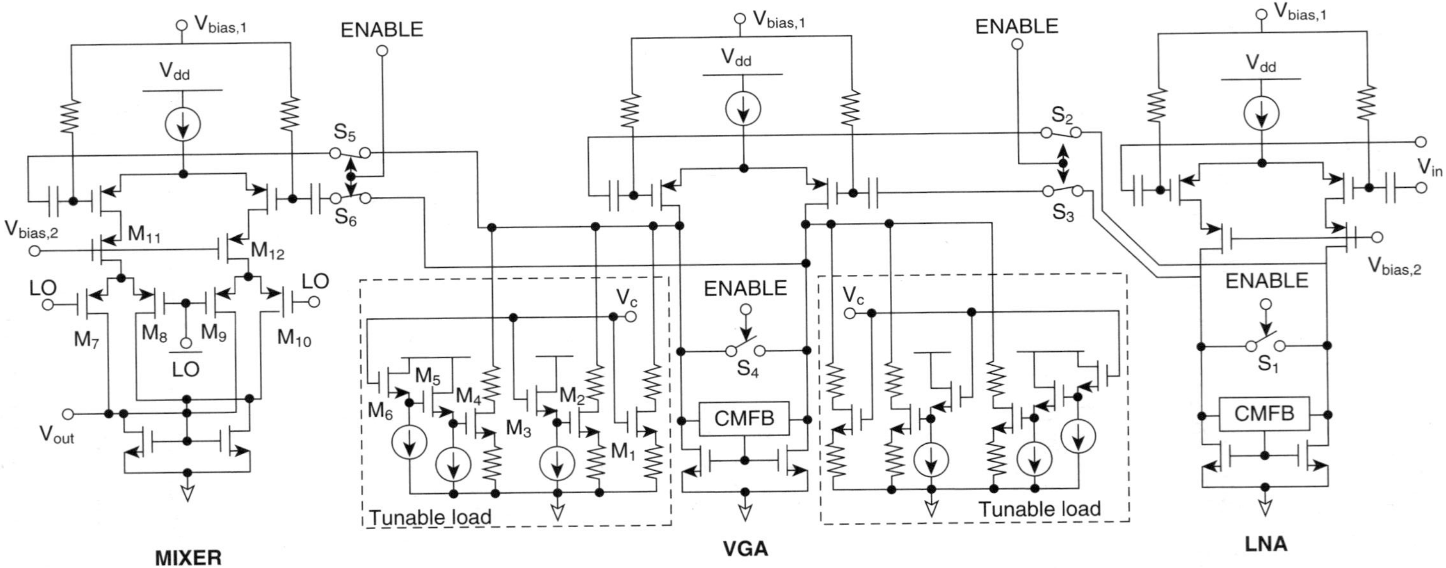

Figure 16.12. Schematics of the LNA, VGA, and mixer, and their connections through a set of switches. (From N. Sun, Y. Liu, H. Lee, R. Weissleder, and D. Ham, "CMOS RF biosensor utilizing nuclear magnetic resonance," *IEEE J. Solid-State Circuits*, 44, pp. 1629–1643, 2009.)

pulse transmissions, for example, Figure 16.11b. In this approach, the large RF signal from the PA will saturate the receiver during the transmission mode. If the receiver is saturated during the transmission mode, restoration to its normal operation at the transition to the receiving phase (e.g., at $t = T_{90°}$) will exhibit a settling behavior over the characteristic time of the receiver, compromising the initial reception of the signal.

To address this issue, during the transmission mode, we reduce the gain of the LNA by closing switch $\mathbf{S}_1$ in Figure 16.12, preventing its saturation, and disconnect the rest of the receiver from the LNA by opening switches $\mathbf{S}_2$ and $\mathbf{S}_3$ of Figure 16.12. Although $\mathbf{S}_2$ and $\mathbf{S}_3$ lie in the RF signal path, they come *after* the LNA substantially amplifies the signal, thus their turn-on resistances hardly contribute to the overall noise figure. Since $\mathbf{S}_2$ and $\mathbf{S}_3$ are not perfect switches and some of the large RF signal could leak to the VGA via their parasitic capacitors during the transmission mode, we also place switch $\mathbf{S}_4$ to reduce the VGA gain during the transmission phase, and switches $\mathbf{S}_5$ and $\mathbf{S}_6$ to cut off the mixer from the VGA.

16.6.3 Mixer

As shown at the left of Figure 16.12, we use a double-balanced active mixer. Although the RF signal is down-converted to a low frequency of $\delta/(2\pi) = 3\text{kHz}$, $1/f$ noise in the mixer does not interfere with the down-converted signal to any significant degree because the signal is amply amplified before the mixer and the local oscillator (LO) generates a square wave. This is quantitatively confirmed, resorting to the study in Darabi and Abidi [14]. The voltage gain of the mixer is 26 dB. The cascode transistors $\mathbf{M}_{11}$ and $\mathbf{M}_{12}$ are used to mitigate the undesired feed through from the LO to the LNA's input path, which could otherwise interfere with the weak signal from the coil.

16.7 EXPERIMENTS

We implemented the NMR transceiver in the 0.18-μm CMOS technology. The IC occupies an area of 2.0 mm by 1.9 mm (Fig. 16.1). It contains, among components of Figure 16.5, digital circuits for the pulse transmission, and most components associated with the heterodyning, that is, the LNA, VGA, mixers, and local oscillators including the frequency dividers. The matching network, PA, and ADCs were implemented with discrete components.

As we deal with weak echo signals, the most important aspect of our design is the receiver noise figure, which we obtained by directly measuring the receiver input referred noise. After feeding an RF signal to the receiver input (node ②, Fig. 16.5), we measured the output at node ③ using a spectrum analyzer. From this, we extracted the gain and output noise of the receiver. By dividing the output noise by the gain while factoring out image noise, we obtained the input referred noise of $1.8\,\text{nV}/\sqrt{\text{Hz}}$. Since the LNA has a voltage gain of 41 dB, the input-referred noise is primarily contributed by the LNA. This measured input-referred noise is larger than the SPICE-simulated value of $13\,\text{nV}/\sqrt{\text{Hz}}$. The extra amount originates mostly from the power supply and substrate noise, which were not taken into account in the simulation.

From the measured input-referred noise and the measured impedance of the coil, we obtain the noise figure of the receiver. For the resonance noise matching with $\alpha = 16$ (Fig. 16.8c), the noise figure is only 0.7 dB. If we use the configuration of Figure 16.8c but with C_1 and C_2 arranged for 50-Ω matching, $\alpha = 3.4$ and the noise figure is 7.0 dB. These results agree well with our time domain measurements of the echo signal. For $\alpha = 3.4$, we averaged 64 echo signals to obtain a clean echo signal. In contrast, for the resonance matching with $\alpha = 16$, only averaging for four times was sufficient to obtain a clean echo signal, from which we were able to reliably extract T_2. This is an unequivocal proof of substantial reduction of the noise figure via maximum passive amplification.

We assembled the entire NMR system of Figure 16.1 by putting together the RF IC with auxiliary electronics on the printed circuit board, along with the planar microcoil and the magnet. The magnet and planar microcoil shown in Figure 16.1 are borrowed from another work of ours [4], where the microcoil and its interface with sample were fully developed with extensive biosensing experiments but the electronics were at the discrete level with lower sensitivity and larger size. The magnet is a commercial product. The planar microcoil was in-house fabricated using standard photolithography. A 5-μL water sample is held on the microcoil inside a microfluidic container fabricated on top. In the following, we present NMR and biosensing experiments performed with the system.

16.7.1 Nuclear Magnetic Resonance Experiments

We first performed proton NMR experiments in a phosphate buffered saline (PBS) solution, which is essentially biocompatible water. The top of Figure 16.13 shows a down-converted echo signal measured at the output of the ADC in Figure 16.5, which is directly interfaced with a computer. From the exponentially decaying envelope of the echoes, shown as a dotted line, we extract $T_2 = 523$ msec. The repeated large spikes are coupling of the transmitted RF sequence through unidentified coupling paths, but they do not compromise the observation of the conspicuous echo signal, as they occur at different times. As magnetic nanoparticles (Fe) (diameter: 30 nm) are suspended in the PBS with a concentration of 0.17 mM, the measured T_2 is decreased to 60 msec (Fig. 16.13, bottom), as discussed in Section 16.2.3. We remark that the data presented here and in what follows were obtained using an external signal generator instead of the on-chip local oscillator. Although the on-chip oscillator yielded echo signals from which T_2 values could be extracted, they were noisier as the on-chip oscillator was a free-running oscillator.

We discussed that T_2 decreases ($1/T_2$ increases) with an increasing concentration of the magnetic nanoparticles (Section 16.2.3). The NMR theory tells more specifically that $1/T_2$ increases linearly with the concentration [15]. Measured $1/T_2$ with a varying concentration of the magnetic particles in Figure 16.14 conform to the theory, affirming our system's functionality.

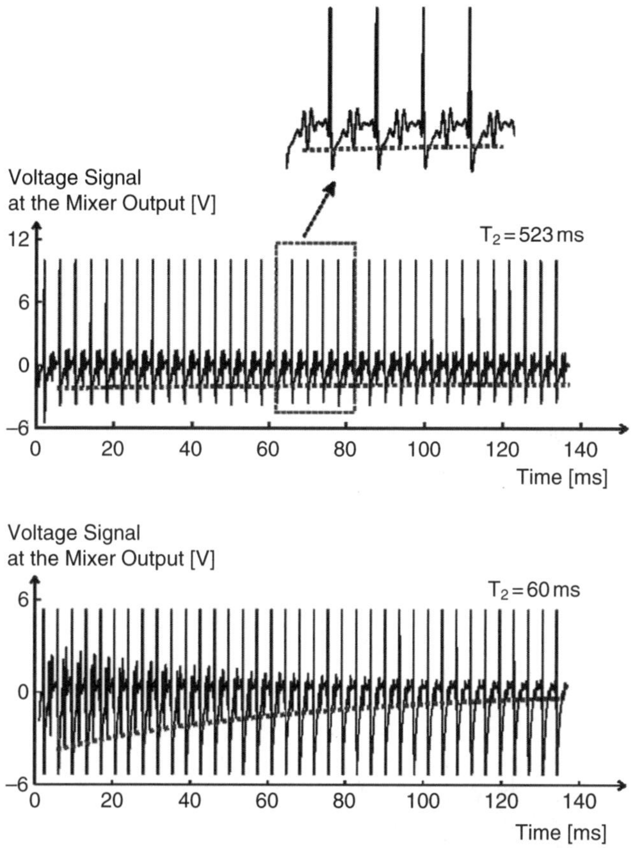

Figure 16.13. Measured, down-converted echo signals: (top) pure PBS; (bottom) PBS with magnetic nanoparticles (0.17 mM). (From N. Sun, Y. Liu, H. Lee, R. Weissleder, and D. Ham, "CMOS RF biosensor utilizing nuclear magnetic resonance," *IEEE J. Solid-State Circuits*, 44, pp. 1629–1643, 2009.)

While the planar microcoil ($Q = 16$) used to obtain Figure 16.13 was ultimately part of our system due to its fabrication advantage, we performed NMR experiments using a higher-Q solenoidal microcoil ($Q = 200$) to show the echo signals more clearly, far beyond the spikes, as shown in Figure 16.15. As the magnetic particle concentration changes from 0.1 mM to 1 mM, T_2 is reduced from 150 msec to 15 msec.

16.7.2 Biomolecular Sensing

Our NMR system with the planar microcoil can be used to sense biomolecules, as described in Section 16.2.3. Consider, for example, the detection of avidin (protein) in the PBS. Magnetic nanoparticles with surfaces coated with biotin (vitamin H) that specifically bind to avidin are suspended in the sample

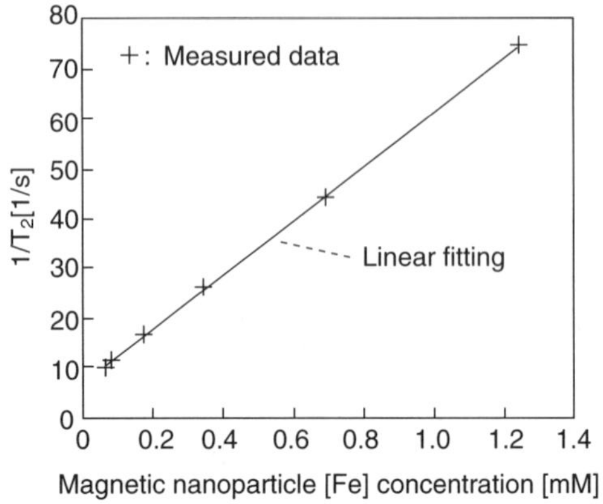

Figure 16.14. Measured $1/T_2$ versus magnetic nanoparticle concentration in PBS. (From N. Sun, Y. Liu, H. Lee, R. Weissleder, and D. Ham, "CMOS RF biosensor utilizing nuclear magnetic resonance," *IEEE J. Solid-State Circuits*, 44, pp. 1629–1643, 2009.)

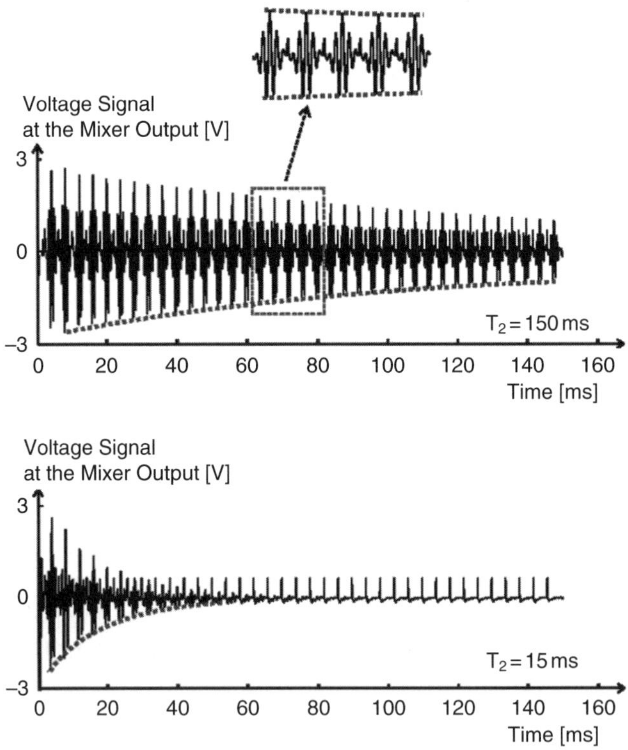

Figure 16.15. Measured, down-converted echo signals for two different concentrations of magnetic nanoparticles; with a solenoidal microcoil. (From N. Sun, Y. Liu, H. Lee, R. Weissleder, and D. Ham, "CMOS RF biosensor utilizing nuclear magnetic resonance," *IEEE J. Solid-State Circuits*, 44, pp. 1629–1643, 2009.)

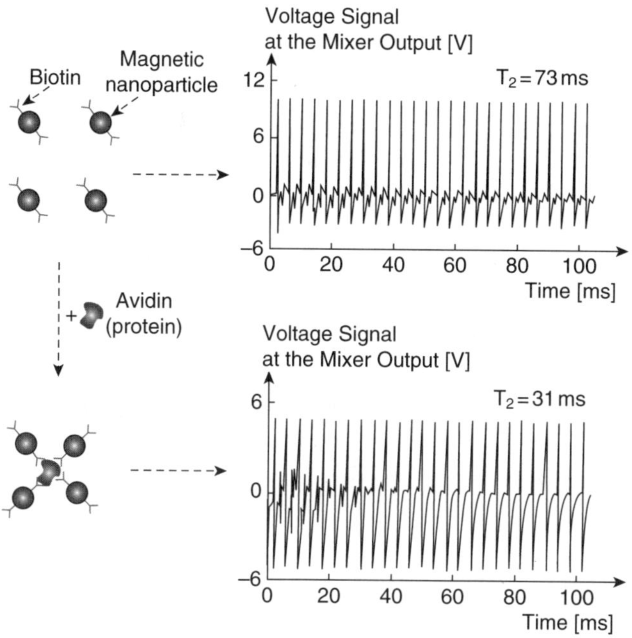

Figure 16.16. Experimental avidin detection. (From N. Sun, Y. Liu, H. Lee, R. Weissleder, and D. Ham, "CMOS RF biosensor utilizing nuclear magnetic resonance," *IEEE J. Solid-State Circuits*, 44, pp. 1629–1643, 2009.)

(Fig. 16.16). When avidin was absent in the sample, the magnetic particles were monodispersed, yielding T_2 of 73 msec (Fig. 16.16, top). When avidin existed in the sample with a concentration of 320 nM, the magnetic particles self-assembled into clusters [2], reducing T_2 to 31 msec (Fig. 16.16, bottom), thus indicating the presence of avidin.

We repeated the experiment with a decreasing concentration of the avidin. The minimum detectable quantity of avidin was 20 fmol and 1.4 ng in the 5-μL sample volume (Fig. 16.17). In comparison to the state-of-the-art commercial benchtop NMR system [3], our system is 60 times more sensitive in terms of the absolute detectable quantity (20 fmol), thus our system is far more advantageous in desired mass-limited sensing applications. Even a better detection threshold can be attained via improvement in the bioassay, for our current detection threshold is limited not by our transceiver, but by the specific bioassay used [4–5]. We also successfully carried out the detection of another biomolecule, folic acid, whose presentation is omitted here due to page limitation.

For more extensive biosensing experiments, we refer readers to another work of ours in *Nature Medicine* [4], where the same magnet and planar microcoil are used while the transceiver electronics are implemented in a discrete form with a lower receiver sensitivity.

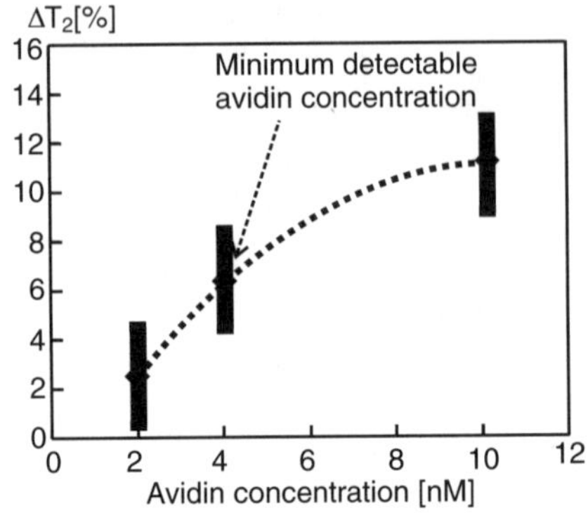

Figure 16.17. Measured ΔT_2 for various avidin concentrations. (From N. Sun, Y. Liu, H. Lee, R. Weissleder, and D. Ham, "CMOS RF biosensor utilizing nuclear magnetic resonance," *IEEE J. Solid-State Circuits*, 44, pp. 1629–1643, 2009.)

16.8 CONCLUSION

Combining RF microelectronics and biotechnology in the unifying framework of NMR, this work showcases how silicon RF ICs can be used not only for cell phones but also for biosensing aimed at improving human health care. The portable, low-cost NMR system, made possible by the development of a highly sensitive CMOS RF transceiver with a noise figure of 0.7 dB, is an addition to the growing library of emerging biosensors aimed at early disease diagnosis and/or at personalized medical care. We are now developing an even smaller, handheld NMR system by using an even smaller magnet: This increases the associated burden on the transceiver IC design, thus more joy for circuit designers.

ACKNOWLEDGMENTS

We thank Matthew Cole of Oxford University, Winfield Hill and Kevin Donoghue of Harvard University, and David Ricketts of Carnegie Mellon University for their valuable suggestions.

REFERENCES

[1] D. Ham and R. M. Westervelt, "The silicon that moves and feels small living things," *IEEE Solid-State Circuits Soc. Newsl.*, 12(4), pp. 4–9, 2007.

[2] J. M. Perez, L. Josephson, T. O'Loughlin, D. Hoegeman, and R. Weissleder, "Magnetic relaxation switches capable of sensing molecular interactions," *Nat. Biotechnol.*, 20, pp. 816–820, 2002.

[3] Bruker Optics, "The Minispec TD-NMR Analyzers," http://www.brukeroptics.com/minispec.html (accessed February 27, 2011).

[4] H. Lee, E. Sun, D. Ham, and R. Weissleder, "Chip-NMR biosensor for detection and molecular analysis of cells," *Nat. Med.*, 14(8), pp. 869–874, 2008.

[5] I. Koh, R. Hong, R. Weissleder, and L. Josephson, "Sensitive NMR sensors detect antibodies to infuenza," *Angew. Chem. Int. Ed. Engl.*, 47(22), pp. 4119–4121, 2008.

[6] D. Canet, *Nuclear Magnetic Resonance: Concepts and Methods*. United Kingdom: John Wiley & Sons, 1996.

[7] H. Lee, Y. Liu, R. M. Westervelt, and D. Ham, "IC/Microfluidic hybrid system for magnetic manipulation of biological cells," *IEEE J. Solid-State Circuits*, 41(6), pp. 1471–1480, 2006.

[8] T. Cherifi, N. Abouchi, G.-N. Lu, L. Bouchet-Fakri, L. Quiquerez, B. Sorli, J.-F. Chateaux, M. Pitaval, and P. Morin, "A CMOS microcoil-associated preamplifier for NMR spectroscopy," *IEEE Trans. Circ. Sys.—I.*, 52(12), pp. 2576–2583, 2005.

[9] R. Meller and D. Hartill, "Pulsed NMR magnetometers for CESR," *Proc. Particle Accelerator Conf.*, pp. 2339–2341, 2003.

[10] T. R. Fox and R. J. Faehnrich, "Signal/noise ratio optimization tuning system," United States Patent, 5,347,222, Sep. 13, 1994.

[11] D. M. Binkley, J. M. Rochelle, B. K. Swann, L. G. Clonts, and R. N. Goble, "A micropower CMOS direct-conversion, VLF receiver chip for magnetic-field wireless applications," *IEEE J. Solid-State Circuits*, 33(3), pp. 344–358, 1998.

[12] T. H. Lee, *The Design of CMOS Radio-Frequency Integrated Circuits*. Cambridge, MA: Cambridge University Press, 1998,

[13] D. K. Shaeffer and T. H. Lee, "A 1.5-V, 1.5-GHz CMOS low noise amplifier," *IEEE J. Solid-State Circuits*, 32(5), pp. 745–759, 1997.

[14] H. Darabi and A. A. Abidi, "Noise in RF-CMOS mixers: a simple physical model," *IEEE J. Solid-State Circuits*, 35(1), pp. 15–25, 2000.

[15] A. Roch, R. N. Muller, and P. Gillis, "Theory of proton relaxation induced by superparamagnetic particles," *J. Chem. Phys.*, 110(11), pp. 5403–5411, 1999.

INDEX

Note: Page numbers in *italics* refer to Figures; those in **bold** to Tables.

CMOS Biomicrosystems: Where Electronics Meet Biology, First Edition. Edited by Krzysztof Iniewski.